T0251000

Sound Reproduction

Sound Reproduction: The Acoustics and Psychoacoustics of Loudspeakers and Rooms, Third Edition explains the physical and perceptual processes that are involved in sound reproduction and demonstrates how to use the processes to create high-quality listening experiences in stereo and multichannel formats.

Understanding the principles of sound production is necessary to achieve the goals of sound reproduction in spaces ranging from recording control rooms and home listening rooms to large cinemas. This revision brings new science-based perspectives on the performance of loudspeakers, room acoustics, measurements and equalization, all of which need to be appropriately used to ensure the accurate delivery of music and movie sound tracks from creators to listeners.

The robust website (www.routledge.com/cw/toole) is the perfect companion to this necessary resource.

Floyd E. Toole was a research scientist at the National Research Council of Canada and, more recently, the corporate vice president of acoustical engineering at Harman International. Now retired, he is a consultant to Harman. He has received the Audio Engineering Society Silver and Gold Medal awards, and lifetime achievement awards from CEDIA and ALMA International. He is a Fellow of the AES, the Acoustical Society of America and CEDIA, and is in the Consumer Technology Hall of Fame.

Audio Engineering Society Presents (AES)

www.aes.org/

Officers of the Technical Council include:

- Francis Rumsey—Chair
- Juergen Herre—Vice Chair
- Michael Kelly—Vice Chair
- Bob Schulein—Vice Chair

Publications:

Handbook for Sound Engineers, Fifth Edition
(9780415842938)
Authored by Glen Ballou
Focal Press, April 1, 2015

Audio Production and Critical Listening, Second Edition
(9781138845947)
Authored by Jason Corey
Routledge, September 1, 2016

Recording Orchestra and Other Classical Music Ensembles
(9781138854536)
Authored by Richard King
Routledge, December 8, 2016

Sound Reproduction

The Acoustics and Psychoacoustics of Loudspeakers and Rooms

Third Edition

Floyd E. Toole

Routledge
Taylor & Francis Group

NEW YORK AND LONDON

Third edition published 2018
by Routledge
711 Third Avenue, New York, NY 10017

and by Routledge
2 Park Square, Milton Park, Abingdon, Oxon OX14 4RN

Routledge is an imprint of the Taylor & Francis Group, an informa business

First edition published by Focal Press 2008
Second edition published by Routledge 2013

Library of Congress Cataloging-in-Publication Data
Names: Toole, Floyd E., author.
Title: Sound reproduction : the acoustics and psychoacoustics of loudspeakers
    and rooms / Floyd Toole.
Description: Third edition. | New York ; London : Routledge, 2017.
Identifiers: LCCN 2017007635| ISBN 9781138921375 (hardback) |
    ISBN 9781138921368 (pbk.)
Subjects: LCSH: Loudspeakers. | Acoustical engineering. | Sound—Recording
    and reproducing—Equipment and supplies. | Psychoacoustics. | Recreation
    rooms.
Classification: LCC TK5983 .T66 2017 | DDC 621.382/84—dc23
LC record available at https://lccn.loc.gov/2017007635

ISBN: 978-1-138-92137-5 (hbk)
ISBN: 978-1-138-92136-8 (pbk)
ISBN: 978-1-315-68642-4 (ebk)

Typeset in Times New Roman
by Apex CoVantage, LLC

Visit the companion website: www.routledge.com/cw/toole

Contents

Acknowledgments

Special thanks are due to the National Research Council of Canada and to Harman International Industries, Inc., who for 27 and 25 years respectively provided funding and facilities for my colleagues and me to investigate the physical and psychoacoustic factors involved in sound reproduction. That we were able to do public presentations of the research findings and to publish them is commendable. Harman engineering staff generously provided many of the measurements seen here. The numerous researchers around the world whose efforts have contributed to the science of audio and to the contents of this book have both my respect and my gratitude.

My thanks to Todd Welti and the late Brad Wood for their constructive criticism of the manuscript, and to my extremely supportive wife Noreen, who shared in the creative process on a daily basis.

I dedicate this book to my father, Harold Osman Toole, who set a high moral standard, taught me the skills to follow in his footsteps as a competent woodworker and handyman, and who insisted that I start life with a good education. He has been a wonderful father and friend—and as this is written, he still is, at 104 years old.

Introduction to the Third Edition

This book is about science in audio, the acoustics and psychoacoustics of loudspeakers, rooms and their combined effects on what listeners hear. It contains many references to work done by researchers all over the world, but among them are references to work done by my research colleagues and me over the years. Consequently, in this introduction to the new edition, I will also introduce myself, my motivations, and my approach to examining aspects of audio.

The first edition of the book was clearly oriented to explaining the science underlying the acoustics and psychoacoustics of loudspeakers, rooms and the listeners who derive pleasure from the combinations. What was called the second edition was a labeling error associated with a change in publishers. The book was unchanged. This edition is substantially new. I have tried to adopt a more linear approach to explaining how the art, technology and science combine to create listening experiences and how we perceive them. Nevertheless, it is a long book with more information than most people need, so I anticipate that it will be read in pieces, dipping into it as appropriate for individual readers. For that reason, certain explanatory details are repeated for clarity.

Readers will find some thoughts that run contrary to conventional audio tradition because new scientific knowledge has superseded incomplete reasoning that provided the original foundations. Some audio folklore needs to be retired. This will not happen overnight, especially when the art is so intertwined with technology and science.

Audio is entertainment, but doing it well may require some homework. Understanding how it works may make the result even more pleasurable.

I was born in 1938 in Moncton, New Brunswick, in eastern Canada. I grew up as a hi-fi enthusiast through the eras of 78s, LPs, open-reel tape, cassettes, tube/valve electronics, and so on. At that time audio was a "participatory" hobby. My father was a consummate do-it-yourselfer, and I followed in his footsteps, building preamps and power amplifiers—from scratch in the beginning, using a lot of war surplus parts, and later from Heath and Eico kits.

My father and I built loudspeaker enclosures in our woodworking shop from designs published in hobbyist magazines. Thick catalogs were full of electronic and audio-related components—so many choices and not a shred of useful data to tell us how they might sound, even if we could understand the data. It was a time of unrestrained opinion, and trial and error. There were aftermarket add-on tweeters, loudspeaker drivers for home-built systems constructed using "universal" enclosures and "universal" crossover networks. The sound quality by today's standards was seriously lacking. The famous acoustician Dr. Leo Beranek once said something like "the sound quality of a home

FIGURE 0.1 *The audio industry of today might wish for this kind of enthusiasm for its mainstream products. Cartoon by Simon Ellinas, www.caricatures.org.uk. Originally published in* Hi-Fi News and Record Review, *1981. Reproduced with permission.*

built loudspeaker increases with the effort put into the hand-rubbed finish." It's true. There were tweaks that claimed to improve the performance of playback hardware and electronics, some of which might even have worked. And always there were maintenance chores, to keep records clean, needles replaced, tubes tested. It was called "hi-fi," high fidelity, a term so abused that it has lost its meaning.

When the term "high fidelity" was coined in the 1930s, it was more a wishful objective than a description of things accomplished—many years would pass before anything resembling it could be achieved. While recreating a live performance was an early goal, and remains one of the several options today, the bulk of recordings quickly drifted into areas of more artistic interpretation.

The essence of high fidelity, the notion of "realism" and the uncolored reproduction of music, dominated almost every discussion of home audio equipment. However, commercial recordings

FIGURE 0.2   *As the author saw himself ca. 1958. The stereo cabinetry was self-styled and built in the home workshop mostly by my craftsman father. The enclosures were small Karlsons (a terrible acoustical design), with Goodmans Axiom 12-inch "whizzer-coned" drivers. A Garrard RC-88 changer with a GE VRII mono cartridge, later replaced with an Acos stereo cartridge, drove a homemade preamp (a variation of a Fisher design, as I recall) and homemade Williamson 6L6 power amplifiers, later modified to use Acrosound Ultralinear output transformers. It was loud, and I was proud.*

themselves betrayed the growing divide between the ideals of high fidelity and the reality of what happened in the recording studio.

(Morton, 2000, p. 39)

I went on to study electrical engineering, first at the University of New Brunswick, Canada, then at the Imperial College of Science and Technology, University of London, from which I graduated with a PhD in 1965. My research topic was multidisciplinary, evolving from discussions with Prof. Colin Cherry, known for his expertise in human communication and creator of the term "cocktail party effect," referring to binaural discrimination in complex listening situations. Stereophonic sound and the directional and spatial effects it yielded intrigued him. He was also an electronics engineer, which explains the origins of my own thesis project: an investigation of binaural hearing—sound localization—employing signal generating, processing and data gathering electronics of my own design. In those days, most psychoacoustics research was done without the benefits of modern electronics and acoustical knowledge, so it was interesting to see where these new experimental capabilities took us. The engineering methodology was greatly advantageous, with new experiments being created in days, not months. A thesis and papers resulted (Sayers and Toole, 1964; Toole and Sayers, 1965a, 1965b). It is gratifying to see the results discussed in what is arguably the reference text on spatial hearing (Blauert, 1996). I became captivated by science, finding answers to questions that interested me, and at the same time that seemed to fill a need for a larger audience. It suited my temperament.

Contemporary science is based on a method of inquiry developed in seventeenth-century Europe. The *scientific method* involves observing the natural world, questioning what is seen (or heard, in this context) and then conducting experiments to gather measurable evidence to provide insights or answers. It is a tedious process, requiring care and repetition to ensure that the data are reliable, and a disciplined approach to designing the experiments to ensure that the data bear on the question being asked without being influenced by extraneous factors, including the person asking the question. The answers need to be free from bias. It is not simple, but as will be seen, the results are worth the effort, yielding insights that allow us to now design and predict many aspects of "good sound."

FIGURE 0.3 *Psychoacoustics research, engineering style. Racks of Toole designed and fabricated germanium transistor and tube electronics, including a four-track pulse-width-modulated (DC—200 Hz) analog tape system for control of randomized experimental parameters, and storing listener responses. Results were printed out on an automated X/Y plotter.*
Imperial College, London, ca. 1963.

It is so much easier just to offer an opinion. The problem is that there are so many of them, and they keep changing. It was not surprising to find that the variations were caused by more than the sound. We humans are very susceptible to non-auditory influences that bias our perceptions. Fortunately non-auditory factors are easily controlled.

Upon graduation I was employed as a research scientist at the National Research Council of Canada (NRCC) in Ottawa. My job was to ask questions and find answers by applying the scientific method. I was in the Applied Physics Division, so the emphasis was on real-world issues. A major performance metric was peer-reviewed publications, but also, for my chosen line of investigation, evidence that industry benefitted from the work—the NRCC was taxpayer funded. There could not have been a better place to engage in this kind of research. First, I worked among, and was tutored by, some of the best acoustical scientists in the world. We had access to excellent anechoic and reverberation chambers, as well as all of the latest measuring equipment necessary to quantify sound. There was also the budget and space to create listening rooms for subjective evaluations. The research was successful, and publications resulted.

The embryonic Canadian loudspeaker industry rented the NRCC measurement and listening facilities to design products, and, importantly, Canadian audio magazines paid for the facilities to perform product reviews—anechoic measurements and double-blind listening tests. The products that were designed and reviewed became part of the database for the research, and everyone benefitted from the knowledge as it emerged. A small staff was hired, and I traveled widely, telling the science story to Audio Engineering Society (AES) audiences and to interested manufacturers. The relatively unknown Canadian loudspeaker manufacturers used the credibility of the NRCC and the research to help gain recognition (some are now well known and respected international suppliers). All was as it should be.

One day in early 1991, the phone rang. It was a headhunter offering the possibility of an interesting job with a major audio corporation, Harman International Industries. After 26 years of research, I was intrigued by this opportunity to get directly involved with applying the science to product development—moving closer to the "real" world. Soon I was hired as the corporate vice president of Acoustical Engineering, but very quickly it became more, because I was able to convince the company leaders that we could afford, and indeed needed, a corporate research group that was not attached to the brands and that did not develop audio products. Knowledge was the product, and obviously some of it would migrate into products if it proved to be of value. Harman generously permitted us to publish freely, following the scientific tradition of free exchange of knowledge at AES conventions, conferences and in the journal (some corporations do not allow this).

Harman spent large sums on improved engineering facilities and innovative listening rooms for product evaluation. The benefits were soon seen in improved consistency and quality of sound from the products. Nevertheless, there were arguments from some sales and marketing people who may not have had the same faith in science as we did. Good sound does not guarantee good sales. There are many factors involved in that aspect: appearance, price, size, marketing and retail distribution. These all fall outside the domain of engineering. Still, there was a resolute effort to ensure that products at all price points were competitive in sound quality.

A program of measurements and double-blind evaluations of competing products was set up and continues. However, it has been difficult to maintain for all products because of tight schedules, the numbers of new products being developed, and the decentralization of design and manufacturing as Harman grew into an enormous worldwide, diversified corporation. When I joined Harman in 1991, sales were about $500M, we had a few thousand employees, and we were primarily an audio company headquartered in California. Things changed. Now sales are about $7B, there are about 26,000 employees worldwide, and audio is just part of what Harman does. It is a different company.

I retired in 2007, but I have remained in a consulting role since then. I continue to learn, publish and teach. I had the great pleasure of turning my audio hobby into my profession, and I can truly say that I feel that I have never had a "job."

SCIENCE IN AUDIO

Early in my career I came to grips with some fundamental truths about the role of science in audio. Scientific explanations of the physical world and new technologies have allowed us to enjoy the emotions and aesthetics of music, whenever or wherever the mood strikes us. Music is art, pure and simple. Composers, performers and creators of musical instruments are artists and craftsmen. Through their skills, we are the grateful recipients of sounds that can create and change moods, that can animate us to dance and sing, and that form an important component of our memories. Music is part of all of us and of our lives.

However, in spite of its many capabilities, science cannot describe music. There is nothing documentary beyond the crude notes and symbols on a sheet of music. Science has no dimensions to measure the evocative elements of a good tune. It cannot technically describe why a famous tenor's voice is so revered, or why the sound of ancient Cremona violins has been held up as an example of how it should be done. Nor can science differentiate, by measurement, the mellifluous qualities of trumpet intonations by a master, and those of a music student who simply hits the notes. Those are distinctions that must be made subjectively, by listening. A lot of scientific effort has gone into understanding musical instruments, and as a result, we are getting better at imitating the desirable aspects of superb instruments in less expensive ones. In fact, recent blind evaluations are indications of success (see the box). We are also getting better at electronically synthesizing the sounds of acoustical instruments. However, the determination of what is aesthetically pleasing remains firmly based in subjectivity.

It is at this point that it is essential to differentiate between the production of a musical event and the subsequent reproduction of that musical event. Subjectivity—pure opinion—is the only measure of whether music is appealing. That will necessarily vary among individuals. Analysis of music involves issues of melody, harmony, lyrics, rhythm, tonal quality of instruments, musicianship and so on. In a recording studio, the

SCIENCE, PSYCHOACOUSTICS, MUSICAL INSTRUMENTS AND MUSICIANSHIP

This book discusses the science of sound reproduction. Others have been applying scientific methods to musical instruments and concert venues. Concert hall acoustical investigations have been numerous and well publicized, but those pertaining to the instruments themselves, much less so. Recent papers have stirred interest and controversy by challenging some widely held beliefs. A paper by Fritz et al. (2014) reports results of elaborate blind evaluations of six new and six Old Italian violins (including five by Stradivari). The players were "significant" professional soloists and the evaluations were done in a rehearsal room and in a small concert hall. The result was that when asked to choose a violin to replace their own for a hypothetical tour, 6 of the 10 soloists chose new violins. In the individual ratings, a single new violin was chosen four times, a single Stradivari three times, and two new violins and a Stradivari once each. Tracking those instruments that soloists rejected as unsuitable, the new violins prevailed by a 6:1 ratio. So, to 10 performers, seven of whom routinely play Old Italian violins, the new—much less expensive—alternatives were very attractive.

Bissinger (2008) delved into the details of violin acoustics, and summarized common remarks about the best violins: "they are more 'even' across the measured range, and strong in the lowest range." As an audio person, I interpret this as "flat frequency response and good bass," which seems reasonable. Campbell (2014) provides additional perspective on scientific contributions to several musical instruments—interesting reading for lovers of music with a technical inclination.

On a very different, but related, topic, Tsay (2013) tested the popular notion that "sound is the most important source of information in evaluating performance in music." He found that both novices and professional musicians were able to identify the winners of prerecorded live music competitions better when viewing a video of the event in silence than when listening only, or viewing and listening together. The evaluation of musical performances was found to be dominated by the visual impact of gesticulations, not audible output. Remarkable. It is no wonder that the visual aesthetics of loudspeakers precondition our reactions to the sounds they produce.

recording engineer is an additional major contributor to the art. All of the many electronic manipulations used to create the final stereo mix are judged subjectively, on the basis of whether it reflects the artists' intent and, of course, how it might appeal to consumers.

The evaluation of reproduced sound should be a matter of evaluating the extent to which any and all of these elements are accurately replicated or attractively reproduced. It is a matter of trying to describe the respects in which audio devices add to or subtract from the desired objective. A non-trivial problem is that we, the listeners, were not in the control room at the time the final mix was approved. We don't know what the creators heard, but we still have opinions about what we like and dislike. We don't know who or what to praise or criticize. Often the playback apparatus carries more than its share of responsibility.

When making audio product evaluations, the terminology appropriate to describing the music itself is either insufficient or inappropriate. A different vocabulary is needed. Most music lovers and audiophiles lack this special capability in critical listening, and as a consequence art is routinely mingled with technology. In subjective equipment reviews, technical audio devices are often imbued with musical capabilities. Some are described as being able to euphonically enhance recordings—others to do the reverse. It is true that characteristics of technical performance of playback devices must be reflected in the musical performance, but the technical performance attributes are fixed, and music is infinitely variable. Consequently the interactions are unpredictable. This does not help our efforts to investigate and improve sound reproduction.

Add to this the popular notion that we all "hear differently," that one person's meat might be another person's poison, and we have a situation where a universally satisfying solution might not be possible. Fortunately reality is not so complex, and although tastes in music are demonstrably highly personal, enormously variable, we discover that recognizing the most common deficiencies in reproduced sounds is a surprisingly universal skill when listeners are given a chance to reveal their unbiased opinions. More good news is that most people can do it, even those who think they have "tin ears." Inexperienced listeners take more time, make more mistakes along the way, but in the end, their opinions generally agree with those of the experts. Only those with hearing loss routinely depart from the norm. To a remarkable extent we seem to be able to separate the evaluation of a reproduction technology from that of the program. It is not necessary to be familiar with or to enjoy the music to be able to recognize that it is or is not well reproduced.

How do listeners approach the problem of judging sound quality? Most likely the dimensions and criteria of subjective evaluation are traceable to a lifetime accumulation of experiences with live sound, even simple conversation. If we hear things in reproduced sound that do not occur in nature, or that defy some kind of perceptual logic, we seem to be able to identify it. By that standard, the best sounding audio product is the one that exhibits the fewest audible flaws. Perhaps this is how we are able to make such insightful comments about sound quality based on recordings that either had no existence as a "live" performance, or that we have no personal experience with.

Figure 0.4 shows that in live performances things are relatively uncomplicated. The musicians radiate sound into a performance space and two ears and a brain interpret the combination. These are "reference" listening experiences—there is literally nothing between you, the listener, and the performance. The very complex sound field of a live performance can be sampled by some number of microphones, taken to a recording studio, and manipulated to sound good through some number of channels and loudspeakers. Naturally the spatial complexity of the live performance cannot be replicated through two channels; multichannel schemes can be more persuasive. It is the skill of the recording engineer that allows these compromised reproductions to sound as good and as realistic as they do.

The center of Figure 0.4 illustrates the origin of most of the popular music and jazz that we enjoy. Performers sing and/or play in a studio, together or separately, and their contributions are stored in "tracks," perhaps many of them. Then the recording engineer(s) and musicians "mix" the final product, adjusting the contributions of individual performers, perhaps altering the timbre of voices and instruments with equalization, and

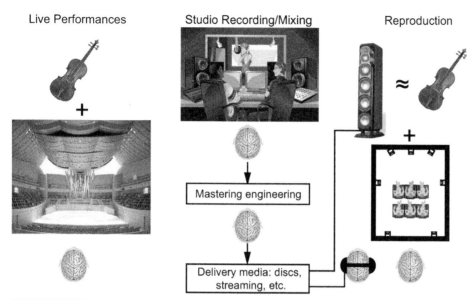

FIGURE 0.4 *A conceptual view of the parallel universes of live performance "originals," and the sound recording and reproduction "replicas," all interpreted by and influenced by two ears and a brain at several points along the way.*
Disney Hall photograph by Federico Zignani.

adding spatial effects: reflections, reverberation and so forth. These days even the pitch of off-tune singing can be corrected. All of this is evaluated as it happens by human listeners—ear/brain combinations—probably several of them, while listening to loudspeakers in a control room. The resulting mix may then be passed to a mastering engineer who applies personal subjective judgment to the musical product through different loudspeakers in a different room, changing it as necessary to fit the chosen delivery format(s). In the days of LPs the manipulations were substantial. The engineer tried to anticipate how the music may sound to the majority of customers in the majority of listening situations while embracing the properties of the delivery media. It would be ideal if there were no limitations to bandwidth and dynamic range, but not all media permit it, and few customers are equipped, or motivated, to enjoy such recordings. Often the audio product is tailored for modest playback equipment, possibly anticipating listening with a certain amount of background noise, and the sound is appropriately preconditioned. Consumers can know none of this, so what we hear from recordings, related to sound and spatial qualities, involves a measure of chance.

On the right of Figure 0.4 is the final step for those fortunate enough to have a listening room. Here is yet another opportunity for changing the audio product: different loudspeakers in different places in different rooms, using different electronics that may impose different signal processing and equalizations to the sounds radiated into the rooms. Again it is the human ear/brain system that generates the perceptions, and the opinions. Of course only a small percentage of people have such playback facilities.

Most make do with modest audio gear, or listen through headphones, the latter yielding a totally different sound experience from stereo recordings that are created for loudspeaker playback.

So, reviewing the process, there are several opportunities for significant personal input into the content of music recordings and how they sound on playback. Serious music lovers should attempt to experience live—unamplified—music performances if they wish to hear "reality." *No* recording, through any number of channels, can perfectly duplicate that reality, however good the playback apparatus. The vast majority of music recordings originate in studios, and before they reach us they have been subject to numerous subjectively guided manipulations, evaluated using loudspeakers of varying pedigree. Even supposedly "pure" classical recordings are massaged in the mixing and mastering processes to make them more pleasant when auditioned through two loudspeakers. The signal from a microphone focused on an acoustically weak instrument or voice may be spatially enhanced so that it fits into the acoustical setting of the entire orchestra. When well done, the trickery is not detected.

This scene may seem hopelessly disorganized, but it works most of the time. Music is very durable, and very likely all of us have spent hours enjoying music through seriously compromised audio systems. However, when a truly good recording is auditioned through a truly good playback system, it can be a spine-chilling experience. It isn't "reality," but it can be absolutely superb music and high-quality entertainment. A general principle for the audio industry should be: if in doubt, at least make it sound good.

Whatever happens in the continuing evolution of audio, it is helpful to understand the basics of the technology, the principles of sound propagation and the psychoacoustics of perception, because it is unlikely that any of those will change. The goal of this book is to provide knowledge and guidance for high-quality reproduction of existing audio formats, and to prepare the way for future developments.

CHAPTER 1

Sound Production vs. Sound Reproduction

Before getting into sound reproduction, the title of this book, it is good to have a look at where it all begins: in live performances—sound production. We may like to think that our audio systems are capable of reconstructing such experiences, but that is simply not possible. Even today, with nearly unlimited bandwidth available, two-channel stereo is the default format. There is no doubt that stereo can be greatly entertaining, and at times can make us feel close to the real experience. But it is sad to say that many recordings can be well described as: left loudspeaker, right loudspeaker and phantom center. This is mono, mono and double mono. Listened to through headphones it becomes left ear, right ear and center of the head. The essence of the music can be conveyed, but any semblance of acoustical space and ambiance is missing. Skillful microphone setup and signal processing by recording engineers can improve things, but at best stereo remains a directionally and spatially deprived format, and an antisocial one, requiring a sweet spot.

We need more channels to capture, store and reproduce even the essential perceptions of three-dimensional sound fields. This is what the movie world has known for decades, and now cinemas have as many as 62 channels in the immersive sound formats. That is excessive for musical needs, but more than two would be nice. Fortunately there are examples of excellent multichannel music, and indications that a binaural version of it will be a part of virtual reality systems. Stay tuned.

On the scientific side, the origins of modern acoustics lie largely in the domain of halls for the performance of classical music. Whether this music appeals to a person or not, the basic perceptions generated by these live performances are generously shared within all recorded music, whatever the genre. Reverberation, spaciousness, envelopment and so on are all simply pleasant perceptual experiences, and recording engineers have been provided with elaborate electronic processors allowing them to be incorporated into any kind of music, adding to the artistic palette. The future is sounding good.

1.1 LIVE CLASSICAL MUSIC PERFORMANCES— SOUND PRODUCTION

I have an approximation of a state-of-the-art sound reproducing system at home, and have always had such systems, gradually improving over the years. As good as they have been, the real thing is a very different and more satisfying auditory event. Dressing up, driving, perhaps eating out, crowds of similar-minded people, the overall visual atmosphere all add to the experience, but the aural components of the experience are the real treat.

I attend about a dozen live concert hall performances a year. Sitting, watching people take their seats and the orchestra assembling on stage, I am aware of a pleasing sense of a large space—I hear the space that my eyes see. This does not happen at home. As the musicians tune up and practice difficult passages, the timbres are enriched by countless reflections—repetitions that give our hearing system more opportunities to hear subtleties. I can localize individual musicians, and even though the sound begins at those locations, the timbres linger in the decaying reverberation all around me. When the music starts, it all magnificently comes together. The sound of the hall is an inseparable part of the performance: rich timbres combined with enveloping space. I am "in" the performance. This is a complex listening experience, allowing one to begin to discover what contributes to an engaging final product. However, it is interesting to consider that with different musicians, instruments, conductors and halls, the "reference" is really not a constant entity.

> In fact, one can correctly assert that a live concert hall performance is what it is at the time, and may never be repeated again. It is sound production.

It is interesting to note that, even in different halls, the essential timbres of voices and musical instruments remain remarkably constant. We have considerable ability to separate the sound of the source from the sound of the hall. In other words, we appear to adapt to the room we are in and "listen through" it to hear the sound sources. A variation on this interpretation is that we engage in what Bregman (1999) calls "auditory scene analysis," and we "stream" the sound of the voices and instruments as significantly separate from the sound of the room. We do this to such an extent that one can focus on the sound from one section of the orchestra, suppressing others. Two ears and a brain are remarkable. If a concert hall performance seems to lack bass, as some do, the inclination is to blame the hall, not the musicians or their instruments. We instinctively know where the blame lies.

Later in the book we will discuss the elements of these auditory events from both perceptual and measurement perspectives. For now, it is sufficient to note that reproducing a concert hall experience means delivering both the timbral and the spatial components. This is not easy.

Achieving satisfactory reproductions of these performances is partially determined by hardware: the electronics, the loudspeakers and the rooms. These are things

FIGURE 1.1 *Los Angeles County Music Center's Walt Disney Concert Hall Auditorium.*
Photo by Federico Zignani.

consumers can choose and manipulate to some extent. But the most important contribution comes from the "system":

- The number of channels and the placement of the loudspeakers and listeners for playback.

- The microphone choice and placement, mixing/sound design, and mastering performed in the creation of the recording.

If the "system" does not allow for certain things to be heard, disappointment is inevitable. The existing system evolved within the audio industry itself and is being implemented by professionals operating within it. These are key factors in our impressions of direction and space, and there is persuasive evidence that spatial perceptions are comparable in importance to timbral quality in our overall subjective assessments of reproduced sound.

Time-domain information, the reverberation, is an important clue to the nature of the performance space. It can be conveyed by a single channel—mono—but it is a spatially "small" experience, with all sound localized to the single loudspeaker. Adding more channels allows for a soundstage, conveyed by the front channels: the lateral spread of the orchestra, with depth. There is also a more subtle component: apparent source width (ASW), or image broadening, wherein sounds acquire dimension, and an acoustical setting; some call it "air" around the instruments. Two channels suffice for a single listener in the symmetrically located "stereo seat," but multiple listeners or a single listener not in the stereo seat require the addition of a center channel to prevent the phantom image soundstage from distorting and ultimately collapsing to the nearer loudspeaker. The perception of being "in" the space with the performers—envelopment—is hinted at in good stereo recordings, absent from many, but is more persuasively delivered by multichannel recordings, with side and other channels providing long-delayed lateral reflections that contribute to both image broadening and envelopment. It is the length of the delays that creates the impressions of large space, something that limits the spatial augmentation possible with multidirectional loudspeakers in small listening rooms. In fact, envelopment is, according to some authorities, the single most important aspect of concert hall performance.

As good as stereo can be, more channels are better for a single listener, and most definitely for multiple listeners. Unfortunately the 5- or 7-channel options almost universally used in movies have been not been commercially successful in the music domain, even though they are capable of more engaging reproductions. The new "immersive" formats, employing even more channels and loudspeakers, were created for movies, where they provide exciting spatial dynamics, but some demonstrations of music programs provide compelling impressions of real concert halls or cathedrals, even as one moves around the listening room. The two-ears, two-channels relationship works for headphones, but is spatially deprived for loudspeaker reproduction. This topic will be discussed in more detail later in Chapter 15.

1.2 LIVE POPULAR MUSIC PERFORMANCES— SOUND PRODUCTION

As pleasurable as the classics are, the majority of our entertainment falls into the numerous subdivisions of what is collectively called "popular" music. The recording methods used are also shared with most jazz we hear. Most of what we listen to is captured by microphones located unnaturally close to voices and musical instruments, or electrically captured without any acoustical connection, and then mixed and manipulated in the control room of a recording studio. The "performance," the art, is what is heard through the monitor loudspeakers during the mixing and mastering operations. The era of large, expensive recording facilities is fading, as more recordings are done in converted bedrooms or garages in homes. The wide availability of sound mixing and processing programs, and the low cost of powerful computers, has given almost anybody access to capabilities that once were the exclusive domain of elaborate recording studio facilities. This paradigm shift is an important factor, expanding the repertoire of recorded music, changing the business model of the music industry itself, and liberating creative instincts that previously had been "damped by dollar signs."

Big-name artists engage in elaborate tours, spanning the globe in some cases. They became popular through recordings created in studios, and reconstructing the essence of that recorded "sound" in a concert situation is sometimes a goal. Occasionally, excerpts of studio recordings creep into the live performances. None of this is a problem because it was always an artificial creation, owing little or nothing to any live unamplified performance. It might seem like cheating, but there are some effects that cannot be duplicated in live performances. In the end, the delivered "art" is what matters.

Figure 1.2 shows that in live, amplified/sound reinforced, popular music performances, the front-of-house (FOH) engineer is in control of the performance heard by the audience. It is an artistic creation in real time, and can vary enormously with different engineers, their tastes and, interestingly enough, how well they hear. I left one show at intermission because the sound was far too loud, and not very good. I learned later that the FOH engineer was known by insiders to have serious hearing loss, but had a long relationship with the performer. Pity. This is a situation in which, for a variety of acoustical, technical and personal reasons, live popular music performances are variable.

One can correctly assert that a live popular music performance is what it is at the time, and may never be repeated again. It is sound production.

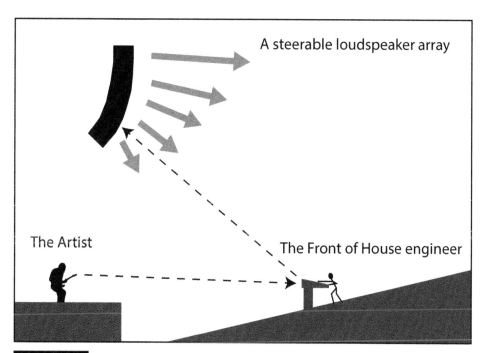

FIGURE 1.2 *A functional diagram of a tour sound system. The microphone and direct-wired inputs from the artists on stage are mixed and manipulated by the front-of-house (FOH) engineer, based on what is heard from the loudspeaker array, the sound of which is shared with the audience.*

1.3 REPRODUCED SOUND—THE AUDIO INDUSTRY

Sound *re*production is different. At some time and place, an original performance occurs, and the objective is to reproduce that performance with as much accuracy as possible whenever and wherever someone chooses to press a "play" button. Most of our listening experiences involve recordings, broadcasts or streaming audio that is reproduced through loudspeakers in a room, loudspeakers in a car or through headphones. This is the audio industry, as shown in Figure 1.3.

Clearly the audio industry is a complex operation, requiring extensive standardization if all of the devices within these different operations and business units are to be compatible with the signals moving through them. More importantly, there is the matter of what listeners in these varied situations hear. Is the result of the mixing and mastering engineering accurately delivered to the customers' ears?

A fallacy: That reference to a "live" sound is the only way to judge sound quality. Reason: microphones capture only a sample of the live sound field that we would hear in a live performance. Parts are missing. Recording engineers can manipulate the mix to sound something like the live sound, to create a totally artificial experience or anything in between.

Figure 1.4 illustrates the enormous contrasts in scale involved in reproduced sound, and it is reasonable to think that the differences might be insurmountable. Can movies created for large cinemas be credible in small home theaters or television sets in tiny apartments? The delivery systems, the film and music formats also differ, and new options continue to evolve. Making it all work involves complex engineering of many

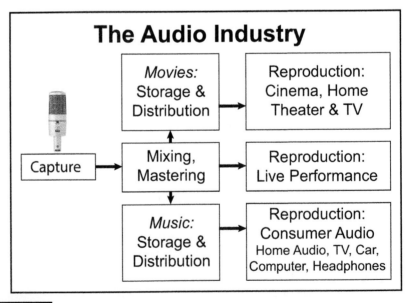

FIGURE 1.3 *The audio industry as we know it.*

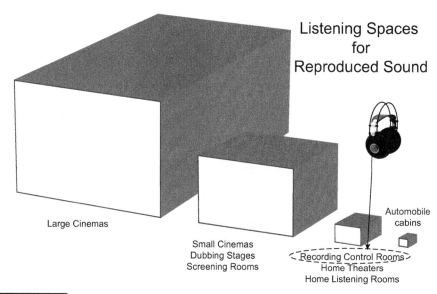

Listening Spaces
for
Reproduced Sound

Large Cinemas

Small Cinemas
Dubbing Stages
Screening Rooms

Automobile
cabins

Recording Control Rooms
Home Theaters
Home Listening Rooms

FIGURE 1.4   *The range of listening spaces within which accurate sound reproduction is desired.*

kinds, and some of it requires research to optimize the processes that deliver sounds to listeners' ears, making them appropriate for the circumstances.

Fortunately, cinemas are designed to have reverberation times that are not very different from domestic rooms. Combined with the large directional loudspeakers normally used, listeners end up in fundamentally similar sound fields and the experiences are acoustically more similar than might have been expected. This is discussed in detail later on.

Headphones normally replay stereo sound tracks created using loudspeakers in recording control rooms. Although it is possible to achieve a good timbral match, the spatial presentation is fundamentally different from that delivered by loudspeakers, sounds being localized primarily within or close to the head. For optimum headphone experiences, binaural (dummy head) recordings are needed, but we seem to have adapted well to the gross spatial distortions commonly experienced. The music survives.

The portrayal in Figure 1.5 could very well be wrong in detail, but there is little doubt that the trend is correct. The circumstances within which we are primarily entertained by audio programs have not had the proportional benefit of scientific research. As a result opinions, traditions and outright folklore have had and still have significant influence in these domains.

Audiophiles seeking excellence in sound reproduction are frequently advised by journalists and reviewers who have little or no awareness of the research that has been done. Opinions substitute for fact-based guidance. Most product reviews are done without the benefit of measurements, a situation that creates great uncertainty. Some publications perform basic measurements, but they are necessarily compromised in accuracy, resolution and comprehensiveness. Nevertheless, I respect their efforts, because they

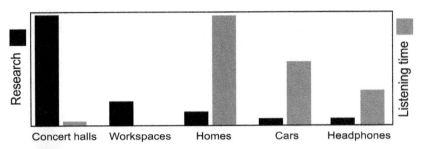

FIGURE 1.5 *A crude estimate of the quantity of scientific research devoted to various acoustical spaces and listening situations compared to the amount of time spent listening in each of them.*

are legitimate attempts to gain some technical insights. Facilities capable of anechoic or useful quasi-anechoic measurements are rare.

Some audio journalists are hostile to the very notion that audio is amenable to scientific investigation—asserting that only subjective opinion, preferably theirs, matters. There is nothing wrong with subjective opinions. As will be seen, it is these that have allowed for deliberate and productive research into what matters in sound reproduction. However, the opinions must be collected under circumstances in which the ever-observant human brain is deprived of information that could bias the opinion about the sound. The tests must be blind—the listeners must not know prices, brands, sizes and so forth. Ideally they should be double-blind, so the experimenter cannot bias the results. There should also be comparison sounds from competing products readily available, because humans are notoriously forgetful when it comes to recalling the details of sound quality. When this is done the subjective data begins to resemble technical measurements, in that they become impressively repeatable and, more importantly, generalizable to a large population. It is this discipline that has been lacking in audio in both the consumer and professional domains.

A motto for the audio industry:

Science in the Service of Art is our Business

Good sound is our product.

Inspiration, invention and trial-and-error experimentation have taken us a long way in audio. But there is a growing body of scientific understanding that can take the industry even further. The great benefit of scientific knowledge is that it allows for predictable, repeatable results. Knowing what matters and what does not matter permits us to optimize the design of audio components that can deliver predictably good sound at affordable prices, and excellent sound at higher prices. If we acknowledge the message in the preceding motto, it is the *art* that matters, and because it is audio art, sound quality is an essential deliverable. Nobody gets chills down the spine from the metal box containing an amplifier, or the hand-rubbed finish on an expensive loudspeaker, but the perception of truly excellent sound can be a greatly pleasurable experience. That is our product. Providing the means for successful creation and delivery is the challenge.

1.4 PRESERVING THE ART—THE CIRCLE OF CONFUSION

If "good sound" is our product, how do we know what is "good"? Trying to answer that question plunges us immediately into what I have called "the circle of confusion." A little thought tells us that if consumers are to hear what the artists, musicians and recording engineers created, they should have similar loudspeakers and rooms. If not, they will be hearing something different. With no standards for monitor loudspeakers in recording control rooms, and no standards for consumer loudspeakers, sound quality is not predictable—it is a gamble.

When we listen to recorded music, we are listening to the cumulative influences of every artistic decision and every technical device in the audio chain. Many years ago I created the cartoon shown in Figure 1.6 for my tutorial presentations, illustrating how the never-ending cycle of subjectivity can be broken.

The presumption implicit in this illustration is that it is possible to create measurements that can describe or predict how listeners might react to sounds produced by the device being tested. There was a time when this presumption seemed improbable, and even now some people claim that we cannot measure what we hear. The reality is that with research and the development of newer and better measurement tools, it has been possible to move the hands of the "doomsday clock" to the point where detonation is imminent. In fact, it would be correct to say that the explosion has begun. Some aspects of audible sound are now more reliably revealed by technical data than by the *normal kinds* of subjective evaluations.

The consequences of this circle of confusion are apparent in both loudspeakers and microphones. I recently read of a microphone that had been "voiced" by a well-known industry person, but there was no mention of what loudspeaker he was listening to when voicing, so what does this mean? Many years ago, a loudspeaker manufacturer had a product that was favored as a monitor by a well-known classical music label. The loudspeaker was not neutral, and those recordings revealed a distinctive coloration when reproduced through more neutrally balanced loudspeakers. For a period of time, most of the loudspeakers made by this

FIGURE 1.6   *The first version of the "circle of confusion" illustrating the key role of loudspeakers in determining how recordings sound, and of recordings in determining our impressions of how loudspeakers sound. The central cartoon suggests that the circle can be broken by using the knowledge of psychoacoustics to advance the clock to the detonation time when the "explosive" power of measurements will be released to break the circle.*

company exhibited the same basic coloration. Of course they flattered those recordings, but did less well with other recordings—the majority. This is clearly not a good situation for professionals or consumers.

Figure 1.7 expresses the ideas of the "circle of confusion" in a form that more accurately reflects the impact on the audio industry. In it, the role of the recording engineer and his/her "tool kit" of microphones, equalizers and the vast array of electronic signal processing algorithms is acknowledged. What is done during the tracking, mixing and mastering of the recorded sounds is influenced by what is heard in the control room. There is nothing wrong with any of it, because what is being created is audio art.

It is naïve to think that it is possible to capture a "pure" sound from a musical instrument and to reproduce it without loss. Musical instruments, especially stringed instruments, radiate directionally complex sound fields—the sound quality varies substantially depending on where the microphone is located (Meyer, 2009). Capturing the total sound as one hears it in a live classical performance is simply not possible with a single microphone or even a small number of microphones.

The situation was lucidly summarized by Brittain (1953). He said:

In any form of reproduction, visible or audible, there are nearly always some limitations inherent in the medium employed. It is the purpose of the scientist to reduce those limitations to a minimum. It is the purpose of the artist, in picture or sound, to circumvent these limitations, and even to use them for his own aesthetic purpose. . . . Sound reproduction is very seldom "Sound Recreation" and there are many cases where art can help science to give greater pleasure.

He goes on:

The sound reproducing system is often blamed for shortcomings which are inherent in the medium. These limitations are to be found in the mind of the performer, right through the chain of events, and also in the mind of the listener.

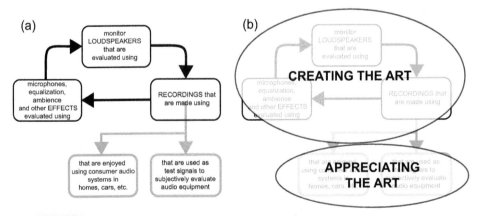

FIGURE 1.7 (a) The "circle of confusion" as it relates to the audio industry showing the important role of loudspeakers. (b) Unless there are significant common factors in the two domains, where the art is created and where it is appreciated, the art is not preserved.

This was written 64 years ago, and it is still true.

Benade (1985) notes that microphone placement and the directivity of musical instruments conspire to distort sounds before they ever make it into a recording. The lack of sufficient early-reflected sounds cannot be corrected with equalization in the mix.

For this sort of music, anything we might try to record has the humble-sounding but difficult task of *reminding* us of what we might hear at a live concert presented in a good concert hall. The task is ultimately made more difficult because what we are trying to produce is an imitation, and imitations are notoriously difficult to carry out successfully.

He goes on to point out the fundamental difference between this and popular music, where technology and signal manipulations are integral parts of the performance. The only "reality" was the final mix session in the recording control or mastering room.

On playback, the restriction to two channels severely limits the sound fields that can be delivered to listeners. However, tasteful mixing and signal manipulation can create a pleasing likeness, which we all have come to enjoy, even if it is not a reconstruction of reality. Recordings are studio creations, and skillful recording and mastering engineers are key artists whose names may or may not appear in the fine print in the liner notes.

Long ago, respected audio engineer John Eargle (1973) recognized the consequences of mismatches between monitoring and home systems in what is done during the recording process. Børja (1977) noted "dramatic and easily audible" differences in recorded sound quality caused by recording engineers compensating for spectral defects in control room monitor loudspeakers. He found that the inverse of the frequency response of the monitor loudspeakers could be seen in the spectra of the recordings. If the monitor loudspeaker had a response peak, the recording exhibited a spectrum dip—the engineer simply did what was necessary to make it sound right in the control room. This is not a new problem. Neither has the problem gone away. In an interview reported by Gardiner (2010) UK producer Alan Moulder, discussing the popular Yamaha NS-10M small monitor (see Section 12.5.1), said that "if you don't do anything they sound kinda boxy. They definitely make you work hard to make things sound right. You have to carve a lot out frequency-wise to make a track sound hi-fi." So, the process is to use a flawed loudspeaker and then equalize the mix to sound "hi-fi" to the mixer. Why? This mix can then only sound similar if reproduced through similarly flawed loudspeakers. This product and its distorted spectrum became so accepted by recording engineers that at least one modern—fundamentally neutral—monitor loudspeaker has built in a switchable equalization to replicate it (http://barefootsound.com/technology) where they describe: "The 'Old School' setting captures the essence of the ubiquitous NS-10M, rolling off the sub-bass and top-end information, and bringing forward mid-range presence." So, the aberrant sound of the discontinued NS-10M may never go away.

And unless the mastering engineer intervenes, the spectrally "carved up" recording is what gets delivered to our playback systems—which, if they are spectrally neutral will clearly expose the errors in the original. If they are not neutral, almost anything is possible. The circle of confusion is aggravated by some of the people who make recordings; see Figure 1.8.

This seems like an unacceptable situation, yet, to this day, the professional audio industry has no meaningful standards relating to the sound quality of loudspeakers used in control rooms. The few standards that have been written for broadcast, control and music listening rooms employed questionable measurements and criteria (e.g., Section 13.2.2). Many years ago, the author participated in the creation of certain of those standards and can report that the inadequacies were not malicious, only the result of not having better information to work with. The film industry has long had standards relating to the performance of loudspeakers used in sound-mixing stages and cinemas. These too were created many years ago, and are deficient, as will be discussed in Chapter 11. Several practices in audio are in need of revision to align with new knowledge and to take advantage of new technologies.

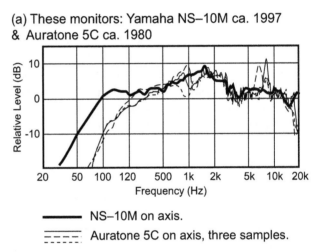

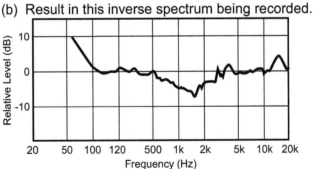

FIGURE 1.8 *An extreme view of what can happen when a recording engineer compensates for the characteristics of the monitor loudspeaker. (a) Two popular reference loudspeakers of the past turn out to be remarkably similar outside of the bass region. The variable performances of the Auratone samples cannot have been advantageous. (b) The inverse of the NS-10M, the equalized spectrum required to make the sound appear to be neutrally balanced to the engineer.*

A consequence of this lack of standardization and control is that recordings vary in sound quality, spectral balances, and imaging. Proof of this is seen each time a person chooses a recording to demonstrate the audio system they want to show off. The choice is not random. Only certain recordings are on the "demo" list, and each will have favorite tracks. This is because the excitement comes from more than just the music—the tune, lyrics or musical interpretation. The ability to exercise and demonstrate the positive attributes of the system is a key factor.

Many of the variations in recordings take forms that can be addressed by old-fashioned bass and treble tone controls. Unfortunately, tone controls are frowned upon by audio purists, who think that they somehow degrade the performance. As a result not all equipment has them, and the ability to compensate for quite common and simple variations like too much or too little bass is lost. The notion that recordings are inherently flawless is seriously misguided, as is the notion that a "perfect" loudspeaker will sound good with all recordings.

We need to be more realistic, acknowledging that "loudspeaker" music is not the same as live music. There are fundamental similarities to be sure, but there are also gross differences. No magic wire, loudspeaker stand or spike, power line filter or trinket on a wall will make it right.

1.5 MUSIC AND MOVIES—THE STATE OF AFFAIRS

Looking at the music side of sound reproduction, the circle of confusion can be reduced to the elements shown in Figure 1.9.

There are problems, all deriving from a lack of uniform loudspeaker performance standards throughout the industry. Sound reproducing systems and devices are highly variable in sound quality, and misguided room equalization practices contribute to that variability. In the consumer world there is now a document, ANSI/CTA-2034-A (2015), which describes measurements on loudspeakers that embrace the spinorama-style data acquisition and presentation described in Section 5.3. As will be made clear in this book, these data are reliable indicators of potential sound quality and, to a limited extent, of what we measure in rooms. However, they are hard to find, partly because manufacturers are content with the present situation in which the bulk of published specifications reveal nothing truly useful, as explained in Section 12.1.1. With rare exceptions, consumers and professionals are deprived of useful data, and loudspeakers in the marketplace continue to exhibit all possible variations in sound quality.

One might think that the professional audio world would be different, but even there it is rare to find usefully descriptive measurements in the specifications for monitor loudspeakers. To a significant extent it is perpetuated by a mistrust of measurements, and a misplaced trust in opinions, including their own. Most subjective evaluations are uncontrolled, and therefore subject to bias.

Some professionals and academics employ ITU-R BS.1116–3 (2015) as a basis for evaluating and establishing system performance. Unfortunately, as will be explained in Section 13.2.2, it embraces a performance target that is different from the default

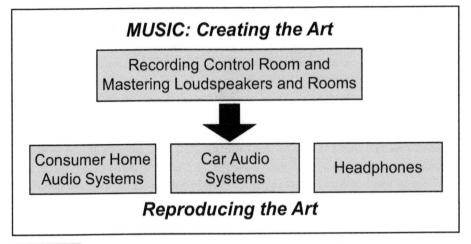

FIGURE 1.9 *The elements that need to possess important common qualities if music recordings are to be reliably delivered to listeners.*

standard throughout the domestic audio industry, and that does not align with guidance provided by decades of controlled subjective evaluations.

Room equalization/correction schemes simply muddy the waters. In spite of advertised claims, these systems are not capable of doing what is stated. Rooms cannot be changed by equalization, and with no knowledge of either the loudspeaker or the room, it is not possible to interpret a measured room curve. Some problems can be addressed by equalization, but many cannot. The sound quality can be changed, just as it can be with simple equalizers, but while it may be possible to improve the situation with seriously flawed loudspeakers, they can never be elevated to high performance standards, and worse, it is possible for the inherent excellence of truly good loudspeakers to be degraded. This topic will come up in several future chapters.

It will be shown that the knowledge exists to change this. However, fortunately, there is a decades-old unwritten rule that loudspeakers should have a "flattish" on-axis frequency response. The idea is that the first sound to arrive at a listener's ears, the direct sound, should be a close replica of the timbre of the recording. The rule is occasionally broken, and frequently distorted, but overall, there is a demonstrable similarity among both professional monitor and consumer loudspeakers in this respect.

In the movie sound domain things are different. The Society of Motion Picture and Television Engineers (SMPTE) and the International Organization for Standardization (ISO—yes, the letters have been rearranged) have basically the same standard for calibrating the B-chain (the equalizers, power amplifiers, loudspeakers and room) in dubbing stages and cinemas. The target is the well-known X-curve, which is fundamentally different from the frequency-response objective employed in all other areas of the audio industry. This creates problems when sound tracks are replayed in home theaters, over TV sets and so on, but there should at least be consistency within the domain of dubbing stages where sound tracks are created, and cinemas where they are reproduced for audiences. Figure 1.10 illustrates the situation.

It would be wrong to assume that all X-curve calibrated systems measure and sound the same. It is a long story, discussed in Chapter 11. When I say "mostly" in Figure 1.9, there are no statistics on the thousands of cinemas distributed around the world (150,000+ in major countries), and it is easy to imagine that many of these subscribe to no standards whatsoever. Some knowledgeable people in the business believe that these sound systems should be more closely allied with those in the rest of the audio industry, and have been acting on this belief. So, standards exist, but standardized sound quality does not. I have experienced some movie industry "reference" cinemas, calibrated according to the appropriate standard, and the sound quality was not impressive. In one dramatic case, turning off the calibrated equalization in the cinema dramatically improved the sound quality—and this was not just my opinion. This is one more reason to watch movies in my excellent home theater (another reason is the "pause" control, and one more is subtitles that reveal dialog rendered ambiguous by theatrical mumbling, or masked by multidirectional music and sound effects).

In an earlier edition of this book, I drew attention to these problems (section 18.2.6), observing: "It seems that no real science has been done." A film sound professional read this, contacted me, and set about trying to improve the situation. The result was a SMPTE investigatory committee to look into the situation. Problems were found

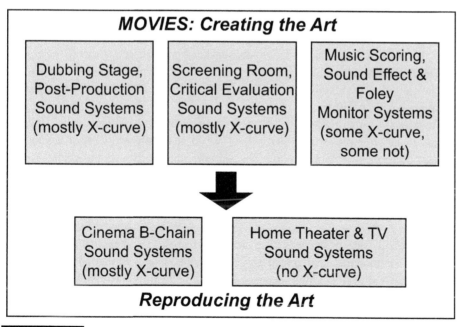

FIGURE 1.10   *The top row shows the elements employed in creating a movie sound track and evaluating it before approval for display. The bottom row shows the circumstances within which consumers get to experience the sound tracks. There are significant differences.*

and a standards committee was set up. Many measurements were made in cinemas and dubbing stages, and a major report was published detailing the findings (SMPTE TC-25CSS, 2014). At time of writing, there are SMPTE and AES standards groups looking at ways to improve the calibration of sound reproduction facilities, and I wrote a paper attempting to summarize the existing science to assist in the process (Toole, 2015b). All of this is discussed in detail in Chapter 11.

1.6 THE ROLE OF LOUDSPEAKERS AND ROOMS

As has been shown, loudspeakers and the rooms they are in influence the "art" as it is being created, and then, again, as it is reproduced. In fact, it will be shown that loudspeakers are the single most important element in sound reproduction. Electronic devices, analog and digital, are also in the signal paths, but it is not difficult to demonstrate that in competently designed products, any effects they may have are small if they are not driven into gross distortion or clipping. In fact, their effects are usually vanishingly small compared to the electro-acoustical and acoustical factors. Tests of these effects quickly become exercises in "is there or is there not *any* difference?" This was the origin of the well-known ABX test, which has shown with monotonous regularity that well-designed power amplifiers, loudspeaker wires and the like are not responsible for offensive sounds. Occasionally a test may show that a difference was observed at a level of statistical significance. This is of importance only if the observers can state a preference—which one is more real or more accurate? It is human nature to think that hearing any difference is associated with an improvement, which is a reason that A vs. B tests need to be randomly balanced with B vs. A tests. Conducting meaningful listening tests is a science unto itself (see Chapter 3).

If we are to set an objective for loudspeaker performance, it should be timbral neutrality. They should not add or subtract from the art being reproduced whether the audience is the mixer creating the product, or the consumer enjoying it. Loudspeakers should not be *part* of the art.

Traditionally, loudspeakers have been chosen "by ear"—subjectively. Underlying this is the widespread assumption that "we all hear differently," so it has to be a personal decision. It is definitely true that we are individualistic in our preferences of "wine, persons and song," but sound quality turns out to be different. In fact, learning this was a turning point in my career when, back in the mid-1960s, I ran some crude listening tests on a few highly regarded loudspeakers (described in Section 18.1). They sounded very different from each other, and the anechoic chamber measurements on them confirmed large differences. But, at the end of the blind evaluation two things were clear: (1) most listeners agreed on what they liked, and (2) the loudspeakers that they liked had the best looking (i.e., smoothest and flattest) frequency responses. To all present, this was a revelation. Nothing has changed since then, but much more detail has been added to the story.

Now we can identify "neutrality" by looking at the right set of measurements. Very "neutral" sounding loudspeakers are available to both professionals and consumers, and many of them are at affordable prices. The difficulty is in identifying such products because comprehensive and trustworthy performance data is rarely exhibited. In the early 1980s, when I was at the National Research Council of Canada, I developed a program of measurements for loudspeakers. During my tenure at Harman International it evolved into what is now informally known as the "spinorama." A wealth of research results support its usefulness, and is now embodied in an industry standard for loudspeaker measurements. It is seen in published specifications for a few products, but time alone will tell if it becomes widespread. Most manufacturers are not able to perform the measurements in-house, and many would be embarrassed to see the results if they saw them. So, business being business, it is probably wise to be skeptical of the future for such candor. Nevertheless, the reality is that sound quality in loudspeakers can very substantially be predicted from the right set of measured data. Chapter 5 discusses this in great detail.

Rooms are the final link in the sound reproduction chain, and a significant industry exists to provide acoustical devices and materials of many kinds. They dispense advice that ranges from transparent salesmanship to well-founded acoustical guidance. Some of the products are useful, others merely decorative.

If one is constructing a dedicated listening room, control room or home theater, custom acoustical treatments are likely to be needed. Because it is a "designed" space, there can be no excuse for bad sound.

In contrast, if one is installing a stereo or multichannel system into a normal living space, there may be a reluctance to deviate from typical furnishings and décor. Fortunately, with a little thought, a perfectly satisfactory listening space can be created using absorption provided by heavy, lined drapes; clipped-pile carpet over felt underlay; upholstered furniture; and what I call the paraphernalia of life: bookcases, cabinets, tables, lamps and so on as scattering objects. Irregular wall profiles created by fireplaces, sculpture niches, and the like add value acoustically and visually. In the end, it is important to realize that good sound begins with good loudspeakers. If that is accomplished, the rest is a lot easier. If it is not, no amount of "treatment" or "room EQ" will make it right.

The exception to all of this is what happens at low frequencies in small rooms. Room resonances cause bass notes to vary in level, and bass transients to boom, and it happens differently in different rooms and different seats in the same room. Equalization can bring some relief to a solitary listener, but sharing similarly good bass is difficult, even with intrusive arrays of bass traps. Chapters 8 and 9 address the topic in detail.

1.7 HUMAN ADAPTATION, A REALITY THAT CANNOT BE IGNORED

All of us have spent time listening to sound systems that were seriously deficient. Many still do. Sound quality may be imperfect, but somehow the music survives to provide enjoyment. Simple pleasure does not need perfection in an audio system. Yet, when we hear excellence, it is recognizable and desirable. But it is significant that we are able, to some extent at least, to adapt to an imperfect situation and derive pleasure.

As with all acclimatization processes, there are limits. In vision, we are usually not aware of the color temperature variations as we move from incandescent, to fluorescent, to direct sunlight, to shade—skin tones seem unchanged until we see photographs taken in each of those situations. However, strong colorations in lighting are easily visible. So, as we examine our abilities to hear various virtues and defects in sound, it is important to keep in mind that absolute perfection may not be necessary. But "defects" should be kept either below the threshold of detectability or below the limits of adaptation.

Adaptation affects the results of listening tests. If we compare loudspeakers within the same room, the product ratings are likely to be almost identical to those arrived at in tests performed in a different room. We adapt to the room we are in to an extent that allows us to make reasonable judgments of relative sound quality. If we quickly switch from one loudspeaker in one room to a different loudspeaker in a different room, adaptation is not achieved and the room becomes a confounding, if not the dominant, factor. This is particularly evident in the increasingly popular binaural listening methods. Section 7.6.2 discusses this.

I have noticed this effect when I return home after a trip. I turn my system on and it may sound a little "different." It doesn't last long, and in a few minutes I am back in the familiar context. Nothing has "broken in"; it was me readapting. This is the common situation of product reviewers noting differences in sound of a new product introduced into their listening room, but over time adapting to its characteristics. Apart from tiny changes in woofer resonance frequency, loudspeakers do not "break in"—that is a physical fact—but listeners certainly do.

BREAKING IN

In parts of the audio industry there is a belief that all components – wires, electronics and loudspeakers – need to "break in." Out of the box, it is assumed that they will not be performing at their best. Proponents vehemently deny that this process has anything to do with adaptation, writing extensively about changes in performance that they claim are easily audible in several aspects of device performance. Yet, the author is not aware of any *controlled* test in which any *consequential* audible differences were found, even in loudspeakers, where there would seem to be some opportunities for material changes. A few years ago, to satisfy a determined marketing person, the research group performed a test using samples of a loudspeaker that was claimed to benefit from "breaking in." Measurements before and after the recommended break-in showed no differences in frequency response, except a very tiny change around 30–40 Hz in the one area where break-in effects could be expected: woofer compliance. Careful listening tests revealed no audible differences. None of this was surprising to the engineering staff. It is not clear whether the marketing person was satisfied by the finding.

To all of us, this has to be very reassuring because it means that the performance of loudspeakers is stable, except for the known small change in woofer compliance caused by exercising the suspension and the deterioration—breaking *down*—of foam surrounds and some diaphragm materials with time, moisture, and atmospheric pollutants. See also Sanfilipo, 2005. It is important to note that "breaking-in" seems always to result in an *improvement* in performance. Why? Do all mechanical and electrical devices and materials acquire a musical aptitude that is missing in their virgin state? Why is it never reversed, getting worse with use? The reality is that engineers seek out materials, components, and construction methods that do *not* change with time. Suppose that the sound did improve over time as something broke in. What then? Would it eventually decline, just as wine goes "over the hill"? One can imagine an advertisement for a vintage loudspeaker: "An audiophile dream. Model XX, manufactured 2004, broken in with Mozart, Schubert, and acoustic jazz. Has never played anything more aggressive than the Beatles. Originally $1700/pair. Now at their performance peak—a steal at $3200!"

1.8 HUMAN SUGGESTIBILITY

One of the remarkable, but frustrating, attributes of human behavior is the extent to which we can demonstrate "mind over matter" dominance in some things. In the medical world, it has been demonstrated many, many times that placebos sometimes function as well as prescription drugs. The power of belief can alleviate symptoms, and even influence a cure. In wine tasting, a well-known label or a high price predisposes tasters to like a product. Blind and double-blind procedures are widely used in both of these activities in order to get closer to the truth.

In audio there are numerous parallel examples of listeners hearing qualities in sounds that are not, or simply cannot, be there. If one believes that there will be a difference, there very likely will be a difference. The sound waves impinging on the eardrums have not changed, but the perceptual process of the brain has decided that there is a difference. Double-blind tests may show that there is no difference, but such is the strength of belief that some people argue that it is the test that is at fault, not the reality that no change in sound existed. Some audio journalists promote these ideas, and products possessing these mystical powers come and go. Thus has evolved what some call "faith-based audio." It exactly parallels the spiritual version.

> In science, contrary evidence causes one to question a theory. In religion, contrary evidence causes one to question the evidence.

If there is to be a resolution to this matter, it may require the recruitment of resources outside the domains of physics and engineering.

Claims of accurate sound quality reproduction began early. Thomas Edison, in 1901, claimed that the phonograph had no "tone" of its own. To prove it, he mounted a traveling show in which his phonograph was demonstrated in "tone tests" that consisted of presentations with a live performer. Morton (2000) reports, "Edison carefully chose singers, usually women, who could imitate the sound of their recordings and only allowed musicians to use the limited group of instruments that recorded best for demonstrations" (p. 23). Of a 1916 demonstration in Carnegie Hall before a capacity audience of "musically cultured and musically critical" listeners, the *New York Evening Mail* reported that "the ear could not tell when it was listening to the phonograph alone, and when to actual voice and reproduction together. Only the eye could discover the truth by noting when the singer's mouth was open or closed" (quoted in Harvith and Harvith, 1987, p. 12).

Figure 1.11 shows that according to marketing claims, perfection in sound was achieved a century ago.

Singers had to be careful not to be louder than the machine, to learn to imitate the sound of the machine, and to sing without vibrato, which Edison (apparently a musically uncultured person) did not like. There were other consequences of these tests on recordings. The low sensitivity of the mechanical recording device made it necessary for the performers to crowd around the mouth of the horn and find instruments that

FIGURE 1.11   *Singer Frieda Hempel stages a tone test at the Edison studios in New York City, 1918. Care was taken to ensure that the test was "blind," but it is amusing to see that some of the blindfolds also cover the ears.*
Courtesy of Edison National Historic Site, National Park Service, US Department of the Interior.

could play especially loud. Because the promotional "tone tests" were of solo voices and instruments, any acoustical cues from the recording venue would reveal the recording as being different from the live performer in the demonstration room. Consequently, in addition to employing what was probably the first "close microphone" recording technique, Edison's studios were acoustically dead (Read and Welsh, 1959, p. 205). All of this is quite an elaborate deception at the very origin of the audio industry, but for listeners there had to be an element of amazement that anything approximately musical could be recorded and reproduced. It is probable that today's listeners might be less easily impressed.

Live vs. reproduced comparison demonstrations of loudspeakers were conducted by RCA in 1947 (using a full symphony orchestra [Olson, 1957, p. 606]) and Wharfedale in the 1950s (Briggs, 1958, p. 302). The purpose was to persuade audiences that near perfection in sound reproduction had arrived. Many thought it had, but others were not convinced. Acoustic Research mounted several live vs. reproduced comparisons in the late 1950s and 1960s using a string quartet and a guitar.

These were not the "party trick" events that Edison staged, but serious demonstrations conducted by knowledgeable people in respected companies. Villchur (1964b) provides great insight into the process. Not everyone was persuaded that those loudspeakers were good enough, but many were. Have we come much further half a century later? Are today's loudspeakers significantly better sounding than those of decades past? For the most part, the answer is "Yes!" Would the person on the street or even a "musically cultured" listener discern the improvement in such a demonstration using the best of today's loudspeakers? Are we now wiser, more aurally acute, and less likely to be taken in by a good demonstration?

Obviously there were careful preparations to improve the chances of success. Great care was taken in the microphone placement to ensure a good reproduction, along with judicious use of equalization (Villchur, 1964b). The need for close microphones to avoid picking up room sound at the recording site made matters more difficult. This means that the tiny sample of the total radiated sound of the instrument would be used to represent the whole sound field.

We may never know all that was done to tailor the recordings that were created for any of these events. But, from one perspective, all of them had the advantage of avoiding the "circle of confusion": recordings were created for reproduction through specific loudspeakers. Most of the comparison demonstrations took place in concert halls or auditoriums and therefore had a distinct advantage. As discussed in Section 1.1, the hall sound embellishes live performances, and in these tests that embellishment would be there for the reproduced sounds too. Villchur (1964b) recognized this, saying: "Playback at the recording site may not sound as similar to the live music as at the actual concert, because the reverberation which will cloak both live and reproduced music equally is missing." In evaluations of concert halls, the sense of envelopment is the major factor in how they are rated. Now there is evidence that perceptions of spaciousness may be comparable to timbral accuracy in providing listener satisfaction in loudspeaker listening tests (Section 7.4.4).

In any event, loudspeakers that by today's standards would be criticized in properly conducted listening tests using commercial recordings were judged to sound fine half a century ago. A cynic might conclude that striving for what used to be called "high fidelity" is futile. I am reminded of the Flanders and Swann (a clever British musical comedy team) "Song of Reproduction" from 1957 that, with a few changes in technical jargon, is still true (it is available on the Internet).

Nevertheless, when confronted with a choice of sound qualities in double-blind tests, listeners in general, even those claiming to have "tin ears," can recognize and like neutral loudspeakers. We know how to describe them in measurements. We all should have them.

CHAPTER 2

A Scientific Perspective on Audio

With its origins based in musical arts, audio has evolved with a very substantial influence from opinion, and even emotion. In the early days, measurements were primitive, and it was decades before there was a capability to measure the performance of audio components, especially loudspeakers, in a meaningful way. There was a long period of subjective vs. objective debating, during which it became an unwritten "truth" that "we cannot measure what we hear." In the beginning there was truth in the statement, but then technology improved and the scientific method began to be applied.

For many in audio, this was unpopular; it was a challenge to the way things had been done. It is so much easier just to listen to something and offer an opinion. The more "qualified" the person, the better the opinion. Musicians were, by definition, qualified. Laypersons—audiophiles—could simply assert that they qualified by virtue of dedication, and some went on to become reviewers.

We can look back now and see that musicians who were not also audiophiles are really not very observant listeners; they listen to the music and can accept quite colored renderings of it as "valid interpretations." From the perspective of an audience member, the impression might be quite different.

Audio journalism is a business, with income, expenses, writers and reviewers to be paid, advertisers to be found and satisfied, and next month's issue to be prepared. Consequently, there are compromises, most of them understandable. Product reviews are a big attraction for readers, but most often there are limited or no proper measurement facilities, measurements or data analysis, no dedicated venues for controlled double-blind listening, and loudness-controlled comparisons. With such little opportunity to conduct meaningful subjective and objective evaluations, the usual result is pages of articulate, often highly literate and colorful prose—opinions. When challenged to produce validation for those opinions, there is pushback.

FIGURE 2.1 *A cartoon I drew after spending time at a high-end audiophile show.*

The hard-core subjectivists created a straw man: a scientist/engineer in a white lab coat, dedicated to numbers and graphs. It is asserted that such people could not care about the "truths" buried within reproduced sounds: the subtle sonic details, soundstage nuances, the emotion and so on. It was an imagined battle between "cold" data and poetic descriptions of reactions to recorded sounds through a product under review. If these people believed in science, and had access to useful, accurate measurements, this situation might not have come about. Both the reviewers and the readers would benefit.

Science is true, whether you believe in it or not.

—Neil deGrasse Tyson

In one case at least, *Stereophile* magazine, useful measurements are made (full credit to John Atkinson), but then, by my observation, these data are rarely incorporated in the discussion by the subjective reviewers who practice the "take-it-home-and-listen-to-it" method. Adaptation sets in, and numerous non-auditory influences are at play. There are many quotable passages in these reviews, but one sticks in my memory. It pertained to a loudspeaker with significant faults, clearly revealed in Atkinson's measurements (and ours), and one can only presume heard by the reviewer. The subjective reviewer, perhaps believing in "breaking in," persisted and stated that after living with the speaker for about two months he experienced an epiphany, and he suddenly heard wonderful things. So, did the loudspeaker "break in"? No. Did the reviewer come to an accommodation with the faults in it? Yes. In the absence of comparison listening to other products, the reviewer will never know what really happened. Readers were encouraged to go spend a lot of money on a mediocre product.

This pattern of the (usually praiseworthy) subjective reviews and the evidence of flaws in the measurements was oft repeated. This said, Atkinson is on record as saying: "as I pointed out in my 1997 AES convention paper, our reviewers' ratings generally follow Floyd's findings: that people tended to prefer speakers with flat response and controlled dispersion" (Atkinson, 2011). This may be his personal synopsis of events, but it would be more convincing to see quantitative evidence of it collected from the many subjective and objective evaluations his publication conducts. The measurements are adequate, although the presentation format could be improved knowing what we do now. The subjective data are the result of undisciplined, sighted listening by essayist reviewers.

The 45th anniversary issue of *Stereophile* magazine (November 2007) contained an editorial by John Atkinson, who interviewed J. Gordon Holt, the man who created the magazine. Holt commented as follows:

As far as the real world is concerned, high-end audio lost its credibility during the 1980s, when it flatly refused to submit to the kind of basic honesty controls (double-blind testing, for example) that had legitimized every other serious scientific endeavor since Pascal. [This refusal] is a source of endless derisive amusement among rational people and of perpetual embarrassment for me, because I am associated by so many people with the mess my disciples made of spreading my gospel.

As I said, at least Atkinson does useful measurements, so technically savvy readers can discern something close to the true performance of the product. Another source of reliable loudspeaker measurements is www.soundstagenetwork.com, in their "measurements library." There one will find anechoic measurements made in the facility I created while working at the National Research Council of Canada (NRCC). The now defunct magazines *AudioScene Canada* and *Sound Canada* paid for NRCC anechoic measurements, and the editorial staff traveled to Ottawa to participate in double-blind listening tests in the IEC (International Electrotechnical Commission) standard listening room. They got to see the measurements only after the listening was done. That was a powerful combination, providing the public with what I believe to be the best published loudspeaker reviews at the time, and perhaps ever. But that, unfortunately, is history. All of this data was incorporated into my research. Through these activities, I was able to experience a steady flow of contemporary loudspeakers, all adding to the scientific sub-jective/objective database. Thank you!

Perhaps things will change, but perhaps not, because reviewing is a business, and doing it right costs money and time that budgets and schedules don't normally permit. A few publications perform simplified measurements, a step in the right direction, but those that eschew all measurements offer no possibility for knowledgeable readers to form their own opinions. It requires the readers to place their faith in the opinions of audio journalists, some of whom claim to hear things that, sometimes, the laws of physics don't permit. Layered over all opinions is a "fog" of uncertainty related to the mostly imagined influence of wires, power cords and conditioners, spikes and numerous other audiophile beliefs. I think I created the following quote—apologies if I am wrong, but it sums up a lot of what happens in audio, where nowadays some of it is clearly what I call "faith based."

In science, contrary evidence causes one to question a theory. In religion, contrary evidence causes one to question the evidence.

—Floyd Toole

As long as it remains a hobby interest that does not compromise a family budget, no harm is done. The real truth is that no progress could have been made in understanding the psychoacoustics of audio without beginning with subjective evaluations. What we hear is the definitive event; our challenge is to find measurements that correlate with subjective opinions. Psychoacoustics is the relationship between what we hear and what we measure—it is an interplay of subjective data and measured data. Over the years we have become very good at acquiring useful data of both kinds.

It is therefore especially disappointing that there is more evidence of closing the subjective/objective loop in the consumer product domain than there is in the profes-sional audio domain. Professionals place great importance in having their products "translate" to different scenarios. Having experienced many of those "translations" it is evident that it is not a high bar to clear. Accuracy in sound quality is not a prime factor.

I see measurements in several publications that review consumer loudspeakers and, as I said earlier, some are better than others. But I see none in publications reviewing monitor loudspeakers. Professionals should be leading the way. What they do is per-manently embedded in the recordings. I have known many audio professionals over the years, and some of them speak disparagingly about consumer audio, while clearly not knowing much of consequence about it.

I have spent most of my professional career in scientific research. Given that research is based on learning, sometimes from mistakes, it was interesting that a science-based, entrepreneurial person I greatly respect, Elon Musk, CEO and visionary for Tesla, Space-X and other high-technology enterprises, said in a 2016 interview (and I may be paraphrasing slightly): "I am always to some degree wrong. The aspiration is to be less wrong . . . and trying to minimize the wrongheadedness of a time. I believe in that philosophy."

So do I. Being less wrong makes one a better scientist or engineer. But the larger issue is the "wrongheadedness" of a significant part of the populace that currently lacks respect for science and, it seems in some respects, facts.

2.1 REQUIREMENTS FOR SCIENTIFIC INVESTIGATIONS

Electrical and acoustical measurements have been with us a long time, and we became very good at doing them a long time ago. Now we can do them faster and less expensively, and we are able to measure more dimensions, but good data from decades ago is as trustworthy as good data from today. The problem in audio was that the data were limited, in analog form and not easy to interpret. Now, with more data, all in digital form, it is possible to manipulate it in ways that were impossible or impractical in the analog era. It is possible to create versions of the data that bear a closer relationship to what we hear than the raw data itself. Interpretation with the eyes becomes more intuitive.

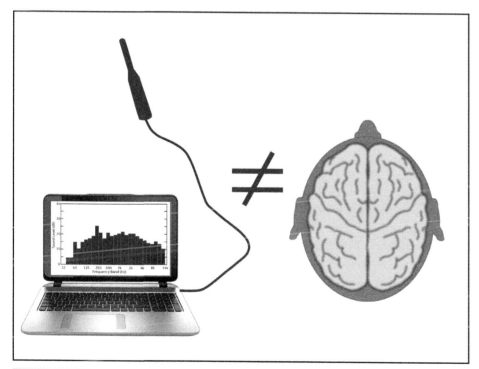

FIGURE 2.2 *There is a fundamental difference between simple RTA measurements and human perception. They are not totally unrelated, but they are not equivalent.*

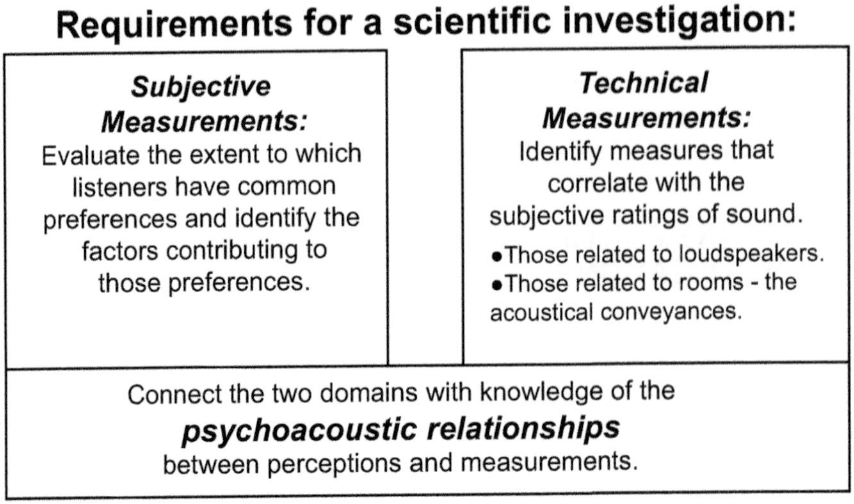

Requirements for a scientific investigation:

Subjective Measurements: Evaluate the extent to which listeners have common preferences and identify the factors contributing to those preferences.	***Technical Measurements:*** Identify measures that correlate with the subjective ratings of sound. • Those related to loudspeakers. • Those related to rooms - the acoustical conveyances.

Connect the two domains with knowledge of the
psychoacoustic relationships
between perceptions and measurements.

FIGURE 2.3 *The elements of scientific investigations in audio.*

Historically, sound reproducing systems were evaluated, in part at least, using an omnidirectional microphone and a 1/3-octave real-time analyzer (RTA). However, when immersed in the complex sound field of a normally reflective room, two ears and a brain are far more analytical than such a simple system. A microphone responds to all sounds from all directions at all times, without discrimination. A human listener does discriminate, and does so in terms of both sound quality/timbre and spatial attributes. Figure 2.2 illustrates the situation.

The key to understanding technical measurements is having reliable, repeatable, subjective data. In contrast to the assertions of some audio subjectivists that "scientists" don't listen or cannot hear, the fact is that subjective evaluations are at the very heart of the investigations. Without them the meaning of measurements cannot be revealed. With a foundation of scientific knowledge guiding engineering efforts, good-sounding products can become more numerous and more cost-effective. Figure 2.3 illustrates the elements of the scientific method as it applies to audio.

The title of the top left block, "Subjective Measurements," may seem odd, but when one experiences the abilities of listeners to report impressively repeatable subjective ratings in double-blind listening tests, the description is appropriate. Technical measurements, properly done, have always been repeatable. The problem has been in knowing what they mean in subjective terms. That is the role of psychoacoustics: connecting the two domains.

CHAPTER 3

Subjective Measurements—Turning Opinion into Fact

First, let me express a personal view on word usage. It has to do with the common expression "subjective listening tests." Of course, they are "subjective" because they involve "listening." But staying with word usage, it may seem oxymoronic to place the words *subjective* and *measurements* in such proximity. However, when it is possible to generate numerical data from listening tests, and those numbers exhibit relatively small variations and are highly repeatable, the description seems to fit. I recall noting this in my very early double-blind listening tests when I employed the simplest metric of variation: standard deviation. In those days, using a 10-point scale of "fidelity," the very best listeners were repeating their ratings over extended periods of time, with standard deviations around 5%. I found this remarkable, because at that time analog VOM (volt-ohm-milliammeter) multi-testers had similar specifications. Yes, these were measurements. Humans can be remarkable "measuring instruments" if you give them a chance.

This, more than anything else, has allowed the exploration of correlations between technical measurements and subjective opinions. Technical measurements, after all, don't change, but however accurate and repeatable they may be, they are useless without a method of interpreting them. We need a way to present them to our eyes so they relate more closely to what we hear with our ears: perceptions.

Subjective measurements provide the entry point to understanding the psychoacoustic relationships. The key to getting useful data from listening tests is in controlling or eliminating all factors that can influence opinions other than the sound itself. This is the part that has been missing from most of the audio industry, contributing to the belief that we cannot measure what we hear.

In the early days, it was generally thought that listeners were recognizing excellence and rejecting inferiority when judging sound quality. Regular concertgoers and musicians were assumed to be inherently more skilled as listeners. As logical as this seemed, it was soon thrown into question when listeners in the tests showed that they could rate products just as well with studio-created popular music as they could with classical music painstakingly captured with simple microphone setups, and sometimes even better. How could this be possible? None of us had any idea how the studio creations should sound, with all of the multitrack, close-miked, pan-potted image building and signal processing that went into them. The explanation was in the comments written by the listeners. They commented extravagantly on the problems in the poorer products, heaping scorn rich in adjectives and occasional expletives on things that were not right about the sound. In contrast, high-scoring products received only a few words of simple praise.

And what about musicians? Their performances in listening tests were not distinctive. Some had the occupational handicap of hearing loss; others were clearly more interested in the music than in the sound, and could find "valid interpretations" of instrumental sound in even quite poor loudspeakers. The best among them often had audio as a hobby (Gabrielsson and Sjögren, 1979; Toole, 1985).

People seemed to be able to separate what the loudspeakers were doing to the sound from the sound itself. The fact that, from the beginning, all the tests were of the "multiple-comparison" type may have been responsible. Listeners were able to freely switch the signal among three or four different products while listening to the music. Thus, the "personalities" of the loudspeakers were revealed through the ways the program changed. In a single-stimulus, take-it-home-and-listen-to-it kind of test, this would not be obvious. Isolated A vs. B comparisons fail to reveal problems that are common to both test sounds. The experimental method matters.

Humans are remarkably observant creatures, and we use all our sensory inputs to remain oriented in a world of ever-changing circumstances. So when asked how a loudspeaker sounds, it is reasonable that we instinctively grasp for any relevant information to put ourselves in a position of strength. In an extreme example, an audio-savvy person could look at the loudspeaker, recognize the brand and perhaps even the model, remember hearing it on a previous occasion and the opinion formed at that time, perhaps recall a review in an audio magazine and, of course, would have at least an approximate idea of the cost. Who among us has the self-control to ignore all of that and to form a new opinion simply based on the sound?

It is not a mystery that knowledge of the products being evaluated is a powerful source of psychological bias. In comparison tests of many kinds, especially in wine tasting and drug testing, considerable effort is expended to ensure the anonymity of the devices or substances being evaluated. If the mind thinks that something is real, the appropriate perceptions or bodily reactions can follow. In audio, many otherwise serious people persist in the belief that they can ignore such non-auditory factors as price, size, brand and so on.

This is especially true in the few "great debate" issues, where it is not so much a question of how large a difference there is but whether there *is* a difference (Clark, 1981, 1991; Lipshitz, 1990; Lipshitz and Vanderkooy, 1981; Nousaine, 1990; Self,

1988). In controlled listening tests and in measurements, electronic devices in general, speaker wire and audio-frequency interconnection cables are found to exhibit small to non-existent differences. Yet, some reviewers are able to write pages of descriptive text about audible qualities in detailed, absolute terms. The evaluations reported on were usually done without controls because it is believed that disguising the product identity *prevents* listeners from hearing differences. In fact, some audiophile publications heap scorn on double-blind tests that fail to confirm their opinions—the data must be wrong.

Now high-resolution audio has become a "great debate" issue, where it has come down to a statistical analysis of blind test results to reveal whether there is an audible difference and if that difference results in a preference. Meanwhile, some "authorities" claim that it is an "obvious" improvement. As I review the numerous objective tests and controlled subjective results, if there is a difference, it is very, very small, and only a few people can hear it, even in a concentrated comparison test. A difference does not ensure a superior listening experience—that is another level of inquiry. At this point, I lose interest because there are other unsolved problems that are easily demonstrated to be audible, and even annoying. I have dedicated my efforts to solving the "10 dB" problems first, the "5 dB" problems next, and so on. There are still some big problems left. So, unless audiophiles *en masse* are spending money that should go to feeding their families, these are all simply "great debates" that will energize discussions for years to come.

Science is routinely set up as a "straw man," with conjured images of wrongheaded, lab-coated nerds who would rather look at graphs than listen to music. In a lifetime of doing audio research, I have yet to find such a person. Rational researchers fully acknowledge that the subjective experience is "what it is all about," and they use their scientific and technical skills to find ways to deliver rewarding experiences to more people in more places. If something sounds "wrong," it *is* wrong. The task is to identify measurements that explain *why* it sounds wrong and, more importantly, to identify those measurements that contribute to it sounding "right."

In the category of loudspeakers and rooms, however, there is no doubt that differences exist and are clearly audible. Because of this, most reviewers and loudspeaker designers feel that it is not necessary to go to the additional trouble of setting up blind evaluations of loudspeakers. This attitude has been tested.

3.1 IS BLIND LISTENING NECESSARY?

When I joined Harman International, listening tests were casual affairs, as they were and still are in many audio companies. In fact, a widespread practice in the industry has been that a senior executive would have to sign off on a new product, whether they were good listeners or not. After engineers had sweated the design details to get a good product, they might be told, "more bass, more treble, and can you make it louder?" Simply sounding "different" on a showroom floor was considered to be an advantage, and sounding louder was a real advantage given that few stores did equal-loudness comparisons. Some managers even had imagined insights into what the purchasing public would be listening for in the upcoming shopping season. In certain markets a prominent product reviewer might be invited to participate (usually for pay) in a "voicing" session to ensure a good launch of a new product. The final product may or may not have incorporated the advice, and it is unlikely that the deception would ever be discovered.

It was a grossly unscientific process, unjust to the efforts of competent engineers and disrespectful of the existing science, and it contributed to unnecessary variations in product "sounds." Many consumers think that there is a specific "sound" associated with a specific brand. That can happen, but it is not universal and not really very consistent, even if that is what was intended. If you reflect on the circle of confusion (Section 1.4), this does nothing to improve the chances of good sound being delivered to customers. There has to be a better way.

With the demise of so many enthusiast audio stores, consumers have few opportunities to listen for themselves, and almost never in circumstances that might lead to a fair appraisal of products. Shopping is migrating to the Internet, and consumers increasingly

LOUDSPEAKER "VOICING"

Music composition and arrangement involves voicing to combine various instruments, notes and chords to achieve specific timbres. Musical instruments are voiced to produce timbres that distinguish the makers. Pianos and organs are voiced in the process of tuning, to achieve a tonal quality that appeals to the tuner or that is more appropriate to the musical repertoire. This is all very well, but what has it to do with loudspeakers that are expected to accurately reproduce those tones and timbres?

It shouldn't be necessary if the circle of confusion did not exist, and all monitor and reproducing loudspeakers were "neutral" in their timbres. However, that is not the case, and so the final stage in loudspeaker development often involves a "voicing" session in which the tonal balance is manipulated to achieve what is hoped to be a satisfactory compromise for a selection of recordings expected to be played by the target audience. There are the "everybody loves (too much)

bass" voices, the time-tested boom and tizz "happy-face" voices, the "slightly depressed upper-midrange voices" (compensating for overly bright close-miked recordings, and strident string tone in some classical recordings), the daringly honest "tell it as it is" neutral voices and so on. It is a guessing game, and some people are better at it than others. It is these spectral/timbral tendencies that, consciously or unconsciously, become the signature sounds of certain brands. Until the circle of confusion is eliminated, the guessing game will continue, to the everlasting gratitude of product reviewers, and to the frustration of critical listeners. It is important for consumers to realize that it is not a crime to use tone controls. Instead, it is an intelligent and practical way to compensate for inevitable variations in recordings, that is, to "revoice" the reproduction if and when necessary. At the present time, no loudspeaker can sound perfectly balanced for all recordings.

rely on Internet audio forum opinions and product reviews. Clearly, there is a need for minimally biased subjective evaluations, and for useful measurements that consumers might eventually begin to trust.

Continuing the Harman story, a blind test facility was set up and the engineers were encouraged to incorporate competitive evaluations into the product development routine. Some went along with the plan of striving to be "best in class," using victory in the listening test as the indicator that the design was complete. However, a few engineers regarded the process as an unnecessary interruption of their work, and a challenge to their own judgment. At a certain point it seemed appropriate to conduct a test—a demonstration that there was a problem. It would be based on two listening evaluations that were identical, except one was blind and one was sighted (Toole and Olive, 1994).

Forty listeners participated in a test of their abilities to maintain objectivity in the face of visible information about products. All were Harman employees, so brand loyalty would be a bias in the sighted tests. They were about equally divided between experienced listeners, those who had previously participated in controlled listening tests, and inexperienced listeners, those who had not.

Figure 3.1 shows that in the blind tests, there were two pairs of statistically indistinguishable loudspeakers: the two European "voicings" of the same hardware and the other two products. In the sighted version of the test, loyal employees gave the big, attractive Harman products even higher scores. However, the little, inexpensive sub/sat system dropped in the ratings; apparently its unprepossessing appearance overcame employee loyalty. Obviously, something small and made of plastic cannot compete with

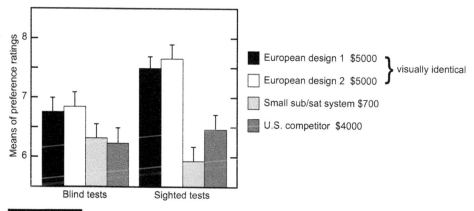

FIGURE 3.1 *The results of blind and sighted listening tests on four loudspeakers, three Harman products and a competing product highly rated in reviews. The two "European design" products differed only in the crossover networks. The sub/sat system was a new design that benefited from excellent engineering, even though it was small and plastic. The lines on top of the vertical bars are the 95% confidence error bars. The large bars must differ by more than this in order for the difference not to be attributable to random factors. From Toole and Olive (1994).*

something large and stylishly crafted of highly polished wood. The large, attractive competitor improved its rating but not enough to win out over the local product. It all seemed very predictable. From the Harman perspective, the good news was that two products were absolutely not necessary for the European marketing regions. (So much for intense arguments that such a sound could not possibly be sold in [pick a country].) In general, seeing the loudspeakers changed what listeners thought they heard.

The effects of room position on sound quantity and quality at low frequencies are well documented. It would be remarkable if these did not reveal themselves in subjective evaluations. This was tested in a second experiment where the loudspeakers were auditioned in two locations that would yield quite different sound signatures. Figure 3.2 shows that listeners responded to the differences in blind evaluations: adjacent bars of the same color have different heights, showing the different ratings when the loudspeaker was in position 1 or 2. In contrast, in the sighted tests, things are very different. First, the ratings assume the same pattern that was evident in the first experiment; listeners obviously recognized the loudspeakers and recalled the ratings they had been given in the first experiment (Figure 3.1). Second, they did not respond to the previously audible differences attributable to position in the room; adjacent bars have closely similar heights. Third, some of the error bars are quite short; no thought is required when you know what you are listening to. Interestingly, the error bars for the two visually identical "European" models (on the left) were longer because the eyes did not convey all of the necessary information.

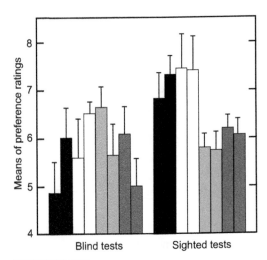

FIGURE 3.2 *A comparison of ratings with loudspeakers in two different locations within the listening room. Adjacent bars of the same shade show results in the two locations for tests conducted under blind and sighted conditions.*

It is normal to expect an interaction between the preference ratings and individual programs. In Figure 3.3 this is seen in the results of the blind tests, where the data have been arranged to show declining scores for programs toward the right of the figure. In the sighted versions of the tests, there are no such changes. Again, it seems that the listeners had their minds made up by what they saw and were not in a mood to change, even if the sound required it. In a special way, they went "deaf." The effect is not subtle.

Dissecting the data and looking at results for listeners of different genders and levels of experience, Figure 3.4 shows that experienced males (there were no females who had participated in previous tests) distinguished themselves by delivering lower scores for all of the loudspeakers. This

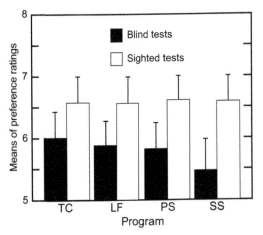

FIGURE 3.3 *Listeners in sighted tests were not responsive to differences caused by different recorded programs.*

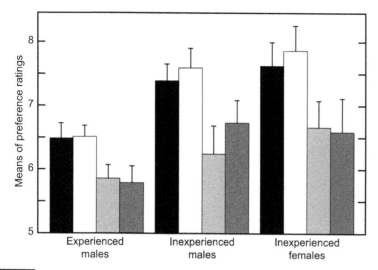

FIGURE 3.4 *The effects of listening experience and sex on subjective ratings of loudspeakers.*

is a common trend among experienced listeners. Otherwise, the pattern of the ratings was very similar to those provided by inexperienced males and females. Over the years, female listeners have consistently done well in listening tests, one reason being that they tend to have closer to normal hearing than males (noisy work environments, or hobbies?). Lack of experience in both sexes shows up mainly in elevated levels of variability in responses (note the longer error bars), but the responses themselves, when averaged, reveal patterns similar to those of more experienced listeners. With experienced listeners, statistically reliable data can be obtained in less time.

Summarizing, it is clear that knowing the identities of the loudspeakers under test can change subjective ratings.

- They can change the ratings to correspond to presumed capabilities of the product, based on price, size or reputation.

- So strong is that attachment of "perceived" sound quality to the identity of the product that in sighted tests, listeners substantially ignored easily audible problems associated with loudspeaker location in the room and interactions with different programs. In sighted tests, it is true to say that listeners heard *less* than they did in blind tests.

These findings mean that if one wishes to obtain candid opinions about how a loudspeaker sounds, the tests *must* be done blind. The answer to the question posed in the title of this section is "yes." The good news is that if the appropriate controls are in place, experienced and inexperienced listeners of both genders are able to deliver useful opinions. Inexperienced listeners simply take longer and more repetitions to produce the same confidence levels in their ratings.

(a)

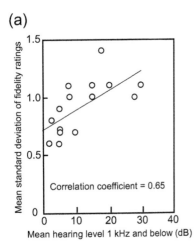

Mean hearing level 1 kHz and below (dB)

(b)

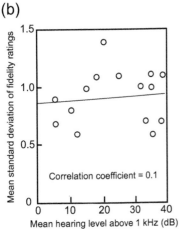

Mean hearing level above 1 kHz (dB)

FIGURE 3.5 *Variability in "fidelity rating" judgments (mean standard deviation) as a function of hearing level (0 dB indicates normal hearing, and higher levels indicate that hearing thresholds have been elevated by that amount. Sounds at lower levels are not audible by those persons. Audiometric measurements were made at the standard frequencies of 250 Hz, 500 Hz, 1 kHz, 2 kHz, 4 kHz and 8 kHz. Hearing levels are the amounts by which the measured thresholds for the listeners are higher than the statistical normal hearing threshold. It is a measure of hearing loss. In (a), hearing levels at 1 kHz and below were averaged. In (b), hearing levels above 1 kHz were averaged. From Toole (1985), figure 8.*

3.2 HEARING ABILITY AND LISTENER PERFORMANCE

Opinions of sound quality are based on what we hear, so, if hearing is degraded we simply don't hear everything, and opinions will change. Hearing loss is depressingly common, and the symptoms vary greatly from person to person, so opinions from such persons are simply their own, and they may or may not be relevant to others.

The first clear evidence of the issue with hearing came in 1982 during an extensive set of loudspeaker evaluations conducted for the Canadian Broadcasting Corporation (CBC), in which many of the participants were audio professionals: recording engineers, producers, musicians. Sadly for them, hearing loss is an occupational hazard (discussed in Chapter 17). During analysis of the data, it was clear that some listeners delivered remarkably consistent overall sound quality ratings (then called fidelity ratings, reported on a scale ranging from 0 to 10) over numerous repeated presentations of the same loudspeaker. Others were less good, and still others were extremely variable—liking a product in the morning and disliking it in the afternoon, for example. The explanation was not hard to find. Separating listeners according to their audiometric performances, it was apparent that those listeners with hearing levels closest to the norm (0 dB) had the smallest variations in their judgments. Figure 3.5 shows examples of the results. Surprisingly, it was not the high-frequency hearing level that correlated with the judgment variability but that at frequencies at or below 1 kHz. Bech (1989) confirmed the trend.

Noise-induced hearing loss is characterized by elevated thresholds around 4 kHz. Presbycusis, the hearing deterioration that occurs with age, starts at the highest frequencies and progresses downward. These data showed that by itself, high-frequency hearing loss did not correlate well with trends in judgment variability (see Figure 3.5b). Instead, it was hearing level at lower frequencies that showed the correlation (Figure 3.5a). Some listeners with high-frequency loss had normal hearing at lower frequencies, but all listeners with low-frequency hearing loss also had loss at high frequencies—in other words, it was a broadband problem. We now know that elevated hearing thresholds are often accompanied by reduced ability to separate sounds in space, which would include separating the listening room from the sound of the loudspeaker, which adds significant difficulty to the task. This is very likely a broadband effect.

Illustrating the situation more clearly is Figure 3.6, which displays the hearing performance of the author as a function of age, and a collection of threshold measurements for the high-variability listeners in Figure 3.5. The hearing threshold data are shown superimposed on the family of equal-loudness contours, the sound levels at which different pure tones

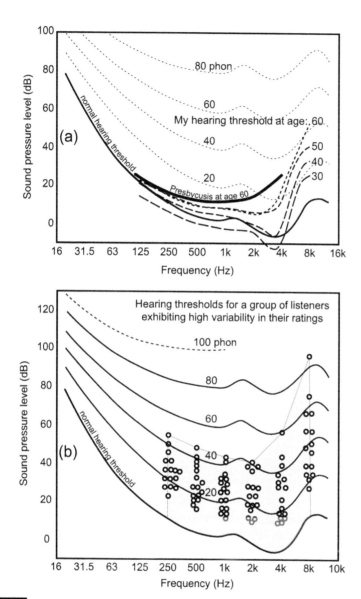

FIGURE 3.6 *(a) A history of the author's hearing threshold compared with the statistical normal hearing threshold and with the normally anticipated hearing loss as a function of age (presbycusis). (b) The hearing thresholds for listeners who had high variations in repeated sound quality ratings of loudspeakers. Adapted from Toole (1985), figure 7. Equal loudness contours are from ISO 226 (2003). © ISO. This material is adapted from ISO 226.2003 with permission of the American National Standards Institute (ANSI) on behalf of the International Organization for Standardization (ISO). All rights reserved.*

are judged to be equally loud. The bottom curve is the hearing threshold, below which nothing is heard. A detailed explanation is in Section 4.4.

My thresholds at age 30 were slightly below (better than) the statistical average. With age, and acoustical abuses of various kinds, the hearing level rose gradually, most dramatically at high frequencies. Even so, my deterioration was less than the population average presbycusis for age 60. My broadband auditory dynamic range was degraded by only 10 dB or so over those years, except that high frequencies disappeared at a higher rate.

In contrast, the professional audio people whose audiometric data are shown in Figure 3.6b were unable to hear large portions of the lower part of the dynamic range. Without the ability to hear small acoustical details in music and voices (good things) or low-level noises and distortions (bad things), these people were handicapped in their abilities to make good or consistent judgments of sound quality, and this was revealed in their inconsistent ratings of products. It did not prevent them from having opinions, deriving pleasure from music, or being able to write articulate dissertations on what they heard, but their opinions were not the same as those of normal hearing persons. Consequently, future listeners underwent audiometric screening, and those with broadband hearing levels in excess of about 20 dB were discouraged from participating.

All of this emphasis on normal hearing seems to imply that a criterion excluding listeners with a greater than 15–20 dB hearing level may be elitist. According to USPHS (US Public Health Service) data, about 75% of the adult population should qualify. However, there is some concern that the upcoming generations may not fare so well because of widespread exposure to high sound levels through headphones and other noisy recreational activities.

Hearing loss occurs as a result of age and accumulated abuse over the years. Whatever the underlying causes, Figure 3.7 shows that in terms of our ability to make reliable judgments of sound quality, we do not age gracefully. A couple of the data points indicate that young persons are not immune to hearing loss. It certainly is not that we don't have opinions or the ability to articulate them in great detail; it is that the opinions themselves are less consistent and possibly not of much use to anyone but ourselves. In my younger years, I was an excellent listener—one of the best, in fact. However, listening tests as they are done now track not only the performance of loudspeakers but of listeners—the metric shown in Figure 3.9. At about age 60, it was clear that it was time for me to retire from the active listening panel. Variability had climbed and, frankly, I found it to be a noticeably more difficult task. It is a younger person's pursuit. Music is still a great pleasure, but my opinions are now my own. When graybeards expound on the relative merits of audio products, they may or may not be relevant. But be polite—the egos are still intact.

Figure 3.8 shows results for individual listeners who evaluated four loudspeakers in a monophonic multiple-comparison blind test. Why mono? The explanation is in the next section. Multiple-comparison test? All four loudspeakers were available to be heard by the listener, switched at will at the push of a button. Loudness levels were normalized. Each dot is the mean of several judgments on each of the products, made by one listener. To illustrate the effect of hearing level on ratings, listeners were grouped according

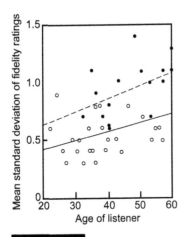

FIGURE 3.7 *Variations in "fidelity" ratings versus the age of the listener. The results are for two sets of experiments. The black dots and dashed regression line represent an early test in which all controls were not in place. The open circles and solid regression line are from well-controlled tests and therefore show lower variability. However, both sets of data display the same overall trend of increasing variability with increasing age.*
From Toole (1985), figure 9.

to the variability in their judgments: those who exhibited standard deviations below 0.5 scale unit and those above that. The low-variability listener results are shown in the white areas and the high-variability results are in the shaded areas. Obviously, the inference is that the ratings shown in the shaded areas are from listeners with less than normal hearing.

In Figure 3.8, it is seen that fidelity ratings for loudspeakers D and U are closely grouped by both sets of listeners, but the ratings in the shaded areas are simply lower. Things change for loudspeakers V and X, where the close groupings of the low-variability listeners contrast with the widely dispersed ratings by the high-variability listeners. This is a case where an averaged rating does not reveal what is happening. Listener ratings simply dispersed to cover the available range of values; some listeners thought it was not very good (fidelity ratings below 6), whereas others thought it was among the best loudspeakers they had ever heard (fidelity ratings above 8). Listeners simply exhibited strongly individualistic opinions. Hearing loss is very likely involved. Be careful whose opinions you trust.

Other investigations agree. Bech (1992) observed that hearing levels of listeners should not exceed 15 dB at any audiometric frequency and that training is essential. He noted that most subjects reached a plateau of performance after only four training sessions. At that point, the test statistic F_L should be used to identify the best listeners. Olive (2003) compiled data on 268 listeners and found no important differences between the ratings of carefully selected (for normal hearing) and trained (to be able to recognize resonances) listeners and those from several other backgrounds—some in audio, some not, some with listening experience, some with none. There were, as shown in Figure 3.9, huge differences in the variability and scaling of the ratings, so selection and training have substantial benefits in time savings. Rumsey et al. (2005) also found strong similarities in ratings of audio quality, anticipating only a 10% error in predicting ratings of naïve listeners from those of experienced listeners.

The good news for the audio industry is that if a loudspeaker is well designed, ordinary customers may recognize it—if given a reasonable opportunity to judge. The pity is that there are few opportunities to make valid listening comparisons in retail establishments, and the author is not aware of an independent source of such *unbiased* listening test data for customers to go to for help in making purchasing decisions.

It is paradoxical that opinions of reviewers are held in special esteem. Why are these people in positions of such trust? The listening tests they perform violate the most basic

FIGURE 3.8 *Averaged sound quality rating judgments for listeners classified according to the variability in their ratings. The small variability/more normal hearing listener results are shown in the white columns. The large variability/less-than-normal hearing listener results are shown in the shaded columns. For interest: D was a PSB Passif II, U was a Luxman LX-105, V was an Acoustic Research AR58s, and X was a Quad ESL-63. From Toole (1985), figure 16.*

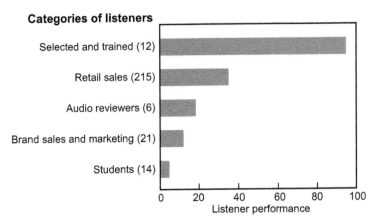

Categories of listeners

FIGURE 3.9 *A metric of listener performance related to the consistency of repeated judgments and the strength of rating differentiation of products having different sound quality levels (based on the F_L statistic). In spite of these differences, averaged product ratings were very similar for all groups.*
From Olive (2003).

rules of good practice for eliminating bias. They offer us no credentials, no proofs of performance, not even an audiogram to tell us that their hearing is not impaired. Adding to the problem, most reviews offer no meaningful technical measurements so that readers might form their own impressions.

Fortunately, it turns out that in the right circumstances, most of us, including reviewers, possess "the gift": the ability to form useful opinions about sound and to express them in ways that have real meaning. All that is needed to liberate the skill is the opportunity to listen in an unbiased frame of mind.

3.3 STRESS AND STRAIN

Subjectivists have criticized double-blind testing because it puts extraordinary stress on listeners, thereby preventing their full concentration on the task. This is presumed to explain why these people don't hear differences that show up in casual listening. Years of doing these tests with numerous listeners have shown that, after a brief introductory period, they simply settle into the task. Regular listeners have no anxiety. Years ago, while at the National Research Council of Canada (NRCC), we routinely contracted to perform anechoic measurements and to conduct blind loudspeaker comparisons for Canadian audio magazines. The journalists traveled to Ottawa, arriving the night before the tests to give their ears a chance to rest, and spent a day or more doing double-blind subjective evaluations before they got to see the measurements on the products they submitted. On one occasion, due to unforeseen circumstances, the acoustically transparent, visually opaque screen was not available—it was on a truck a few hours away. Rather than waste time, it was decided to start the tests without the screen—sighted. A few rounds into the randomized, repetitive tests our customers called a halt to the proceedings—they felt that seeing the loudspeakers prevented them from being objective—they were stressed! They waited for the truck to arrive.

Those of us who have done many of these tests agree. In the end, it is simpler if there is only one task: to judge the sound. There is no need to wonder if one is being influenced by what is seen, noting that certain familiar brand names might not be performing as anticipated, thinking you might be wrong, and so on. And all of us long ago gave up trying to guess what we are listening to—it is another distraction that usually proves to be wrong. In multiple comparison tests it is advisable to add one or more products that are not known to the listeners.

Additional criticisms had to do with notions that the sound samples might be too short for critical analysis, and that a time limit can be stressful. For the past 25+ years the process at Harman has involved single listeners, who select which of the three or four unidentified loudspeakers is to be heard at any time, and there is no time limit—switch at will, and take as long as is needed. The musical excerpts are short, but repeated in nicely segued loops to allow for detailed examination. Stressful?

3.4 HOW MANY CHANNELS?

We listen to almost all music in stereo; that is the norm. So, it makes sense to do double-blind listening tests in stereo too, right? Wrong. It all came to a head back in 1985, when I decided to examine the effects of loudspeaker directivity and adjacent side-wall reflections (Toole, 1985, described in Section 7.4.2). In the process I did tests using one and both of the loudspeakers. Because stereo imaging and soundstage issues were involved, listeners were interrogated on many aspects of spatial and directional interest. To our great surprise, listeners had strong opinions about imaging when listening in mono. Not only that, but they were very much more strongly opinionated about the relative sound quality merits of the loudspeakers in the mono tests than in the stereo tests.

Over the years mono vs. stereo tests have been done to satisfy various doubters. Each time the loudspeaker that won the mono tests also won the stereo tests, but not as convincingly, because in stereo everything tended to sound better. And what about imaging? It turns out to be dependent on the recording and the loudspeaker; there is a strong interaction. Classical recordings, with their high content of "ambiance" (i.e., uncorrelated L and R channel information), tended to be quite unaffected by the loudspeaker. However, some popular and jazz recordings, with close-miked, panned and equalized mixes, exhibited some loudspeaker interactions—as might be expected because they were control-room creations, not capturing a live event. But there were interactions involving specific recordings with specific loudspeakers.

More recently, tests were done that included mono, stereo and multichannel presentations (Olive et al., 2008). The results, shown in Figure 3.10, indicate that when there was a perceived preference among the optional sound qualities, it was most clearly revealed in monophonic listening. As active channels were added to the presentations, the ability to distinguish between sounds of different timbres appeared to deteriorate. Condition "B" was by far the worst sounding option on monophonic listening, with

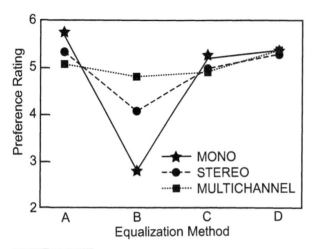

Results of tests in which listeners compared different forms of equalization applied to identical loudspeakers in monophonic, stereo and 5.1 multichannel configurations. From Olive et al. (2008).

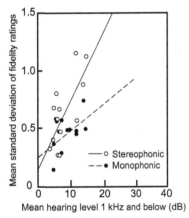

FIGURE 3.11 *A comparison of judgment variability as a function of hearing level for tests done in mono and stereo. Even though these listeners had been selected for relatively normal hearing, only those with very close to zero hearing level performed similarly well in both tests.*

significant resonant colorations, yet, in stereo and multichannel presentations, it was judged to be close to the truly superior systems.

A persuasive justification for monophonic evaluations is the reality that in movies and television programs, the center channel does most of the work—a monophonic sound source. The center channel is arguably the most important loudspeaker in a multichannel system. In popular music and jazz stereo recordings, images are often hard-panned to the left or right loudspeakers, again monophonic sources, and the phantom center image is the product of double mono, identical sounds emerging from L and R loudspeakers.

Finally, the data of Figure 3.11 adds more strength to the argument that monophonic listening has merits. It shows that only those listeners with hearing levels very close to the statistical normal level (0 dB) are able to perform similarly well in both stereo and mono tests. Even a modest elevation in hearing level causes judgment variations in stereo tests to rise dramatically, yielding less-trustworthy data. Age and hearing loss affect one's ability to separate multiple sounds in space. We may notice this in a crowd or in a restaurant setting, but we may not be aware of the disability when listening to a stereo soundstage (see Section 17.2). The directional and spatial complexity of stereo appears to interfere with our ability to isolate the sound of the loudspeakers.

For program material, some people insist on using one of a stereo pair of channels, but summing the channels is possible with many stereo programs; listen, and find those that are mono compatible. Because complexity of the program is a positive attribute of test music, abandoning a channel is counterproductive.

In summary, while it is undeniable that good stereo and multichannel recordings are more entertaining than monophonic versions, it is an inconvenient truth that monophonic listening provides circumstances in which the strongest perceptual differentiation of sound quality occurs. If the purpose of a listening test is to evaluate the intrinsic sound quality from loudspeakers, listen in mono. For pleasure, or to impress someone, listen in stereo or multichannel modes.

When multiple loudspeakers are simultaneously active, the uncorrelated spatial sounds in the recordings modify the ability of listeners to judge timbral quality. As noted in Section 7.4.2, the sound quality and spatial quality ratings were very similar. Is it possible that these perceived attributes couldn't be completely separated? As discussed in Section 7.4.4, Klippel found that 50% of listeners' ratings of "naturalness" were attributed to "feeling of space"; with "pleasantness," it was 70%. So spatial effects are important—more important than the author anticipated—and forming opinions about sound quality in stereo and multichannel listening is obviously a complex affair.

But aren't there qualities of a loudspeaker that *only* show up in stereo listening, because of the many subtle dimensions of "imaging"? One can conceive of loudspeakers with grossly inferior spectral response and/or irregular off-axis performance generating asymmetries in left and right channel sounds. So, possibly there are, but with competently designed loudspeakers the spatial effects we hear appear to be dominated by the information in the recordings themselves. The important localization and sound-stage information is the responsibility of the recording engineer, not the loudspeaker, and loudspeakers with problems large enough to interfere with those intentions should be easily recognizable in technical measurements or from their gross timbral distortions. For stereo, the first requirement is identical left and right loudspeakers and left-right symmetry in the room. That requirement is not always met, in which case many forms of audible imperfections may be heard, and the reasons may not be obvious.

The circle of confusion applies also to stereo imaging.

3.5 CONTROLLING THE VARIABLES IN SUBJECTIVE EVALUATIONS

Any measurement requires controls on nuisance variables that can influence the outcome. Some can be completely eliminated, but others can only be controlled in the sense that they are limited to a few options (such as loudspeaker and listener positions) and therefore can be randomized in repeated tests, or they can be held as constant factors. Much has been written on the topic (Toole, 1982, 1990; Toole and Olive, 2001; Bech and Zacharov, 2006). The following is a summary.

3.5.1 Controlling the Physical Variables

As audio enthusiasts know, and as explained in detail in Chapters 8 and 9, rooms have a massive influence on what we hear at low frequencies. There are also effects at higher frequencies, in the reflection of off-axis sounds described by loudspeaker directivity. Therefore, if comparing loudspeakers, listen to all of them in the same room, taking time to let the listeners adapt to the acoustical setting. Listeners have the ability to adapt to many aspects of room acoustics, as long as they are not extreme problems. As explained in Section 8.1.1, there are no "ideal" dimensions for rooms, in spite of claims to the contrary. There are practical dimensions, however, that allow for the proper placement of a stereo pair of loudspeakers and a listener, or of a multichannel array of loudspeakers and a group of listeners.

3.5.1.1 The Listening Room—Making Tests Blind

The importance of blind listening was established in Section 3.1, but how, in practice, is it achieved? An acoustically transparent screen is a remarkable device for revealing truths, but it is essential to test the acoustical transmission loss of the material by performing a simple on-axis frequency response measurement, then place the material between the loudspeaker and microphone, and measure again. There will be a difference, but it should be less than 1 dB, even at the highest frequencies, as it is in a good grille cloth. Polyester double-knit is a commonly available possibility. However, care must be taken to ensure that the fabric when tested and used is under slight tension to open the weave—it is this that is responsible for the acoustical transparency. Because fabrics with low acoustical loss tend to be very porous (easy to blow air through, and to see glints of daylight through) they can reveal outlines and sometimes more of the loudspeakers being evaluated. Minimize the light behind the screen, and emphasize the lighting levels on the audience side to alleviate this problem—directional lighting helps.

Should you listen with the grilles on or off? It depends on your audience. In reality, only audiophile enthusiasts are likely to want to, or can get away with, leaving the transducers exposed. In good products the differences should be vanishingly small, but this is not always so. Conscientious loudspeaker designers include the grille cloth and frame (often the bigger problem) into the design. In the real world, visual aesthetics count for something, and grille cloths are a deterrent to probing fingers.

3.5.1.2 Real-Time Loudspeaker Comparison Methods

Ideally, use positional substitution: bring the active loudspeakers to the same position so that the coupling to the room modes is constant. This can be expensive, and requires a dedicated facility, something beyond the means of hobbyists, or even audio publications and their reviewers. But, with the expenditure of some time and effort, simpler methods can yield very reliable results. Figure 3.12 shows the listening room used in my original research at the NRCC in the 1980s. In this, three or four loudspeakers were located at the numbered locations, and some number of listeners occupied the seats. Rigorous tests used only a single listener at a time. After each sequence of music, during which

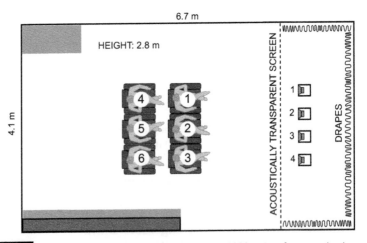

FIGURE 3.12 *The original listening room at the NRCC, ca. 1982, set up for monophonic comparisons of four loudspeakers. In the course of the evaluations, the loudspeakers and listeners would be randomly rearranged to average out the strong influences of standing waves in the room.*

listeners made notes and decided on "fidelity" ratings, the loudspeakers were randomly repositioned and the listeners took different seats. This process was repeated until all loudspeakers had been heard in all locations. It was time-consuming, but it reduced the biasing influence of room position, which in some cases could be the deciding factor.

The results were good enough to provide a solid foundation for the research that followed. The setup needs to be calibrated by measurements from each loudspeaker location to each listener location, and positional adjustments made so as to avoid accidental contamination by one or more obtrusive room resonances. The appendix of Toole (1982) describes the process for one room, and Olive et al. (1998), Olive (2009) and McMullin et al. (2015) document others. The room cannot be eliminated, but it should be a relatively constant, and innocuous, factor. If the result is to be an evaluation relevant to ordinary audiophiles or consumers, deliberate acoustical treatment must be moderated. Imitating an "average" domestic listening room is a good objective, and the most contentious decision is likely to be what to do with the first side wall reflection. My initial decision, many years ago, was to let it reflect—a flat, untreated surface, as happens in many domestic rooms and retail demonstrations—so that the off-axis behavior of the loudspeaker would be revealed. This is a demanding test, as it probably should be for loudspeaker evaluations intended to apply to a general population. But, for one's personal recreational listening there are other factors to consider, as is discussed in Chapter 7.

Figures 3.13 and 3.14 show an evolution of those listening tests, the Harman International Multichannel Listening Laboratory (MLL) (Olive et al., 1998). In this, a positional substitution process has been automated using computer-controlled pneumatic rams to move the entire platform left and right, and to move the desired loudspeaker(s) to a predetermined forward location, storing the silent loudspeakers at the rear. The amount of forward movement is programmable, and stereo and L, C, R arrangements can be compared. The positional exchanges take about 3 s, and take place quietly. The single listener controls the process, choosing the active loudspeaker (which has a coded identity) and deciding when it is time to move on to the next musical selection. When the music changes (a random process), the coded identities are randomly changed, so in effect it becomes a new experiment. There is no time limit. All of these functions are computer randomized to make it a true double-blind test.

There are other ingenious ways to achieve positional substitution for in-wall or on-wall loudspeakers, sound bars and so forth. If these tests are to be a way of life, any are worth considering because of the excellent results they yield and the time savings involved. If none of this is

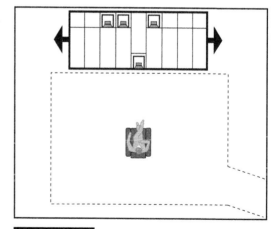

FIGURE 3.13   *The Harman International Multichannel Listening Laboratory (MLL) in its original configuration. Later, the side walls were moved closer to center locations. The programmable mover can compare up to four single loudspeakers at one time, up to four stereo pairs, or up to three L, C, R combinations. The single loudspeakers can be compared at L, C or R locations at varied distances from the side walls. The distance from the front wall is also a controlled variable.*

FIGURE 3.14 *The MLL with and without the curtain, showing three loudspeakers under evaluation, one of them in the forward, active location.*

possible, a variation of Figure 3.12 certainly works, costs nothing, but involves more work and more time. A simple turntable is another cost-effective and practical solution. It can be operated by a patient volunteer behind the screen (Figure 7.11) or automated (McMullin et al., 2015, figure 14). Versions of these automated movers can be built into walls for evaluating flat-panel TV audio systems, sound bars, on- and in-wall/ceiling loudspeakers, and so forth (Olive, 2009; McMullin et al., 2015).

3.5.1.3 Binaural Record/Replay Loudspeaker Comparisons

Many situations do not lend themselves to real-time in situ listening tests. Binaural recordings reproduced at a later time provide an obvious alternative, but many people have a problem accepting the idea that headphone reproduction is a satisfactory substitute for natural listening. The author also had doubts in the beginning, but the appearance of the Etymötic Research ER-1 insert earphones provided what seemed to be a solution to the largest problem: delivering a highly predictable signal to the listeners' eardrums. Toole (1991) describes the problems in detail, and shows results of early tests in which there was good agreement between real-time and binaural listening tests of some loudspeakers. The same basic process was used in experiments comparing results of listening tests done in different rooms (Section 7.6.2 and Olive et al., 1995). Again the results of real-time and binaurally time-shifted tests yielded closely similar results.

These tests employed a static KEMAR head and torso simulator (an anthropometric mannequin) as a binaural microphone, and a problem was that much or all of the time the sound was perceived to be in-head localized. Playing the binaural recordings in the same room that they were recorded in significantly alleviated this, allowing listeners to see and hear the room before the recorded version took over. In such instances the sounds were often credibly externalized. However, the illusion deteriorated when the listening was done in a different room and what the eyes saw did not match what the ears heard.

A more elaborate scheme called Binaural Room Scanning (BRS) involves measuring binaural room impulse responses (BRIR) at each of many angular orientations of the mannequin in the test environment. For playback, the music signal is convolved with the BRIR filters as it is reproduced through headphones attached to a low-latency head-tracking system. By this means, as the listener's head rotates, the sounds at the ears are modified so as to maintain a stable spatial relationship with the real room. The result is a much more persuasive externalization of the sound and the associated spatial cues. Olive et al., 2007; Olive and Welti 2009; Welti and Olive, 2013; Gedemer and Welti, 2013; and Gedemer, 2015b, 2016 describe such experiments in domestic rooms, cars and cinemas.

A literature search will reveal many more examples, because binaural techniques are now accepted in audio and acoustical research. A fact not recognized by many skeptics is that this requires a precise knowledge of the sounds that reach the listeners' eardrums. A carefully calibrated system is required, and headphones or insert earphones are chosen for the ability to deliver predictable sounds to the eardrums.

3.5.1.4 Listener Position and Seating

Listen in the same room, using a single listener in the same location. If there are multiple listeners, on successive evaluations it is necessary to rotate listeners through all of the listening positions. The problem with multiple listeners is avoiding a "group" response. Some people have difficulty muting verbal sounds signaling their feelings, and others may exhibit body language conveying the same message. This is especially true when anyone in the group is respected for his or her opinion. Such listeners sometimes deliberately reveal their opinions verbally or non-verbally, in which case the test is invalid. Single listeners are much preferred and enforce a "no discussion" rule between tests.

Seats should have backs that rise no higher than the shoulders to avoid reflections and shadowing from multichannel surround loudspeakers.

3.5.1.5 Relative Loudness

In any listening comparison, louder sounds are perceptually recognizable, which is an advantage for persons looking for a reason to express an opinion. It is a well-known sales strategy in stores. Perceived loudness depends on both sound level and frequency, as seen in the well-known equal-loudness contours (Figure 4.5). Consequently, something as basic as perceived spectral balance is different at different playback levels, especially the important low-bass frequencies. If the frequency responses of the devices being compared are identical, as in most electronic devices, loudness balancing is a task easily accomplished with a simple signal like a pure tone and a voltmeter. Loudspeakers are generally not flat, and individually they are not flat in many different ways. They also radiate a three-dimensional sound field into a reflective space, meaning that it is probably impossible to achieve perfect loudness equality for all possible elements (e.g., transient and sustained) of a musical program.

There has been a long quest for a perfect "loudness" meter. Some of the offerings have been exceptionally complicated, expensive and cumbersome to use, requiring narrow-band spectral analysis and computer-based loudness-summing software. The topic is discussed in Section 14.2.

The practical problem is that if the frequency responses of the loudspeakers (or headphones) being compared are very different, a loudness balance achieved with one kind of signal would not apply to a signal with a different spectrum. Fortunately, as loudspeakers have improved and are now more similar, the problem has lessened, although it has not disappeared. In all of these cases, if in doubt, turn the instruments off and listen; a subjective test is the final authority.

3.5.1.6 Absolute Loudness—Playback Sound Levels

Spectral balance is affected by playback sound levels, and so are several other perceptual dimensions: fullness, spaciousness, softness, nearness, and the audibility of distortions and extraneous sounds, according to Gabrielsson et al. (1991). Higher sound levels permit more of the lower-level sounds to be heard. Therefore, in repeated sessions of the same program material, the sound levels must be the same. Allowing listeners to find their own "comfortable" playback level for each listening session may be democratic, but it is bad science. Of course, background noises can mask lower-level sounds. This can be a matter of concern in automotive audio, where perceptual effects related to timbre and space can change dramatically in going from the parking lot to highway speeds.

Preferred listening levels for different listeners in different circumstances have been studied, with interesting results. Somerville and Brownless (1949), working with sound sources and reproducers that were much inferior to present ones, concluded that the general public preferred lower levels than musicians, and both were lower than professional audio engineers.

Recent investigations found that conventional TV dialog is satisfying at an average level of 58 dBA in a typical domestic situation, but at 65 dBA in a home theater setting. In home theaters, different listeners expressed preferences ranging from about 55 to 75 dBA (Benjamin, 2004). Using a variety of source material, Dash et al. (2012) concluded that a median level of 62 dB SPLL was preferred (using ITU-R BS.1770 weighting; it can be approximated by B- or C-weighting). The preferences ranged from 52 to 71 dB SPLL. These two studies are in basic agreement, meaning that satisfying everyone in an audience may be impossible. Frequency weighting is explained in Section 14.2.

These are not high sound levels compared to those used by many audiophiles or movie viewers in foreground listening mode. Cinemas are calibrated to a 0 dB reference of 85 dB (C-weighted), which, allowing for 20 dB of headroom in the soundtrack, permits peak sound levels of 105 dB—for each of the front channels. In cinemas this can be *very* loud. Low-Frequency Effects (LFE) channel output can be even higher, so it is not surprising that sustained levels of this combined sound are unpleasant to some. Patrons have walked out of movies, causing some cinema operators to turn the volume down. This compromises the art and degrades dialog intelligibility. It is a serious industry problem.

This same calibration process is used in many home theater installations. Many (most?) listeners find movies played at "0" volume setting to be unpleasantly loud.

Home theater designers and installers in my CEDIA (Custom Electronic Design and Installation Association) classes routinely reported that their customers play movies at −10 dB or lower. That is a power reduction of a factor of 10—a significant cost factor when equipping such a theater. Even a reduction of 3 dB is a factor of two in power.

Recommendation ITU-R BS.1116–3 (2015) chose to standardize the *combined* playback level from some number of active channels at 85 dBA, allowing 18 dB before digital clipping. Bech and Zacharov (2006) comment that they find this calibration level to be excessively high by 5–10 dB for subjects listening to typical program material.

So, the recommended calibrations for playback sound levels are frequently found to be excessively high, as judged by moviegoers and in private home theaters. The point of this discussion is that the choice of a playback sound level for product evaluation is a serious matter. It affects what can be heard, and how it is heard (see the equal-loudness contours in Figure 4.5). Ideally, comparative listening should be done at a sound level that matters to you, or to the audience for your opinions. Olive et al. (2013) used program at an average level of 82 dBC, slow. This was for an audience of serious listeners doing foreground listening of a demanding kind. Most tests do not reveal their listening levels, so we never really can be certain what the results mean.

As a generalization, music and movies are often mixed while listening at high sound levels. When I have walked into these sessions I find myself quickly putting my fingers in my ears and retreating to the back of the room. I value my hearing. Naturally, when these recordings are played at "civilized" sound levels, the bass is perceptually attenuated because of the equal-loudness contours. A bit of bass boost may be required to restore satisfaction (see Section 12.3). This is why I monotonously recommend having easy access to tone controls, especially a bass control.

3.5.1.7 Choosing Program Material

The ability to hear differences in sound quality is greatly influenced by the choice of program material. The audibility of resonances is affected by the repetitive nature of the signal, including reflections and reverberation in the recording and in the playback environment (Toole and Olive, 1988). The poor frequency response of some well-known microphones is enough to disqualify recordings made with them, whatever the merits of the program itself (Olive and Toole, 1989b; Olive, 1990). Some of the historically popular large diaphragm units have frequency responses at high frequencies that would be unacceptable in inexpensive tweeters.

Olive (1994) shows how, in training listeners to hear differences in loudspeakers, it is possible to identify the programs that are revealing of differences and those that are merely entertaining. Figure 3.15 displays the programs that Olive used in those tests, and the effectiveness of those programs in allowing listeners to correctly recognize spectral errors, bass and treble tilts, and mid-frequency boosts and cuts, introduced into the playback. The important information is not specific to the recordings themselves, because more modern recordings should perform similarly, but to the nature of the program itself. The more complex and spectrally balanced the program, the more clearly the spectral errors or differences were revealed to listeners. It is surprising that some

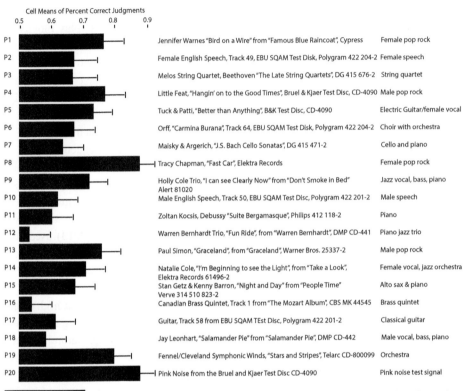

Cell Means of Percent Correct Judgments

P1	Jennifer Warnes "Bird on a Wire" from "Famous Blue Raincoat", Cypress — Female pop rock
P2	Female English Speech, Track 49, EBU SQAM Test Disk, Polygram 422 204-2 — Female speech
P3	Melos String Quartet, Beethoven "The Late String Quartets", DG 415 676-2 — String quartet
P4	Little Feat, "Hangin' on to the Good Times", Bruel & Kjaer Test Disc, CD-4090 — Male pop rock
P5	Tuck & Patti, "Better than Anything", B&K Test Disc, CD-4090 — Electric Guitar/female vocal
P6	Orff, "Carmina Burana", Track 64, EBU SQAM Test Disk, Polygram 422 204-2 — Choir with orchestra
P7	Maisky & Argerich, "J.S. Bach Cello Sonatas", DG 415 471-2 — Cello and piano
P8	Tracy Chapman, "Fast Car", Elektra Records — Female pop rock
P9	Holly Cole Trio, "I can see Clearly Now" from "Don't Smoke in Bed" Alert 81020 — Jazz vocal, bass, piano
P10	Male English Speech, Track 50, EBU SQAM Test Disc, Polygram 422 201-2 — Male speech
P11	Zoltan Kocsis, Debussy "Suite Bergamasque", Philips 412 118-2 — Piano
P12	Warren Bernhardt Trio, "Fun Ride", from "Warren Bernhardt", DMP CD-441 — Piano jazz trio
P13	Paul Simon, "Graceland", from "Graceland", Warner Bros. 25337-2 — Male pop rock
P14	Natalie Cole, "I'm Beginning to see the Light", from "Take a Look", Elektra Records 61496-2 — Female vocal, jazz orchestra
P15	Stan Getz & Kenny Barron, "Night and Day" from "People Time" Verve 314 510 823-2 — Alto sax & piano
P16	Canadian Brass Quintet, Track 1 from "The Mozart Album", CBS MK 44545 — Brass quintet
P17	Guitar, Track 58 from EBU SQAM TEst Disc, Polygram 422 201-2 — Classical guitar
P18	Jay Leonhart, "Salamander Pie" from "Salamander Pie", DMP CD-442 — Male vocal, bass, piano
P19	Fennel/Cleveland Symphonic Winds, "Stars and Stripes", Telarc CD-800099 — Orchestra
P20	Pink Noise from the Bruel and Kjaer Test Disc CD-4090 — Pink noise test signal

FIGURE 3.15 *The ability of different kinds of program to reveal spectral errors in reproduced sound. From data in Olive (1994).*

types of program that are highly pleasurable, or important to a genre—such as speech in movies and broadcasting—are simply not very revealing, and therefore not very demanding of, spectral accuracy.

Figure 3.16 shows that an important requirement for good program material is a long-term spectrum that well represents most of the audible frequencies, presumably giving listeners more opportunity to judge. What they were judging in these tests was the ability to hear narrow and broadband spectral variations in loudspeakers, arguably the problems most disruptive of sound quality/timbre.

The inversion of the bass contents of Groups 3 and 4 breaks what seems like a trend, indicating that there is more to this than long-term spectral content. Vocal/instrumental composition is another factor, some of which can be seen in the program listings in Figure 3.15. And, finally, there is the inescapable factor of the "circle of confusion"—we don't know what spectral irregularities are incorporated into the recordings. More specific details are in Olive (1994).

Group 1: pink noise; symphonic wind orchestra; male pop rock; female pop rock.
Group 2: male pop rock; female vocal with jazz orchestra; alto sax and piano jazz; electric guitar and female pop vocal; female vocal, bass, piano jazz.

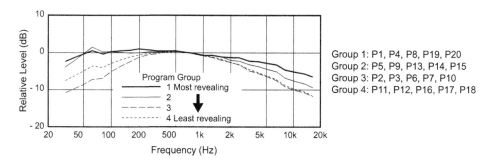

FIGURE 3.16 *Long-term 1/3-octave spectral averages of programs ranked according to their abilities to reveal differences in loudspeakers.*
From Olive (1994).

Group 3: male speech; female speech; string quartet; choir and orchestra; cello and
    piano.
Group 4: solo piano; piano jazz trio; brass quintet; classical guitar; male vocal with
    piano and bass jazz.

Clearly there are no hard rules but, in general, complex productions with broadband, relatively constant spectra aid listeners in finding problems. The notion that "acoustical," especially classical, music has inherent superiority was not supported. Solo instruments and voices appear not to be very helpful, however comfortable we may feel with them. A variety of programs should be used. Amusingly, a program that is good for demonstrations (e.g., selling) can be different from that which is needed for evaluation (e.g., buying).

3.5.1.8 Power Amplifiers, Wire and So Forth

The maturity of this technology means that problems with linear or non-linear distortions in the signal path are normally not expected. However, they happen, so some simple tests are in order. It is essential to confirm that the power amplifier(s) being used are operating within their designed operating ranges. Power amplifiers must have very low output impedances (high damping factor), otherwise they will cause changes to the frequency responses of the loudspeakers (see Chapter 16). This eliminates almost all vacuum-tube and some lesser class-D power amplifiers, although high-quality ones are fine. Class AB power amplifiers are a safe choice. Non-linearities in competently designed amplifiers are insignificant to the outcome of loudspeaker evaluations, so long as voltage or current limiting and overload protection are avoided. It is wise to measure the frequency-dependent impedances of loudspeakers under test to ensure that they do not drop below values that the amplifiers can safely drive. Some of these requirements will be difficult to find for many amplifiers as they are not always specified. Adequately sized speaker wire (Chapter 16) and audio frequency analog interconnects are transparent—in spite of advertising and folklore to the contrary. Digital communication and interfacing is a different and increasingly complex matter—and outside my comfort zone to discuss.

3.5.2 Controlling the Psychological Variables

3.5.2.1 Knowledge of the Products

This is the primary reason for blind tests. Section 3.1 shows persuasive evidence that humans are notoriously easy to bias with foreknowledge of what they are listening to. An acoustically transparent screen is a remarkable device for revealing the truth about what is or is not audible.

3.5.2.2 Familiarity with the Program

The efficiency and effectiveness of listening tests improves as listeners learn which aspects of a performance to listen for during different portions of familiar programs. Select a number of short excerpts from music known to be revealing, and use them repeatedly in prerecorded randomized sequences. They can be edited and assembled on computers. It is not entertainment, but the selections must have some musical merit or they will be annoying. After many, many repetitions the music becomes just a test signal, but because of the familiarity, it is extremely informative. Because the circle of confusion is a reality, it is essential to use several selections from different sources, recording engineers, labels and so forth. By paying attention, it will soon become clear which selections are helpful and which are just taking up time. Ask your listeners what they use to form their opinions. It has been known for listeners to have a very narrow focus—like using only the kick drum, or ignoring anything classical, or ignoring anything popular, or listening only to voices. Such listeners should not be invited back. It is also not uncommon for the peculiarities of certain programs to be adapted to. For example, one pop/rock excerpt that was very revealing of spectral features was expected to have slightly too much low bass; others might be known to be slightly bright. All of this relates to the circle of confusion, but these characteristics can recognized and accepted only when auditioned in the company of several other musical selections.

3.5.2.3 Familiarity with the Room

We adapt to the space we are in and appear to be able to "listen through" it to discern qualities about the source that would otherwise be masked. It takes time to acquire that ability, so schedule a warm-up session before serious listening begins or discard the results from early rounds.

3.5.2.4 Familiarity with the Task

For most people, critical listening of this intensity is a first-time experience. They may feel unprepared for it and are anxious. For audiophiles with opinions, pride is involved, leading to a different kind of tension. Seeing some preliminary test results showing that their opinions are not random is very confidence inspiring. This is the normal situation unless serious hearing loss is involved. Don't place much importance on the results of the first few test sessions. However, most listeners settle in quickly, and become quite comfortable with the circumstances. Experienced listeners are operational immediately, and some actually enjoy it—I did.

3.5.2.5 Listening Aptitude and Training

Not all of us are good listeners, just as not all of us can dance or sing well. If one is establishing a population of listeners from which to draw over the long term, it will be necessary to monitor their decisions, looking for those who (a) exhibit small variations in repeated judgments and (b) differentiate their opinions of products by using a large numerical range of ratings (Toole, 1985; Olive, 2001, 2003). An interesting side note to this is that the interests, experience and, indeed, occupations of listeners are factors. Musicians have long been assumed to have superior abilities to judge sound quality. Certainly, they know music, and they tend to be able to articulate opinions about sound. But, what about the opinions themselves? Does living "in the band" develop an ability to judge sound from the audience's perspective? Does understanding the structure of music and how it should be played enable a superior analysis of sound quality? When put to the test, Gabrielsson and Sjögren (1979) found that the listeners who were the most reliable and also the most observant of differences between test sounds were persons he identified as hi-fi enthusiasts, a population that also included some musicians. The worst were those who had no hi-fi interests. In the middle were musicians who were not hi-fi-oriented. This corresponds with the author's own observations over many years.

Perhaps the most detailed analysis of listener abilities was in an EIA-J (Electronic Industries Association of Japan) document (EIA-J, 1979), which is shown in Figure 3.17. It is not known how much statistical rigor was applied to the analysis of listener performance, but it was the result of several years of observation and a very large number of listening tests (personal communication from a committee member).

	Categories of listeners/audio	Sound analytical ability	Knowledge of reproduced sound	Attitude towards reproduced sound	Application objective	Reliability
1	General public	mediocre	mediocre	biased	various general investigations	
2	General public interested in audio equipment	mediocre	mediocre	biased	various general investigations	
3	General public strongly interested in audio equipment	fairly	some	strong self-assertion tendency	semi-specialized studies and measurements by specific group	low
4	Audio equipment engineers	high	sufficient	most correct	specialized studies and high precision measurements	high
5	Experienced acoustic experts	adequately high	excellent	correct	various studies and high precision measurements	high
6	Musicians, including students	adequately high	some	over-rigorous or non-interest	valuable opinions but unsuited for measurement	
7	General public strongly interested in music	high	some	mostly uninterested	unsuited for study & measurements (opinions are valuable)	
8	General public interested in music	somewhat high	mediocre	biased	various studies as representative of the general public	
9	General public strongly interested in recorded music	fairly high	fair	roughly correct	various studies and measurements	fairly high

FIGURE 3.17 *A chart describing several aspects of listener capabilities and their suitability for different kinds of subjective audio evaluations.*
From EIA-J, 1979.

Olive (2003) analyzed the opinions of 268 listeners who participated in evaluations of the same four loudspeakers. Twelve of those listeners had been selected and trained and had participated in numerous double-blind listening tests. The others were visitors to the Harman International research lab, as a part of dealer or distributor training or in promotional tours. The results shown earlier in Figure 3.9 reveal that the trained, veteran listeners distinguished themselves by having the highest performance rating by far, but obviously years of experience selling and listening on store floors has had a positive effect on the retail sales personnel.

Olive (1994) describes the listener training process, which is also a part of the selection process; those lacking the aptitude do not improve beyond a moderate level. It teaches listeners to recognize and describe resonances, the most common defect in loudspeakers. It does not train them to recognize any specific "voicing," as some have suspected. Instead, by a process of eliminating timbre-modifying resonances, the preferences gravitate to "neutral" loudspeakers, through which listeners get to hear the unmodified recordings.

3.5.2.6 Culture, Age and Other Biases

It is common to think that persons from different countries or cultures might prefer certain spectral balances. It is also common to think that people of different ages, having grown up with different music might have preferences that are not shared by young people. Harman is a large international company, with audio business in all product categories—consumer, professional and automotive—so it is important to know if such biases exist, and if they do, to be able to design products that appeal to specific clienteles. Over the years several investigations have shown that, if such biases exist at all, they are not revealed in blind testing. Figure 3.1 shows two "voicings" of the same loudspeaker, one done by the Danish design team and another by a consultant hired by the German distributor to modify the product to fit the expectations of his clients. The two versions were very similar in sound, and after extended double-blind listening by an international listener pool, including Germans, there was no evidence of a significant preference. Olive has conducted numerous sound quality evaluations with listeners from Europe, Canada, Japan and China, and has yet to find any evidence of cultural bias. Figure 3.18 shows the result for headphone evaluations. Similar results have been found for loudspeakers and car audio. The same is true with respect of age—except, as shown in Figure 3.7, older listeners tend to exhibit greater variation in their opinions because of degraded hearing.

3.5.2.7 Hearing Ability

Section 3.2 has shown that there are significant detrimental effects of hearing loss. Audiometric thresholds within about 20 dB of normal are desirable.

3.5.2.8 Listener Interaction

Listeners in groups are sometimes observed to vote as a group, following the body language, subtle noises, or verbalizations of one of the number who is judged to be the

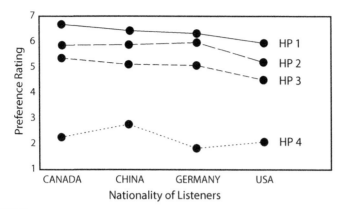

FIGURE 3.18 *Subjective preference ratings for four headphones, as evaluated by listeners in Canada, China, Germany and the US.*
From Olive et al. (2014).

most likely to "know." For this reason, controlled tests should be done with a single listener, but this is not always possible. In such cases, firm instructions must be issued to avoid all verbal and non-verbal communication. Opinionated persons and self-appointed gurus tend to want to "share" by any means possible. Video surveillance is a deterrent.

3.5.2.9 Recognition

The premise of controlled listening tests is that each new sound presentation will be judged in the same context as all others. This theoretical ideal sometimes goes astray when one or more products in a group being evaluated exhibit features that are distinctive enough to be recognized. This may or may not alter the rating given to the product, but it surely will affect the variations in that rating that would normally occur. If listeners are aware of the products in the test group, even though they are presented "blind," it is almost inevitable that they will attempt to guess the identities. The test is no longer a fair and balanced assessment. For this reason it is good practice to add one or more "unknowns" to the test population and to let that fact be known in an attempt to preclude a distracting guessing game.

3.5.3 How to Do the Test

Most people realize that the way a test is conducted can affect its outcome. That is why we do *double*-blind tests—one "blind" is for the listeners, so they don't know what they are listening to; the second "blind" is to prevent the experimenter from influencing the outcome of the test.

The simplest of all tests, and a method much used by product reviewers, is the "take-it-home-and-listen-to-it," or *single-stimulus*, method. In addition to being "sighted," and therefore subject to all possible non-auditory influences, it allows for adaptation. Humans are superb at normalizing many otherwise audible factors. This means that characteristics that might be perceived as virtues or flaws in a different kind of test can go unnoticed.

The scientific method prefers more controls, fewer opportunities for non-auditory factors to influence an auditory decision. It is what we hear that matters, not what we think we hear. Critics complain that rapid comparisons don't allow enough time for listeners to settle into a comfort zone within which small problems might become apparent. Of course, a time limit is not a requirement for a controlled test. It could go on for days. Listeners can listen to any of the optional sounds whenever they like. All that is required is that the listeners remain ignorant of the identity of the products—and that seems to be the principal problem for many reviewers.

Having followed subjective reviews for many years, in several well-known magazines and webzines, I find myself being amazed at how much praise is heaped on products that by any rational subjective or objective evaluation would be classified as significantly flawed. No, the single-stimulus, take-it-home-and-listen-to-it method is at best excusable as an expedient solution for budget and/or time-constrained reviewers. At worst it is a "feel-good" palliative for the reviewer. At no time is it a test procedure for generating useful guidance for readers.

A simple A versus B *paired comparison* is an important step in the right direction. The problem with a solitary paired comparison is that problems shared by both products may not be noticed. There is evidence that in such a comparison the first and second sounds in the sequence are not treated equally, so it is necessary to randomly mix A-B and B-A presentations, and always to disguise the identities.

Several paired comparisons among more than two products selected and presented in randomized order is a much better method, but it is very time-consuming. If a computer is not used for randomization in real time, it is possible to generate a randomized presentation schedule in advance and follow it.

A multiple comparison with three or four products simultaneously available for comparison is how I conducted my very first listening tests, back in the 1960s, and it is still the preferred method. It is efficient, highly repeatable, and provides listeners with an opportunity to separate the variables: the loudspeaker, the room and the program. If positional substitution is not used, remember to randomize the locations of the loudspeakers between musical selections.

In addition, there are important issues of what questions to ask the listeners, scaling the results, statistical analysis and so on. It has become a science unto itself. (See Toole and Olive, 2001; Bech and Zacharov, 2006, for elaborations of experimental procedures and statistical analysis of results.) However, as has been found, even rudimentary experimental controls and elementary statistics can take one a long way.

3.5.3.1 Is It Preference or Accuracy That Is Evaluated?

These tests have been criticized for many reasons, and I would like to think that all have been satisfactorily put to rest. Perhaps the most persistent challenge is the belief that as long as the term "preference" is involved, the results merely reflect the likes, dislikes and tastes of an inexpert audience, not the analytical criticism of listeners who know what good sound really is. They assert that the objective should be "realism," "accuracy," "fidelity," "truth" and the like.

As it turns out, this is where my investigations started, many years ago. I required listeners to report their summary opinions, their "Gestalt," as a number on a 0 to 10 "Fidelity" scale. The number 10 represented the most perfect sound they can recall, and 0 was unrecognizable rubbish. When these tests began, it is fair to say that some of the loudspeakers approached "rubbish" and none came close to perfection. Consequently, scores extended over a significant range. To help listeners retain an impression of what constituted good and bad sound, the Fidelity scale was anchored by always including in the population of test loudspeakers at least one that was at the low end of the scale and one that was at the high end, usually the one we dubbed the "king of the hill," the best of the test population to that date. A parallel scale was provided, called "preference," thinking that there might be a difference, but the two ratings "fidelity" and "preference" simply tracked each other.

Figure 3.19 illustrates the cumulative results from our NRCC tests in the early 1980s—a slice of the history of audio. It is obvious that there are many loudspeakers clustered near the top of the scale. As time passed more products were rated highly, and

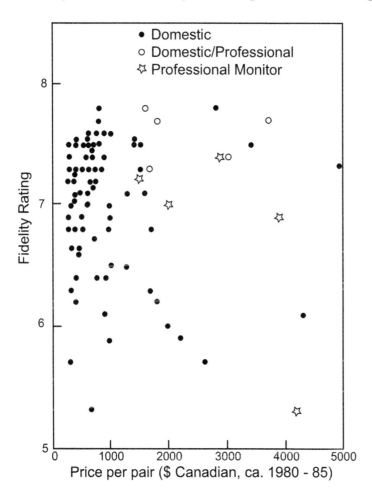

FIGURE 3.19 *The cumulative subjective ratings of loudspeakers evaluated in double-blind listening tests at the NRCC over the years 1980 to 1985. These would have been evaluated in groups of four, randomly selected from a larger number that would include "anchor" products representing low scores and high scores. Each product was auditioned several times in the randomized groups and these are the averaged ratings for all listeners. The "Domestic/Professional" loudspeakers were used in both applications. See Toole (1982, 1985, 1986) for more details.*

listeners complained about wasting time listening to bad sound. I stopped anchoring the scale at the low end, and an interesting thing happened: most listeners continued to use the same numerical range to describe what they heard. The scores rarely went above 8, which is a common subjective scaling phenomenon—people like to leave some "head-room" in case something better comes along. The advantage was that it was now easier to distinguish the rankings of the remaining loudspeakers because they were more widely separated on the scale. However, it was evident that they were no longer judging on the same scale—the upper end of it had stretched. When listening to just a few of the top-rated loudspeakers, which in Figure 3.19 would have been crowded together, statistically difficult to separate, the same product ratings would cover a much larger portion of the scale and be easy to separate and rank. Listeners were still judging fidelity, but it was rated on an elastic scale. I chose to call it preference, and it still is.

Listeners obviously have a "preference" for high "fidelity," also known as "accuracy," "realism" and so on. It is really one scale.

Figure 3.19 is somewhat provocative in that it shows inexpensive loudspeakers being competitive with much more expensive ones. At moderate sound levels that can be true, especially if one is willing to accept limited bass extension. But most of the small inexpensive products misbehave at high sound levels, and almost certainly they do not have the compelling industrial design and exotic finishes that more expensive products have. It is possible to see a slight upward step in the ratings around $1,000 per pair, when the price could justify a woofer in a floor-standing product. The dedicated professional monitor loudspeakers in these tests had some problems. Over the years everything has improved from these examples, and the best of today's products would surpass all of the products in these tests, although poor ones persist.

CHAPTER 4

The Perceptual and Physical Dimensions of Sound

The following chapters will be heavily involved with technical measurements of sounds radiating from loudspeakers and of sound fields in rooms, both of which combine in our two ears and brain to form perceptions. It is important at the outset to have a basic understanding of the units of measurement on the one hand, and of the perceptual consequences of the measured data on the other hand. Some readers may already be equipped with this knowledge, in which case this chapter can be skimmed or skipped. Others, though, may find the information useful and interesting because not all of the psychoacoustic relationships between what we measure and what we hear are intuitive.

The tutorial will begin with explanations of the vertical and horizontal axes of the common amplitude vs. frequency (aka frequency response) graphs, and then will move on to the audibility of variations in those graphs. What is it we are looking for, and how do we interpret it if we find it?

4.1 THE FREQUENCY DOMAIN

Figure 4.1 shows the horizontal axis of the frequency responses shown in this book. It displays frequency on a logarithmic scale because that is an approximation to many of the ways humans perceive sound. Only for special engineering purposes would a linear scale be used.

In this figure a standard piano keyboard is shown, indicating the fundamental frequencies of the keys. An organ keyboard can cover the entire audible frequency range. Other common musical instruments share portions of this range. Of course, harmonics and other overtones extend the frequency spectrum of most instruments to the upper limit of this scale and beyond. In live performances, these very high frequency sounds are normally much attenuated at listening locations because of atmospheric absorption, a phenomenon that progressively attenuates frequencies above about 2 kHz, and

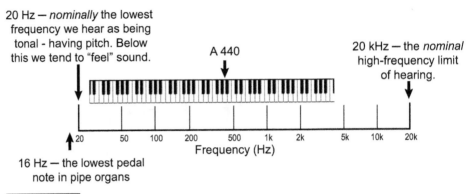

20 Hz — *nominally* the lowest frequency we hear as being tonal - having pitch. Below this we tend to "feel" sound.

A 440

20 kHz — the *nominal* high-frequency limit of hearing.

16 Hz — the lowest pedal note in pipe organs

Frequency (Hz)

FIGURE 4.1 *The horizontal axis of typical frequency-response data.*

that increases with propagation distance (see Figure 10.12). The loss of high frequencies with distance is itself a perceptual clue to the distance of familiar sound sources. In concert venues, reflected sounds travel farther than direct sounds, thereby causing the reverberant sound field to rapidly lose high-frequency energy—live concerts are therefore noticeably more "mellow" toward the rear of the auditorium. In amplified concerts, directional loudspeakers avoid most reflected sound and are often equalized with compensating high-frequency boosts for more distant audience members, thereby bringing them perceptually closer to the performers.

In close-microphone recordings, some of these overtones can be captured if the microphones have the necessary bandwidth, which many of the newer ones do. This has cultivated a belief that extending the recorded bandwidth is necessary, now that digital audio allows for it to be delivered to consumers. We will return to the topic later.

At the low-frequency end of the scale, things are complicated because humans have a tactile perception of high-level low frequencies and a pitch/tonal perception. The former is a pleasant addition to music, while the latter *is* the music. The dual perceptions begin in the upper bass region, around 100 Hz. Pitch cues fade and tactile cues get stronger at very low frequencies. Disproportionately boosted bass is common at rock concerts because it is dramatic to literally have your breath modulated by the kick drum and bass guitar—something not experienced at home. These can be "whole body" experiences. Fortunately, low frequencies are not high-risk sounds for hearing loss.

In perceptual terms, the correlate of frequency is pitch. It is not a perfect relationship because most musical instruments radiate complex spectra and waveforms, not pure tones. The judgment of pitch is therefore somewhat dependent on the relationship of the fundamental frequency and the overtones. The common example of this is the piano, in which, because of insufficiently flexible strings, not all of the overtones are harmonically related to the fundamental. Consequently, "stretched" tuning is necessary to make the highest notes perceptually acceptable, and in smaller pianos, with shorter strings, this is required at low frequencies as well. Pitch is also dependent on sound level, with low notes sounding slightly lower and high notes slightly higher at increased volumes. The consequence of this to sound reproduction is obvious: the music itself changes with sound level, so a somewhat realistic playback level is a requirement if "realism" in all of its manifestations is the expectation.

4.2 THE AMPLITUDE DOMAIN

Humans respond to sounds ranging over a sound pressure range of about one million to one. When interpreted as sound intensity (energy flow per unit area), it is an even more impressive one trillion to one. This enormous range extends from sounds that are just audible at middle frequencies, to those that generate discomfort or pain at the higher limit. Because such large numbers are impractical to work with, a logarithmic scale of sound pressure level is used to describe measured and audible acoustic events. The units are decibels (dB). Figure 4.2 introduces the scale.

The decibel scale can be used to compare any two sounds, disregarding the standardized reference 0 dB. In such cases it is correct to use "sound level" or "relative sound level" instead of sound pressure level as the notation on the vertical axis if the graph displays the differences in sound levels. However, SPL is the choice when discussing *absolute* levels of different sounds. In loudspeaker measurements, it is used

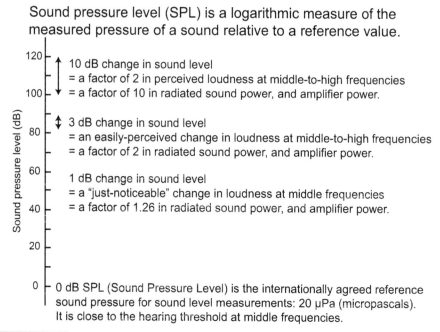

Sound pressure level (SPL) is a logarithmic measure of the measured pressure of a sound relative to a reference value.

10 dB change in sound level
= a factor of 2 in perceived loudness at middle-to-high frequencies
= a factor of 10 in radiated sound power, and amplifier power.

3 dB change in sound level
= an easily-perceived change in loudness at middle-to-high frequencies
= a factor of 2 in radiated sound power, and amplifier power.

1 dB change in sound level
= a "just-noticeable" change in loudness at middle frequencies
= a factor of 1.26 in radiated sound power, and amplifier power.

0 dB SPL (Sound Pressure Level) is the internationally agreed reference sound pressure for sound level measurements: 20 µPa (micropascals). It is close to the hearing threshold at middle frequencies.

FIGURE 4.2 *The basics of the sound pressure level (SPL) measurement scale. The indicated changes are not related to any specific SPL. These differences apply to broadband program material, not narrow-band or bandpass filtered material. Emphasis over a portion of the frequency range, as in the case of resonances, yields different results, as discussed in Section 4.6.2.*

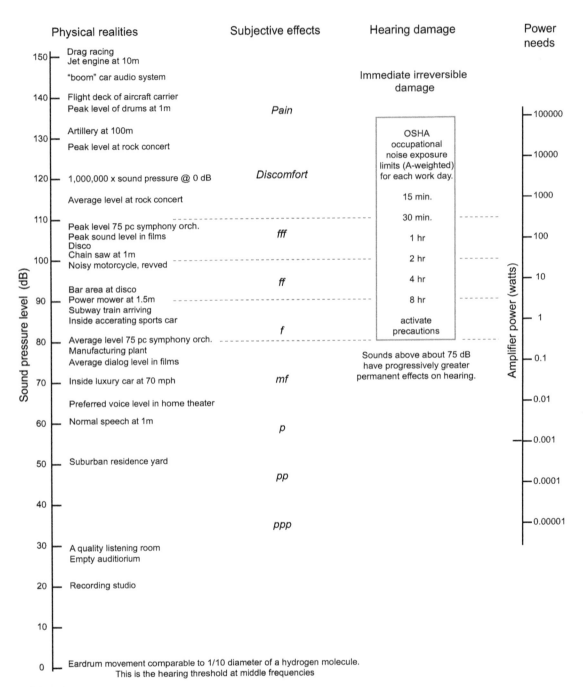

FIGURE 4.3 *Examples of sounds generating various sound pressure levels are shown on the left. In the middle are descriptions of a few subjective effects and considerations related to hearing conservation. On the right are idealized estimates of amplifier power required to produce these sound levels in typical domestic rooms using medium-sensitivity loudspeakers. All of these estimates and descriptors are approximate because of uncertainties in how the measurements were made and normal acoustical variations in real-world circumstances.*

because it is an indicator of absolute sound pressure level at a standardized distance (1 m) for a standardized voltage input (2.83 V) to the loudspeaker—its sensitivity.

Figure 4.3 shows the approximate sound pressure levels of different sound sources, and other relevant information. The sound sources identified with sound pressure levels in the left-hand column have been collected over many years; not all were measured with the same frequency weighting, and many of the sources identified are extremely variable, so these are only estimates.

The occupational hearing conservation guidelines are important, but are often misinterpreted in that it is assumed that they prevent hearing loss. As discussed in more detail in Chapter 17, that is not the case. They permit hearing loss to occur, aiming only to conserve enough at the end of a working life to permit carrying on a generally intelligible conversation at a distance of 1 m. Appreciating the subtleties of live or reproduced sounds would be a pleasure long gone. To these workplace noise exposures must be added non-occupational noise exposures: rock concerts, shooting, motorcycles, lawnmowers, power tools, automobile air bag deployment and so on, all of which contribute to the life-long accumulation of hearing damage. Hearing that is considered "normal" by your audiologist, whose criteria are based only on speech, may already exhibit significant losses when it comes to appreciating or evaluating the many dimensions of music and movies.

The column at the far right is a guess at what amplifier power may be needed to deliver various sound levels in typical rooms. Obviously, this is fundamentally dependent on loudspeaker sensitivity, how many are operating, the frequencies being radiated, and the acoustical properties of the room. In spite of these uncertainties, the *relative* quantities are correct—the overall vertical scale may slide up or down by a few dB. This is shown to give an appreciation for the reality that a 3 dB change in sound level requires a doubling or halving of power, and a 10 dB change translates into 10 times the power. When one is talking about high sound levels from domestic loudspeakers having typically moderate sensitivity (85–90 dB SPL @ 2.83 V @ 1 m), one is also talking about very large amounts of power.

4.3 AMPLITUDE AND FREQUENCY TOGETHER: FREQUENCY RESPONSE

Many important measured quantities, electrical and acoustical, are displayed as an amplitude/magnitude vs. frequency curve. The one of greatest importance in audio is the amplitude of electrical signal or sound level vs. frequency, or as it is commonly known: the frequency response. It is the principal determinant of sound quality. Knowing what is "correct" in examining a frequency response curve is important, so Figure 4.4 identifies some of the key factors that can be revealed in these curves. The example is a loudspeaker, driven with the standard test signal level of 2.83 V, and the sound pressure levels shown are corrected to the standardized measurement distance of 1 m, so it is possible to see the *sensitivity* of the loudspeaker as a function of frequency (dB SPL @ 2.83 V @ 1 m). Curves (a) and (b) were measured on the reference axis of the loudspeaker, which would normally be directed at the listener, providing the direct

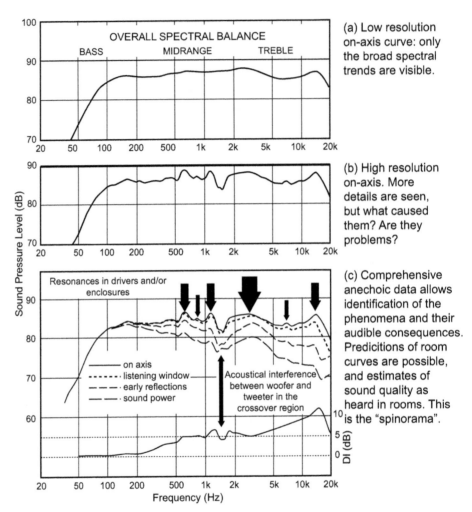

(a) Low resolution on-axis curve: only the broad spectral trends are visible.

(b) High resolution on-axis. More details are seen, but what caused them? Are they problems?

(c) Comprehensive anechoic data allows identification of the phenomena and their audible consequences. Predicitions of room curves are possible, and estimates of sound quality as heard in rooms. This is the "spinorama".

FIGURE 4.4 *An illustration of some information that can be revealed in frequency-response curves. In these examples the sound pressure level at 1 m is shown when the loudspeaker is driven by a constant 2.83 V. This is an industry standard format. Figure 4.4a shows a spectrally smoothed curve revealing upper and lower frequency limits and general trends in sound output. (b) is a measurement with higher frequency resolution that reveals other possibly audible timbral degradations. (c) shows the family of curves informally called the "spinorama" that has proved to be useful in identifying the mechanisms underlying the features in frequency-response curves, and quantifying them in a manner that permits assessment of their potential audibility in rooms.*

sound in a room. However, loudspeakers radiate sound in all directions, and that sound, less any that is absorbed, eventually reaches listeners as reflections, so many more measurements must be made, as indicated in the family of curves shown in (c). All of this will be explained in detail in Chapter 5.

Figure 4.4a provides a starting point for speculation about the sound quality of this loudspeaker. This is an "artistic interpretation" that often appears in product specifications and sometimes in product reviews. It is widely assumed in marketing circles that the public is not ready to see something close to the truth. It can be seen that the bass output is rapidly attenuated below about 80 Hz, meaning that a subwoofer would be needed for satisfying bass, and 80 Hz is a very convenient crossover frequency for bass management systems (Section 8.5). The sensitivity of the loudspeaker over the important middle portion of the frequency range is somewhat frequency dependent, which is why such a specification is usually based on the sound output averaged over a range of frequencies, such as 300 Hz to 3 kHz. The slight upward trend in output from about 100 Hz to 3 kHz suggests that the sound may be somewhat thin, bright or hard depending on the program. Perceived "brightness" is often associated with frequencies much lower than is commonly thought. The depression at higher frequencies deprives the listener of balanced instrumental overtones (see Figure 4.1), possibly leading to a loss of openness and articulation, and perhaps even a hint of dullness.

Figure 4.4b shows more details, but additional information is needed in order to identify what they are with any certainty. The spinorama shown in Figure 4.4c is an example of such data—measures of sounds radiated in other directions. The specifics of these curves will be discussed in Chapter 5. Most of what we hear in typical rooms is reflected sound, and the lower curves reveal increasing amounts of reflected sounds. It is a property of transducer resonances that they tend to be widely radiated into rooms, which means that the bumps associated with resonances will be evident in all of the curves, and that they are likely to be audible in rooms. Because of their wide dispersion, resonances may not be evident in the directivity index (DI).

Resonances that occur within the normal operating ranges of transducers are *minimum-phase* phenomena, meaning that a bump in an amplitude response is sufficient evidence to predict that there will be a localized phase anomaly and ringing in the time domain. Section 4.6.2 discusses this in detail. It turns out that the shape and height of the amplitude response bump is the most reliable indicator of audibility. This and the fact that these resonances can be attenuated by precise equalization mean that these anechoic data are very important. As is explained in Section 4.6.2, the most audible resonances are the low-Q (i.e., spectrally broad) ones, such as that seen here around 3 kHz. Eliminating just this one resonance would greatly improve the sound quality of this loudspeaker. The resonance near the high-frequency limit is unwanted, but in normal program is not likely to be annoying. Those in the middle frequencies are more obvious because they affect the timbre of voices and many musical instruments.

Acoustical interference can arise in the interactions between transducers radiating simultaneously in crossover regions, and also from direct sound interfering with sound diffracted from edges of enclosures and the like. These peak and/or dip shapes change with direction, sometimes averaging out over a range of angles, and at other times, as in the present example, being sustained over a range of angles but diminishing in the ultimate sound power spatial integration. This is why it is visible in the directivity index, the difference between the listening window and normalized sound power curves (explained in Section 5.3). Interference effects tend to be less audible than resonances.

Good engineering design can reduce or eliminate any of these problems, but it is important first to know what they are. It is obvious that a simple product specification like ±3 dB 80 Hz to 18 kHz contains almost no useful information for consumers, especially in the common situation where there is no tolerance and no description of how or where it was measured. Section 12.1.1 discusses this.

4.4 AMPLITUDE AND FREQUENCY TOGETHER: EQUAL-LOUDNESS CONTOURS

Loudness is the perceptual correlate of sound level. But it is also dependent on frequency, incident angle, the duration of the signal and its temporal envelope. It is one of many psychoacoustic relationships that defy a simple description. Yet, one must start somewhere, and so most investigators focused on pure tones in acoustically simple circumstances, like headphone or anechoic listening.

Fletcher and Munson (1933) were among the first to evaluate the sound levels at which we judged different frequencies to be equal in loudness. Starting with a reference pure tone at 1 kHz, other tones at different frequencies were adjusted until they appeared to be equally loud. With enough of these measurements it was possible to draw a contour of equal loudness identified by the sound pressure level of the reference tone, but called *phons* rather than decibels, to denote its subjective basis. For each SPL of the 1 kHz reference tone, curves were derived, collectively called the equal-loudness contours. Fletcher and Munson used headphones, and there has been concern about the way they were calibrated, so while we give them credit for their pioneering work, their data are no longer trusted.

Several more recent studies, notably the work of Robinson and Dadson (1956), used pure tones presented to listeners in an anechoic chamber; their results were the ISO standard for several years. Even then, errors were identified in the contours at low frequencies. Pollack (1952) evaluated a set of equal-loudness contours for bands of noise, but the circumstances of the listening were not well specified: "Listening was monaural. All tests were conducted in a quiet room." If they were monaural (one ear), headphones had to be used—or did he mean monophonic (one loudspeaker)? If headphones, how were they calibrated? "A quiet room" does not describe an acoustical environment. Stevens (1957) moved closer to typical listening conditions by employing fractional-octave bands of steady-state sounds in reflective spaces, which he used as a basis for calculating the loudness of complex sounds (Stevens, 1961). The Stevens curves shown in Figure 4.5 are from Toole (1973), scaled to the present format.

Figure 4.5a shows a sample from the current ISO standard (ISO 226, 2003), the result of an international collaboration initiated to address errors in the 1987 edition of the standard that was based on Robinson and Dadson (1956) data. Figure 4.5b shows a comparison of it and contours from four earlier studies showing how very different they

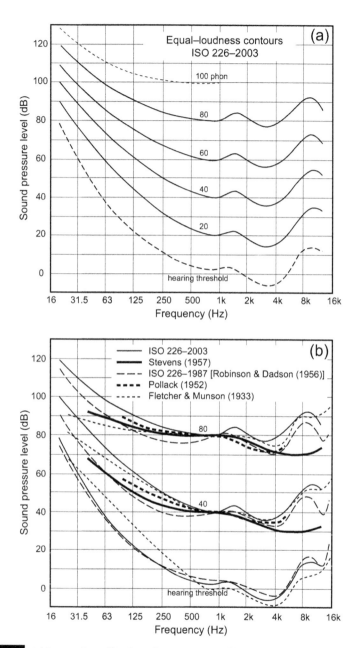

FIGURE 4.5   *(a) Curves of equal loudness for pure tones auditioned in an anechoic space according to the most recent internationally agreed standard: ISO 226 (2003). Each curve is created with reference to a pure tone at 1 kHz, and is identified by the sound pressure level at that frequency, but expressed in "phons." (b) A comparison of the current standard with three other determinations of equal-loudness contours.*
The curves from ISO 226 (2003) © ISO are reproduced with permission. See details in the caption for Figure 3.6. All other curves are reproduced with permissions from the respective sources.

are in shape. These are obviously *not* hard engineering data. The curves are averages over many listeners and there were large inter-subject differences.

Nevertheless the curves all share two important characteristics:

- Our sensitivity to sound reduces dramatically as frequency is reduced. Bass sounds must be much higher in sound level to be perceived to be as loud as mid- and high-frequency sounds.

- The curves progressively crowd together at lower frequencies, meaning that smaller changes in sound level (dB) are required to yield the same change in perceived loudness (phons).

This is responsible for the very familiar and easily audible rapid growth and decline of loudness at bass frequencies when the overall volume is changed. It can be seen that the contours are very similar at frequencies above about 500 Hz, with a relatively constant separation. At these frequencies a change of 10 dB in sound level corresponds to a change of 10 phons—which is a roughly a factor of two in perceived loudness. However, at lower frequencies the curves move closer to each other. At 63 Hz the relationship is about 7 dB for 10 phons, and at 31.5 Hz it is about 5 dB for double loudness. Small changes in sound level at bass frequencies are audibly consequential.

4.4.1 Loudness Controls and Tone Controls—Do They Work, Are They Necessary?

"Loudness" controls have been in audio equipment for decades. The attractive premise is that as the volume is reduced, automatic adjustments to frequency response compensate for the disproportionately reduced bass. For no obvious reason, some of them also boost the treble. This is wrong. It is also wrong to think that a simple variable bass boost can compensate for all components in a program that has dynamic range; loud portions need only slight bass boost, while quieter portions will need much more boosting. If the loudness compensation addresses the low-level portions, the high-level portions will have greatly excessive bass and so on.

The equal-loudness contours describe a characteristic of our hearing systems that is always there and unchanged, whether we are at a live musical performance, having a conversation, or listening to music or movies at home. They are part of the human listener. They are typically measured using pure tones at different frequencies that are compared to a reference tone at 1 kHz and adjusted to be equally loud. It is a simple process using simple sounds, and the key concern is the extent to which these curves reliably relate to the complex continuous spectra of music. It is an issue to which there is no definitive answer.

Nevertheless, they cannot be ignored in sound reproduction. In order for a sound reproducing system to accurately portray what was heard in a live performance it must have a uniformly *flat* frequency response, and reproduce the sounds at or close to the original sound levels, so that our built-in loudness processes operate on the sound in the same way. This is where complications set in, because that is rarely the case. Even if our sound systems have flat frequency responses, the playback sound levels are not likely to be what they were at live performances or in recording control rooms where important artistic decisions were made. For recreational listening, sound levels are almost always lower, often much lower, than the "original" sound.

To accommodate this issue it is necessary to look closely at the family of contours and pay attention not to the shapes of the curves, but to the *differences* in their shapes at different sound levels. As playback sound levels are reduced, not only will the bass frequencies be disproportionately reduced in apparent loudness, but progressively more low frequencies will fall below the hearing threshold, becoming completely inaudible. Olive (2004a, 2004b) found that in the overall ratings of sound quality, bass accounted for about 30% of the total factor weighting. It is important to get it right, and it cannot be right if some of it is missing.

Figure 4.6 shows results of an exercise performed by the author over 40 years ago (Toole, 1973) to elucidate the problem: if the loudness of music is reduced by, say 10 or 20 phons, what new frequency response must the playback system have in order that:

1. The timbre or spectral balance of the original will be preserved in the less loud reproduction?

2. Any frequency audible in the original will be audible in the less-loud reproduction?

Condition 1 cannot be perfectly satisfied because music has dynamic range and different components of the program at different levels will require different compensations. Music has large dynamic range, *ppp* to *fff*, as shown in Figure 4.3. The ear is a non-linear device generating distortions, pitch shifts and more that are sound-level dependent. These cannot be replaced. Condition 2 is a concession to human nature—we don't want to lose any of the music.

Figure 4.6 shows results for three loudness reductions. A 10 phon reduction from a high "reference" level (a) is a substantial change (approximately half loudness), but one that is not at all uncommon. Most people would still consider this to be "foreground" listening. A 20 phon reduction (b) yields "background" music, and a 30 phon reduction (c) is for "ambiance." It is immediately obvious that no single loudness compensation curve can work for music that has significant dynamic range. The low frequency boosts are substantial to maintain audibility of low-level sounds, and these would render high level sounds grossly bass heavy. Clearly condition (a) will not be satisfied, but it may be worth considering a compromise solution, and some musical realities may moderate what is needed. First, low bass sounds are most often in high-level musical passages, because even in live performances the sound levels must be high in order for these sounds to be audible (Figure 4.5a). It therefore seems reasonable to focus the compromise compensation on the higher-level components, such as the 60 and 80 dB SPL portions.

Figure 4.6d shows the result of plotting compromise curves between the 60 and 80 dB curves in each of (a), (b) and (c). These look quite practical. It is important to note that no great changes are required at high frequencies. Many loudness

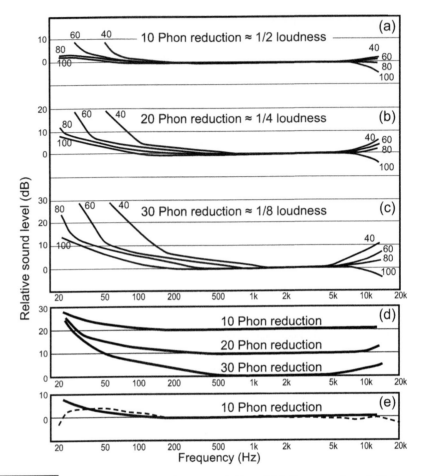

FIGURE 4.6 *The frequency-response changes needed to maintain the apparent spectral balance and audibility of sounds having flat spectra at original sound pressure levels of 40, 60, 80 and 100 dB SPL when the loudness level is reduced by (a) 10 phons, (b) 20 phons and (c) 30 phons. Derived from Robinson and Dadson (1956) equal-loudness contours. Figure 4.6d shows compromise curves drawn between the 60 and 80 dB predictions. Data shown in (a) to (d) are from Toole (1973). Figure 4.6e shows a comparison of the 10 phon reduction compromise curve and the tone control settings that average listeners preferred in loudspeaker listening tests done by Olive et al. (2013). It is a reasonable assumption that the playback level was lower than that in the recording control room, but by an unknown amount.*

NOTE: all of these curves would change slightly if they were derived from the current ISO equal-loudness contours. See Figure 4.5.

compensation devices over the years have significantly boosted both bass and treble as sound level was reduced. In the author's experience with such controls, sound quality was diminished—a result of misinterpreting the equal-loudness contours. With 10 or 20 phon reductions, essentially no high-frequency compensation is needed.

Finally, Figure 4.6e shows data adapted from work by Olive et al. (2013), as shown in Figure 14 of Toole (2015b). Here listeners in double-blind tests adjusted bass and treble controls to arrive at preferred spectral balances for several programs. A highly rated loudspeaker with a flattish on-axis frequency response in the Harman reference room was used. The curve shown here is the average tone control setting for all listeners; it is the difference between the steady-state room curves before and after adjustment. The listening was done at "foreground" recreational listening level, which is probably lower than recording studio listening level. In any event, a small bass boost was preferred—not very different from the 10 phon reduction curve. The similarity existed at high frequencies as well. This could be interpreted as a real-world validation of a theoretical prediction, but at the very least it is an interesting comparison.

In summary, listening at sound levels below the "original" or intended level causes timbral changes and a progressive loss of low bass unless corrective measures are taken. Unfortunately, because music has dynamic range, the corrections cannot be correct for all components of the original sounds. Strongly amplitude compressed music would respond more favorably to such correction, though, meaning that much modern music would certainly qualify. It is possible to choose a compromise loudness correction, catering to the needs of the louder portions of programs. When this is done, the substantial loudness compensations are at low frequencies—essentially it is a bass tone control. In fact, because we have no idea what the original sound level was, and no sound level calibrations are typically done at the playback locations, the loudness compensations cannot be accurately predicted. All that can be said is that some amount of bass boost will be needed, and the amount of the bass boost will vary from place to place and among different recordings.

Movies are mixed and reproduced in cinemas at sound levels that are at least somewhat standardized. Home theaters are often calibrated to meet a similar sound level for 0 dB volume control setting. This is a situation where a calibrated loudness compensation could function predictably. The reality is that playback at 0 dB in a home theater is commonly regarded as being uncomfortably loud and volume settings are often reduced by 10 dB or more. An interesting current issue with movie sound in cinemas is that many customers find some of them too loud and complain to the management. The problem has reached a point where some jurisdictions have talked about placing legal limits on the playback sound levels in cinemas. Some cinema owners have reduced the playback sound levels by as much as 10 dB—half loudness! For those few blockbuster films where directors have insisted on sustained, penetrating, loud sound (compensation for an anemic plot?), this is understandable. But the dialog is reduced along with the annoyingly loud passages, and intelligibility suffers. This is a serious situation because the "art" is being compromised. Some sounds, gunshots, explosions and so forth *must* be loud to be persuasive and to convey appropriate dramatic punctuation. However, prolonged loud sounds that seem unnecessary for the events portrayed on the screen are simply tiresome and annoying. A loudness meter cannot solve this problem. Good taste is the answer. Hearing loss is not a key issue for casual moviegoers, although one should be concerned for the sound mixers.

Music programs are totally uncalibrated in loudness, requiring music lovers to listen and adjust a bass control until the sound is pleasantly balanced spectrally. The curves in Figure 4.6 do not call for a treble control (although there may be other reasons to use one), and loudness compensators that automatically boost the treble along with the bass are to be avoided—they are ill-conceived.

It has become a part of high-end audiophile folklore that tone controls damage the "purity" of sound, and some equipment proudly has no such controls. The underlying assumption has to be that recordings are flawless, but from common experience this is simply not the case. Some recordings have insufficient bass, others too much. Some are too bright, others too dull, and there are other audible maladies. Recording studios are not uniformly calibrated—see the "circle of confusion" in Figure 1.7 and the enormous variations in bass response found in a survey of recording studios (Mäkivirta and Anet, 2001). Add to that the fact that we may wish to listen at lower than "reference" sound levels at times, and it is hard to avoid the notion that a bass tone control can be an important contributor to achieving rewarding listening experiences. Easily accessed tone controls allow listeners to find their own compromise settings in dealing with the characteristics of their audio systems and rooms, and in real time, following the needs of individual programs.

Peter Walker was the founder of Quad, and I had the pleasure of knowing him. He gave us what have long been respected as relatively neutral sounding electrostatic loudspeakers, implying that no "equalization" was needed. Nevertheless he incorporated a "tilt" tone control in his Quad 34 Preamplifier (1982). It was basically bass and treble controls linked, so that as one increased the other decreased. Because the entire spectrum was "tilted," the audible effects were easily heard, so the boost/cut was limited to about 3 dB. There was also a separate low bass control. The explanation was simple: not all rooms are the same and personal tastes vary, so find the spectral balance you most like. He went on to explain that recordings are not consistent in their bass/treble balance, excessive brightness being a common problem at the time, and this could be compensated for. Then for classical works, one may wish to assume a different perspective on the orchestra than the recording engineer provided, and a tilt control gave the listener some ability to change this. Altogether it was a very sensible addition to an audio system, adding value—and participation—to the listening experience. Much later, Lexicon incorporated bass, treble and tilt controls in its surround processors. Others may also have done so. It is a good idea. The author has used them extensively over the years.

Not all playback electronics have these controls, and some of those that do require entering a menu and/or waiting for a digital reconfiguration, which are disincentives to using them. "Old fashioned" knobs or the real-time digital equivalents are user-friendly.

4.5 THE BOUNDARIES OF WHAT WE CAN HEAR

Figure 4.7 summarizes some aspects of sound and hearing that collectively bear on sound reproduction. The bottom of the display begins with a shaded area representing sounds that are below the threshold of audibility. The short wavelengths at high frequencies make it difficult to be sure of the sound pressure level at the eardrum when using conventional audiometric methods, so audiometric measurements rarely go above 8 kHz. However, there have been some independent investigations focusing on

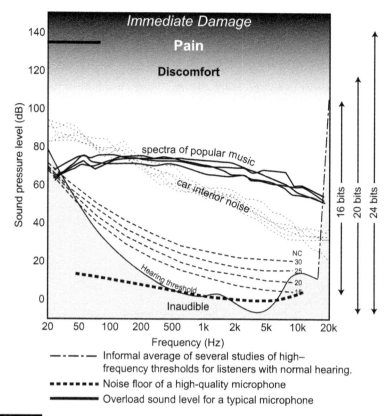

- —·—·— Informal average of several studies of high–
  frequency thresholds for listeners with normal hearing.
- ······· Noise floor of a high-quality microphone
- —————— Overload sound level for a typical microphone

FIGURE 4.7 *A simplified two-dimensional display of what we can record, store and hear. At the bottom is the hearing threshold, and an example of the noise floor of a good microphone. Above this are some examples of typical background sounds in recording and listening environments, including cars at highway speeds, to high sound levels where things become uncomfortable and then permanently damaging. Running across the middle of the white area, representing the useful listening region, are examples of long-term average spectra of popular music at foreground listening levels (from Olive, 1994). Peak levels will be higher. On the right are shown estimated dynamic ranges of perfect digital record/playback systems.*

these frequencies. The dot-dashed curve on the right side is my attempt to summarize results of efforts to measure hearing thresholds at high frequencies. There is considerable individual variation among subjects and some issues related to calibrations of the measurements—anyone seriously interested in this topic should to go to sources such as Stelmachowicz et al. (1989) and Ashihara (2007) and references cited therein. Also at the bottom is an example of the noise floor of a good microphone, indicating that it is possible to capture most of the smallest sounds the ears can hear.

Moving up the display are some examples of NC (Noise Criteria) curves used for evaluating background noise levels. These represent background sound spectra that might find their way into recordings, or that might be present during their playback. Here the curves have been lowered by 4.8 dB to make the octave-band NC curves compatible with the 1/3-octave filtered music and car interior noise. The true NC curves are shown in Figure 4.8.

In the middle of the display is a collection of background noise measurements made inside cars at highway speeds, which, when combined with simultaneous masking, explains why under these circumstances music loses much of its bass, timbral subtlety and spatial envelopment. Only in the parking lot or in stop-and-go traffic can good car-audio systems reveal their true excellence. Quiet cars are highly desirable. It also explains why car audio systems are often balanced to have rather more bass than would

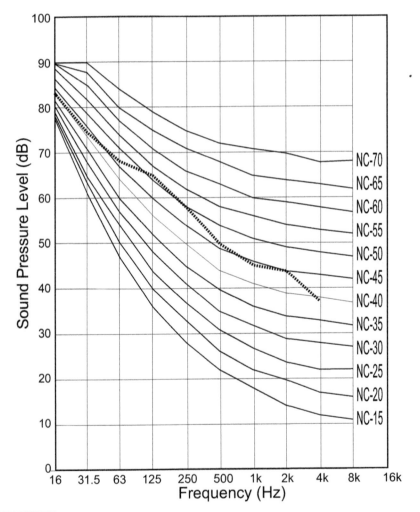

FIGURE 4.8 *The new NC contours showing a measured spectrum registering NC-51 according to the tangency criterion. The NC rating of a spectrum is designated as the value of the highest NC curve "touched" by the measured octave-band spectrum.*
Reprinted from ANSI/ASA S12.2:2008 American National Standard on Criteria for Evaluating Room Noise, © 2008, with the permission of the Acoustical Society of America, 1305 Walt Whitman Road, Suite 300, Melville, NY 11747.

be usual in the home, and why elaborate ones incorporate bass boost activated according to road speed and/or interior noise. Surround channel levels would similarly benefit from that kind of automated adjustment.

All of this is made directly relevant by the set of music spectra slicing through the middle of the display. These come from measurements done by Olive (1994) and represent four programs exhibiting unusually flat and extended spectra. These are long-term average spectra, which ignore whatever dynamic range the program had. Today, in much of the music, movies and television we are exposed to, dynamic range is diminishing. Part of it has to do with highly compressed audio delivery formats that sacrifice both bandwidth and dynamic range, and some of it is that programs are tailored for where we listen—cars, earphones on the street, in buses, the subway and so on. Every once in a while we need to sit down in quiet surroundings, put on an old-fashioned (relatively) uncompressed music source, and be reminded what dynamic range sounds like. It is not "all loud all the time," which so much of available programming seems to be. Sadly, it is part of the "dumbing down" of audio. Compressed programs sound tolerable when played through inexpensive audio systems that are incapable of playing loud. In this technology intensive era it seems that it should be possible to supply programs and playback devices that could satisfy a number of different audiences or the same audience in different circumstances—at home, in the car, walking the dog and so forth (metadata?).

At the top of Figure 4.7 are shown regions of discomfort and pain and the absolute barrier of instantaneous severe hearing loss or deafness. These are the reasons why turning up the volume is not the answer to elevated background noise levels. It becomes tiresome, and ultimately damaging. In this region even microphones reach their limits.

On the right are shown the dynamic ranges of some digital record/playback systems, showing that the conventional 16-bit CD, if well used, has the potential to be a satisfactory delivery format for most program material. Twenty bits would be better, allowing leeway for some imperfection in the process. Twenty-four-bit systems permit the encoding of everything from "I can't hear it" through to "I'll never hear again." Taking full advantage of wide system dynamics requires some serious power from amplifiers and loudspeakers, as is shown in Figure 4.3. In reality, we make do with (and enjoy) much less dynamic range than is possible. The reason that occasional loud sounds emerge from modest audio systems is that we can tolerate up to 6 dB of clean amplifier clipping with remarkably little complaint; 6 dB is a factor of four in power (Voishvillo, 2006). Those experiments used music. In movies, *very* loud events are mostly sound effects with no real "fidelity" issues, because the original could possibly be the sound of an elephant trumpeting, equalized, and played at quarter-speed, backwards.

4.5.1 What Is Acceptable Background Noise?

Over the years various criteria have evolved aimed at enabling acoustical designers to set goals for background noise levels that are acceptable. Most targeted commercial spaces, and the criterion was speech: how well could normal speech be understood against a background of HVAC and other building noises? Although simple weighted sound level meter measurements can provide a rough guide, it is usual to start with an octave-band analysis of the noise spectrum, and then to compare this spectrum to one of several criterion curves purporting to describe the acceptability of the background noise for different specific purposes.

Measures driven by such concerns remain at the core of the popular criteria used in setting acceptable background sound levels in listening and recording spaces, even though speech interference is not a problem. In those situations one is likely to be more concerned with whether the noises are annoying or pleasant. In North America, the Noise Criteria (NC) curves are widely applied to define acceptable background noise levels for audio environments. "[The NC curves] are intended to be octave band noise levels that just permit satisfactory speech communication without being annoying"—so said the creator of these curves, Beranek, as reported in Tocci (2000). Warnock (1985) states:

NC contours are not ideal background spectra to be sought after in rooms to guarantee occupant satisfaction, but are primarily a method of rating noise level. There is no generally accepted method of rating the subjective acceptability of the spectral and time-varying characteristics of ambient sounds. In fact, ambient noise having exactly an NC spectrum is likely to be described as both rumbly and hissy, and will probably cause some annoyance.

This does not sound like an objective for an expensive studio facility or recreational listening space.

Tocci (2000) and Broner (2004) provide lucid surveys of the numerous families of contours proposed by acousticians for various purposes. If the sound quality of the background noise is important—and this should be the case in listening environments—there is agreement that it is worth considering something other than the traditional NC contours where the shape of the spectrum is not evaluated, only the penetration of the highest point in the spectrum into the family of curves (the tangency criterion). For audio purposes the noise spectra should be as smooth as possible. The latest version of the NC contours is shown in Figure 4.8.

NC criteria levels that are generally considered acceptable for various sound reproduction venues are listed here. They are approximations. Obviously lower levels are preferred if they can be achieved.

> Recording studios—distant microphone pickup, NC-5 to NC-10.
> Recording and broadcast studios—close microphone pickup, NC-15 to NC-25.
> Concert halls and other live performance spaces, NC-15 to NC-20.
> Home theaters, suburban homes, NC-15 to NC-25.
> Movie theaters, NC-25 to NC 35.
> Urban residences and apartments, NC-30 to NC-40.

Commercial or multiple-dwelling buildings with massive central HVAC systems present the greatest problems. When making distant microphone recordings in some venues it may be necessary to shut down the HVAC to reduce the background noise. In these cases two NC ratings should be provided, one with HVAC and one without.

After a detailed investigation, the ultimate goal for recording studios was concluded to be as close to the hearing threshold as possible (Cohen and Fielder, 1992). Everything else leaves a residual background that inevitably will have a character of its own, and which might be captured in recordings and reproduced in quiet listening rooms. Figure 4.9 shows this perspective on background noise, which includes some real-world data on home listening rooms. Clearly, with rare exceptions, any sound isolation in home theaters is most likely to be needed to keep good sounds in, rather than bad sounds out.

A very quiet room is very impressive, and customers like to demonstrate the silence, whether or not it matters once the movie or music is under way. Home theaters that are acoustically isolated from the rest of the house to prevent the escape of theatrical dynamics almost always end up being very quiet—the sound transmission loss works both ways.

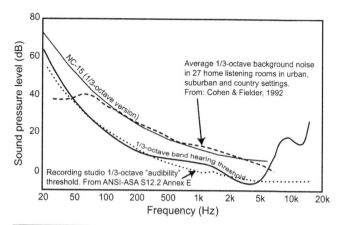

FIGURE 4.9 An interesting comparison of data originating in the Cohen and Fielder (1992) paper. The bottom "audibility" curve was motivated by this paper and underlines the need to apply more restrictive background noise requirements to audio recording and listening environments than has been traditional. Fortunately, many homes are adequately quiet. The NC and ANSI-ASA curves are reproduced with permission of the Acoustical Society of America.

4.6 LINEAR DISTORTIONS: AMPLITUDE AND PHASE VS. FREQUENCY

Audio devices should ideally yield outputs that are perfect replicas of the signals that went in, or perfectly scaled versions of those signals if amplification is involved. Transducers (microphones and loudspeakers) convert energy from one form to another, and again the requirement is that the output is a perfectly scaled transduction of the input signal (e.g., voltage in and sound level out). The success of this is evaluated by measurements comparing inputs and outputs, looking for discrepancies—distortions.

Linear distortions are those deviations from perfection that are independent of sound level. The shape of a waveform (amplitude vs. time)—the music—may be linearly distorted, but that distortion is the same at all sound levels. Linear distortions can change the relative levels (amplitude) or timing (phase) of frequency components in the complex music signal, either or both of which can alter the shape of the waveform. The combination of amplitude and phase as a function of frequency is commonly known as the transfer function that constitutes the Fourier transform of the impulse response; they are equivalent information. This equivalence of frequency-domain and time-domain information is not widely appreciated in audio, but it is fundamentally important. As will be seen, humans are very sensitive to variations in amplitude vs. frequency, and very insensitive to phase shifts. In other words, we do not "hear" waveforms *per se*, but we are very sensitive to their spectral content.

If the waveform changes as a function of signal or sound level, those changes are non-linear distortions—these non-linear mechanisms *add* frequencies to the signal that were not there originally and are especially objectionable if they are high enough to be audible.

The amplitude vs. frequency curve, which is known as the frequency response, is the single most important measurable parameter of any audio device, and the general objective is a smooth flat curve extending over the audible frequency range. However, that goal has yet to be universally achieved in our imperfect physical world. Consequently, especially in loudspeakers, we must learn how to evaluate the audible consequences of different kinds of deviation from the ideal—the audibility of tilts, peaks, dips, bumps and wiggles. In what follows, we will look closely at the audibility of deviations from what might be considered to be target performances.

4.6.1 Spectral Tilt

The simplest deviation from flat is probably a spectral tilt. There is evidence that we can detect slopes of about 0.1 to 0.2 dB/octave, which translates into a 1 to 2 dB tilt from 20 Hz to 20 kHz (Kommamura and Mori, 1983). Such a spectral error, if small, is likely to be quite benign and subject to adaptation: we simply would get used to it. However, if large, it is very audible, and causes those of us who have them to reach for the bass, treble or tilt control to restore a more normal spectral balance.

However, the most common complaint in subjective evaluations of loudspeakers is coloration due to resonances, so it is worth dwelling on this topic.

4.6.2 Resonances Viewed in Frequency and Time

Simple resonances exhibit their presence in the frequency and time domains. Things get more complicated with room resonances, discussed in Chapter 9, because what you hear also depends greatly on the room, and where you and the sound sources are located in it—"space" becomes a factor.

Simple resonances are found in electronic, acoustical or mechanical devices and constitute "lumped elements" in a resonating system: a mass component (inductance), a compliance (capacitance) component and a damping (resistance) component. The mass and compliance determine the frequency of the resonance. The damping determines the amount of energy loss in the system, which defines the quality factor, or Q. This determines the bandwidth of the resonance in the frequency domain and the duration of the buildup and decay of energy in the time domain. High-Q resonances have a small footprint in the frequency domain (a narrow, sharpish spike in a frequency response) and a long buildup and decay in the time domain. As Q falls, the resonance gets wider in the frequency domain and occupies less space in the time domain. Eventually, the frequency response gets to be "flat," there is no evidence of a resonance, and the time-domain misbehavior disappears. So it is inevitable that all systems designed to minimize resonances have flattish, smooth frequency responses.

Resonances in program material (all voices and musical instruments are complex collections of resonances) will be propagated to all parts of a listening room, and they constitute the essential sound quality, or timbre, of the sound sources. The details in the resonances differentiate the sounds of voices and musical instruments. These resonances are good. It is curious that psychoacoustic researchers have not put more effort into understanding the perception of resonances. It is reasonable to assume that humans have evolved to have special capabilities in detecting them. Instead, effort has been put into the detection of abstract spectral anomalies, like "profile analysis," discussed in Moore (2003, pp. 105–107). Perhaps this exemplifies the difference between (practical) engineers and (theoretical) psychoacousticians.

If resonances exist in loudspeakers, either in the transducers or in the enclosures, they are added to those in the program, changing the timbre. These changes are monotonously added to all reproduced sounds: voices, musical instruments and so forth. Such resonances are unwanted, and it is one of the prime challenges of loudspeaker designers to eliminate them because they radiate throughout the room, about equally to all listeners. An interesting fact is that reflected sounds are perceived as "repetitions" of the direct sound, and the result of the accumulated "looks" at the sound is that low- and medium-Q resonances become more audible. Consequently two contrasting events follow: flaws in loudspeakers are more obvious (bad), and subtle timbral cues in music are better revealed (good) (Toole and Olive, 1988). Listening through headphones or in a dominant direct sound field (a dead room) make us less sensitive to low- and middle-Q resonances, possibly explaining why some headphones with unimpressive measurements are tolerable, at least with pop music.

Büchlein (1962) conducted one of the first investigations into the audibility of spectral irregularities. His subjects listened through headphones (which, as just noted, is not the most revealing circumstance), and he used peaks and dips with equivalent, but inverted, shapes. It is tidy, but it does not acknowledge the physical mechanisms that cause such shapes to occur in the real world. The most common mechanism for generating peaks in loudspeaker frequency response curves is resonances. If a dip were to have the same shape, but inverted, it would indicate the presence of something that functions as a powerful resonant absorber of energy. Such occasions are extremely rare. More likely as a cause of a dip is a destructive acoustical interference, in which case the dip will not have the appearance of an inverted "hump" like a resonance, but rather a very sharp, possibly very deep, dip at the frequency where the interfering sounds cancel. Büchlein concluded that dips are less noticeable than peaks and that narrow interference dips would be the least noticeable of all. Wide peaks and dips were easier to hear than narrow ones. He also found that they were difficult to hear with solo instruments as test sounds, because they were audible only when the frequency of the defect and the musical tones coincided. Broadband sounds, such as noise, were more revealing of these irregularities in frequency response.

Fryer (1975, 1977) reported detection thresholds for resonances of different Q and frequency added to different kinds of program material. He found that detection thresholds fell with decreasing Q, and that frequency (at least from 130 Hz to 10 kHz) was not a strong factor. As for program, the lowest thresholds were found with white noise, the next lowest for symphonic music, and the highest thresholds for a female vocalist with jazz combo. It seems that spectral complexity matters: the more dense the spectrum and the more continuous the sound, the lower the thresholds.

Toole and Olive (1988) repeated some of these tests, confirming the results, and went further. Continuous sounds are more revealing of resonances than isolated transient sounds. The repetitions (reflections) necessary to lower thresholds (increase our sensitivity to resonances) can either be in the recorded sounds themselves or contributed by the listening room. The conclusion reported there was that we appear not to be sensitive to the ringing in the time domain (at frequencies above about 200 Hz at least), but to the spectral feature—the peak. As discussed earlier, at low bass frequencies we have the potential to hear both the spectral feature and the ringing. Section 8.3 discusses this in detail.

The least revealing sounds are, as Büchlein found, solo instruments and voices, especially those recorded in an acoustically "dry" or close-miked situation (lacking "repetition"). Olive (1994) found simple vocal and instrumental recordings not to be the most revealing of differences between loudspeakers—see Section 3.5.1.7. Although film and broadcast people place great importance on voices, for good reason, they happen to be among the sounds that are least revealing of imperfections in sound reproducing systems. High speech intelligibility is possible through systems that are significantly flawed when judged from a timbre/sound quality perspective. The most useful recordings had broadband, dense spectra, such as those shown in Figures 3.16 and 4.7. This confirms in a different way our observation over many years that if a loudspeaker has an audible problem, there is a high likelihood that resonances are involved. It also suggests a strategy for choosing program material: for demonstrations intended to impress, use simple sounds—solo voice, guitar, small combos and so on, especially with little reverberation—and, if possible, use a relatively dead room. To look for problems, listen to complex instrumental combinations with wide bandwidth and reverberation—and listen in a room with some reflections.

Those experiments looking into the audibility of resonances provided important insights into the perceptual process, and helped to set performance objectives for loudspeaker designers. Figure 4.10 shows some resonances precisely at the thresholds of detectability; any more and they become more easily audible; any less and you may be wasting money on engineering and materials. These data are valuable assets when designing loudspeakers at the low end of the price scale.

The key idea being communicated in Figure 4.10 is that neither the amplitude of the peak in the frequency response nor the duration of ringing is a reliable clue to the audibility of the resonances, at least above the transition frequency. It is counterintuitive. Adding damping reduces the Q, and therefore the ringing. It also reduces the amplitude of the peak in the frequency response, all of which would seem to be good. But the peak is now much wider and is therefore more frequently energized by components of the program. This means that it will be more often heard; the threshold ends up being lower—this well-damped resonance is easily heard. In contrast, the Q = 50 resonance is energized only when a musical component is precisely at the right frequency and remains there long enough to allow it to accumulate energy—to build up—which is easily seen in the center column of data. The amplitude of the ringing is of the same order as the tone burst itself, but it happens relatively infrequently. The right-hand column shows a very different situation for impulses. Because the spectrum of a brief transient

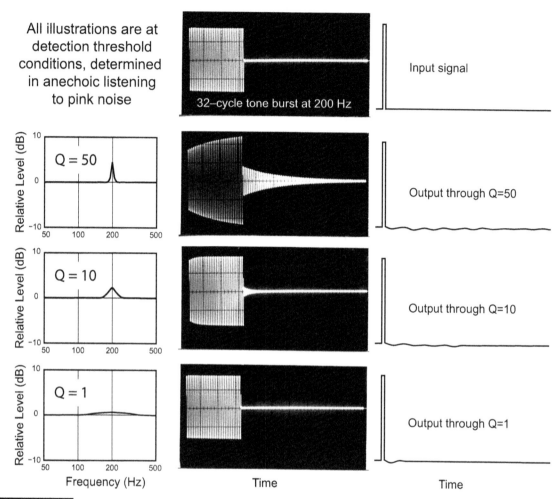

FIGURE 4.10 *On the left are shown frequency responses of resonances when they are at the threshold of detection when listening to pink noise, the most revealing of all program material. In the center column are oscilloscope pictures of how these resonances respond to a tone burst at the center frequency of the resonance. The right column shows how these resonances respond to brief impulses. Toole (2008), figure 19.10, adapted from Toole and Olive (1988).*

is very broadband, with little energy at any one frequency, there is ringing, but at a very much lower amplitude than the peak of the impulse. Forward temporal masking will reduce the perception of the ringing even further.

A key question is "what do we hear: the peak in the frequency response or the ring-ing?" For frequencies above about 200 Hz, the answer seems to be the spectral peak. Therefore the audibility of resonances will be best described by the frequency response curve. This topic is discussed in more detail in Toole and Olive (1988).

So what do just-audible resonances look like in frequency response curves? Fig-ure 4.11 shows examples of deviations from flat for high- (50), medium- (10) and

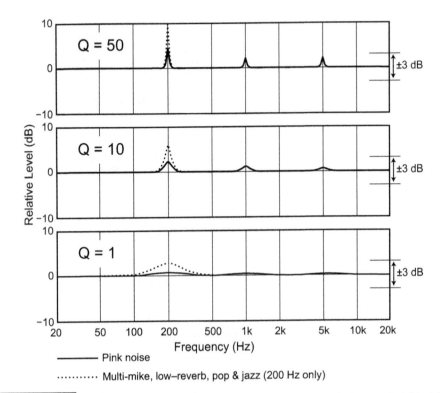

FIGURE 4.11 *Deviations from flat caused by resonances of three different Qs that were just detectable when listening to pink noise (all three frequencies) and to close-miked, low-reverberation, pop/jazz recordings (200 Hz only). Also shown is a ±3 dB tolerance, which can be seen to be too tolerant of medium- and low-Q resonances and unnecessarily restrictive of some high-Q resonances.*
Adapted from Toole and Olive (1988).

low- (1) Q resonances at three frequencies when they were adjusted to the audible threshold levels using pink noise in an anechoic chamber, and for the 200 Hz resonances detected when listening to typical close-miked, low-reverberation pop and jazz. In these simple recordings the absence of repetition (reflections/reverberation in the mix), and possibly a lack of spectral complexity, has relaxed the requirements.

These data show that we are able to hear the presence of very narrow (high-Q), low-amplitude, spectral aberrations. Measuring these, especially at the lower frequencies, requires free-field, anechoic circumstances. The popular digital FFT/TDS measuring systems that are based on time-windowed data can have severe limitations when used in reflective spaces, as shown in Figure 4.12.

Figure 4.12 shows data from related to a real experience with some loudspeaker engineers who were used to doing 10 ms time-windowed measurements in their listening rooms. This was common at that time, around 1991. I was summoned to comment on a problem they could hear, but could find no explanation for. Everything they played through the inexpensive prototype loudspeaker sounded acceptable, except for one female vocalist when she sustained a specific note. There was an unmistakable and unpleasant change in timbre. We did a quick pitch-matching to a pure tone, and found that the problem was around 280 Hz. It was clearly a very high-Q problem and only became audible because this singer had a vocal formant that coincided exactly with the resonant frequency. When she hit the right note, we were hearing what is shown in the Q = 50 tone-burst measurement in Figure 4.9. There was no evidence of a problem in the low resolution (100 Hz) gated data, but was visible in the anechoic data subsequently gathered. It was an enclosure resonance that was easily subdued.

The probability of any high-Q resonance being heard is low because exact frequency matches with some component of the program are required. However, it *does* happen. Anyone who holds negative opinions about resonances in metal diaphragms has heard very high-Q, very-narrow-bandwidth effects. Superb metal diaphragms exist, and high-resolution measurements confirm their excellence.

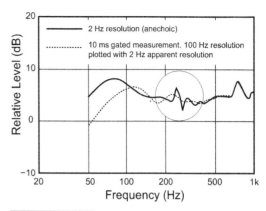

FIGURE 4.12 *Measurements of a loudspeaker with an audible high-Q problem at 280 Hz. One measurement used a 10 ms time-windowed FFT, as done in a normal room (dotted), and the other measurement was performed in an anechoic chamber using 2 Hz resolution.*

In short, if you wish to foresee *any* audible problems in a loudspeaker, high-resolution measurements cannot be avoided. Howard (2005) was aware of this limitation of in-room measurements and offers insightful options yielding better data. If equalization of a loudspeaker based on anechoic data is contemplated—as in active loudspeakers—it is necessary to have high-resolution data in order to identify center frequency, Q and amplitude of the resonances, even those of moderate Q.

In summary, some facts need to be emphasized:

1. The amplitudes of the resonances shown in these frequency responses are the *steady-state* measured changes in the playback system caused by the presence of the resonances that have been adjusted to the detection-threshold level while listening to different kinds of program in an anechoic environment. This is *not* the amplitude of the output from the resonance when listening to musical program material because music is not a steady state signal. That amplitude will probably be lower. The fact that the resonant peaks are higher for the chosen pop/jazz example is a reflection of the fact that the program material exhibits a lower probability of exciting the resonance than pink noise, a spectrally dense, steady-state signal. Not shown, but interpretable from the Fryer (1975) data, shown in Toole and Olive (1988), is the fact that symphonic music, which is both spectrally complex and reverberant, exhibits thresholds that are between the two shown in Figure 4.10 at 200 Hz.

2. High-resolution measurements are necessary to reveal high-Q features in the frequency responses. Because they exist at all frequencies, including the lower-frequency regions, it means that anechoic, or long-window simulated-anechoic measurements, are necessary to reveal them.

3. Any tolerance applied to a frequency-response curve needs to take into consideration the bandwidth/Q of the deviations that are being described. The

conventional ±3 dB, or other tolerances, have no meaning without being able to see the curve(s) that are being numerically described.

4. Finally, all of these threshold determinations were done in anechoic listening conditions. As mentioned, these thresholds may be even lower when listening in reflective rooms. However, if the thresholds were determined using signals incorporating significant repetitions (e.g., reverberation), the effect of the listening environment is minimal.

4.6.3 Finding and Fixing Resonances

The fact that loudspeaker transducers are basically minimum-phase devices means that parametric equalization can be used to address a resonance in such a device, and by reducing the amplitude, the audibility of the resonance is reduced, as is the ringing. In technical terms minimum-phase systems are "invertible." This is a powerful argument for loudspeakers with dedicated electronics, including DSP (digital signal processing). All of this evidence confirms the logic of designing loudspeakers to have smooth and flat on-axis frequency responses, which, as it turns out, is what listeners in double-blind tests have told us for decades.

The challenge is to identify what is or is not a resonance in the fluctuating frequency response curves that also contain peaks and dips associated with acoustical interference, a non-minimum-phase problem that is not invertible—that is, not correctable by simple equalization. Fortunately, it is a much less audible problem. In order to properly equalize a resonance in a loudspeaker, the center frequency, Q and amplitude must be known. This requires a moderate amount of resolution in the measurements and a parametric equalization filter that can be appropriately tuned.

Figure 4.13 shows a loudspeaker with several resonances that are easily identified because they appear in all of the spatially averaged curves. If these spinorama curves are puzzling, look ahead to Section 5.3 for an explanation. Resonances radiate widely into rooms. Interference effects change with location. In Figure 4.13a it can be seen that there are four strong resonances. Keen eyes may find a couple of others, but they are so low in amplitude that they are not audible problems. The most significant resonance is the broad low-Q hump around 3 kHz. The frequency response examples in Figures 4.10 and 4.11 provide guidance, even though they were for isolated resonances in an otherwise flat response curve. It can be seen that this bump is several times higher than the detection threshold for the $Q = 1$ resonance, the closest match. The other medium-Q resonances in (a) are around the anechoic detection threshold, and in a reflective room the threshold will be even lower, so these may be audible in program.

The loudspeaker in its original state is an undistinguished product, not rating very high in the normal monophonic double-blind evaluations done at Harman. Here it has been used as a medium for evaluating different ways to try to improve its sound using equalization. The two methods shown here are the obvious ones:

1. Equalize a spatially averaged room curve to be reasonably flat and smooth.

2. Equalize the loudspeaker to have a reasonably flat and smooth on-axis frequency response in an anechoic chamber.

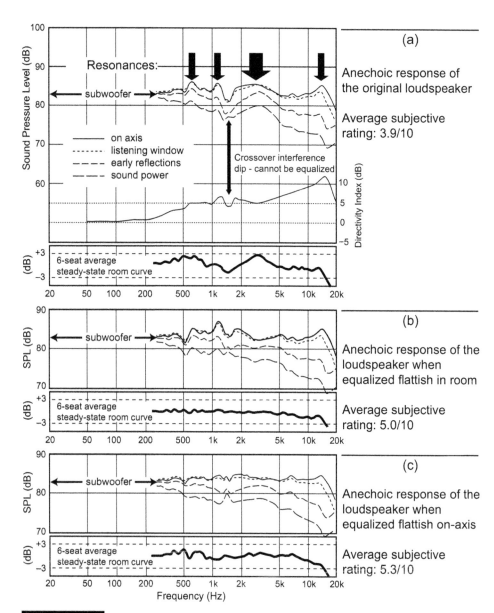

FIGURE 4.13 *Measurements on a loudspeaker in an anechoic chamber and also in a listening room. The loudspeaker has been equalized above about 400 Hz in two different ways, and here it is possible to see the performance of the loudspeaker measured in both ways for each circumstance. These data have been extracted from Olive et al. (2008), and detailed information about the tests may be found there.*

The loudspeaker seemed like a good candidate for equalization because of its generally well-behaved directivity. It will be noted, however, that there is an interference dip around 1.5 kHz caused by a flaw in the crossover. Because it is acoustical interference, it changes form with increasing spatial averaging, and almost disappears in the sound power curve. Because of this it shows up in the directivity index. This is a most often a non-minimum-phase phenomenon, and in those cases should not be equalized—but, as will be seen, it has been in these experiments that employed an automatic EQ algorithm.

The special, and very rare, feature of these data is that we get to see the performance of the loudspeaker itself in comprehensive anechoic data, and as it performs in a room as revealed by the steady-state frequency response. We also get to relate both of these data sets to double-blind subjective evaluations. This is a powerful experiment, of which more need to be done.

Based on these results it appears that any form of equalization was beneficial, elevating the subjective ratings from an unimpressive 3.9/10 for the original loudspeaker, to 5.0 for the in-room equalized version and 5.3 for the anechoic equalized version. It is clear that the medium-Q resonances do not show up in the room curve in (a), therefore, when the room performance was equalized as in (b), only the low-Q resonance was attenuated. The subjective ratings went up substantially because that was the most serious problem. The medium-Q resonances were still there, only slightly changed.

In the anechoic chamber all of the problems are on display, and all are addressed by the on-axis equalization. The result is a slightly higher subjective rating. Obviously, the anechoic chamber manipulation improved the loudspeaker, suggesting that if one had started with a comparably good loudspeaker, no in-room equalization would be necessary. The motivation is clearly for manufacturers to build good loudspeakers, and for customers to seek out and purchase such loudspeakers. Equalization to achieve an immaculately smooth room curve is not sufficient. The residual problem in the listening room is then related to the bass frequencies only, where equalization can be very helpful.

Not to be overlooked is the fact that in (c) the equalization process also filled the interference dip in the on-axis curve. This resulted in the addition of a resonance to the loudspeaker, which now shows up in the sound power curve. We will never know if the subjective scores would have been different if that had not happened.

Another feature of these tests, and one of the motivations, was to compare subjective rating performances when listening in mono, stereo and 5-channel multichannel modes. Figure 4.14 shows the result, and it is very clear, as discussed in Section 3.4 and 7.4.2, that adding more active channels to a listening experience reduces one's ability to discern details of loudspeaker performance.

As has been mentioned, monophonic components are prominent in music (the featured artist) and movies (the workhorse center channel), so the performance of loudspeakers needs to be evaluated using the most demanding test method. See Section 7.4.2 for more elaboration on this topic.

4.6.4 A Persistent Problem: Differentiating between Evidence of Resonances and Acoustical Interference

Figure 4.13 shows that resonances are recognizable in the family of curves that constitute the spinorama. A loudspeaker exhibiting strong resonances in these curves may generate corresponding evidence in steady-state room curves, but clearly this is not always the case.

If the only data one has is a steady-state room curve, it is important to differentiate between peaks and dips caused by resonances and anti-resonances, and those caused by constructive and destructive acoustical interference resulting from the summation of sounds arriving at the microphone at different times from different incident angles. The perceptual effects are very different. In general, resonances alter the timbre of sounds, while the delayed arrivals of reflected sounds that create acoustical interference are perceived as spaciousness—the sound of a room.

If the anechoic loudspeaker data do not reveal evidence of resonances, fluctuations in steady-state room curves are almost certainly attributable to acoustical interference. It is highly improbable that the propagation path from the loudspeaker to the microphone/listener contains a potent acoustical resonance mechanism above the transition frequency. Section 14.1.3 discusses the interpretation and processing of steady-state room curves.

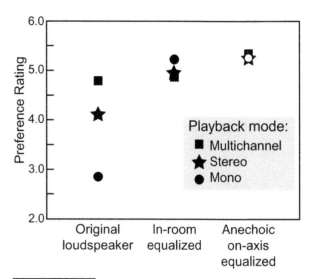

FIGURE 4.14 *The subjective ratings of the loudspeaker auditioned in three playback modes. Only in mono mode were the rating differences statistically significant.*

4.6.5 Critical Bands, ERB$_N$s and the "Resolution" of the Hearing System

One of the traditional justifications for 1/3-octave spectral analysis has been that this is a rough approximation of perceptual "critical bands" over much of the middle and high frequency range. Some people even have argued that we "hear" in critical bands, that this is the "resolution" of the hearing system. Such statements are simplistic and misleading.

The critical band is not the *resolution* of the hearing system, but it is related to the *discrimination* of tones and how multiple tonal components interact with each other, thereby entering the domain of timbre. Roederer (1995) discusses this in detail, summarizing that "It plays an important role in the perception of tone quality and provides the basis for a theory of consonance and dissonance of musical intervals."

Readers wishing to understand details of the process are referred to Roederer (1995) and Moore (2003), where the latter leads us to a new definition, the equivalent rectangular bandwidth, the ERB$_N$ (see Figure 4.15).

These bands define the bandwidth over which spectral information is summed for estimates of loudness and in the simultaneous masking of tonal signals by broadband noise. But they also define the separation required for two adjacent tones to be individually identifiable. Within these bands multiple tones (which can be fundamentals and overtones of musical sounds) beat with each other and also generate a perceptual quality called "roughness" (see Figure 4.16). Beats and differences in roughness contribute to the distinctiveness of sounds—their timbre—and so timbre can be changed if the reproducing system

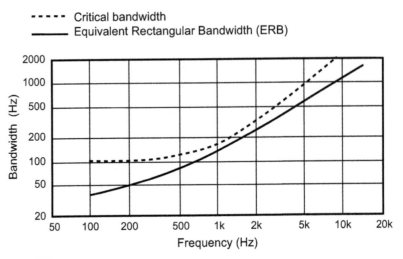

FIGURE 4.15   *The critical bandwidth and ERB$_N$, shown as a function of frequency.*

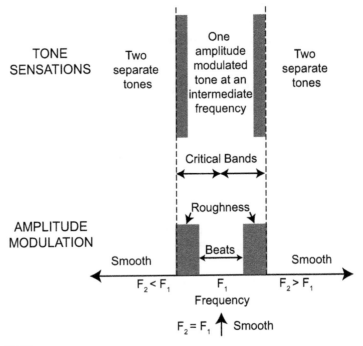

FIGURE 4.16   *An explanation of the pitch and timbre-modifying events that occur within and around critical bands.*
Inspired by Roederer (1995), figure 2.12.

has spectral variations occurring within a single critical band or ERB$_N$. The bandwidths of these bands seen in Figure 4.15 indicate that this is easily possible.

A full understanding of the performance of loudspeakers requires measurements with resolution higher than a critical band or ERB$_N$. Most of the anechoic measurements in this book are 1/20-octave, because that was technically convenient, not that it has a unique qualification. There are situations where spectrally smoothed measurements are useful for portraying general trends in spectral balance, as in room curves, but for analytical data collection contemporary understanding suggests that 1/3-octave bands are not optimal (Moore and Tan (2004). The availability of inexpensive or free measurement software that can run on portable devices means that these requirements are not burdensome.

Moore (personal communication, 2008) summarizes: "The auditory filter bandwidths (ERB$_N$ values) are about 1/6- to 1/4-octave at medium to high frequencies. And . . . within-band irregularities in response can have perceptual effects." As has been found in other aspects of perception, it is difficult to find a "one-curve" description for what we hear. These observations emphasize the importance of having high-resolution measurement data that may reveal fluctuations that should ideally be smoothed.

Finally, it almost goes without saying that all of these perceptual factors are changed in hearing-impaired listeners. With hearing loss, the critical bands get wider, and even our perception of music is altered. It is not just a matter of hearing or not hearing; it affects how we perceive the sounds that we are able to hear.

4.7 AMPLITUDE, FREQUENCY AND TIME TOGETHER: WATERFALL DIAGRAMS

Waterfall diagrams are very photogenic and intellectually appealing because they combine the three domains of amplitude, frequency, and time. The problem is that the audio environment is awash with waterfall diagrams that are inadequately described. They may show what is claimed; they may not. These displays have been used to inform, to impress and, unfortunately, also to deceive. The observer needs some help in order to know what the display is revealing. This provocative introduction needs explanation.

Ideally we want to be able to see high-resolution data in the frequency and time domains simultaneously. Unfortunately this is not always possible; in general you can have one or the other, not both. Figure 4.17 illustrates some examples of the consequences of making different choices in frequency and time resolutions.

Figure 4.17a shows a very high resolution (about 3 Hz) steady-state frequency response. Many measuring devices, including smart-phone, tablet and laptop-based packages, can generate this kind of data. It has excellent detail in the frequency domain, but it reveals nothing *directly* about what is happening in the time domain. However, because we know that low-frequency room resonances generally behave in a minimum-phase manner, we know that, if there are no prominent peaks protruding above the average spectrum level, there will not be prominent ringing in the time domain. It is this *indirect*, inferential knowledge that permits us to confidently use frequency responses as a primary source of information about room behavior at low frequencies. Figure 4.16b, c and d show the *same* situation displayed in time and frequency, as a waterfall, but employing different resolutions.

In Figure 4.17b the frequency response at the back, near, time = 0, is closely similar to the curve in (a), because a narrow 7 Hz bandwidth was chosen. However, the

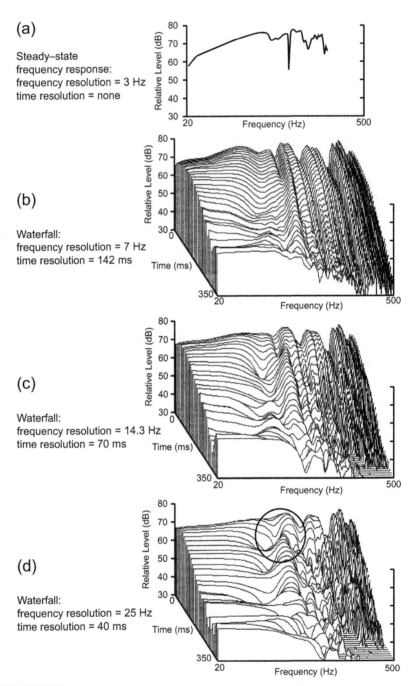

(a)

Steady–state
frequency response:
frequency resolution = 3 Hz
time resolution = none

(b)

Waterfall:
frequency resolution = 7 Hz
time resolution = 142 ms

(c)

Waterfall:
frequency resolution = 14.3 Hz
time resolution = 70 ms

(d)

Waterfall:
frequency resolution = 25 Hz
time resolution = 40 ms

FIGURE 4.17 *Examples of differing resolutions in time and frequency domains, for the same acoustical situation. (a) a very-high-resolution steady-state frequency response. (b), (c) and (d) show waterfall diagrams with different frequency and time resolutions.*

successive curves making up the waterfall all look very much like it because in achieving detail—resolution—in the frequency domain, the time-domain resolution has been sacrificed; it is 142 ms. Within the total 500 ms decay that is shown, nothing much changes because every successive curve moving forward is an average of events within a 142 ms window. The first several curves after $t = 0$ are almost identical, and those following simply drop smoothly in amplitude. It looks as though this room has uncontrolled ringing at many frequencies, but this is wrong—the data are misleading.

In (c) the frequency resolution has been reduced; now it has an effective resolution of about 14.3 Hz. The top frequency response is smoother, revealing less detail but, because of the improved 70 ms time resolution, we now can see some detailed changes in the decay curves. All of the curves decay more rapidly and there is evidence of what seems to be a frequency shift in one of the resonances in the early part of the decay. At frequencies approaching 500 Hz, it can be seen that the output has dropped below the lowest measured level before 350 ms is reached.

In (d) the top frequency response, and all that follow it, are much smoothed by the further reduced frequency resolution, now 25 Hz. However, because the time resolution is a much-improved 40 ms, it is possible to see more of what is happening in the time domain. Notice in the circled area that what looked like a single decaying resonance in (a) is actually two adjacent resonances with the higher amplitude one decaying very quickly, leaving a higher Q but much lower amplitude one to ring longer. All of the other resonances that looked so alarming in (a) can now be seen to decay very quickly. In fact, the room appears to have only one resonance with prolonged decay and its amplitude is more than 10 dB down at the outset—not likely to be a dominant audible factor.

The system looks as though it should have good bass performance, and it did—it was my first home theater (Figures 7.18 and 8.14). This same conclusion can be drawn from the relatively well-behaved steady-state room curve in (a). Good news.

In audio forum discussions and manufacturers' data, there have been numerous examples of erroneous conclusions being drawn from data of this kind. Usually the authors don't reveal the parameters of the measurement; in some cases one suspects that they did not know, or know that it mattered. Other mathematical methods exist that reveal different aspects of the time/frequency tradeoff (e.g., Wavelets, Wigner-Ville and Gabor distributions), but each requires skilled interpretation, and none that I am aware of has any supportive psychoacoustical data.

4.8 PHASE AND POLARITY—DO WE HEAR WAVEFORMS?

A very long time ago, as part of my self-education about the hearing system, I learned that the basic transduction process in the inner ear performed as a half-wave rectifier. This alone gives one reason to think that acoustic compressions will be heard as different from acoustic rarefactions. We should be sensitive to details in the acoustic waveform as it generates movement of the eardrum, and thus initiates the hearing process. Based on this simple logic, phase shifts and absolute polarity should be audible phenomena.

Naturally I did some tests, reversing polarities of loudspeakers, and introducing phase shift to distort musical waveforms—listening for big differences. They weren't there. At least not in the music I was listening to, through the loudspeakers I was using, from the musical sources I employed. Maybe I was simply unable to hear these things. Yes, there were times when I thought I heard things, but they were subtle, and hard to repeat. Changing loudspeakers made huge differences. Changing recording companies or engineers made huge differences. But, the anticipated "dramatic" event of inverting polarity appeared to be missing, in spite of how appealing the idea of waveform integrity is from an engineering perspective. Are they audible or not, and if so, does it matter in real-world listening?

4.8.1 The Audibility of Phase Shift and Group Delay

The combination of amplitude vs. frequency (frequency response) and phase vs. frequency (phase response) totally define the linear (amplitude independent) behavior of loudspeakers. The Fourier transform allows this information to be converted into the impulse response and, of course, the reverse can be done. So, there are two equivalent representations of the linear behavior of systems, one in the frequency domain (amplitude and phase) and the other in the time domain (impulse response).

An enormous amount of evidence indicates that listeners are attracted to linear (flat and smooth) amplitude vs. frequency characteristics; more to be shown later. Figure 7 in Toole (1986) and the excerpts shown here in Figure 5.2 indicate that listeners showed a clear preference for loudspeakers with smooth and flat frequency responses. Figure 5.2 also shows phase responses for those same loudspeakers. It is difficult to see any reliable relationship to listener preference, except that those with the highest ratings had the smoothest curves, but linearity did not appear to be a factor. Listeners were attracted to loudspeakers with minimal evidence of resonances because resonances show themselves as bumps in frequency response curves and rapid up-down deviations in phase response curves. The most desirable frequency responses were approximations to horizontal straight lines. The corresponding phase responses had no special shape, other than the smoothness. This suggests that we like flat amplitude spectra, we don't like resonances, but we seem to be insensitive to general phase shift, meaning that waveform fidelity is not a requirement.

If one chooses to design a loudspeaker system having linear phase, there will be only a very limited range of positions in space over which it will apply. This constraint can be accommodated for the direct sound from a loudspeaker, but even a single reflection destroys the relationship. Therefore it seems that (a) because of reflections in the recording environment, there is little possibility of phase integrity in the recorded signal, (b) there are challenges in designing loudspeakers that can deliver a signal with phase integrity over a large angular range, and (c) there is no hope of it reaching a listener in a normally reflective room. All is not lost, though, because two ears and a brain seem not to care.

Many investigators over many years attempted to determine whether phase shift mattered to sound quality (e.g., Hansen and Madsen, 1974a, 1974b; Lipshitz et al., 1982; Van Keulen, 1991; Greenfield and Hawksford, 1990). In every case it has been shown that, if it is audible, it is a subtle effect, most easily heard through headphones or in an anechoic chamber, using carefully chosen or contrived signals. There is quite general agreement that with music, reproduced through loudspeakers in normally reflective rooms, phase shift is substantially or completely inaudible. When it has been audible as a difference, when it is switched in and out, it is not clear that listeners had a preference.

Others looked at the audibility of group delay. Blauert and Laws (1978), Deer et al. (1985), Bilsen and Kievits (1989), Krauss (1990), Flanagan et al. (2005), Møller et al. (2007) found that the detection threshold is in the range 1.6 to 2 ms, and more in reflective spaces. These numbers are not exceeded by normal domestic and monitor loudspeakers.

Lipshitz et al. (1982) conclude: "All of the effects described can reasonably be classified as subtle. We are *not*, in our present state of knowledge, advocating that phase linear transducers are a requirement for high-quality sound reproduction." Greenfield and Hawksford (1990) observe that phase effects in rooms are "very subtle effects indeed," and seem mostly to be spatial rather than timbral. As to whether phase corrections are needed, without a phase correct recording process, any listener opinions are of personal preference, not the recognition of "accurate" reproduction.

4.8.2 Phase Shift at Low Frequencies: A Special Case

In the recording and reproduction of bass frequencies, there is an accumulation of phase shift at low frequencies that arises whenever a high-pass filter characteristic is inserted into the signal path. It happens at the very first step, in the microphone, then in various electronic devices that are used to attenuate unwanted rumbles in the recording environments. More is added in the mixing process, storage systems and in playback devices that simply don't respond to DC. All are in some way high-pass filtered. One of the most potent phase shifters is the analog tape recorder. Finally, at the end of all this is the loudspeaker, which cannot respond to DC and must be limited in its downward frequency extension. I don't know if anyone has added up all of the possible contributions, but it must be enormous. Obviously, what we hear at low frequencies is unrecognizably corrupted by phase shift. The questions of the moment are, how much of this is contributed by the woofer/subwoofer, is it audible and, if so, can anything practical be done about it? Oh yes, and if so, can we hear it through a room?

Fincham (1985) reported that the contribution of the loudspeaker alone could be heard with specially recorded music and a contrived signal, but that it was "quite subtle." The author heard this demonstration and can concur. Craven and Gerzon (1992) stated that the phase distortion caused by the high-pass response is audible, even if the cutoff frequency is reduced to 5 Hz. They say it causes the bass to lack "tightness" and become "woolly." Phase equalization of the bass, they say, subjectively extends the effective bass response by the order of half an octave. Howard (2006) discusses this work and the abandoned product that was to come from it. There was disagreement about how audible the effect was. Howard describes some work of his own, measurements and a casual listening test. With a custom recording of a bass guitar, having minimal inherent phase shift, he felt that there was a useful difference when the loudspeaker phase shift was compensated for. None of these exercises reported controlled, double-blind listening tests, which would have delivered a statistical perspective on what might or might not be audible, and whether a preference for one condition or the other was indicated.

The upshot of all of this is that, even when the program material might allow for an effect to be heard, there are differences of opinion. It all assumes that the program material is pristine, which it patently is not, nor is it likely to be in the foreseeable future. It also assumes that the listening room is a neutral factor, which, as Chapter 8 explains, it certainly is not. However, if it can be arranged that these other factors can be brought under control, the technology exists to solve the residual loudspeaker issue.

4.8.3 The Audibility of Absolute Polarity—Which Way Is "Up"?

Finally, there is the matter of simple polarity, which is where this discussion started. Johnsen (1991), a strong advocate of the importance of polarity, offers an extensive survey of literature, much of it anecdotal, and conducts experiments that are claimed to be definitive evidence of its audibility. Unfortunately, from my perspective, there was a serious flaw in the choice of source material for the subjective tests. Johnsen said: "Sources were LP records exclusively, as the CD medium was found (in collateral, unreported tests) to reveal polarity rather less well." I have an intimate understanding of the LP medium, and its lengthy list of non-linear behaviors. It is easy to understand that the asymmetrical distortions added by the playback medium would indeed make polarity changes more easily audible than the demonstrably more linear CD. Were the listeners responding to the medium or the message?

In contrast, Greiner and Melton (1994), who used CDs as sources, took precautions to avoid contamination of the signals by both source and delivery devices. They employed both synthesized asymmetrical waveforms and naturally asymmetrical sounds. Their conclusion was:

It is certain from our listening tests that inversion of acoustic polarity is clearly audible for some instruments played in some styles and for some listening conditions. It is not likely that the effects observed in this work were an artifact of the record reproduce system because of the considerable care taken to maintain waveform integrity.

While polarity inversion is not easily heard with normal complex musical program material, as our large-scale listening tests showed, it is audible in many select and simplified musical settings. Thus it would seem sensible to keep track of polarity and to play the signal back with the correct polarity to ensure the most accurate reproduction of the original acoustic waveform.

Greiner went on to say in a subsequent "letters to the editor" exchange that: "Neither the quality of the transducers nor the loudness level, again within reason, has in my experience a great effect on the audibility of acoustic polarity inversion."

So, as speculated at the outset, the asymmetrical detection process in the ear can yield audible polarity effects if the signal and the listening conditions are right. The problem is that there appear to be no audio or film industry standards that ensure the delivery of such absolute polarity sounds from the microphones, through the extensive electronic manipulations in control rooms or dubbing stages, which may be different for different components of a mix, and finally through the playback electronics and loud-speakers at home. If there is a polarity switch in the playback chain, one may occasionally find a preferred setting, but there can be no knowing which setting is correct, and it may be different for different voices and instruments in an ensemble.

4.9 NON-LINEAR DISTORTION

The memory of distortions in 78 rpm recordings and early LPs is still vivid in my memory. They ranged from inaudible to intolerable, and all recordings had them. LPs and their playback devices improved, and at their best they became highly enjoyable—except for pesky inner-groove crescendos, the exciting climax of a symphony or opera that one has carefully been prepared for by the composer.

I recall testing phonograph cartridges as part of the effort to improve playback quality (Toole, 1972). This exercise is substantially a test of test records, which really is a test of the entire LP mastering, pressing and playback process. I participated in the creation of a test record, creating some of the sounds and signals recorded on the master tape. Comparisons of the master tape to the LP playback revealed that it was impossible to replay from an LP the same signal that was delivered to the (carefully chosen) mastering lab. When using pure tones or bands of noise, distortions were easily measured and easily audible. They registered in whole (sometimes high) percentages much of the time, and this applied to both harmonic and intermodulation versions. Yet, when the signal was music, the experience was mostly enjoyable. What is going on here?

Masking, in which a smaller sound (the distortion) is rendered less audible by a larger sound (the same musical signal that created the distortion), as shown in Figure 4.18. Simple test signals leave the distortion products spectrally exposed so that they can be measured, and sometimes enough of them are unmasked that we can hear them. The wide-bandwidth, dense spectrum of music is a much more effective masking signal, in spite of the fact that it is at the same time a generator of much more complex distortion products. It is also almost useless as a measurement test signal.

The masking contours shown in Figure 4.18b are especially revealing in that a loud bass sound can have an effect on sounds covering much of the audible spectrum, including the critical voice-frequency range. This may explain why it is sometimes difficult to understand the lyrics in a loud, bass-heavy rock concert, or the dialog in a movie when loud bass rumbles are used for dramatic effect.

To understand what is going on, first consider what the real problem is: non-linear behavior in the audio device. This means that the relationship between the input signal and the corresponding output from the device changes with level. If the input increases by a certain percentage, the output should change by the same percentage—no more, no less. However, the percentage does change; the shapes of audio waveforms are altered

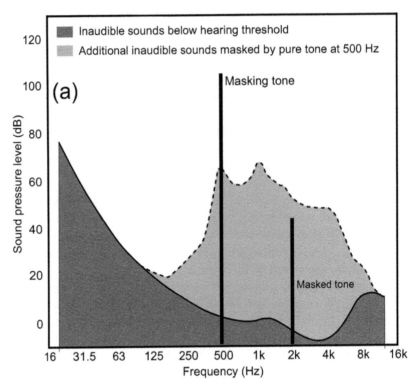

FIGURE 4.18
(a) a simple view of simultaneous masking of a tone at 2 kHz by another, louder one, at 500 Hz. It shows that the masking effect spreads substantially upward in frequency and only slightly downward. The detailed shape of the dashed masking curve is highly variable, depending on masker level and differing among studies, but the general effect is as shown. At lower sound levels, the masking effect exhibits less spreading. (b) shows the substantial masking effects of very low frequencies, especially at high sound levels.
From Fielder, chapter 1.5, figure 1.60 in Talbot-Smith (1999). Reproduced by permission of Taylor & Francis Group, LLC, a division of Informa plc.

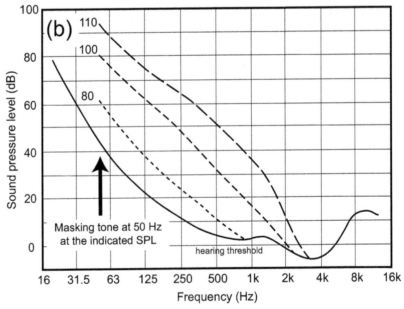

from a little to a lot. When we hear these distorted waveforms, we notice that new sounds (distortion products) have been created. When the non-linear device is driven by a known, well-specified signal, we can measure how different the output is from the input, and obtain a measure of the magnitude of the distortion—for that specific signal. But, instead of measuring the nature of the non-linearity itself—as an input-to-output transfer relationship—we have chosen to try to quantify what it does to audio signals. Put a known signal into the device and spectrally analyze what comes out.

Engineers have invented several ways to quantify the magnitude of the non-linearity by measuring the distortion products and expressing them as a percentage of the original input signal, or of that signal plus distortion and noise. If the input signal is a pure tone, the distortion products show up as harmonic overtones of that signal: harmonic distortion. The problem with that method is that as the fundamental frequency gets higher, the higher harmonics move beyond the hearing range so it ceases to be a relevant metric. Combining two closely spaced tones and scanning them to high frequencies is more useful. Each of the tones will generate a set of harmonic distortions, but they will also interact with each other in the non-linearity and create combination tones— intermodulation distortions. These are sum and difference multiples of the two input signals and their harmonics that extend upwards and downwards in frequency. Because masking is much more effective in the upward frequency direction (see Figure 4.18), it is the difference frequencies that end up being more audible and, because they are inherently unmusical, more annoying.

Thus began the legend that harmonic distortions are relatively benign and intermodulation distortions are bad. It is true in the context of these test signals, but they are both simply different ways of quantifying the *same* problem (the non-linearity in the audio component), and neither test signal (one tone or two) is even a crude approximation of human voices or music. Contributing to the mismatch between perception and measurement is the fact that such a technical measurement totally ignores masking. Included in the numbers generated by the measurements are distortion components that are partially or completely masked. Some of what is measured is inaudible. The numbers are wrong.

The result of this is that traditional measures of harmonic or intermodulation distortion are almost meaningless. They do not quantify distortion in a way that can, with any reliability, predict a human response to it while listening to music or movies. They do not correlate because they ignore any characteristics of the human receptor, itself an outrageously non-linear device. The excessive simplicity of the signals also remains a problem. Music and movies offer an infinite variety of input signals and therefore an infinite variety of distorted outputs. The only meaningful target for conventional distortion metrics is "zero." Above that, somebody, sometime, listening to something, may be aware of distortion, but we cannot define it in advance.

A recent listening test proved its worth when it revealed that a loudspeaker having excellent looking spinorama data (Section 5.3), which normally is sufficient to describe sound quality, was not rated highly as expected. The problem was found to be intermodulation distortion, an extremely rare event, associated with the way sounds from a woofer and tweeter combined in a concentric arrangement—so constant vigilance and listening are essential.

Taking advantage of advances in psychoacoustic understanding, and utilizing the analytical and modeling capabilities of computers, some new investigations are attempting to identify some of the underlying perceptual mechanisms, and develop better test methods. Cabot (1984) provides a good historical perspective; Voishvillo (2006, 2007) provides an excellent overview of the past, present and possible future of distortion measurements. Geddes and Lee (2003), Lee and Geddes (2003, 2006), Moore et al. (2004), and Temme and Dennis (2016) provide additional useful perspectives.

In loudspeakers it is fortunate that distortion is something that normally does not become obvious until devices are driven close to or into some limiting condition. In large-venue professional devices, this is a situation that can occur frequently. In the general population of consumer loudspeakers, it has been very rare for distortion to be identified as a factor in the overall subjective ratings. This is not because distortion is not there, or not measurable, but it is low enough that it is not an obvious factor in judgments of sound quality at normal listening levels.

4.10 WAVELENGTH, THE KEY TO UNDERSTANDING MUCH IN AUDIO

In audio most discussions involve frequency, and events in the frequency domain. But to understand why many of these things happen, it is essential to understand the concept of wavelength. The explanation begins with the speed of sound, which, at sea level and normal room temperature (72°F/22°C), is:

- 1,131 ft/s (771 mi/hr)

- 345 m/s (1,240 km/hr).

All audio frequencies travel at the same speed, so the distance traveled in one cycle of sound at a given frequency is the speed of sound divided by the frequency.

Wavelength = Speed of Sound ÷ Frequency

This gives the wavelengths for a few frequencies as shown in Figure 4.19.

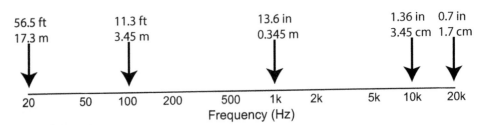

FIGURE 4.19 *Wavelengths at various frequencies.*

4.10.1 Loudspeaker Directivity

Let us consider what this enormous range of wavelengths implies in sound repro-duction. Sound radiated from a source, a loudspeaker, a musical instrument, a voice and so forth is omnidirectional—equal in all directions—when the wavelength of the sound it is radiating is long compared to the dimensions of the source. So, for a 12-inch (0.3 m) subwoofer operating below 80 Hz, the wavelengths are longer than 14.1 ft (4.3 m) and the sound will radiate omnidirectionally. It therefore doesn't mat-ter which way a subwoofer diaphragm faces—up, down or sideways—the radiated sound will not change.

However, at higher frequencies—shorter wavelengths—the radiated sound will tend to become more focused in the forward direction of the diaphragm, radiating in a progressively narrowing beam. This is illustrated in a grossly simplified manner in Figure 4.20a; when the wavelengths approach and eventually become smaller than the

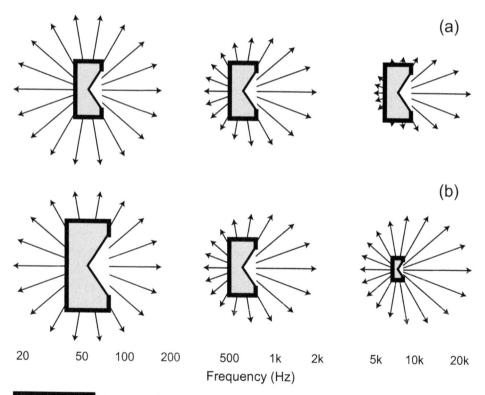

FIGURE 4.20   *A very simplified illustration of sound radiation patterns from loudspeaker drivers of the same size (a) and when the sizes are graduated to somewhat correspond to the wavelengths (frequencies) being radiated. The goal is relatively constant directivity at frequencies above the transition frequency. Nevertheless, the lowest frequencies will trend to omnidirectionality and the highest frequencies from a conventional tweeter will progressively beam.*

diaphragm of the loudspeaker, it becomes more directional. This is a problem for two reasons:

1. The off-axis, room reflected sound cannot have the same spectral balance as the direct sound. This is a problem for sound quality and sound localization.

2. Large diaphragms don't behave well when driven at high frequencies. They acquire breakup modes—resonances—that color the sound.

However, if the size of the loudspeaker diaphragm is reduced as frequency increases, something closer to constant directivity can be achieved, along with much improved frequency and time-domain performance. This is illustrated in Figure 4.20b. The residual problem is that at very low frequencies the bass is radiated omnidirectionally, so the directivity control becomes effective in the low hundreds of Hz. In practice this is not an issue, because below the transition frequency what we hear is dominated by room modes, as will be discussed in Chapter 9. The choice of crossover frequency is based on both frequency response and directivity. Ideally, at the crossover frequency the directivity of the low-passed driver and the high-passed driver should be similar.

Interestingly, dividing the frequency range is advantageous for distortion reduction. Spectrally complex sounds are shared among multiple drivers, thereby reducing the opportunity for intermodulation distortion. More crossovers complicate matters, but improved measurements, waveguides to better match directivities, and computerized crossover design aids have all but eliminated audible artifacts, as can be seen in measurements in later chapters.

What about large-panel loudspeakers? Because directivity is determined by the wavelength relative to the size of the radiating surface, a large surface radiating the full audio bandwidth must become progressively more directional with increasing frequency. The limited displacement of electrostatic and electromagnetic panels requires that they be large in order to produce enough sound to be interesting. The usual solution for the directivity problem is to subdivide the panel into progressively smaller areas, radiating successively higher frequencies. Many employed a simple narrow tweeter strip. Perhaps the most ambitious example of this approach was the Quad ESL63, which drove the diaphragm with an arrangement of concentric rings, expanding the source area as frequency decreased, with the aim of generating the wavefront of a point source. Others curve the surface, but the nature of the drive mechanism allows only a gentle curvature.

There are practical reasons why cones and domes became and remain the prevalent loudspeaker paradigm.

4.10.2 Room Resonance Basics

Sounds of all frequencies and wavelengths reflect around rooms in a somewhat chaotic manner, but when the wavelengths are of the same order as the dimensions of a room some special events occur. Between parallel surfaces that are half a wavelength apart, or multiples of half-wavelengths apart, the sound traveling in both directions add constructively, reinforcing each other, and a standing wave develops. This is the basis of a room mode or resonance. Figure 4.21 illustrates the basics.

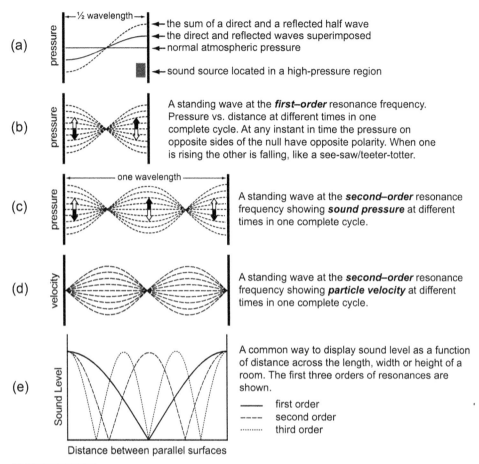

(a)

½ wavelength
- the sum of a direct and a reflected half wave
- the direct and reflected waves superimposed
- normal atmospheric pressure

- sound source located in a high-pressure region

(b)

A standing wave at the **first–order** resonance frequency. Pressure vs. distance at different times in one complete cycle. At any instant in time the pressure on opposite sides of the null have opposite polarity. When one is rising the other is falling, like a see-saw/teeter-totter.

(c)

one wavelength

A standing wave at the **second–order** resonance frequency showing **sound pressure** at different times in one complete cycle.

(d)

A standing wave at the **second–order** resonance frequency showing **particle velocity** at different times in one complete cycle.

(e)

A common way to display sound level as a function of distance across the length, width or height of a room. The first three orders of resonances are shown.

—— first order
---- second order
········ third order

Distance between parallel surfaces

FIGURE 4.21   *(a) The mechanism by which a standing wave is formed, illustrated in a "stop-action" view. At the frequency for which the distance between the walls is exactly one-half wavelength, the direct sound wave traveling toward one wall is precisely replicated by the reflected sound traveling away from it. They add, producing a higher amplitude resultant, which is called a "standing wave" because it has a constant form. In (b) is an attempt to show what happens in a running situation. The "stop-action" resultant waveform shown in (a) cycles up and down with time, once per period of the signal, and it is shown at several points in the cycle. In this first-order standing wave along one axis of a room there is one null, one location where it is possible to stand and hear almost nothing—in the center of the room. Moving toward either wall the sound gets louder. It is very important to note that, at any instant in time, at any one of the "stop-action" curves, when the sound pressure on one side of the null is increasing, the sound pressure on the other side is decreasing (shown by the white and black arrows). (c) shows the situation when the distance between the walls is one wavelength. This pattern can also exist between the walls in (a), but at double the frequency. (d) shows the distribution of particle velocity as a function of distance. This is important to understanding resistive absorbers. (e) shows a common manner of representing the sound pressure distribution across a room that is graphically simpler, but it must be remembered that there is a polarity reversal at each null. This will be important in multiple-subwoofer mode-attenuating schemes.*

At the frequencies for which the wall separation is a multiple of one-half wavelength there will be a resonance and a standing wave. These are called axial modes, because they exist along each of the principal axes of a rectangular room: length, width and height.

When there is a standing wave, it is obvious that the sound level at the frequency of the resonance will change as one moves around the room. In fact, a classic demonstration of the phenomenon is to set up a loudspeaker placed against an end wall radiating a pure tone at the first-order resonance frequency along that dimension of the room, and have listeners walk from one end to the other. If all is well, the sound will be about equally loud at both ends of the room and will almost disappear at the halfway point. A variation of this is to have the listeners stand at a single point and scan the signal generator through a range of low frequencies—there will be huge fluctuations in loudness, high at some frequencies, low at others. It is a simple and persuasive demonstration of the problem confronting us.

This topic will be elaborated on in Chapter 8, but for now it is sufficient to understand that it is the wavelength of sound relative to the dimensions of rooms that determine whether and at what frequencies there will be resonances and the accompanying standing waves. The resonant modes illustrated here occur between two parallel surfaces and exist along the principal axes of rooms: length, width and height. They are called axial modes, and although other kinds of modes will be shown to exist, these are the dominant factors in most rooms.

It may seem odd that Figure 4.21d shows "particle velocity" when what we hear and normally measure is sound pressure. It turns out that it is important in understanding how resistive absorbers (fibrous tangles and acoustical foam) work—the next topic.

4.10.3 Resistive/Porous Absorbers and Membrane/ Diaphragmatic Absorbers

Most of the absorbers we use are resistive: messy tangles of glass, cellulose or mineral wool fibers, or leached foam that creates a similarly complex path for sound to travel through. Sound is a pressure wave that propagates through an elastic medium, the compressible fluid air in this case, creating compressions and rarefactions as it passes. The air molecules move back and forth, oscillating in place, in the direction of sound propagation, but not traveling. The wave motion travels at the speed of sound, but the molecules themselves just vibrate. If this movement is forced to happen inside a resistive absorber, some energy is lost in the turbulence of moving within the fibrous tangle—it is absorbed, converted to heat. Hence the description of the material (porous) and the mechanism of absorption (resistance). But the resistive losses can happen only if the air molecules are in motion and the porous material has significant flow resistance.

Looking at Figure 4.21c and d, it is evident that pressure and particle velocity are inverses of each other. It is intuitively easy to understand that air molecules that are against a hard wall have nowhere to go, and particle velocity is zero. Equally instinctive is the notion that a sound pressure wave arriving at a rigid wall will build up pressure, as it is unsuccessful at moving the wall and is reflected. So, at a reflecting boundary, pressure is at a maximum, and particle velocity is at a minimum.

Thus we reach the first level of understanding how best to absorb sound. If resistive absorption is to be useful, it must be at a distance from the room boundary—the portion of it that is adjacent to the boundary is doing nothing, and the effectiveness increases with distance from the boundary, or the thickness of the material, up to a maximum at the 1/4-wavelength distance where the particle velocity is maximum (Figure 4.21d). Look at Figure 4.19 and divide those wavelengths by four to get an idea of why resistive absorbers are not attractive solutions at low frequencies—even at a mid-bass frequency of 100 Hz the quarter-wavelength is 2.8 ft (0.86 m), which seriously intrudes on expensive real estate. Fortunately, at any distance from the boundary the resistive absorber is doing something, even if not its best effort, which explains why it is the solution of choice at middle and high frequencies.

The popular alternative at low frequencies is membrane absorption. When the pressure builds up at the room boundary, if the boundary is flexible it will move, thereby extracting energy from the sound field and reflecting less back into the room. Again sound energy is turned into heat in the frictional losses of the moving wall surface. When you feel the bass in the floor or walls, membrane absorption is happening. Naturally, if the room boundaries are not adequately absorbing, membrane absorbers must be added. Most of these are tuned to absorb sounds over specific bands of frequencies, sometimes narrow bands, sometimes wider bands. To know what to build or buy, it is important to know what frequencies are in need of attenuation so that (a) the right membrane absorber is used, and (b) it is possible to find the high-pressure regions for that frequency in the standing wave patterns of the room, and place the absorber where it will be most effective.

All of this involves the concept of wavelength.

4.10.4 Diffusers and Other Sound-Scattering Devices

Sound scattering, diffusing surfaces are necessary in performance spaces where there is a fixed amount of sound energy from voices and musical instruments and it is necessary to deliver that sound to all parts of the auditorium. In sound reproduction systems there is a volume control, and the record/reproduction process attempts to incorporate the important directional and spatial cues, in the stereo or multichannel media.

Do we want a diffuse sound field in our listening rooms? No, and even if we did, normally furnished rooms have so much absorption that we could not get it. In small listening rooms and home theaters, sound-scattering/diffusing surfaces are used as an alternative to absorption in breaking up strong reflections, or flutter echo between parallel surfaces.

The classic demonstration is the "handclap" test. Stand between two reflective parallel walls and clap the hands; especially to the person clapping, there may be a disturbing flutter echo: the clap rattling back and forth between the walls. This is an interesting demonstration, but the real problem is when the "clap" emerges from a loudspeaker and is heard as a flutter echo to a person sitting in the audience. So the first test involves someone moving from speaker to speaker in a system, clapping the hands, while someone else moves around the listening area to see if there is a problem. Then, if there is, find the offending reflecting surface and treat it. Absorption is one solution; scattering or diffusing the energy is an alternative.

Handclaps tend to exhibit a spectral maximum around 2 kHz and higher, so it takes very little to damp them. From the preceding discussion, 1″ (25 mm) of resistive absorber will do the job. If a diffusing surface is decided on, there is again a decision about the thickness, and again the answer will be based on wavelength. Figure 4.22 illustrates the basic principle. Each diffuser, depending on its design and depth, will have a frequency below which it does not scatter the sound, meaning that listeners exposed to only the scattered portion of the sound will hear a high-pass filtered version of the original sound. It is therefore desirable that the scattered/diffused sound should cover the bandwidth above the transition frequency, to avoid distorting the spectrum of the reflected sound field.

A crude guideline is that the depth of a geometrical (e.g., semi-cylinders, multifaceted shapes) diffuser should be not less than about a foot (300 mm), and that of an engineered (e.g., Schroeder, QRD) diffuser should be not less than about 8 inches (200 mm) deep. Gilford (1959) concluded that to be useful, projections from walls must be 1/7-wavelengh or greater at the lowest frequency of interest. At 300 Hz, that is about 7 inches (180 mm). For handclaps (say 1 kHz and up), 2 inches (50 mm) will suffice—but we listen to more than handclaps. Textured paint or plaster are visually interesting, but acoustically of no value.

If, for example, the diffuser has a lower limit of 1 kHz, it means that the sound sent off in the direction of the main (specular) reflection arrow will have a lot of energy up to the diffusion limit, and then the high frequencies will be attenuated. In other directions,

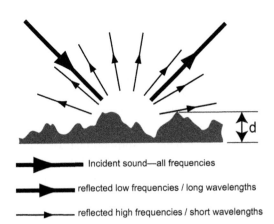

Incident sound—all frequencies

reflected low frequencies / long wavelengths

reflected high frequencies / short wavelengths

FIGURE 4.22 *A sketch showing the basic concept of a scattering/diffusing surface. At frequencies where the wavelength is long compared to the depth "d" of the protruding features of the irregular surface, there is no scattering and the sound is simply reflected as if the surface were flat. At shorter wavelengths, sounds are scattered in many directions simultaneously. Because the sound is spread over a large angular range, the sound sent in any single direction is substantially attenuated compared to the original incident sound.*

listeners will hear the reverse: high frequencies with insufficient spectral energy in the lower frequencies. It all depends on the specific circumstances, to be sure, but one cannot ignore the fact that only a portion of the spectrum is being manipulated by these devices. As the experts (e.g., Cox and D'Antonio (2009)) caution us, there is more to the acoustical performance of these devices than meets the eye. When sellers offer only dimensions measurable with a tape measure as "specifications," one needs to be suspicious.

CHAPTER 5

Characterizing Loudspeakers—Can We Describe What Is Good?

The three components—loudspeaker, room and listener—comprise a system, functioning together. Combined with prerecorded music or movie sound tracks, the result is a listening experience—a subjective event. It is not possible to optimize the design of loudspeakers without knowing certain physical properties of the room through which that sound is propagated to the listener. It is not possible to optimize the design of the listening room without knowing certain physical properties of the loudspeaker. It is not possible to optimize either of those components without knowledge of the perceptual process—psychoacoustics—that identifies and prioritizes the physical variables contributing to satisfying listening experiences.

In real-world experiences, there is no way that these factors can be fully considered, so what one hears is substantially determined by chance. These chance factors also apply to the circumstances within which the program material was created—the recording studios and movie dubbing stages. The art itself is the result of listening to loudspeakers in rooms—different loudspeakers and different rooms—the "circle of confusion" discussed in Section 1.4. Fortunately humans are very adaptable, and from these varying acoustical experiences we manage to extract key elements of the music and the great pleasure it offers. But, could it be better? Are there ways to reduce the variability in listening experiences?

The following is an attempt to develop evaluation and design criteria that increase the probability of good sound in normal listening environments. As it turns out, certain characteristics of loudspeaker design appear to be able to improve the delivery of good sound in a variety of rooms. This is good news for those of us who value visual aesthetics in our living and listening spaces. For those with the desire and budget to create custom, acoustically treated, listening rooms or home theaters, good things are possible, but the point of diminishing returns is quickly reached. In truth, high-quality listening is possible in rooms that appear to be normally furnished, and this is the largest audience. Chapter 7 discusses some options.

The dominant problem in small rooms is bass response, but with some knowledge and lateral thinking, it is possible to achieve high levels of sound reproduction quality employing what might be called "stealth treatments," substituting technology-based solutions for the traditional passive ones, completely or in combination. Chapter 8 discusses these.

In the end, the fact is that we now know enough to be able to have options to consider in our individual paths to rewarding listening experiences. But, as will be seen, it is essential to begin with good loudspeakers—that is not optional.

5.1 THE WISDOM OF THE ANCIENTS

Reviewing the very old literature is a humbling activity. One can only wonder at some of the remarkable insights possessed by our predecessors. Hugh Brittain (1936–1937)—80(!) years ago—listed the "common imperfections of electroacoustic systems . . . roughly in order of importance" (some of the author's terminology has been updated). My comments are in parentheses.

- Amplitude/frequency response. [Yes!]

- Harmonic distortion. [Yes]

- Spurious noises and intermodulation distortion. [Yes]

- Frequency shift. [I'm not sure what this means, but it could very well be reference to the abundant resonances in old transducers that pitch-shifted and spectrally colored everything, in which case an emphatic "yes." If he meant the frequency-modulation effects of Doppler distortion, then "no."]

- Dynamic range compression. [Yes]

- Transient distortion. [Yes, but now we know that most of this information is in the amplitude response because transducers are minimum-phase devices.]

- Phase distortion. [We now know that phase shifts are essentially inaudible in music reproduced in rooms. Brittain knew it too, saying that phase distortion only became apparent when "accompanied by some other phenomenon"—a resonance, perhaps?]

- Group delay. [Only if it is quite large, more than about 2 ms, which typically does not happen in domestic and studio monitor loudspeakers.]

- Electroacoustic efficiency. [Back then, absolutely yes. There were no large power amplifiers. Not a major concern now except in large-venue sound reinforcement or reproduction systems.]

- Power-handling capacity. [Yes]

- Constancy of performance. [Yes. Today, performance changes as a function of time are not problematic, but if we let "constancy" include the consistency of manufacturing, it is still a problem.]

Brittain was obviously a very intelligent and thoughtful man to have created this comprehensive list when all audio technology was in its infancy, the moving coil loudspeaker having been invented in 1925. The ability to measure was seriously limited, and program material greatly compromised. When I was in England as a post-graduate student (1960–1965), Brittain's novel GEC metal cone loudspeaker from the 1950s was still respected as a significant achievement. Metal cones are common nowadays, especially in high-end loudspeakers.

Another contributor to early audio science was Harry Olson, along with his colleagues at RCA. In 1954, some 60 years ago, he knew that "Of all the performance characteristics of a loudspeaker, the response-frequency characteristic is the most useful and important because it conveys the most information." He goes on to say that "a smooth response-frequency characteristic and a broad directivity pattern" are necessary. "These loudspeakers provide directivity patterns as broad and uniform as possible and still retain uniform response, low distortion and adequate power-handling capacity." Looking into the time domain, he did tone-burst tests to evaluate transient response, which would also ferret out bothersome resonances. All of this was combined in cabinets designed "to eliminate the deleterious effects of diffraction on the response-frequency characteristics" (Olson et al., 1954). These are the same objectives that we would apply today. The difference is that with modern technology we can succeed at it better than he was able to.

But these thoughts were either not widely known or believed, because later publications and products indicated that the matter was not settled. In preparation for one of my early papers (Toole, 1986), I did an extensive literature search and summarized where I thought things stood at that time. There were dramatic differences in the measured parameters that were favored by different people in different places. They could not all be right. There were four separable schools of thought that differentiated the engineering approaches to loudspeaker design.

1. Anechoic on-axis frequency response (focus solely on the direct sound from the loudspeaker).

2. On- and off-axis anechoic data in some kind of weighted assessment (more detailed information on the loudspeaker, anticipating reflected sounds in rooms).

3. Total radiated sound power (less detail about the loudspeaker, anticipating a diffuse-field room response).

4. In-room, listening position, steady-state frequency response (the loudspeaker and the room are inseparably combined).

If one looks at those who put their beliefs on paper in the 1970s and '80s, some regional biases are evident. From top to bottom on the preceding list, one would find most of the kindred spirits in (1) the UK, (2) the UK with some US and Canadian associates, (3) the East Coast of the US with a couple of Scandinavian associates, and (4) the US with a Danish associate. A few authors were clearly hedging their bets, but it is interesting when one thinks about what might have been the influential factors. I am not naming names, because with time some of the people changed their views, as I did. That is the scientific method.

We now know that no one of these targets is a complete descriptor, but all contain some useful evidence of performance. Most products of that period had some things right, and some things not so right, the result of measurement and engineering limitations at the time, and company opinions about what the public wanted. Chapter 18 has examples.

5.2 IDENTIFYING THE IMPORTANT VARIABLES— WHAT DO WE MEASURE?

After about 50 years of doing research in and around audio and sound reproduction, it is evident that we have learned a lot. Looking back, it is clear that I progressed through several phases: thinking I knew a lot about audio, realizing that I didn't, designing and conducting experiments to find out what might be true, discovering what seems to be true, testing it many times to prove it, and then writing about it and teaching it to as many people who will read and listen to me.

Some of the knowledge is what some of us expected, but the confirmation from unbiased tests makes it impersonal, allowing us to cease debating and move on to other topics. But, there are other aspects of measured data that surprised many of us. In the audibility of resonances it turns out that high-Q resonances, as revealed in frequency response measurements, are less audible than lower-Q phenomena. High-Q resonances are the ones that produce narrow spikes in curves and that energetically ring on after the driving signal has stopped. They show up as prominent features of the attractive "waterfall" diagram measurements, but they are not as bothersome to our ears as our eyes might suggest. Damping a resonance changes it, but may not eliminate it as an audible factor. Intuition failed us, but there is an explanation (Section 4.6.2).

Phase shift, half of the amplitude and phase duo in a transfer function, is an important engineering metric. It is essential data when designing loudspeaker crossovers and arrays where the outputs from multiple drivers must acoustically sum in an orderly manner. However, once the sound has left the loudspeaker system, phase, by itself, turns out not to be a significant audible factor. Because both amplitude and phase information are needed to describe sound pressure as a function of time, this means that we really don't hear waveforms. Incredible?

Phase shifts and even complete polarity inversions turn out to be "is it there or not?" subjective issues when tested in blind circumstances. Using contrived signals in headphone or anechoic listening, some listeners report hearing a "difference," but a reliably stated "preference" has been elusive. In normally reflective rooms using music, the differences, if real, become vanishingly small or non-existent. Intuition fails again (Section 4.8).

Loudspeaker transducers were found to be minimum-phase devices within their operating bandwidths, meaning that the time-domain, transient response behavior is predictable from the amplitude response. "Massless" film diaphragms don't move faster than conventional cones and domes covering the same bandwidth—it is the strength of the motor that makes the difference. I now have a massive electric luxury sedan that accelerates more rapidly than most many-cylinder turbocharged lightweight sports cars—it's a horsepower/torque-to-mass matter. The high frequency limit revealed by the amplitude response defines the rise time. Instinct is wrong yet again.

If the high-resolution amplitude response of a transducer is smooth and flat, there are no audible resonances, and no audible ringing—it is not necessary to do "waterfall" diagrams, however visually appealing they are. Properly designed electronic equalization can address specific resonances and attenuate them, making active loudspeakers, or loudspeakers with dedicated digital electronics, distinctly advantageous.

The "ancients" discussed in Section 5.1 realized that frequency response was the most important measurable parameter. They were right, but loudspeakers radiate a three-dimensional sound, propagated through a room, generating a sound field at the listening location that varies with frequency, time and direction of arrival. Two ears and a brain are sensitive to all of these variables. Therefore no single curve is a sufficient descriptor of loudspeaker frequency response. It is essential to measure all of the sound that is radiated and subsequently delivered to the listeners both directly and by reflections. Adding them together in a steady-state room curve discards information. The importance of this reflected sound is the part that has been disputed for decades, but now, it seems that it can be reconciled with subjective opinions.

More details will emerge from the discussions to come, but without accurate and comprehensive measurements and corresponding controlled listening tests, none of this would have been unambiguously revealed. Having an understanding of these things allows us to move on to topics that we know less about.

5.3 ANECHOIC MEASUREMENTS—THE SPINORAMA EVOLVES

From the beginning, my loudspeaker measurements at the National Research Council of Canada (NRCC) were analog paper-and-ink curves using state-of-the-art Brüel and Kjær chart recorders. They were accurate enough, but data in that form cannot be easily post-processed. Even simple sum or difference curves were tedious manual operations to be avoided. The appearance of digital computers changed what was possible. In 1983, I initiated a new loudspeaker characterization scheme that included high-resolution (1/20-octave) anechoic frequency-response measurements at 2 m—that is, in or close to the acoustical far field for most consumer and monitor loudspeakers. The chamber was anechoic down to about 80 Hz, and lower-frequency data were corrected for anechoic chamber errors (standing waves) by fixing the locations of the loudspeaker reference point and microphone and comparing the chamber measurements with those done at the identical distance on a 10 m tower. A correction was incorporated into the data. Measurements were made on horizontal and vertical orbits, at 15° increments in the front hemisphere and 30° increments in the rear hemisphere—34 measurements in all (see Figure 5.1). The data were acquired using a programmable oscillator and voltmeter controlled by a DEC PDP-11/03, which stored the data and performed post processing. The rack-mounted computer with its peripheral equipment was not easily moved, so many useful measurement opportunities were lost. It was a reasoned speculation as to what may be required to describe the acoustical performance of loudspeakers as they are auditioned in normal rooms. It was a start.

The anechoic measurements from that system were post processed to display the following indicators of linear behavior:

1. On-axis amplitude vs. frequency (frequency response)

2. Individual curves at each of the 33 off-axis measurement locations

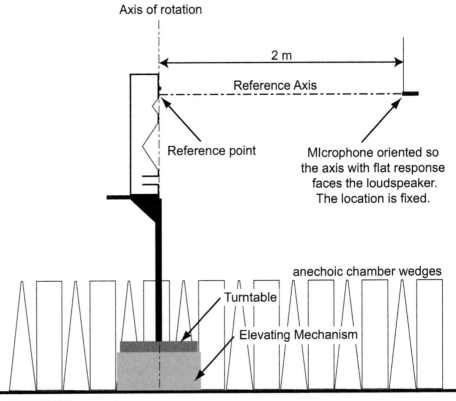

FIGURE 5.1 *The basic measurement configuration inside the anechoic chamber. This has experienced several versions of drive and elevation mechanisms over the years. The setup shown measures the horizontal orbit. Placing the loudspeaker on its side, and elevating it to realign the reference point and the reference axis, permits measurements on the vertical axis. By these means the microphone and the reference point of the loudspeaker are fixed in location, thereby allowing low-frequency chamber calibrations to reduce errors. This illustration was created by the author, who contributed it to ANSI/CTA-2034-A (2015).*

3. Mean amplitude response in the front hemisphere

4. Mean amplitude response in the total sphere

5. Mean amplitude response in the ±15° "listening window"

6. Mean amplitude response in the ±30°–45° annulus

7. Mean amplitude response in the ±60°–75° annulus

8. Total sound power calculated from individual measurements weighted according to their contributions to the total power radiated

9. Directivity index

10. Phase response.

More details can be found in Toole (1986), part 2, sections 5.2 and 5.3. The remainder of that paper shows many examples of the measurements displayed according to the subjective ratings of sound quality by listeners in double-blind tests. There is no doubt from these displays that there was an intimate relationship between the smoothness and flatness of these curves and the perceived sound quality (then called "fidelity") ratings of the loudspeakers. *Fidelity* was rated on a scale of 0 to 10, where 10 represented a reproduction that is perfectly faithful to the ideal, and no improvement is possible, whereas 0 represented the worst reproduction imaginable.

A *Pleasantness* scale was also provided, but eventually it was dropped because the two ratings simply tracked one another. The tests involved double-blind randomized comparisons of four loudspeakers at a time, with several repeated presentations of the same loudspeaker in different groupings. Listeners were professional colleagues, friends interested in audio, and a few audio journalists. Some were amateur musicians. None was a "trained" listener, but all who participated on a regular basis became very adept at the task. Switching among the loudspeakers was instantaneous. Many tests were self-administered by single listeners—the best method. Others involved two or three at a time who shared control. This was not ideal because the switching sequence quickly revealed which loudspeakers the person in control was favoring, and which were being ignored. An operator following a randomized sequence, and switching at suitable intervals was preferred. Several musical selections were used: classical, jazz and popular. There was no time limit, but it became clear that with experience, decisions were made quickly. A selection of programs that listeners found useful also emerged. Figure 5.2 shows an example of the results.

Some trends are worth noting in Figure 5.2:

■ There is an underlying "flat" trend in these clusters of frequency-response curves. The variations, even the larger ones, seem to be fluctuations around a horizontal line for the on-axis groups, and around gently sloping lines for the off-axis groups. Crossover imperfections can be seen in several of the curves.

■ The low-cutoff frequency progressively decreases as the fidelity rating increases. The listeners liked bass that extended to lower frequencies, not more bass, in the sense that it was boosted.

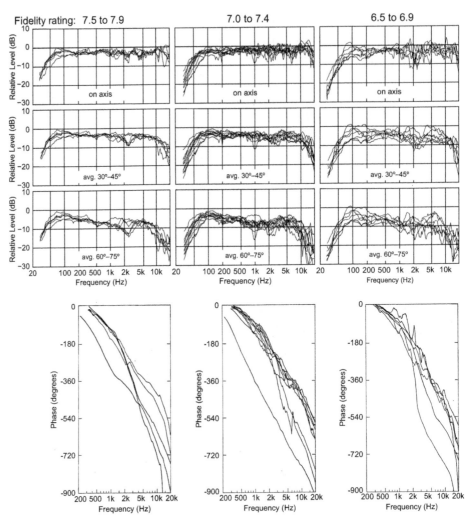

FIGURE 5.2 *A sample of results from Toole (1986), part 2, figures 7 and 13, showing loudspeakers grouped according to subjective "fidelity" ratings in three categories. There were 6 loudspeakers awarded ratings between 7.5 and 7.9, 11 loudspeakers in the range 7.0 to 7.4, and 7 loudspeakers in the range 6.5 to 6.9. The original data include a fourth, lower category that adds nothing to the discussion. The measurements are unsmoothed, 200-point, log-spaced stepped-tone anechoic measurements (1/20-octave). To eliminate the effects of loudspeaker sensitivity, the vertical positions of the frequency-response curves were normalized to the mean sound level in the 300–3,000 Hz band. This same frequency band was used to normalize listening levels during subjective evaluations. The phase responses may include a small residual time-delay error that could affect the slopes of the curves but not the contours and fine structure.*

- The best sounding loudspeakers all had smooth, gently undulating phase responses. Lower ratings were given to loudspeakers with resonances that showed up as discontinuities in the curves; high-Q resonances as sharp discontinuities. As explained in Section 4.6, the most useful evidence of resonances is in the frequency response, and Section 4.8 indicates that phase response *per se* is not an audible factor. The lack of any persuasive visual correlation between phase response and Fidelity rating caused it to be relegated to the category of "optional" measurements. Of course, for system design it is essential to the design of properly functioning crossovers; the contributions from the low-pass and high-pass portions must acoustically sum smoothly.

Clearly a ±3 dB numerical description does not do the highest-rated (7.5 to 7.9) loud-speakers justice. It is evident that smooth and flat was the design objective for all of these loudspeakers. Deviations from this target are seen at woofer frequencies (below 150 Hz), in the off-axis curves in the woofer/midrange to tweeter crossover region (1 to 5 kHz) and in the tweeter diaphragm breakup region above 10 kHz. The variations seen among these six loudspeakers from different designers, manufacturers and countries of origin are smaller than the production tolerances allowed by some manufacturers for a single model. Thirty years later, these loudspeakers would be competitive in comparisons with many products in today's marketplace.

The apparent preference for extended bass motivated Figure 5.3, where the low-cutoff frequencies for all of the loudspeakers were determined at two levels relative to the reference 300–3,000 Hz band. The normal −3 dB "half-power" level was also tried, but it yielded no relationship, which was anticipated because of the substantial effects of

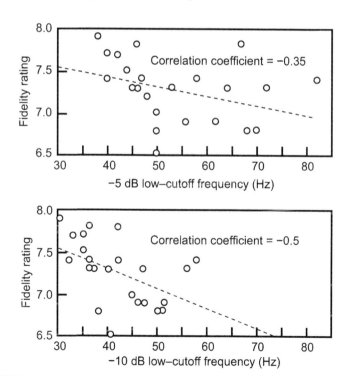

FIGURE 5.3 Plots of the low-cutoff frequencies determined at −5 dB and −10 dB relative to the average sound level over the 300–3,000 Hz band.
From Toole (1986), part 2, figure 10.

solid angle gains (Chapter 9) and bass resonances (Chapter 8) in the room in which the listening tests took place (or indeed any room). Therefore it was not a total surprise to find that the relationship between "fidelity" rating and low-frequency cutoff reached a maximum correlation for cutoff frequencies determined at the −10 dB level. Bearing in mind that this is a correlation achieved with all other factors varying indicates that bass extension is a *very* important factor in overall sound quality evaluations. This was confirmed many years later in work by Olive (2004a, 2004b), discussed in Section 5.7, which revealed that bass performance—smoothness and extension—accounted for about 30% of the factor weighting in double-blind loudspeaker evaluations of sound quality.

These results were extremely gratifying, indicating that, given a listening situation without non-auditory biasing factors, listeners were able to agree on which loudspeakers sounded "good." The preferences exhibited a simple visual correlation with the anechoic data—smooth and flattish on- and off-axis seeming to be the desired pattern, combined with low-bass extension. Because this was done in a listening room, it was logical to think that these data must be intimately related to measurements in the listening room.

Figure 5.4a shows anechoic loudspeaker measurements in which an excellent on-axis performance is clear, as is the irregular off-axis performance. Similar undulations can be anticipated in the total sound power output, which is dominated by off-axis radiated energy.

Figure 5.4b shows the principal components of the sound fields as they were estimated to be at the prime listening location in the NRCC (IEC 268–13 1985) listening room. In addition to the direct sound (the on-axis response) there is an energy sum of first reflections from the floor, ceiling and side walls derived from the appropriate anechoic off-axis responses, assuming spectrally perfect reflections and including inverse-square-law attenuation. There is also an estimate of the contribution made by the total radiated sound power, calculated from an area-weighted sum of measurements made on 360° horizontal and vertical orbits around the loudspeaker, modified according to the frequency dependent absorption in the room revealed in reverberation-time measurements.

The first observation is that if one wishes to anticipate how this loudspeaker might sound in a room, it is necessary to pay attention to sound power at low frequencies—it is the highest curve in Figure 5.9b—and the direct sound at highest frequencies for the same reason. In the middle frequencies all three components contribute significantly, so all three need to be measured. The on-axis curve by itself is insufficient data. Full 360° data, appropriately processed, is important information. An energy sum of the measurements contributing to these curves yields an estimate of a steady-state room curve, shown shifted upwards by 10 dB.

The loudspeaker was then placed at three possible stereo-left locations in the listening room, and averaged measurements at six seats yielded the curves shown in Figure 5.4c with the prediction from Figure 5.4b superimposed. Above about 400 Hz, the curves are essentially identical and the prediction aligns well with the measurements.

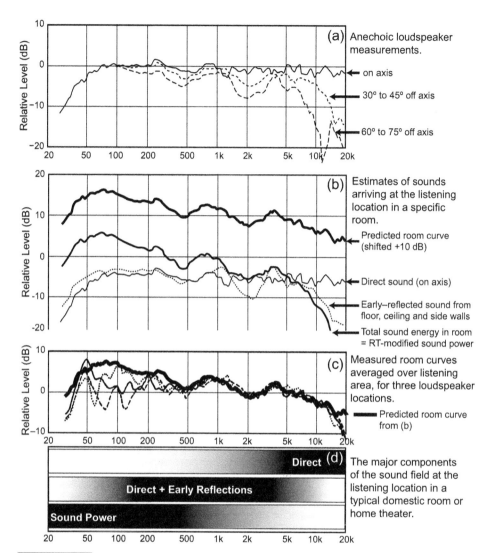

FIGURE 5.4   *A prediction of a room curve from anechoic measurements and a generalized description of the sound field at the listening location for a KEF 105.2 loudspeaker. Descriptions are in the text.*
Data from Toole (1986), part 2, as adapted for Toole (2015b), figure 4.

Below this, the effects of standing waves in the room dominate, and the locations of the loudspeaker and the listener/microphone determine the acoustical coupling at different frequencies. The predicted curve provides an estimate of the upper limit of the steady-state sound levels, but destructive interference in the standing waves substantially reduces overall bass energy. Below the transition/Schroeder frequency—around 300 Hz here—the room is the dominant factor; above it, the loudspeaker is substantially in control.

Figure 5.4d attempts to illustrate the sounds arriving at the listening location in a typical domestic listening room or home theater—different portions of the radiated sound dominate at different frequencies, determined by the frequency-dependent directivity of the loudspeaker and the reflective nature of the room. Obviously the frequencies at which the transitions occur will change with different loudspeaker designs and with different room acoustical configurations. In general, as loudspeakers become more directional and/or rooms become less reflective, the transitions move down in frequency.

These data illustrate some fundamentally important concepts:

■ With sufficient anechoic data on a loudspeaker, it is possible to predict with reasonable precision middle- and high-frequency acoustical events in a listening space with known properties.

■ There is a difference between the spectrum of the direct sound arriving at a listener and that of the steady-state sound level that is achieved after reflected sounds arrive. The shape of a steady-state room curve is determined by the sound radiated by the loudspeaker modified by the geometry and frequency-dependent reflectivity of the room. In an acoustically dead room, the room curve will be identical to the on-axis response of the loudspeaker. As reflections within the room increase, the room curve will rise toward the predicted room curve, as the off-axis sounds add to the result. The bass and midrange sound levels will build up over a short time interval, affecting what is measured and heard. At very high frequencies, the direct sound becomes progressively dominant. Therefore, with no knowledge of the loudspeaker and no knowledge of the room acoustical properties, a steady-state room curve conveys ambiguous information.

■ In normal rooms, the on-axis frequency response is not the dominant physical factor. However, the direct sound has a high priority in perception, establishing a reference to which later arrivals are compared in determining such important perceptions as precedence effect (localization), spatial effects and timbre. In this example, the poor off-axis performance dominated the in-room measurements and in listening tests caused audible timbral degradation. Equalization of the room curve will degrade the best performance attribute of the loudspeaker: the on-axis/direct sound response. Equalization cannot change loudspeaker directivity; the remedy is a better loudspeaker. Adequate anechoic data on the loudspeaker would have revealed the problem in advance of measurements or listening.

■ Below the transition/Schroeder frequency, the room resonances and the associated standing waves are the dominant factors in what is measured and heard. These are unique to each room, and are strongly location-dependent. Only on-site measurements can reveal what is happening, and different loudspeaker and listener locations will result in different bass sound quality and quantity.

■ In domestic listening rooms and home theaters or in any acoustically well-damped room like a cinema, traditional diffuse reverberation is significantly absent. As discussed in Toole (2006), Toole (2008, chap. 4), Toole (2015b) and

Chapter 10 here, the sound fields in low-reverberation rooms are different from those described in classical acoustics.

Different sound sources in different rooms will change the pattern of sound fields shown in Figure 5.4d, moving the transition regions up or down the frequency scale, but the basic principles hold.

With this evidence it was a short step to what has now come to be known as the spinorama. With faster data acquisition more data can be acquired in a reasonable time, thus the angular resolution is now 10° for the entire rotations. The new version of the two-axis measurement scheme is shown in Figure 5.5.

The low-frequency calibrations were referenced to both ground plane and tower measurements and are most accurate with closed-box woofer systems. Multiple-woofer tower systems, and reflex systems with ports on other than the front baffle, can show anomalies, especially when the woofers of horizontally positioned tower loudspeakers swing far from the rotational axis shown in Figure 5.1. The problem is that the chamber standing waves being corrected for are energized differently by the different woofer configurations. Often the best estimate of bass response is the sound power calculation, embracing all 70 curves. Relocating the woofers of tall loudspeakers to be closer to the axis of rotation can yield more precise data at low frequencies, which then can be spliced to the mid-/high-frequency data. Doing low-frequency measurements outdoors is the only way to avoid such complications, but that has its own difficulties. In reality, listening rooms are the dominant factors in the bass region, so small errors at low frequencies are unlikely to be traceable to listening experiences in rooms, certainly if room equalization is involved.

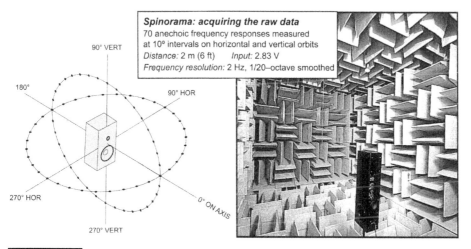

FIGURE 5.5 *The process used in acquiring the raw data used in the spinorama. The chamber shown has 4 ft (1.2 m) wedges and anechoic ±0.5 dB at 1/20-octave from 60 Hz to beyond 20 kHz, and has been calibrated to be ±0.5 dB from 20 Hz to 60 Hz at 1/10-octave resolution for the reference microphone and loudspeaker locations.*

After the data are accumulated, the next step is to process it for visual display. Figure 5.6 illustrates the basic method. The curves shown are:

■ The **on-axis frequency response** is the universal starting point for loudspeaker evaluation and, in many situations it is a fair representation of the first sound to arrive. However, as shown in the Devantier (2002) survey, over half of those investigated had the prime listening position 10° to 20° off-axis, providing

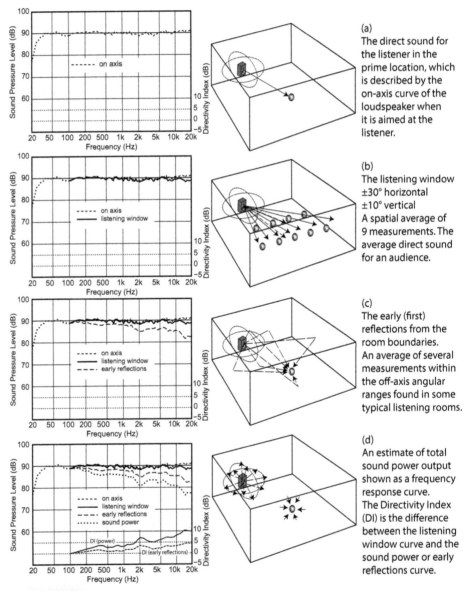

(a)
The direct sound for the listener in the prime location, which is described by the on-axis curve of the loudspeaker when it is aimed at the listener.

(b)
The listening window ±30° horizontal ±10° vertical A spatial average of 9 measurements. The average direct sound for an audience.

(c)
The early (first) reflections from the room boundaries. An average of several measurements within the off-axis angular ranges found in some typical listening rooms.

(d)
An estimate of total sound power output shown as a frequency response curve. The Directivity Index (DI) is the difference between the listening window curve and the sound power or early reflections curve.

FIGURE 5.6 *A pictorial description of the individual curves shown in a spinorama.*

a justification for the following measure; a modified version of the original listening window used in the NRCC data, which was a combination of five curves: 0° and 15° off-axis vertically and horizontally.

- The **listening window** is a spatial average of the nine frequency responses in the ±10° vertical and ±30° horizontal angular range. This embraces those listeners who sit within a typical home theater audience and those who disregard the normal rules when listening alone. Because it is a spatial average, this curve attenuates small fluctuations caused by acoustical interference (something far more offensive to the eye than to the ear) and reveals evidence of resonances (something the ear is very sensitive to). Interference effects change with microphone position and are attenuated by the spatial averaging, while resonances tend to radiate similarly over large angular ranges, and remain after averaging. Bumps in spatially averaged curves tend to be caused by resonances (see Figure 4.13).

- The **early reflections** curve is an estimate of all single-bounce first reflections in a typical listening room. Measurements were made of early reflection "rays" in 15 domestic listening rooms. Figure 5.7 shows an example of the horizontal reflections in one of the rooms.

From these data, a formula was developed for combining selected data from the 70 measurements in order to develop an estimate of the first reflections arriving at the listening location in an "average" room (Devantier, 2002). It is the average of the following:
- Floor bounce: average of 20°, 30°, 40° down
- Ceiling bounce: average of 40°, 50°, 60° up
- Side wall bounces: average of ±40°, ±50°, ±60°, ±70°, ±80° horizontal
- Front wall bounce: average of 0°, ±10°, ±20°, ±30° horizontal
- Rear wall bounces: average of 180°, ±90° horizontal

The summary illustration for the important forward hemisphere of a forward-firing loudspeaker is shown in Figure 5.8.

The number of averages mentioned in that description may make it seem as though anything useful would be lost in statistics. However, this turns out to be a very useful metric for two reasons. First, it provides an estimate of the contribution to direct and early-reflected sounds arriving at a listener in a typical room. These are very influential in establishing timbral and spatial qualities. Second, being a substantial spatial average, a bump that appears in this curve and in other curves is strong evidence of a resonance. It is also, as will be seen, the basis for a good prediction of what is measured in rooms.

- **Sound power** is a measure of the total acoustical energy radiating through an imaginary spherical surface surrounding the loudspeaker. In the present context it is something of limited relevance because all of the sound radiated by a loudspeaker is not equally important to what is measured and heard at a listening position in a rectangular semi-reflective room: a normal listening room. Ideally,

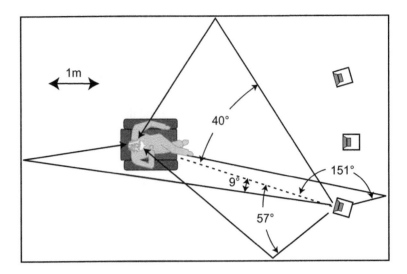

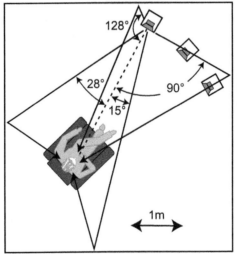

Angles of early-reflected sounds relative to the on-axis direct sound.

FIGURE 5.7 *Examples of horizontal-plane early (first) reflections in listening rooms arranged with the primary listening axis along a parallel room axis and also on an oblique room axis. The angles shown are relative to the direct sound axis shown as a dashed line. It should be noted that in the parallel arrangement the first lateral reflections are associated with moderately large off-axis angles (i.e., significant mid- and high-frequency sounds from most loudspeakers). The sound reflected from the wall behind the loudspeaker will have little mid-high frequency energy from forward-firing loudspeakers. The sound from the wall behind the listener is close to the on-axis (i.e., direct) sound, and will be a significant factor. In the oblique arrangement the first "lateral" reflections are associated with very large off-axis angles (i.e., mid and high frequencies are greatly attenuated), while the "rear" reflections fall within the listening window angles (strong mid and high frequencies). This arrangement in a large room, with the walls behind the listener at a greater distance, leaves the listener in a predominantly direct sound field at mid to high frequencies. It is a good option, albeit not a popular one, perhaps for aesthetic reasons. Two of my personal listening rooms have used it.*

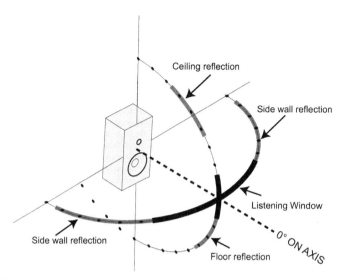

Side wall reflection

Ceiling reflection

Side wall reflection

Floor reflection

Listening Window

0° ON AXIS

FIGURE 5.8 *A visual portrayal of the angular ranges considered in calculating the listening window and early reflections in the front hemisphere of a forward-firing loudspeaker. The early-reflections calculation includes the horizontal portion of the listening window.*

such a measurement would be made at many closely spaced points over the entire surface of the sphere. In the spinorama the sound power is estimated by calculating an energy sum of the 70 measurements on two circular orbits, with individual measurements weighted according to the portion of the spherical surface that they represent. The weighting coefficients are in ANSI/CTA-2034-A (2015). Thus the on-axis curve has very low weighting because it is in the middle of other, closely adjacent measurement points (see the perspective sketch in Figure 5.5), and measurements further off-axis have higher weighting because of the larger surface area that is represented by each of those measurements (the area between the orbits). Because it is an energy sum, the final curve must be computed from sound pressures, converted from dB SPL, which are squared, weighted, summed and then converted back to dB. The result could be expressed in acoustic watts, the true measure of sound power, but here is left as a frequency-response curve having the same shape and normalized to the other curves at low frequencies. This serves the present purposes more directly. Any bump that shows up in the other curves, and persists through to this ultimate spatial average, is a notable resonance.

■ **Directivity index (DI)**. Traditional DI is defined as the difference between the on-axis curve and the normalized sound-power curve. It is thus a measure of the degree of forward bias—directivity—in the sound radiated by the loudspeaker. It was decided to depart from this convention because it is often found that, because of symmetry in the layout of transducers on baffles, the on-axis frequency response contains acoustical interference artifacts, due to diffraction, that do not appear in any other measurement. It seems fundamentally wrong to burden the directivity index with irregularities that can have no consequential effects in real listening circumstances. Therefore, the DI has been redefined as the difference between the listening window curve and the sound power. In most loudspeakers the difference is small; in highly directional systems it can be significant. In any event, for the curious, the raw evidence for the classic DI is there to inspect. Obviously, a DI of 0 dB indicates omnidirectional radiation. The larger the DI, the more directional the loudspeaker in the direction of the reference axis. This is illustrated in Figure 5.9. Because of the special importance of early reflections in what is measured and heard in rooms, a second DI is calculated, the Early Reflections DI, which is the difference between the listening window curve and the early-reflections curve. Because of the importance of early reflections in common sound reproduction venues, this is arguably the more important metric.

All of this is summarized in Figure 5.6.

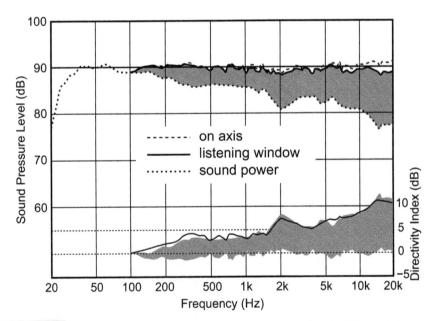

FIGURE 5.9 *An explanation of the origin of directivity index (DI) based on the difference between the listening window curve and the sound power curve. Also shown is the difference area inverted and superimposed on the DI, demonstrating that the inverted DI is equivalent to the sound power of this loudspeaker if it were equalized to be flat in the listening window. This can be visualized by imagining that the bottom of the shaded area in the lower illustration is adjusted to be flat and smooth—the top of the shaded area would then align with the DI curve. The same process can be applied to the Early-Reflections DI.*

5.4 TOTAL SOUND POWER AS A MEASURED PARAMETER

Historically, sound power was measured in reverberation chambers: hard, irregularly shaped epoxy-painted concrete rooms with suspended, and sometimes rotating, panels or vanes to assist in creating a diffuse sound field. Sounds radiated in all directions from sound sources such as appliances, power tools and, yes, loudspeakers, were combined in the very reverberant sound field. These measurements are not high precision exercises because of the difficulty in establishing a truly diffuse sound field. They are traditionally done in 1/3-octave bands. The data from these measurements are for general consumption, in that no consideration is given to the environment within which these noise/sound sources are heard. Obviously, a power tool listened to outdoors sounds very different from the same device in a furnished room, or in an unfurnished room—the latter being closest to the reverberation chamber situation. Nor is there any data pertaining to the frequency-dependent directivity of the radiated sound. Varying the orientation of the device might significantly change the perceived sound timbre and level in these different spaces. Therefore the metric is really of use only in a relative sense, allowing "noise rating" comparisons between different, but similar, products, but not providing assurances of specific sound levels or timbre in any specific environment. The same applies to loudspeakers measured in this way, even though the metric has had its proponents, as discussed in Section 5.1. Sound power can be reliable only if the listening space is highly reverberant and the sound field highly diffuse.

That said, it is nevertheless used as part of the process of defining loudspeaker directivity index: classically the difference between on-axis and sound power curves. As such, it is convenient as a rating scheme, but clearly cannot ensure audible characteristics in conventional semi-reflective venues, especially above a few hundred hertz (Toole, 2015b). From a rational perspective, a better metric is needed.

In the present case, an anechoic chamber provides a controlled space within which sound power radiated by loudspeakers can be estimated by making measurements at numerous points on an imaginary sphere surrounding the device. The more points, and the more closely spaced the points, the more accurate will be the result. Celestinos et al. (2015) performed simulated and real comparisons among several spatial sampling schemes for estimating total sound power. With a 7,082-point measurement providing a defendable reference, it is no surprise that there were differences between it and the original NRCC (34-point) and the spinorama (70-point) two-orbit simplifications. The simulations used a pair of theoretical monopoles oriented along x-, y- and z-axes, and clear differences were observed. However, real loudspeakers exhibit substantially different radiation characteristics than this theoretical source, so anechoic measurements were performed on a two-way loudspeaker. These showed very different error patterns, indicating that the two-orbit methods deviated from the ideal by exhibiting a 1 dB bump between 200 Hz and 300 Hz, which was thought to be provoked by port tube positioning, and an approximately 2 dB bump at 2.2 kHz, in the crossover frequency region of the sample loudspeaker. It is interesting to note that the difference in the errors between the original NRCC method and the spinorama was very small. The NRCC method simplified the rear hemisphere measurements to 30° increments because it was believed, and these data appear to confirm, that there is little useful information in that hemisphere when the loudspeaker is a conventional forward-firing design.

To evaluate the audible consequences of these errors, one must consider the sound propagation paths to a listener in a room. With the exception of the direct sound, all of the radiated sound arrives at the listening position by means of reflections, which means that both the directivity of the loudspeaker and the geometry and reflectivity of the listening room determine what is measured and heard. A requirement for sound power measurements is that equal value is put on sounds radiating in all directions. This is not compatible with the way loudspeakers are used. Normal listening rooms have perpendicular, vertical and horizontal, boundaries, and reflected sounds from these surfaces are the second loudest sounds arriving at a listening location following the direct sound—the first/early reflections described earlier. The two-orbit, vertical and horizontal, measurement addresses this practical reality, which was a factor in choosing it for the original NRCC simplified measurement scheme. Sounds originating at all other off-axis angles will be subject to longer propagation distances (greater inverse-square-law attenuation), and interactions with multiple reflecting surfaces, each reflection possibly resulting in some energy loss, as they travel to the listening location. These facts need to be combined with the common reality that loudspeakers become more directional with increasing frequency, and that normal listening rooms are less reflective and the sound field less diffused with increasing frequency. This means that sound power, *per se*, contributes progressively less to what is measured and heard in a room at middle and high frequencies. Errors found in the technically correct sound power measurements in this frequency range may or may not be evident in what is measured or heard in a room. It is a measurement that is best adapted to omni- or multi-directional loudspeakers that may be auditioned in relatively reflective spaces, from any arbitrary listening perspective. By themselves, the on-axis curve and the sound power are imperfect data for anticipating sounds arriving at a listening location. This of course implies that the DI is a simplistic metric for typical loudspeakers in typically absorptive listening venues. That said, they are both useful indicators, albeit not accurate predictors, of real physical events in listening rooms.

5.5 WHY DO WE MEASURE WHAT WE DO? ARE THERE BETTER WAYS?

As we have learned increasingly more about the details of the perceptual process, relating what is measured to what is heard, there have been ever-present questions: what is it that needs to be measured and how do we interpret the results? Let us examine the components of the present spinorama as described in Section 5.3.

■ **On-axis and listening window.** These curves attempt to describe the direct (first) sound to arrive at a listener seated on the axis, or up to 30° off-axis horizontally and 10° vertically. It is a small fraction of the sound energy in the room, but the direct sound is perceptually very important because it is responsible for determining localization—the azimuth and elevation of the sounds we hear: the soundstage. It also provides a basis for comparison of later, reflected sounds that participate in the precedence effect, and that generate perceptions of spaciousness, envelopment and timbre. Smooth and flat on-axis/listening window curves are indicators of potential good sound, but not a guarantee, because the off-axis (reflected) sound is strong over most of the frequency range. These are important metrics.

■ **Early reflections.** In spinorama, the early reflections being evaluated are the first reflections only, having had only a single reflection from a room boundary. They are the second loudest sounds to arrive at the listening location and on average will arrive within 2 ms to 15 ms after the direct sound (Devantier, 2002, figure 12). They also are propagated along vertical and horizontal axes from the loudspeaker because most rooms have vertical and horizontal boundaries. The present two-orbit measurement scheme addresses most situations. A possible improvement would involve prior knowledge or measurements of the specific off-axis angles in particular listening rooms, thus avoiding the imprecision of averaging over a range of angles for each of the boundary reflections. Of course, the acoustical characteristics existing at each reflecting point are a factor, contributing frequency dependent attenuation and possibly scattering. Unless these properties are incorporated into the model of the loudspeaker in a particular room, the predictions will have errors. However, evidence thus far has indicated that the predictions of the physical sound fields in rooms are commonly much better than anticipated. The perceptions arising from early reflected sounds are a more complicated matter. The simple-minded interpretation of these delayed sounds is that they create destructive comb filtering, and therefore all strong reflections should be absorbed. Reality is very different because two ears and a brain respond very differently from an omnidirectional microphone. This important topic will be discussed in detail in Chapter 7. For the moment, it is fair to state that the more similar the early reflections curve is to the direct sound curve, the more favorable is the perceptual result in both timbral and spatial respects.

■ **Sound power** is most relevant to loudspeakers designed to be omnidirectional, or widely dispersing, radiating sound into reflective venues. There it is a good prediction of a steady-state room curve (Figure 7.20) and an indication of spectral uniformity. However, it is less useful for the common relatively directional forward-firing loudspeakers, radiating sound into relatively non-reflective rooms. As seen in Figure 5.4, it is most revealing at low to middle frequencies. It is best known as a component in defining Directivity Index (Figure 5.9). In measuring loudspeakers, those with multiple woofers and/or certain port configurations can radiate sound in ways that are incompatible with the low-frequency calibrations of anechoic chambers. In such cases the sound power metric can be a more accurate assessment of bass performance. In the real world of small rooms, that accuracy is lost among the gross modifications caused by room boundary absorption, resonances and the associated standing waves. Alone, it is a metric of limited utility, but it usefully fits into the pattern of the spinorama curves as the ultimate spatial average.

- **A combination of all curves** from "on-axis" through "listening window" and "early reflections" to "sound power" is highly revealing. This is the spinorama shown in Figures 5.5 and 5.6. From an engineering perspective it quickly allows the separation of acoustical interference effects (which are different in different microphone locations) and resonances (which exhibit peaks that are consistent over large angular ranges). If a peak persists through all of the curves, it is undoubtedly a resonance that needs to be addressed. Naturally high-resolution data are necessary for this to work.

- **Non-linear distortion** is routinely measured, but as discussed in Section 4.9 the data are not easily interpreted in meaningful ways. The good news is that it is not normally a factor in full-sized domestic and professional monitor loudspeakers. But, as described in that section, it does happen. It happens frequently, and indeed is expected, in small portable loudspeakers that are challenged in bandwidth, amplifier power and transducer capabilities. In all cases, subjective evaluation is the only certain metric.

5.6 PREDICTING ROOM CURVES FROM ANECHOIC DATA—AN EXERCISE IN CURVE MATCHING

In the author's opinion, it is extremely valuable to have the collection of curves ranging from a single on-axis curve through increasingly comprehensive spatial averages to the sound power curve. Resonances are quickly revealed as bumps that persist through many or all of the curves, and resonances are a major contributor to unwanted coloration. It also provides indications of frequency-dependent directional characteristics that may be heard as spectral colorations in reflected sounds. These are clues that do not require numerical analysis or engineering degrees to be understood. However, it is human nature to strive for simplification: a single curve descriptor.

The idea that there might be a better metric than sound power is not new. It begins with the notion that what is heard in a normal listening room by two ears and a brain puts greater importance on early-arriving, louder sounds than later-arriving, quieter sounds. The physical circumstances are such that with conventional forward-firing loudspeakers, the frontal hemisphere measurements contribute most to what is measured and heard in rooms.

Gee and Shorter (1955) developed a sound power measurement system for loudspeakers and microphones, and speculated:

it might be possible, by weighting the results obtained in different zones [areas of the sphere], to give more prominence to the front response, and so to arrive at an empirical figure, intermediate between the axial and mean spherical responses, which would give a useful approximation to the overall effect obtained in the average listening room.

Unaware of this work at the time, the author ended up sharing the opinion (Toole, 1986, section 7.2). The prediction exercise shown in Figure 5.4 showed the importance of the direct and early-reflected sounds, which led to an intuitively inspired alternative metric. Quoting from Toole (1986, p. 340):

It is simply the mean amplitude response in the front hemisphere calculated as the energy average of the 25 measurements made at 15° intervals between ±90° horizontal and vertical. The spatial distribution of these measurements over the surface of the front hemisphere causes this curve to be the equivalent of an axially weighted hemispheric sound-power measure. (A true sound-power calculation involves weighting the individual measurements according to the solid angle represented by each one.) Thus the high frequencies will tend to be elevated slightly, compared to the true sound-power measurement, and because the low-frequency rear radiation is not included, the low-frequency portion of the curve will be lowered. The result, fortuitously, is a simple but effective estimate of the mean room response that seems to work well for loudspeakers in general, as long as they are of the conventional forward-facing variety.

Figure 5.10 shows some interesting comparisons.

Obviously, the mean amplitude response in the front hemisphere is a very capable predictor of steady-state room curves above about 600 Hz. Below that, room modes/standing waves create progressively greater disorder, with much energy being lost in destructive acoustical interference. The front-hemisphere prediction curves in both Figure 5.10a and b are too high by about 3 dB at frequencies below 600 Hz. The available acoustical mechanisms for this disparity appear to be the abundant room-mode destructive interference dips, such as those seen in (a), and energy loss to the room boundaries. The mismatch can be expected to be different in different rooms; all of which underlines the importance of in-situ measurements to determine what is really happening at low frequencies—see Chapter 8.

Perhaps more remarkable is the agreement between this curve, calculated from 25 measurements, and the average amplitude response in the 30° to 45° horizontal and vertical windows, calculated from only eight measurements (extracted from Figure 5.4a). The underestimation at the very highest frequencies is clearly due to the absence of direct sound in the data. This is confirmed in the curve showing only the floor, ceiling and side wall reflections (four measurements), which captures the essence of the room curve, but exaggerates the undulations because the direct sound, front- and rear-wall reflections are not included in the summation. These are flatter curves and would moderate the undulations (as will be seen in the upcoming spinorama predictions). It appears that for this loudspeaker in this room the dominant sounds contributing to the room curves relate to the early reflections.

Phase 1: What has been learned? Different forward-firing loudspeakers in the same room generate steady-state room curves that are well predicted at middle and high frequencies by data from only the front hemisphere. Of that data the sounds radiating at angles appropriate to early reflections contribute the bulk of the information. Full spherical data, sound power, is not necessary for loudspeakers of this configuration.

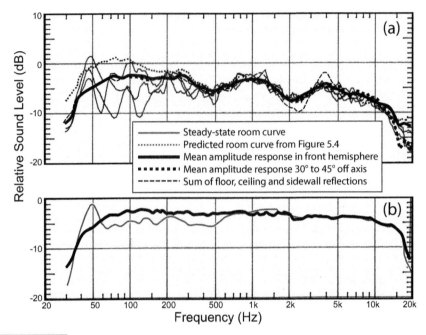

FIGURE 5.10   *(a) illustrates data from Figure 5.4 with the mean amplitude response in the front hemisphere superimposed. The steady-state room curves are six-seat averages for three different loudspeaker locations. Also shown is the average amplitude response 30° to 45° off-axis from Figure 5.4a and four of the first reflections from Figure 5.4b. Data are from Toole (1986). (b) shows averaged results for three highly rated loudspeakers, comparing the mean amplitude response in the front hemisphere to the mean listening room curves. The same room was used throughout. Data from Toole (1986), figure 21.*

Moving to the present, spinorama data provides another version of this approach. Figure 5.4d indicates that for conventional loudspeakers in conventional rooms, sound power is the dominant influence at the lowest frequencies, yielding at middle frequencies to a combination of direct and early-reflected energy and at the highest frequencies to the direct sound itself. Figure 5.8 shows that the measurements used to compute the early-reflections curve cover a substantial part of the front hemisphere, with the listening window almost completing the coverage. Figure 5.11 shows sound power and early-reflections data superimposed on steady-state room curves for two very different loudspeakers. Both exhibit imperfections, but the example in (a) comes closer to the ideal.

As indicated in Figure 5.4d, sound power matters progressively less with rising frequency. This explains why the frontal hemisphere data, exemplified by the early-reflections curve, is a better match to the room curve. The sound power curve overestimates the low frequency room curves, and both it and the early-reflections curve underestimate the high frequencies. This is a reason to blend in at least high-frequency data from the listening window to improve the frontal weighting of the data. Figure 5.10

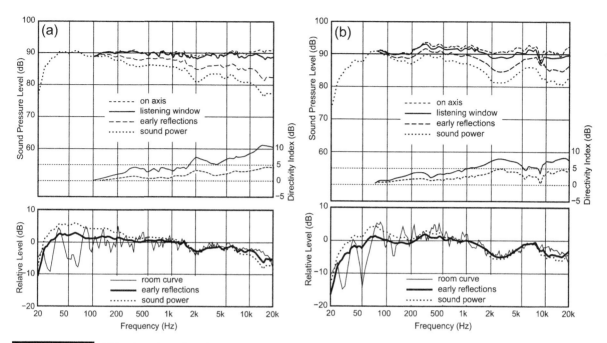

FIGURE 5.11   *(a) Loudspeaker: Infinity Prelude MTS prototype, ca. 2000. Top: the full anechoic spinorama. Bottom: steady-state in-room measurements with the loudspeaker in the front-left location averaged over six head locations within a typical listening area in a typical rectangular room. As a means of understanding which sounds from the loudspeaker contribute to this measurement, the early-reflections and sound-power curves from the anechoic data set at the top have been superimposed. Obviously, it is the off-axis performance of the loudspeaker that is the dominant factor in determining the sound energy at the listening location and, in the several comparisons that have been done, the early-reflections curve has been a better fit than the sound power curve, which is excessively tilted. As noted in Figure 5.4b the on-axis curve is the dominant factor at very high frequencies, so both of the overlaid curves underestimate the steady-state room curves at high frequencies. More direct sound is needed in the prediction. As is expected, the standing waves in the room take control at low frequencies and the prediction fails.*
(b) Loudspeaker: B&W 802N. This loudspeaker has both frequency-response and directivity issues. In the DI curves, one can see the directivity of the woofer rising with frequency, then falling in the crossover around 350 Hz to a rather large (6-inch, 150 mm) midrange driver that exhibits increasing directivity up to around 2 kHz before crossing over to an unbaffled tweeter with wide dispersion, which by 5 kHz has taken over. In the lower box of curves the pronounced mid- to upper-frequency undulations seen in the spatially averaged room curve are clearly associated with the off-axis behavior of this loudspeaker system. The shape of the room curve is clearly signaled in the shapes of both the anechoic early-reflections and sound-power curves.

shows that mean amplitude response in the front hemisphere provides a good estimate of a room curve above about 600 Hz, but it overestimates the room response at low frequencies. In all cases, the on-axis frequency response by itself is not a useful indicator of how loudspeakers will measure in rooms, although it is critical to anticipating how loudspeakers sound in rooms. It bears repeating: two ears and a brain are not equivalent to an omnidirectional microphone and an analyzer.

An obvious message from Figure 5.11 is that measuring a steady-state room curve and equalizing it cannot guarantee excellent sound. Equalization can only change frequency response. In the loudspeaker in Figure 5.11b, the dominant problem is non-uniform directivity as a function of frequency, and equalization cannot change directivity. If the only information available is a room curve, this fact is not known. Comprehensive anechoic data on the loudspeaker would have revealed this in advance.

All of the loudspeakers examined so far have been cone/dome forward-firing designs. It is in their nature to exhibit directivity that changes with frequency. Figure 5.12 shows performance data on a horn loudspeaker with constant directivity, a JBL Professional M2, a monitor loudspeaker consisting of a 15-inch (381 mm) woofer and an innovative compression driver and horn. It is an active loudspeaker with dedicated outboard electronics, equalized to be relatively flat in the listening window. As discussed in Figure 5.9, for a loudspeaker with flat listening window response, the sound power curve and the DI are inverses of each other. This is shown to be true here. Also of interest is the very constant directivity of the horn over most of its frequency range, revealed here in the similarly flat spinorama curves from about 1 to 8 kHz. Inevitably, the steady-state room curve is comparably flat. Later, in Figure 11.11 it is seen that this performance is substantially unchanged when measured in several cinema venues ranging up to 516 seats.

Phase 2: What has been learned? It can be seen that the early-reflections curve is a good match for a steady-state room curve from approximately 500 Hz to 8–10 kHz in the loudspeakers tested, but it is slightly high at lower frequencies, and slightly low at

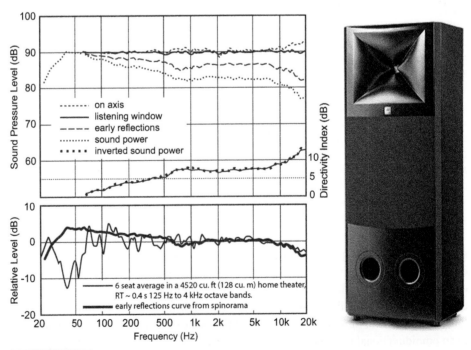

FIGURE 5.12 *The JBL Professional M2 spinorama curves and the same loudspeaker in a home-theater setting.*

high frequencies. As anticipated in Figure 5.4, more direct sound needs to be blended in at high frequencies for a better match. The sound power curve overestimates the bass and underestimates the high treble by even more. However, at mid- and high frequencies a constant-directivity horn is impressively consistent. At low frequencies the room is the dominant factor, most dramatically below about 100 Hz. Once again it is evident that the details of low-frequency performance must be evaluated and compensated in situ.

Finally, another approach to predicting room curves was proposed by Devantier (2002), where in figure 18 there is a prediction based on a combination of three spin-orama curves. This process is also included in ANSI/CTA-2034-A (2015). Figure 5.13 shows the loudspeaker used as an example in that document, which states:

It [the prediction] shall be comprised of a weighted average of 12% Listening Window, 44%, Early Reflections, and 44% Sound Power. The sound pressure levels shall be converted to squared pressure values prior to the weighting and summation. After the weightings have been applied and the squared pressure values summed they shall be converted back to sound pressure levels.

As Figure 5.13 shows, the agreement between the calculated prediction and the spatially averaged steady-state room curve is good. However, with access to the original spin-orama data it can be seen that, for this loudspeaker in this room, the early-reflections curve by itself is a similarly good predictor.

Phase 3: What has been learned? From the perspective of anticipating room curves for forward-firing loudspeakers, considerable simplification is possible. Any predictive scheme involving sound power requires the full spinorama of 70 curves. If the loudspeaker under test is a horizontally symmetrical forward-firing design the early-reflections calculation involves only 18 measurements to embrace the floor, ceiling, one side wall, rear and front wall reflection estimates. Asymmetrical designs require both side

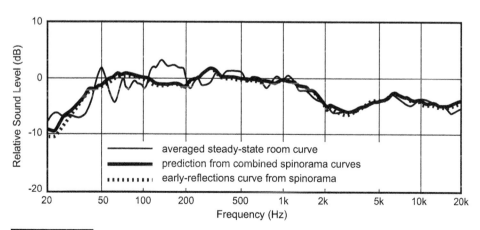

FIGURE 5.13   *Measured and predicted room curves. One curve (heavy solid line) is based on a combination of 12% listening window, 44% early reflections and 44% sound power. The other curve (dotted) is the early-reflections curve from the spinorama.*

walls and both left and right front-wall data points, bringing the sum to 26 measurements. All but one of these (the 180° measurement) are in the front hemisphere.

This conclusion was foreshadowed in Figure 5.4b, from the 1986 NRCC tests, where the four first reflections from the floor, ceiling and side walls were determined for a specific room. In Figure 5.10 the sum of these four data points is compared to more elaborate predictions. That curve is a reasonable overall match, but it exaggerates the off-axis imperfections of the loudspeaker, needing the addition of direct sound and front and rear reflected sounds to moderate the undulations. But it is clear that, for a specific room whose geometry is known, useful predictions can be obtained with as few as six angular measurements: floor, ceiling, two side wall, front- and rear-wall reflections. It is possible that more reliable predictions are possible if some direct sound is added, weighted to have a high-frequency bias. That needs experimental confirmation. However, even then the total is seven measurements. When attempting a generic room curve prediction for "statistically typical" rooms, as in the original spinorama, and the ANSI/CTA-2034-A context, many more data points are required. Sound power *per se* is not required for this purpose although some of the measurements that contribute to it are useful.

As mentioned at the beginning of this section, in 1955 Gee and Shorter speculated that, rather than sound power, it was necessary "to give more prominence to the front response, and so to arrive at an empirical figure, intermediate between the axial and mean spherical responses, which would give a useful approximation to the overall effect obtained in the average listening room." They were right.

> But all of this assumes that the room curve is a definitive statement of potential sound quality. As shown in Figure 5.2, that is not true, anechoic data prevail. We measure room curves because it is easy to do and most of the time we operate without sufficient measured data on the loudspeakers themselves. One is forced to get information wherever possible, even if it is imperfect data.

5.6.1 A Message about Sound Absorption and Scattering

The dominant factor in the shape of the room curves, and therefore a factor in what we hear, is the off-axis sound, the first reflections paramount among them. This observation sends a clear message that in order to preserve whatever excellence exists in the loudspeakers, the room boundaries must not change the spectrum of the reflected sound. This means that absorbing and scattering/diffusing surfaces must have constant performance over the frequency range above the transition frequency around 300 Hz, below which the room resonances become the dominant factor. As was discussed in Section 4.10.4, such devices are significantly larger and thicker than some of those in common use.

5.6.2 Why Do We Care about Room Curves?

The notion that a single curve resulting from sounds picked up by an omnidirectional microphone is an adequate representation of perceptions by two ears and a brain is preposterous. It is related, but it is incomplete data. A prediction of a room curve from anechoic data allows one to identify what the room is doing to the sound. If a real room curve differs from the prediction, it is possible that the geometry of the setup or the acoustical properties of the room need adjusting. However, a room curve tells us very little that is reliable about the loudspeaker itself, although at very low frequencies in-room measurements are very instructive when it comes to dealing with room modes and standing waves. When we get into the domain of discrete reflections, however many there may be, the rules change. Listeners have demonstrated an uncanny ability to "listen through" rooms to be able to identify key properties of loudspeaker performance. It is therefore clear that one must begin with comprehensive data on the loudspeakers in order to be confident about the sounds arriving at the listening position in a room. Then, there is the matter of how to interpret those data and to decide what remedies, if any, are necessary. In all cases it is best to start with a well-designed loudspeaker, and in many cases, above a few hundred hertz, that may be all that is required.

5.7 CLOSING THE LOOP: PREDICTING LISTENER PREFERENCES FROM MEASUREMENTS

It is one thing to use anechoic frequency-response data to predict steady-state room curves, but quite another to use that same data to predict the sound quality preferences of two ears and a brain. The following describes some important experiments, with eye-opening results.

Psychoacousticians dream of being able to insert measured numbers into an equation that represents a model of a perceptual function, and accurately predict a subjective response. It has not been straightforward for simple perceptual dimensions, much less for more complex ones. Something as multidimensional and abstract as preference in the sound quality from loudspeakers presents a new level of challenge—one that many people have said is impossible.

There have been a few attempts, all of which have shown some degree of correlation with perceptions. Olive (2004a) discusses a number of them. The oldest investigations suffer from a lack of truly good loudspeakers, and a lack of comprehensive measurements on them. Differences were audible and measurable, but the differences were so large and the misbehavior so gross that almost any measurement would have shown a correlation with perceptions. Loudspeakers have improved, and our understanding of how we perceive their sound has advanced. It is now evident that more than a single curve is necessary to explain or describe how a loudspeaker might sound in a room. It is further evident that 1/3-octave or critical-bandwidth measurements lack the resolution to reveal all audible subtleties.

Figure 5.2 shows results from the author's early investigations (Toole, 1986), and there is no doubt that there is a close relationship between the measured curves and the subjective "fidelity" ratings. It is not a calculated correlation coefficient, but the trends in the families of curves are visually evident—smoother, flatter, wider-bandwidth curves are associated with higher subjective ratings. Good performance at angles well off-axis appears to be an important factor. If we had such data on loudspeakers in the marketplace, choosing a good sounding one would be much easier than it is.

Well before any of this, though, *Consumer Reports*, the publication of Consumers Union (CU) that tests and rates the performance of many consumer products, began to rate loudspeakers (Seligson, 1969). The process was based on a sound power estimate compiled from anechoic measurements in 1/3-octave bands. Avoiding the contentious low-frequency room mode region, sound levels at frequencies from 110 Hz to 14 kHz were converted to loudness units, sones, in an attempt to get closer to what might be perceived. The sone-converted curve is then compared to an "ideal" target and an accuracy score calculated. Listening tests were conducted, and results are claimed to have supported the predicted ratings, although no details of those tests appear to have been published. Altogether it appeared to be a serious-minded attempt at correlating measurements with subjective perceptions. However, the method had several problematic aspects that would gradually be revealed, resulting in revisions to the process.

Allison (1982) criticized the method for not adequately weighting low-frequency performance, including his special interest, adjacent boundary interactions. However, on the matter of the basic method he said:

the most important aspect of a loudspeaker's performance (certainly its most clearly audible aspect) is how its acoustic-power output varies as a function of frequency. There is wide, but certainly not universal, agreement with CU on this point. I am among those who agree.

Evidence discussed earlier in this section points to disagreement.

In 1985 I hosted a loudspeaker measurement seminar at the NRCC, where we showed and demonstrated our measurement capabilities and had lectures on things that were believed to be measurable and important to what we heard and preferred. The ultimate test was a double-blind subjective evaluation—ears and brains determined what was good, not the appearance of a curve. We also demonstrated those. The audience was a who's who of loudspeaker designers and researchers from the US and Canada, and there were animated discussions. Among them were two engineers from *Consumer Reports*. Knowing they were coming, I had an overhead transparency prepared. For four loudspeakers that both of us had evaluated, it showed the subjective "fidelity" rating from our double-blind listening tests, and the "accuracy score" from their publication. The correlation was −0.7 (negative 0.7!). I still have the transparency. There was discussion, some puzzlement on their part, but nothing changed at the magazine. Based on this comparison, readers would have been well advised to invert the page of ratings.

Over the years the CU loudspeaker ratings were frequently at odds with our findings, and discussions with several loudspeaker manufacturers found confirming evidence, all revealing discontent with the situation. "High-end" consumers may put more weight on recommendations from audiophile publications, but masses of general consumers pay attention to these ratings. For me it came to a head after I joined Harman International as corporate vice president of acoustical engineering. In 1999, when some of our products received poor CU ratings, I was challenged by management to explain why.

At the time Harman was in the process of upgrading its measurement facilities and defining a process by which our many products could be evaluated. Taking into account the reputations of the different brands under the Harman umbrella, and satisfying the personalities and marketing stories associated with them, could be challenging. There was a growing confidence that our products were good performers, but there were exceptions. This situation provided an opportunity to move toward a trustworthy performance evaluation process. My suggestion was initiate research aimed at finding a better way of rating products that we all could use. It would require time, manpower and money. Remarkably, the CEO, Bernie Girod, and Dr. Sidney Harman, who started the corporation, agreed. This was an important decision, and these men deserve credit for having faith in the scientific method.

My long-term colleague from NRCC days, Sean Olive, was then in my research group at Harman. We assessed where we stood with respect to psychoacoustic rules that could be applied to numerically evaluating spinorama curves to reveal subjective preferences. It seemed as though a lot was known. Sean then initiated the following exercise, all of which is described in Olive (2004a, 2004b). In parallel with this the spinorama measurement technique was evolving, and it would play a significant role.

5.7.1 The Olive Experiments—Part One

Using the collection of anechoic measurements described in Figure 5.6, an evaluation of 13 loudspeakers that had recently been reviewed by CU was undertaken. All were bookshelf-sized products, and all lacked low bass. It began with a fully balanced, double-blind listening test (every loudspeaker was auditioned against all others the same number of times) by a group of selected and trained listeners. The result is shown in Figure 5.14a compared with the corresponding accuracy scores for the same loudspeakers published by Consumers Union. The contrast between the two sets of data is striking, with loudspeaker 1 exhibiting both "best" and "worst" evaluations. Overall, there was no correlation between the two results, indicating that, if the subjective data are correct, the CU method of evaluation is based on a faulty premise: that a sound power measurement alone contains information necessary to describe listener perceptions of sound quality in small rooms. In fact, the flaw in this logic was indicated in much earlier results by Toole (1986), and in recent data, as discussed in Section 5.6.

Employing the much more comprehensive, higher-resolution data generated routinely in the Harman International tests, Olive identified a model that described the subjective preference ratings in a near-perfect manner, as shown in Figure 5.14b. The precision exhibited by this model implies a complex analysis, and this is indeed the case.

The statistics derived from the different curves of the kind exemplified in Figure 18.6 are as follows:

- AAD: *absolute average deviation* (dB) relative to the mean sound level between 200 and 400 Hz, in 1/20-octave bands from 100 Hz to 16 kHz.

- NBD: average *narrow-band deviation* (dB) in each 1/2-octave band from 100 Hz to 12 kHz.

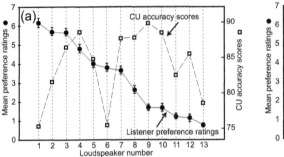
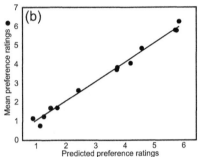

FIGURE 5.14 *(a) shows a comparison of mean preference ratings from double-blind listening tests and accuracy scores from evaluations of the same loudspeaker models published by Consumers Union (CU). The correlation between the two was small and negative (r = –0.22), and not statistically significant (p = 0.46). Confidence intervals of 95% are shown for the subjective preference ratings, indicating high confidence in several of the rating differences, although there were clusters of loudspeakers that were in statistical ties: 2 and 3; 5, 6 and 7; 9 and 10; and 11 and 12. Data from Olive (2004a), figure 3. (b) shows the comparison of the same subjective data and predictions from a model created by Olive, using anechoic data of the kind shown in Figure 5.6. In this case the correlation was 1.0 (r = 0.995) and the statistical significance a very high p ≤ 0.0001.*
Data from Olive (2004b), figure 4.

- SM: *smoothness* (r^2) in amplitude response based on a linear regression line through 100 Hz to 16 kHz.

- SL: *slope* of the best-fit linear regression line (dB).

All of these statistics were applied to all of the measured curves (see Figure 18.6):

- ☐ On-axis (ON)
- ☐ Listening window (LW)
- ☐ Early reflections (ER)
- ☐ Sound power (SP)
- ☐ Both directivity index curves (ERDI and SPDI)
- ☐ Predicted-in-room (PIR), a proportioned combination of ON, ER and SP that approximates measured room curves in several typical rooms.

Then there are two statistics that separately focus on low-frequency performance:

- LFX: low-frequency extension (Hz), the 6 dB down point relative to listening window (LW) sensitivity in the range 300 Hz to 10 kHz.

- LFQ: absolute average deviation (dB) in bass frequency response from LFX to 300 Hz.

When the dominant contributing factors to this model were analyzed they were:

- Smoothness and flatness of the on-axis frequency response: 45%
  - ☐ AAD_ON (18.64%) + SM_ON (26.34%)

- Smoothness of sound power, SM_SP: 30%

- Bass performance: 25%
    - LFX (6.27%) + LFQ (18.64%)

On-axis flatness and on- and off-axis smoothness account for 75% of the preference estimate. Bass, on its own, accounts for 25% of the preference estimate—something not to be dismissed.

These data were shared with CU before public presentation, and they acknowledged that they had problems, and agreed to further discussions. Most importantly, they ceased evaluating loudspeakers. There were a few discussions, and it seemed that improvements were being made to their process, but eventually the personnel involved moved on to other projects.

5.7.2 The Olive Experiments—Part Two

As impressive as this result is, there is a common factor that prevents the model from being generalized—all 13 of the loudspeakers were bookshelf models, a natural basis for a comparative product review in a magazine. All suffered from a lack of low bass, and many of them had common features: enclosure size, driver complement and configuration, and so on. The population needed to be expanded to include larger products, using different driver configurations and types. This was done in a second test that involved 70 loudspeakers from many origins, covering large price and size ranges. The penalty in this test was that the listening had been done in a fragmented manner—19 independent tests over an 18-month period—as happens in the normal course of business. All loudspeakers were evaluated in comparison with some other models, but there was no overall organization to ensure that the comparisons were balanced, each model against all other models. It would be unreasonable to expect the same high precision in the subjective-objective correlations as has just been seen.

Still, the result was impressive: predicted preference ratings correlated with those from listening tests with a correlation of 0.86, with a very high statistical significance ($p \leq 0.0001$). These are remarkable numbers given the opportunities for variation in the listening tests, meaning that the listeners themselves were highly stable "measuring instruments" and that the strategy of always doing multiple (usually four products) comparisons is a good one. The model that, for these products and these subjective data produced the best result was different from the earlier one. Here, the dominant factors were:

- Narrow-band and overall smoothness in the on- and off-axis response: 38%
    - NBD_PIR (20.5%) + SM_PIR (17.5%)
    - Because PIR incorporates on-axis, early-reflection and sound power curves, it is evident that all three must exhibit narrow and broadband smoothness.
- Narrow-band smoothness in on-axis frequency response (NBD_ON): 31.5%

- Low-bass extension (LFX): 30.5%

Again, on- and off-axis smoothness and the lack of narrow-band deviations account for most of the preference estimate (38 + 31.5 = 69.5%). Bass again is a big factor at 30.5%. This time, perhaps because there was a mixture of large and small loudspeakers, listeners were attracted to differences in bass extension. Because the bass smoothness metric does not include the influence of the room (it is based on anechoic data), the issue may be debatable, but extension alone seems fairly straightforward.

5.7.3 The Olive Experiments—Part Three

As part of the Part One tests, listeners were required to "draw" (using sliders on a computer screen) a frequency response, describing what they thought they heard in terms of spectral balance. This is a task that obviously could not be asked of average listeners, but these listeners had been through a training program (Olive, 1994, 2001) and were able to estimate the frequencies at which audible excesses and deficiencies occurred. When these data were compiled the results indicated that all of the highly rated loudspeakers had been judged to have very flat curves. Of great interest, of course, is which one of the several measured curves for these loudspeakers does this correspond to—the logical candidates being sound power (the CU model), the early-reflections curve, which is similar to the PIR (the steady-state room curve that is the basis of room-equalization schemes), or the on-axis curve (not the dominant measured component, but important in perception)?

Figure 5.15 shows a revealing comparison of three loudspeakers chosen because of their ratings in Figure 5.14. The subjectively "drawn" spectrum curves are obviously best matched to the on-axis and listening window frequency responses of the loudspeakers, thus confirming the observation noted from the earliest tests shown in Figure 5.2, that a good axial amplitude response appears to be a requirement for good sound. Beyond that, attention must be given to off-axis behavior.

Let us dig deeper into these examples to see why the Harman and CU results did not always agree.

- Loudspeaker 1, the highest rated in Harman tests, and the lowest rated in CU tests. A straightforward examination of the spinorama reveals a wide bandwidth, flattish and quite smooth on-axis amplitude response. Bass extends smoothly down to about 60 Hz. The smoothness is maintained in the progressive spatial averages, indicating an absence of coloring resonances. The DI is respectably smooth. This should be a good-sounding, highly rated loudspeaker, and it is revealed as such in the Harman double-blind listening tests. However, the CU calculated "accuracy" clearly did not approve, possibly because of the tilted sound power curve.

- Loudspeaker 9, the highest rated in CU tests and one of the low-rated ones in the Harman tests. The rising axial response is a bad start, and the severe resonances seen in all of the curves indicate significant coloration. The bass shows a bump around 100 Hz, lending "punch" to kick drums, but it does not extend as far down as Loudspeaker 1. DI indicates some problems in the crossover region

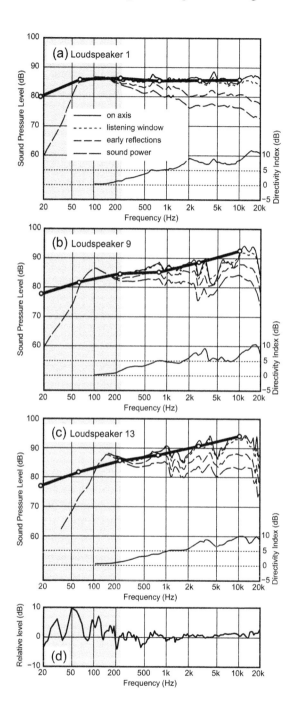

FIGURE 5.15 *Figures (a), (b) and (c) show the spinorama data for 3 of the 13 loudspeakers tested (results in Figure 5.14). The heavy dark line superimposed on each curve is the subjective estimate of the spectral balance given by the listeners while in the listening room. The open circles indicate the center frequencies of bands used by listeners to describe the spectra of what they heard. This is a "dimensionless" curve on the vertical axis, reflecting subjective impressions of boost and cut relative to the personal ideals of the listeners. Consequently, this author has taken the liberty of adjusting the vertical scale and the vertical position of these curves until they appeared to fit the pattern of anechoic data. The chosen vertical scale is used in all three curves. Below about 250 Hz the anechoic data cease to reflect accurately what is heard in the listening room— it will be less smooth and significantly higher in amplitude in the room, and this portion of the curves is shaded. As the tests were done using positional substitution, the contribution of the room to the sound would be quite constant. (d) shows the average error between PIR (from anechoic data) and in-room measurements (from Olive, 2004b, figure 7). The most obvious visual correlation with the subjective spectral balance is with the on-axis or listening window curves.*

between the woofer/midrange and tweeter. The low subjective rating in Harman evaluations seems justified. The CU calculation appeared to like the flatter sound power. The 1/3-octave resolution in the CU sound power data would not reveal the resonances, and the lack of high-resolution anechoic data hinders the identification and evaluation of resonances.

- Loudspeaker 13, the lowest rated in the Harman tests, and also low rated in CU tests. Again, a dramatically rising on-axis response is not good. Evidence of several resonances is not good. A bass rolloff that begins around 120 Hz, following a bass bump, is a significant deficiency. This is not a good-sounding loudspeaker, and both evaluations identified it as such.

In these experiments, trained listeners were able to draw curves of spectral trends—crude frequency responses—describing what they heard in the listening room. All of the high-scoring loudspeakers were described as having flattish spectra (above the transition frequency), a trend that matches the flattish on-axis/listening window curves for all of the corresponding anechoic measurements.

Loudspeakers with lower preference ratings were described as having either insufficient bass or excessive treble, or both. These trends tend to yield a flatter sound power curve, and thereby higher accuracy scores in CU reviews. However, it is a trend that runs contrary to the interests of high sound quality as judged by these listeners and many others. The direct sound has a special place in the hierarchy of sounds arriving at the ears, and flat is a good objective. Figure 5.2 indicates this, and Queen (1973) concludes an investigation by saying: "The results thus far tend to confirm the presupposition that spectrum identification depends principally upon the direct sound of the sound source."

Below the transition frequency, in the gray shaded area in Figure 5.15, it is evident that listeners responded to something closer to the low-frequency response in the room than to the anechoic frequency response. This is the ultimate limitation of all such models, in that the listening room and the arrangements within it determine the bass level and quality. With bass accounting for about 30% of the overall subjective rating, this is a matter of great importance and it implies that if there is to be parity among listening situations—professional and consumer—there needs to be substantial control over the low frequencies in the room. Chapter 8 provides guidance.

5.8 LOUDSPEAKER RESONANCES—DETECTION AND REMEDIES

Section 4.6 discusses this topic in detail, and Figure 4.13 shows how resonances are revealed in a spinorama presentation of anechoic data. The persistence of a bump in the spatially averaged curves is confirmation. There is no need to look at a waterfall diagram, because if there is a resonant peak, there will be ringing; no peak, no ringing. As was reported in Toole and Olive (1988) and discussed here in Section 4.6, it is the spectral bump that is the most reliable indicator of audibility.

Loudspeaker transducers behave as minimum-phase devices, and therefore resonances in them can be attenuated by (minimum-phase) parametric equalization in the electronic signal path. Done correctly, the amplitude and phase response of the loudspeaker is corrected: the resonant peak is reduced, as is the time-domain misbehavior. Ideally, this would be done by the manufacturer, based on high-resolution anechoic measurements, as part of an active loudspeaker or one with outboard dedicated electronics. If a passive loudspeaker has resonances, as did two of the examples in Figure 5.15, the problem is hard for a customer to identify and to repair.

The conclusion is that it is strongly advantageous to begin with comprehensive high-resolution anechoic data from which problems in a loudspeaker can be reliably anticipated. Below the transition/Schroeder frequency, the room is in control, and other remedies are necessary (Chapters 8 and 9).

5.9 SUMMARY AND DISCUSSION

From the simple analog ink lines on graph paper of the pre-digital era, we have moved on to computer-controlled collection of unlimited amounts of data, elaborate digital processing of that data, useful predictions of steady-state room curves, and ultimately to the prediction of listener subjective ratings of sound quality. We may never eliminate the need for subjective evaluations, but it is greatly reassuring to know that there are measurements that add guidance to the evaluation of a product. In fact, one may venture a challenge that an examination of the right set of anechoic measurements may well be more reliable as an indicator of sound quality than a "take it home and listen to it" subjective evaluation, and far more trustworthy than an uncontrolled listening session at an audio show or store. Now, with brick and mortar audio stores difficult to find, and Internet sales becoming more popular, it is highly useful if data like spinoramas, or a reasonable facsimile, are available to shoppers.

It is impressive to think that we now have a technical means of identifying loudspeakers that have the potential of sounding good. Equally impressive is that this is confirmed by listening tests done in a variety of different rooms (Section 7.6.2). Humans have a remarkable ability to "listen through" rooms.

Figure 5.16 illustrates the so-called Central Paradox, in which the communication of sound, live or reproduced, through a room appears acoustically to be a chaotic mess, and yet two ears and a brain make perceptual sense of it. As envisioned by its creator, Dr. Arthur Benade, it applied to live performances, but it turns out to be true for reproduced performances as well. It is wise not to underestimate the power of the binaural perceptual process.

If we wish to, it is possible to construct loudspeakers that are essentially neutral in their timbral signature. However, whether they sound good in a demonstration depends on the recordings. No self-respecting audiophile demonstrates his pride and joy audio system without first carefully selecting the demonstration programs. The reality is that recordings differ in their spectral characteristics in ways that are related to the circumstances in which they were mixed (Børja, 1977). Over the years it has been possible

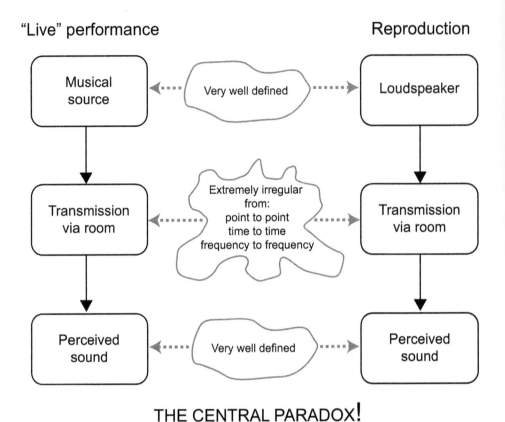

FIGURE 5.16 *The "Central Paradox" of sound in rooms (Benade, 1984), expanded by the author to include reproduced sound. The message is that what we measure in a room may or may not directly relate to what is heard. However, in the case of loudspeakers, an adequate description of performance may provide useful data. This has now been proved.*

to find recordings with spectral properties that flatter many different loudspeakers, but finding the match is a trial-and-error process. The origin of the problem is described in Figure 5.17, a pictorial adaptation of the Benade diagram that introduces some realities of the audio industry.

In live performances we don't normally feel a need to "equalize" (even if we could) the sound from instruments and voices. They are what they are, and for the most part, the timbral signatures are well replicated in different performance venues. Bass balance is a possible exception, but it is interesting that if we sense an excess or deficiency of bass in a live performance, we are not inclined to blame the instrument or the musician, but the acoustics of the venue. This can be interpreted as a case of perceptual streaming, the result of, as Bregman (1999) describes it, auditory scene analysis. Clearly there is significant perceptual analysis, deduction and adaptation going on. This cognitive activity apparently happens subconsciously in everyday listening; we hear and recognize the

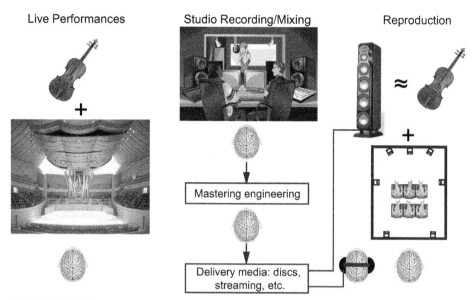

Live Performances Studio Recording/Mixing Reproduction

Mastering engineering

Delivery media: discs, streaming, etc.

FIGURE 5.17 *The Central Paradox elaborated to include the practical considerations of what is involved in sound reproduction. There are several stages, all of which must have some important common physical factors if there is to be the possibility of listeners hearing the art that was created. The technical portions of this chain of events need to be subject to some amount of standardization. Human perceptions and judgments are part of the process, contributing to the artistic content—this is where freedom of choice should have no boundaries. Monitor loudspeakers with distinctive "personalities" should not become part of the art.*

timbral identities and subtleties of the same sound sources, but in different rooms. Two ears and a brain are massively more analytical and adaptable than an omnidirectional microphone and an analyzer.

Whether microphones capture the sound in a concert hall or in a dedicated recording studio, it will be auditioned in a control room, perhaps multiple control rooms, possibly through multiple loudspeakers. During this artistic creation process, changes in balance and timbre may be made based on what is heard. This is the first stage of "voicing" in the sound reproduction process, and the problem is that some characteristics of the specific loudspeakers being used modify the art being created. The recording may then go through a mastering stage, in which it is auditioned by other persons through other loudspeakers in other rooms, and additional changes may be made to create a product that is hoped will please listeners, whoever and wherever they are, within the limitations of the delivery media. This is the second stage of "voicing," and clearly a certain amount of guesswork is involved.

Loudspeaker designers do the third stage of "voicing." Frequently the sales and marketing people collaborate in finding a frequency-response balance and shape that flatters the kind of music they like, that they think their customers may like, or that they think might make their product stand out in a store demonstration. This is another process involving guesswork, and over the years, it has been significantly responsible for some of the distinctive sounds in different brands.

Nowadays there is a fourth stage of "voicing," room equalization, in which automated algorithms make measurements in the listening room and adjust the frequency response to meet a criterion set by someone who knows nothing about the loudspeakers, or the room, or the geometry of the setup. Companies that do not manufacture loudspeakers have designed most of these algorithms and, as will be discussed later, some of the algorithms present a significant possibility of degrading the performance of good loudspeakers.

Clearly, there are too many "voices" in the "voicing" of sound reproducing systems. That things work as well as they do is a credit to the increasing numbers of loudspeaker designers who have come to realize that timbrally "neutral" is a good place to be. The monitor loudspeakers used in the creation of the art should be the equivalent of clear windows through which the artists can view their art. Then if they don't like what they see, they can fix the mix. If consumers have similarly neutral loudspeakers, they have a reasonable chance of hearing the art as it was created. Otherwise it is a gamble.

Consumers who wish to compensate for audible spectral biases should have access to tone controls, but unfortunately in contemporary digital equipment some of these have become inconvenient to use, or in the case of high-end misguidance, non-existent. Pity.

CHAPTER 6

Loudspeaker/Room Systems—An Introduction

He was alone in his private space at the end of a long day, a cool Scotch in one hand and, in the other, the remote control for a new CD player. He pressed "play" and sat back for some enjoyable imaging.

That was my introduction to an article titled "Stereo Imaging and Imagery" that I wrote for *Sound and Vision* magazine in 1986. It was a sarcastic way of introducing a serious discussion of a topic that is as old as stereo itself. In addition to the melody, rhythm, harmonies and lyrics of the music, serious listeners have expectations of a soundstage, with performers arranged across it, playing in an acoustic environment that, with luck, may seem to envelop the listener.

Reading the audiophile literature indicates that imaging is influenced by everything in the playback chain—that the real "truths" are in the recordings, waiting to be revealed by the right combination of interconnects, power cords, amplifiers, speaker wire, loudspeakers, stands, room treatments and so forth. Sarcasm comes easily after reading some of the advertising claims and reviews. There are substantial topics for discussion, no doubt, but some claims stretch credibility and flout physical laws. Here we will stick to mainstream physics and psychoacoustics, and address the final links in the playback chain: loudspeakers and rooms.

Recordings are major determinants of what we hear. Figure 5.17 describes a process with numerous variables that are upstream from the playback medium we are experiencing. The loudspeakers and rooms in the stages of creating a recording may have a greater influence on what is ultimately heard than those at the point of playback. That is the basis for the circle of confusion (Section 1.4). It would also be presumptuous to think that the creators of the art have the same perceptual capabilities and priorities as listeners, so results inevitably vary. We may never know why they vary, but it is useful to understand the key physical factors so that listeners can exercise some control. It would be good if we had some assurance that what we hear is the art that was created.

Over the years, loudspeakers and rooms have been treated as separate entities. To an extent, they must be, but in the end, the more that is known about how they function together, the more likely that the result will be high-quality entertainment.

6.1 ONE ROOM, TWO SOUND FIELDS—THE TRANSITION FREQUENCY

The room curves displayed in Chapter 5 show that the right combination of anechoic measurements can closely anticipate events in typical rooms at middle and high frequencies, but at low frequencies the curves are lumpy and irregular. The acoustical explanation is the dominance of relatively isolated room modes and associated standing waves at low frequencies, and of a complex collection of overlapping modes and reflected sounds at high frequencies. As a result, as frequency increases it becomes progressively less useful to think about regular patterns of standing waves in rooms, but rather to think in terms of irregular patterns of constructive and destructive acoustical interference caused by numerous reflections traveling in many directions.

Between the orderly low-frequency room resonances and the different higher frequency acoustical behavior is a transition zone. The middle of this zone, in large rooms

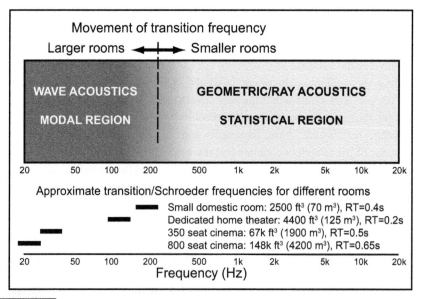

FIGURE 6.1   *An artistic interpretation of the transition between the low-frequency region dominated by room modes and the high-frequency region dominated by reflected sounds. Also shown is the effect of room size on the position of the transition region in the frequency domain.*

such as concert halls and auditoria, would be defined as the Schroeder frequency or, as Schroeder himself calls it, the cross-over frequency f_c (Schroeder, 1954, 1996).

$$f_c = 2000 \sqrt{\frac{T}{V}}$$

Where: T is the reverberation time in seconds, and V is the volume of the enclosure in cubic meters. The multiplier constant changed from the original 4,000 to 2,000 in the 1996 paper, emphasizing that this is an estimate, not a precision calculation.

Calculation of the Schroeder frequency assumes meaningful reverberation times, a strongly diffuse sound field and an unimpeded volume. Small listening rooms and recording control rooms are too absorptive to have a consequential reverberant sound field (see Section 10.1). Therefore, in small rooms, especially those with large furnishings, these are mismatched concepts, so the calculated value may be in error, as noted by Baskind and Polack (2000). For the room used in the measurements of Figure 5.4c, the Schroeder frequency is 129 Hz ($T = 0.32$s, $V = 76.9$ m³). This appears to be on the low side because the large undulations in the curves have not diminished, although some of these variations may be associated with adjacent-boundary effects. However, no matter how it is identified, or what it is called, the transition region is real, and it is necessary to take different approaches to dealing with acoustical phenomena above and below it.

Figure 6.2 gives us more insight into this topic. Here are shown, using an expanded frequency scale, high-resolution frequency-response measurements from each of the five loudspeakers in a surround-sound system at the prime listening position. The room is geometrically symmetrical, but differences in the curves reveal that it is not acoustically symmetrical. A door in one end wall causes it to flex more than the other one at certain frequencies, and a concrete wall behind, but not touching, one of the side walls gives it more stiffness than the opposite wall. The result is asymmetry in the standing wave patterns. Five identical loudspeakers were arranged in the ITU-R BS.775–2 recommended arrangement and measurements were made at the listener's head location.

The standing waves cause huge variations at low frequencies, covering the full 40 dB range of the display. Above about 100 Hz the variations are attenuated and, above about 200 Hz, they seem to settle down even more. Looking at the details, below about 200 Hz, in spite of some obvious variations related to the very different loudspeaker locations, one can see evidence of relatively independent resonant peaks at clearly identifiable frequencies. The frequencies may not be an exact match to the calculated frequencies because the room boundaries were not infinitely stiff, as theory assumes. Above 200 Hz, the pattern becomes much less orderly, and the peak-to-peak variations are smaller. Yet, an underlying trend is visible, including the step at 500 Hz, which is a characteristic of the loudspeaker, a woofer running without a low-pass filter. The real transition frequency for this room appears to be around 200 Hz. The calculated Schroeder (crossover) frequency (f_c) is 111 Hz ($T = 0.4$s, $V = 128$ m³), which plainly seems to be too low. For this room substituting the original constant, 4,000, into the equation would yield a better answer for this small room.

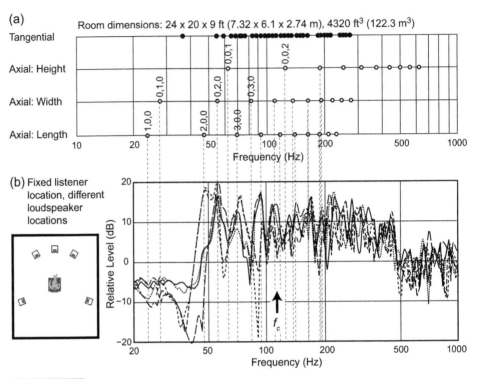

FIGURE 6.2 *(a) Axial and tangential modes calculated for a listening room set up with a listening arrangement according to ITU-R BS.775–2. (b) 1/20-octave steady-state measurements for each of the five loudspeakers, measured at the listening position. The Schroeder crossover frequency (f_c) is shown.*

The following chapters will deal separately with events above and below the transition frequency because they require different analysis and treatment.

6.2 A BRIEF HISTORY OF LOUDSPEAKER/ROOM INTERACTIONS

There is a discussion thread in audio that encourages the belief that the listening room is a problem to be significantly tamed, if not eliminated. First reflections are claimed to be especially problematic, and the primary task is to find the reflection points on the room boundaries and apply absorbent or diffusers. Damping low-frequency room resonances requires massive low-frequency absorbers. There are rooms shown on the Internet in which all sense of normal interior décor is lost. They are literally filled with large absorbers—42 in one room is the highest count I can recall, and this is stated with pride, as though anyone with fewer is less committed to good sound. No doubt audiophiles with families enter into serious debates with their significant others if this option is considered.

If there is a need for social or acoustical isolation, a dedicated home theater is the answer. Then there are few restrictions. However, normally furnished domestic rooms should not have to be unattractive in order to be highly performing listening spaces. That belief has motivated some of the scientific research, looking for alternatives to the traditional acoustical solutions. Fortunately, now there are options that can allow for a variety of preferences to be satisfied, so that good sound can be part of one's lifestyle, whatever it is.

Sound reproduction has endured decades of directional and spatial deprivation. It began with the first sound reproduction technology, monophonic sound, which stripped music of any semblance of soundstage, space and envelopment. This was further aggravated by the need to place microphones close to sources; early microphones had limited dynamic range and high background noise. Adding further to the spatial deprivation was the use of relatively dead recording studios and film soundstages. "Reproduced in the home, where upholstered furniture, drapes and rugs quite often prevented such an acoustical development of ensemble through multiple reflections, the Edison orchestral recordings were often singularly unappealing" (Read and Welsh, 1959, p. 209). Recording engineers soon learned that multiple microphones could be used to simulate the effects of reflecting surfaces, so the natural acoustics of the recording studio were augmented by the techniques of the sound recording process.

Directional microphones gave further control of what sounds were captured. With relatively "dead" source material, it became necessary to add reverberation, and the history of sound recording is significantly about how to use reverberation rooms and electronic or electromechanical simulation devices to add a sense of space. In the past, these effects were used sparingly and "the typical soundtrack of the early 1930s emphasized clarity and intelligibility, not spatial realism" (Thompson, 2002, p. 283).

A coincidental influence was the development of the acoustical materials industry. In the 1930s dozens of companies were manufacturing versions of resistive absorbers—fibrous fluff and panels, to absorb reflected sound and to contribute to acoustical isolation for bothersome noises. Acoustical treatment became synonymous with adding absorption. Dead acoustics were the cultural norm—the "modern" sound—which aligned with recording simplicity, low cost, small studios and profitability (Blesser and Salter, 2007, p. 115). "When reverberation was reconceived as noise, it lost its traditional meaning as the acoustic signature of a space, and the age-old connection between sound and space—a connection as old as architecture itself—was severed" (Thompson, 2002, p. 172).

Read and Welsh (1959, p. 378), recount a discussion in 1951, by popular audio commentator Edward Tatnall Canby, writing in *Saturday Review of Recordings*. He says:

"liveness," the compound effect of multiple room reflections upon played music, is—if you wish—a distortion of "pure" music; but it happens to be a distortion essential to naturalness of sound. Without it music is most graphically described as "dead." Liveness fertilizes musical performance, seasons and blends and rounds out the sound, assembles the raw materials of overtone and fundamental into that somewhat blurred and softened actuality that is normal, in its varying degrees, for all music. Disastrous experiments in "cleaning up" music by removing the all-essential blur long since proved to most recording engineering that musicians do like their music muddied up with itself, reflected. Today recording companies go to extraordinary lengths to acquire studios, churches and auditoriums, (not to mention an assortment of artificial, after-the-recording liveness makers) in order to package that illusively perfect liveness.

This notion that reflections result in a corruption of "pure" music, and the apparent surprise in finding that musicians and ordinary listeners prefer "muddied-up" versions, reappears in audio, even today. We now have quite detailed explanations why this is so, but one can instinctively grasp the reality that, toward the rear of a concert hall, the direct sound (the "pure" music) is not the primary acoustic event. And yet, its subtle presence is strongly influential in what we perceive (Griesinger, 2009). Two ears and a brain comprise a powerful acoustical analysis tool, able to extract enormous resolution, detail and pleasure from circumstances that, when subject to mere technical measurements, seem to be disastrous. Something that in technical terms appears to be impossibly scrambled is perceived as a splendid musical performance (see Figure 5.16).

When Sabine introduced the concept of reverberation time into acoustical discussions of rooms at the turn of the last century, he provided both clarification and a problem. The clarification had to do with adding a technical measure and a corresponding insight to the temporal blurring of musical patterns that occur in large live spaces. The problem appeared when recordings made in spaces that were good for live performances were often perceived to have too much reverberation. A single microphone sampling such a sound field, and subsequently reproduced through a single loudspeaker, simply did not work; it was excessively reverberant. Our two ears, which together allow us to localize sounds in three dimensions, to separate individual conversations at a cocktail party and to discriminate against a background of random reverberation, were not being supplied with the right kind of information. Multiple microphones, conveying information through multiple channels, delivering the appropriate sounds to our binaural hearing system were needed, but not available in the early years.

Habits die hard. The introduction of stereo in the 1950s gave us a left-to-right soundstage, but close-microphone methods, multitracking, and pan-potting did nothing for a sense of envelopment—of being there. The classical music repertoire generally set a higher standard, having the advantage of the reflectivity of a large performance space, but a pair of loudspeakers deployed at ±30° or less is not an optimum arrangement for generating strong perceptions of envelopment (as will be explained later, this needs additional sounds arriving from further to the sides). Perhaps that is why audiophiles have for decades experimented with loudspeakers having wider dispersion (to excite more listening room reflections), with electronic add-ons and more loudspeakers (to generate delayed sounds arriving from the sides and rear). In the control room, audio engineers now employ spatial simulation algorithms to break the monotony of a frontal soundstage by adding some low-correlation spatial sounds, even including binaural crosstalk-cancellation. All have been intended to contribute more of "something that was missing" from the stereo reproduction experience that, even at its best, serves only a single listener.

The real solution is more channels, giving the capability of delivering anything from a single point image through to spatial envelopment and immersion, as appropriate.

Both stereo and multichannel sound reproduction were motivated by movies, not music. I am old enough to remember some audio (then hi-fi) authorities arguing that mono delivered all the musically important sound, and that stereo was not necessary. Incredible! We have had 5.1 and 7.1 options for several years, used mainly in movies, rarely in music. Now there is "immersive" sound, and cinemas are equipped with many (currently up to 60 or more) independently driven loudspeakers located around the walls and across the ceiling.

Someone is paying attention to the perceptual dimensions of direction and space. Most of the use is for sound effects in movies, but not long ago I heard a many-channel Auro-3D immersive demonstration of classical music in a concert hall and also in a cathedral. Both were spine-chillingly realistic, and one could move around the listening room without losing the illusion of being in the real space. Two ears and a brain recognize the difference when sounds arrive from the appropriate directions at the appropriate times. Multidirectional stereo loudspeakers add some of the directions, but not those far to the side and rear, and not with the appropriate time delays to generate true envelopment. Downscaled versions of these multichannel systems for homes offer both thrilling movies and the prospect of engaging music. A state-of-the-art home theater can also be a state-of-the-art concert hall replicator.

An interesting recent study indicated that when listening to a recorded space in a real space—the listening room—it is the larger of the two spaces that is most often perceived (Hughes et al., 2016). So the objective in sound reproduction would appear to be to employ a playback space that sounds "smaller" than any reproduced space in program material; in domestic rooms, home theaters and recording control rooms, that is the case. Cinemas have a problem reproducing the impressions of small spaces.

6.3 TIMBRAL AND SPATIAL EFFECTS ATTRIBUTABLE TO ROOMS

Figure 6.3 is an attempt to summarize the perceptual effects that are experienced when loudspeakers energize a room. The direction and space effects are estimates based on comments in the literature, mostly pertaining to large venues, but the wave-acoustic effects at the bottom are well documented. All will depend to some extent on the specific characteristics of loudspeakers, rooms and program.

Before we get into an examination of experimental evidence of acoustical events in small rooms, let us very briefly summarize the dominant effects of sounds above the bass-frequency range. There is some blending of perceptions in large venues and those reproduced by multichannel systems in small venues—occasionally they are based on the same goals. These topics are discussed in more detail elsewhere in the book, as indicated in the index.

- **Localization** has two principal dimensions:
  - □ **Direction.** Identifying the direction from which sound appears to be coming. Because of the ear locations we are much better at localizing in

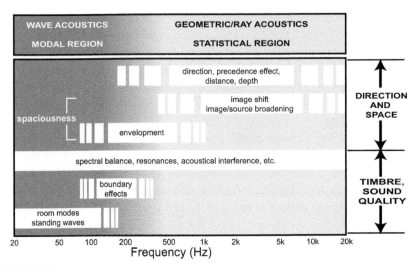

azimuth than we are in elevation, and more precise close to the median plane (a vertical forward/back plane running through the head) than we are to the sides. In rooms the **precedence effect** allows us to localize in the presence of numerous reflections. It is a cognitive effect, occurring at a high level in the brain, meaning that it can be different at different times and places, for different sounds and changing with experience.

- □ **Distance.** Reflections help us to determine distance. Distance perception also has a cognitive component, meaning that we can learn to recognize aspects of the sound field. Because the floor reflection is always with us, and because it is closely similar for all horizontal channels, it is entirely feasible that humans adapt to it (Section 7.4.7). This leaves the multiple channels to deliver cues to different distances.

- ■ **Spaciousness or spatial impression** can be separated into two components:
  - □ **Envelopment and the sense of space.** Also called listener envelopment (LEV), this is the impression of being in a specific acoustical space. It is perhaps the single most important perceived element distinguishing truly good concert halls. It is the one thing I hear at my regular live concert hall experiences that simply is not replicated at home without multichannel audio delivering independent greatly delayed sounds from different directions. The new immersive systems hold much promise in this respect. In movies, where multichannel audio originated, the purpose was to provide the envelopment to "transport" the listener to the space portrayed on the screen.
  - □ **Image size and position.** Strong reflections have the ability to slightly shift the apparent position of a source in the direction of the reflection

and/or to make the source appear larger. In live classical performances this is called ASW (apparent source width), and audiences like it. In sound reproduction there is evidence that the tendency continues. Small positional shifts are innocuous because nobody knows what the intended location really was. Image broadening is what happens in live performances—the "air" around the instruments and voices. It gives an acoustical setting for the sound, rather than having it float as an abstract pinpoint in space—although it seems that some listeners like that enhanced effect.

- **Immersion.** Here I will go out on a limb and invent a definition. The idea of large numbers of loudspeakers around and above the audience in a cinema or home theater is to deliver impressions of specific sounds emerging from specific directions. They can be birds in trees, rusting leaves in a forest, raindrops, overflying aircraft, alien critters and the like. One is surrounded by sounds, but those sounds do not describe a surrounding acoustic space, as in envelopment (LEV). The same loudspeakers can create envelopment, and with so many active loudspeakers delivering strong direct sounds the listening room itself fades in importance.

- **Timbre changes** have two basic components: one can be negative, the other mostly positive. Simply detecting a "difference" is not a sufficient criterion.
  - ☐ **Comb filtering, repetition pitch.** Colorations can be created when a sound is added to a delayed version of itself. When the result is measured we see a repeating pattern of peaks and dips in the frequency response, which is why it is called "comb" filtering. In some circumstances a pitch can be perceived that is associated with a frequency defined by the inverse of the delay. The effect is audibly obvious if it occurs in an electronic signal path, or if there is a single strong vertical (median plane) reflection in an otherwise echo-free environment. However, for delayed sounds arriving from large horizontal angles, and in normally reflective spaces where there are multiple reflections, the effect ceases to be a problem—the perception is simply of a "room."
  - ☐ **The audibility of resonances.** Resonances are the "building blocks" of voices and musical instrument sounds. Reflective spaces enhance our ability to hear resonances, making these sounds timbrally richer and more interesting. In loudspeakers, resonances are huge problems because they monotonously color all reproduced sounds. Listening in reflective rooms is more likely to reveal inferior loudspeakers.

- **Speech intelligibility.** If the perception of speech is unsatisfactory, our ability to be informed and entertained is seriously compromised. Fortunately, the human hearing system is remarkably tolerant of reflections typical of small rooms. In fact, up to a fairly generous level, we make use of them to assist us in understanding speech.

A lot of the pioneering research work was done using speech as the test signal and, while it is fundamentally important, movies and music contain much more complex, wider bandwidth sounds. Similarly, many experiments examined the effects of a single reflection auditioned in an otherwise reflection-free environment. The results are not wrong, but they very likely need to be modified to be applicable in normally reflective spaces. As will be seen, our instincts about what is audible are not always right, and our instincts about what is audibly important may need adjusting as well.

CHAPTER 7

Above the Transition Frequency: Acoustical Events and Perceptions

In typical home listening rooms, home theaters and recording control rooms, the transition frequency is somewhere in the region of 200 to 300 Hz. Below this region we must cope with room resonances and the associated standing waves, which are discussed in Chapter 8, and adjacent boundary effects, which are discussed in Chapter 9. Above this region we need to look at how loudspeakers radiate sounds into rooms, how rooms modify those sounds on the way to listeners, and how listeners respond to the combination.

In monophonic days, the room mattered greatly because reflections within the room provided the only spatial setting for what was reproduced. Recordings contained only reverberation: a time domain phenomenon. Some enthusiasts aimed the loudspeaker toward a wall or into a corner to add more spatial complexity, softening the single "point source" image. Multidirectional loudspeakers are often used in stereo reproduction in part because they soften the hard-panned left and right images in a lot of pop and jazz, and provide slight expansion of a classical music soundstage.

Stereo recordings have the advantage of being able to include sounds that are poorly correlated in the two channels, thereby creating a certain amount of spatial impression in addition to the panorama of phantom-images across the soundstage between the loudspeakers. Phantom images rely on symmetry in the physical setup and the discipline of the listener to find and stay in the sweet spot—lean left or right and they move toward and ultimately collapse into the nearer loudspeaker.

The huge contrasts in music add complications. The "in-your-face" sharply delineated imaging that characterizes much pop and rock is a contrast to the soft imaging and spaciousness that is expected of classical music. Over time, popular music has selectively adopted some of the spatial characteristics of the classics, with some of the components in a mix being treated with signal processing that generates perceptions of reverberant space. Already, it is possible to identify some of the variables, and to anticipate that there will be differences of opinion and preference among music lovers.

In multichannel audio, nowadays mostly movies, all of the stereo issues exist, in multiples. But there is one distinctive feature of movies: the center channel not only conveys almost all the dialog, but, for much of the time, all on-screen sounds. Because of the ventriloquism effect, viewers tend not to notice the directional contradictions—we localize to the moving lips, the slammed door, whatever the sound source is. It is widely used because it keeps costs down and makes editing much easier. Only big-budget films do much in the way of panning sounds across the screen. Other channels can also be employed in this manner, with a voice emanating from left front, conversing with another voice in right front. These are all examples of single loudspeaker, monophonic sound sources, and the room interactions may be significant for them. However, if multiple channels are simultaneously radiating sound, the influence of the room becomes diluted by the recorded sounds.

The default format for music is stereo. I fully understand the economic pressures of the past—multiple inventories of physical media (LPs, CDs, SACDs, DVDs, Blu-ray, etc.), but as streaming entertainment grows, more options are possible. With movies delivering many channels for immersive entertainment, perhaps some of the bandwidth can be devoted to music. I have heard people dismiss multichannel formats for music. It is true that not all such recordings are excellent. But after about 60 years of practice, I still hear stereo recordings that are audibly challenged. Give it time. For now, though, most of the following discussions will focus on conventional stereo. If it is well reproduced, quality entertainment is possible. Additional channels simply add artistic options in direction and space (Chapter 15).

7.1 THE PHYSICAL VARIABLES: EARLY REFLECTIONS

Since the appearance of the first edition of this book, a few people took the position that "Toole" is in favor of side-wall reflections, and let their views be known in various Internet audio forums. I became the "straw man" opponent to their views that, not surprisingly, have been known to encourage the purchase of acoustical products they sell. In fact, if my critics had read to the end of the book, figure 22.3 shows some suggested room treatments, and it is stated that acoustical treatment of those contentious side-wall areas is "optional: absorb, diffuse, reflect"—the choice is left to the designer, with, one would hope, input from the customer.

It has been alleged that I have this commitment to lateral reflections in ignorance of, or dismissal of, decades of professional audio tradition in which these reflections are absorbed in recording control rooms. These people don't know that early in my career I designed a few recording studios as a learning exercise. One of them could hold a 75-piece orchestra, had separate vocal and drum booths, and two control rooms. Many of the high-power monitor loudspeakers at the time were less than impressive, so I also designed the monitor loudspeakers that went into a couple of them—see Section 18.3.1. In those days the powerful main monitor loudspeakers were moderately directional mid- and high-frequency horns, and side walls were usually angled to direct the residual first lateral reflections into the broadband back wall absorber. Recording engineers preferred to be in a strong direct sound field, and that is what they got. Some control rooms of that period had a "dead end" that absorbed all early reflections in the front of the room. These practices in control room design evolved at a time when high-power monitor loudspeakers were simply not very good. Some misbehaved so badly off-axis (see Figures 12.8 and 18.5) that the only way to alleviate the problem was to absorb the off-axis sounds.

However, as a research scientist, I saw a number of interesting questions to be answered, and so, it turns out, did several other investigators over the years. Toole (2003) gives an updated opinion of what I thought might be useful guidelines for control room practice, and Toole (2015a) is an invited "distinguished lecture" to an audience that included recording engineers and tonmeister students at CIRMMT, the Centre for Interdisciplinary Research in Music Media and Technology, housed at the Schulich School of Music at McGill University (available for viewing on YouTube). All of this is also discussed in detail in several chapters of the original book, and a revised and expanded version follows.

It is not a black and white situation, and whatever personal opinions I may have, they are overwhelmed by experimental data from others; I am the messenger. So, read on and draw your own conclusions. It may be that "one size does not fit all."

Before getting into experimental evidence, it is important to understand the variables involved. Our focus will be on the first reflections because these are the second most energetic sounds to reach the listener.

7.1.1 Problems with the Stereo Phantom Center Image

In most of the following tests listeners will be auditioning stereo programs. The center image, the featured artist in most cases, is a prominent component of the soundstage. Whatever else we subjectively respond to, this cannot be ignored.

Figure 7.1 illustrates the basic sets of reflections for the loudspeakers in stereo and the front channels in home theaters. It is evident that the sounds arriving at the listener are dependent on the loudspeaker directivity patterns and the acoustical performance of the reflecting points on the room boundaries. In general, for forward-firing loudspeakers, sounds radiated to the rear are mainly low frequencies. For this reason they are not a significant factor, but for some designs (bipoles, dipoles, omnidirectional) they are. Apparent in this illustration is the dramatic simplification in the reflection pattern of a real center channel loudspeaker compared to the "double-mono" phantom center image. They cannot sound the same in terms of timbre or space, not even considering the interaural crosstalk dip at 2 kHz. The importance of geometric symmetry in stereo setups is clear, including a "sweet spot" seat, if one wishes to hear the recording as it was created.

The stereo phantom center will suffer from a significant dip in the spectrum around 2 kHz because the sound from both loudspeakers reaches both ears at different times. This interaural crosstalk cancellation is the first dip in an acoustical interference comb filter as explained in Figure 7.2.

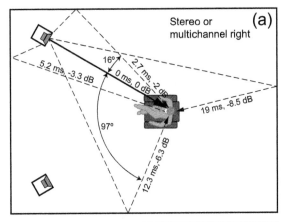

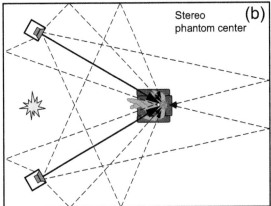

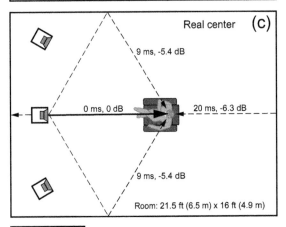

FIGURE 7.1 *First reflection diagrams for stereo and for the front channels in a home theater setup. All of the delays and sound levels shown are relative to the direct sounds. In (c) the rear wall reflection follows the same outgoing path as the direct sound.*

The consequence of this is that the sound quality of the phantom center image (usually the featured artist) in stereo recordings is fundamentally compromised. Those who believe that phase shift is audible and that waveform fidelity is essential must think again, because for this sound image both are seriously corrupted. As an illustration I recommend a simple experiment. Arrange for monophonic pink noise to be delivered at equal level to both loudspeakers. When seated in the symmetrical sweet spot, this should create a well-defined center image floating midway between the loudspeakers. If it does not, something is seriously wrong. If it does, consider what you hear as you lean *very slightly* to the left and to the right of the symmetrical axis: the timbre of the noise changes. It will be more obvious the closer you sit to the loudspeakers placing you in a stronger direct sound field. It will also be more dramatic in acoustically dead rooms. In fact, it is possible to find the exact sweet spot by simply listening to when the sound is dullest. Moving even slightly left or right of the sweet spot causes the sound to get audibly brighter. It is much more exact to find the sweet spot by listening to the timbre change than by trying to judge when the center image is precisely localized in the center position. There is nothing faulty with the equipment or setup; this is stereo as it is: flawed.

In Figure 7.2a a listener receives direct sounds, one per ear, from a center channel loudspeaker. Figure 7.2b is the situation for a phantom center image; there are two sounds per ear, one of which is delayed because of the additional travel distance. Both of these are symmetrical situations, so the sound in both ears is essentially identical. Figure 7.2c shows the frequency responses measured at one ear when sounds were delivered by a center loudspeaker and then by a stereo pair of loudspeakers, set up in an anechoic chamber. Measurement were made using a KEMAR mannequin, an anthropometrically and acoustically precise head and torso with ear canals terminated by microphones and appropriate acoustical impedances. The curves are *very* different. The solid curve is the "correct" one. It shows what a real center sound source delivers to the eardrum of a listener—it has the odd shape it has because of the acoustics of the external auditory system, the head-related

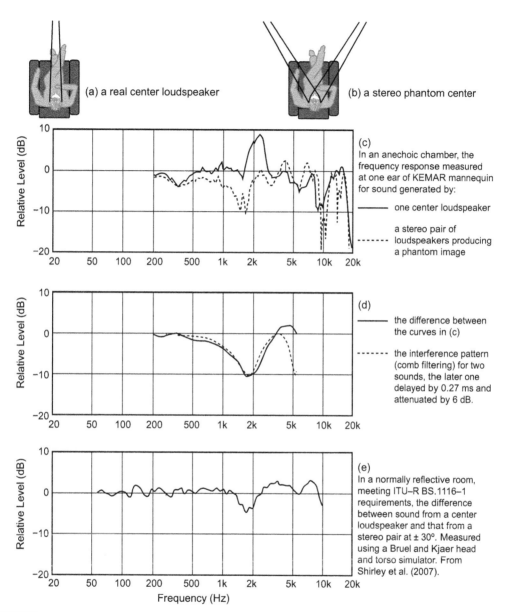

(a) a real center loudspeaker

(b) a stereo phantom center

(c) In an anechoic chamber, the frequency response measured at one ear of KEMAR mannequin for sound generated by:

——— one center loudspeaker

- - - - - - a stereo pair of loudspeakers producing a phantom image

(d)

——— the difference between the curves in (c)

- - - - - - the interference pattern (comb filtering) for two sounds, the later one delayed by 0.27 ms and attenuated by 6 dB.

(e) In a normally reflective room, meeting ITU–R BS.1116–1 requirements, the difference between sound from a center loudspeaker and that from a stereo pair at ± 30°. Measured using a Bruel and Kjaer head and torso simulator. From Shirley et al. (2007).

FIGURE 7.2 *Anechoic frequency-response measurements (c) made at one ear of a KEMAR mannequin for sounds arriving from a real center loudspeaker, shown in (a), and from a stereo pair, shown in (b). The curves contain the axial frequency response of the loudspeaker used in the test, and the head-related transfer functions (HRTFs) for the relevant incident angles for that particular anthropometric mannequin. The important information, therefore, is in the difference between the curves that, around 2 kHz, is substantial. The smoothed difference is shown in (d). Nothing is shown above about 5 kHz because it is difficult to separate the effects of this specific acoustical interference from those of other acoustical effects. The dashed curve is the first interference dip (the first "tooth" of the comb filter) calculated for an interaural delay of 0.27 ms (appropriate for a loudspeaker at 30° left or right of center) and for an attenuation of the delayed sound of 6 dB. (e) is the same kind of measurement done in a normally reflective room, showing that early reflections within the room reduce the depth of the interference dip. Data from Shirley et al. (2007).*

transfer functions (HRTFs). The dashed curve, for a phantom center, includes the effects of acoustical interference caused by the acoustical crosstalk from the two loudspeakers and, the errors in HRTF processing because the incident sounds arrive at the ears from the wrong directions: $\pm30°$ rather than straight ahead. This situation is one in which there is a conflict between the timbre of the sound where it is coming from and that for the location it is perceived to come from. This is not ideal.

Figure 7.2d shows the difference between the curves, revealing the result of acoustical interference. This can be confirmed by a simple calculation. The time differential between the ears for a sound source at 30° away from the frontal axis is about 0.27 ms for an average head. A destructive acoustical interference will occur at the frequency at which this is one-half of a period: 1.85 kHz. It won't be a perfect cancellation, because of a tiny propagation loss, and a significant diffraction effect. The wavelength is just over 7 inches (178 mm) that, because it is similar in dimension to the head, will experience a substantial head-shadowing effect at the ear opposite to the sound source. There will be an interaural amplitude difference of the order of 6 dB in this frequency range. Taking a simplistic view, the dashed curve in Figure 7.2d shows the first cancellation dip (the first "tooth") of a comb filter with these parameters. The fit is quite good. The cause of the "dullness" in the phantom center image is destructive acoustical interference. The rest of the comb is not seen because at higher frequencies the whole situation is muddied by head-related transfer functions and rapidly increasing attenuation of the delayed sound at higher frequencies caused by head shadowing.

By any standards this is a huge spectral distortion, a serious fault because it affects the featured "talent" in most recordings, the person whose picture is on the album cover. Under what circumstances are we likely to hear it as shown here? Obviously, only when the direct sound from the loudspeakers is the dominant sound arriving at the listener's ears. This is the situation in many recording control rooms and custom home theaters, where special care is taken to attenuate early reflections.

In normally reflective rooms, reflections arriving from other directions at different times will help to fill in the spectral hole because there will be no acoustical interference associated with those sounds. Therefore, in normally reflective rooms this will not be as serious as the curve in Figure 7.2d suggests, a fact confirmed by data from Shirley et al. (2007) in the data shown in Figure 7.2e.

It is a clearly audible effect. Plenge (1987) explains it, as done here, and goes on to relate it to some theories of timbre perception, pointing out along the way that most listeners do not sit still, and are obviously not bothered by the consequent changes in sound image location and timbre. However, dedicated audiophiles and recording engineers who gravitate to the stereo seat are exceptions. Augspurger (1990) describes how easy it is to hear the effect using 1/3-octave bands of pink noise, observing a "distinct null at 2 kHz" (p. 177). Pulkki (2001) confirmed that the comb filter was the dominant audible coloration in anechoic listening to amplitude-panned virtual images, but that it was lessened by room reflections. Listeners in experiments by Choisel and Wickelmaier (2007) reported a reduction in brightness when a mono center loudspeaker was replaced by a stereo phantom image (their figure 4). Listeners in the Shirley et al. (2007)

experiments not only heard the dip seen in Figure 7.2e but demonstrated that it had a significant negative effect on speech intelligibility. This is hardly the epitome of sound reproduction technology.

Comb filters are measurable in any room when a direct sound and a reflected version of that sound combine at the microphone. However, there is a great difference between what we measure and what we hear depending on where the two sounds come from. Clark (1983) investigated aspects of this, and Figure 7.3 illustrates one of his experiments.

Starting with a standard stereo arrangement in a normal listening room, frequency response measurements were made in three situations that yielded the same comb-filter interference pattern. Note that the first destructive interference dip moving left from 0 Hz is the 2 kHz dip shown in Figure 7.2. The events described in Figure 7.3 can be explained as follows:

Figure 7.3a The first was a measurement at one ear location, with a microphone 4 inches (100 mm) off the stereo symmetrical axis. In a phantom center image situation, both loudspeakers radiate the same sound, so each ear location receives the direct sound from the nearer loudspeaker and then a slightly delayed version of it from the opposite loudspeaker. In this simple experiment there was no head between the microphone locations, so all of the comb notches are visible. In contrast, Figure 7.2 shows that only the first interference dip is present in the mannequin measurements that include head shadowing. Figure 7.3a shows this situation, with the corresponding waterfall (amplitude vs. frequency vs. time) measurement. The curve at the back (time = 0) is related to the steady-state measurement. The curves moving toward the front show events as a function of time and it is seen that later sound arrivals (reflections) reduce the depth of the first interference dip, as shown in Figure 7.2e. Interestingly, the dominant effects happen in the first 10 ms.

Figure 7.3b The second involved sound from a single loudspeaker that arrived at an ear location directly, and after reflection from a (2 ft × 3 ft, 0.6 m × 1.0 m) panel positioned to produce the same delay in the reflected sound path that occurred in (a). Figure 7.3b.

Figure 7.3c The third situation involved electrically delaying the playback signal and adding it to itself so that the comb filtering took place before the signal was radiated into the room. Figure 7.3c.

The subjective impressions of the three circumstances that generated very similar frequency-response curves (the curve at the back of the waterfall diagrams) were greatly different. According to Clark, listeners found the following:

1. The stereo phantom image: "moderate to pleasing effect."

2. The reflector delay: "very small effect."

3. The electronic delay in the signal path: "greatly degrading effect."

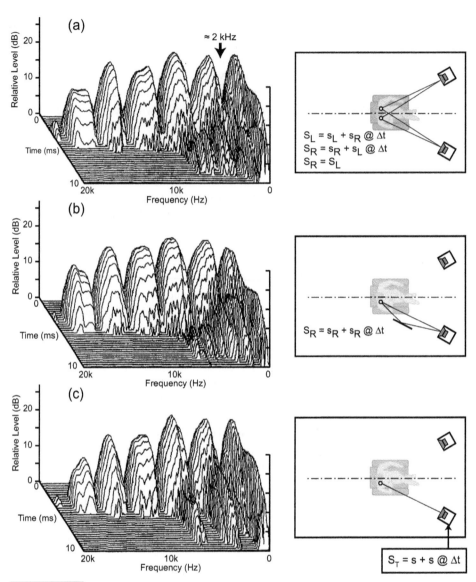

FIGURE 7.3 *(a) shows a waterfall measurement for a microphone located at an ear position (no person was present) when a stereo pair of loudspeakers was radiating the same sound. The comb filter is the result of the direct sounds from both loudspeakers arriving at slightly different times. (b) shows the effect of a single lateral reflection having the same delay as (a). (c) shows the effect of adding the signal to itself, with a delay, before it is radiated into the room. Note that the frequency scale is linear to show the regular comb pattern, and it is reversed to reveal the early reflections at lower frequencies.*
Adapted from Clark (1983).

Returning to Figure 7.3 and observing what happens while the sound is decaying, it can be seen that other early (< 10 ms) room reflections fill in the lower parts of the initial 2 kHz comb notch. The same behavior can be seen in (a) and (b) because both situations allow for room reflections from different directions and times to arrive at the listening position. This does not happen in (c) because the comb filtering occurred in the electrical signal path and the comb notch is in all of the sounds, direct and reflected. This is the worst situation for the audibility of comb filtering, and it occasionally happens in the microphone pickups at live events, and in broadcasts, when a signal accidentally gets routed so that it combines with a delayed version of itself. We hear these occasionally in news broadcasts when the outputs of two microphones (among the sometimes massive collections) are blended. The audible coloration is not subtle and not pleasant.

It is relevant that experiments by other investigators have tended to support these findings. For example, Vickers (2009) experimented with using signal processing to improve the sound quality of the phantom center stereo image, mentioning the Clark experiments as partial inspiration.

It is curious that some people hear enough of a problem with the phantom center stereo image to embark on research to fix it, while others condemn a real center channel loudspeaker as being inferior to a phantom image. It is a situation complicated by inconsistencies in recordings and human adaptation to stereo after decades of exposure. As shown in Figure 10.15a, the radiating pattern, the directivity, of a human voice is very similar to that of a conventional cone and dome loudspeaker. Logically it should do well as a reproducer of a solo voice (Figure 7.1c) compared to the artificially complex sound field associated with a phantom center image (Figure 7.1b). Listening away from the sweet spot eases the phantom image timbre problem, but then the soundstage is distorted from what was intended. A center channel avoids the timbre problem, stabilizes the soundstage for more listeners, and is less sensitive to room effects.

In summary, the phantom center image is spatially and timbrally flawed, even to the point of affecting speech intelligibility, especially if the direct sound is dominant. In the context of the phantom stereo center image, early-reflected sounds appear to be beneficial.

Reflections and reverberation added to the mix will also help. However, the more acoustically "dead" the listening environment and the closer we sit to the loudspeakers, the more dominant will be the direct sound and the more clearly the problem is revealed. This means that the common practice of eliminating reflections in control rooms and the common use of "near-field" loudspeakers, sitting on the meter bridge of a console, both create situations where this problem is more likely to be audible. Taking a positive attitude to the effects of this on recordings, perhaps it will be the motivation to add delayed sounds to the vocal track, filling the spectral hole and in a very tangible manner "sweetening" the mix.

On the other hand, if a recording engineer chooses to "correct" the sound by equalizing the 2 kHz dip, a spectral peak has been added to the recording, which will be audible to anyone sitting away from the stereo sweet spot. Even worse, if the two-channel original is upmixed for multichannel playback, the center channel loudspeaker will not be flattered by the unnatural signal it is supplied with.

All of this should provide reasons to employ a real center channel in recordings, another point made by Augspurger (1990), who notes:

But, no matter what kind of loudspeakers are used in what kind of acoustical space, conventional two-channel stereo cannot produce a center image that sounds the same as that from a discrete center channel, even if it is stable and well defined. This leads to a certain amount of confusion in both the recording and playback processes. A dubbing theater that deals exclusively with [multichannel] motion picture sound does not have to worry about this problem. . . . A dialog track can be panned across the width of the screen without a noticeable change in tonal quality.

One is left to conclude from the absence of outrage from purist audiophiles that we humans have significantly adapted to this situation in loudspeaker stereo. The author, probably because of his listening experiences, has not.

We need multichannel audio. In the present state of affairs, however, the experimental evidence suggests that we look favorably on having some room reflections.

7.2 THE PHYSICAL VARIABLES: LOUDSPEAKER DIRECTIVITY

The next issue is: room reflections from what kind of loudspeaker? Figure 7.4 shows a single reflection from three distinctively different loudspeaker configurations among even more alternatives. The strength of the reflected sound depends on the sound level radiated by the loudspeaker on that particular axis. For this side wall reflection, the omnidirectional loudspeaker will generate the strongest lateral reflection, the conventional forward-firing loudspeaker will produce a moderate reflection, and in this geometry, the dipole will have the least lateral energy arriving at the listening position. Popular bipole configurations can approach omnidirectionality in the low- to mid-frequency range, but drift toward forward/backward-firing designs at high frequencies. The reader

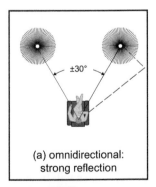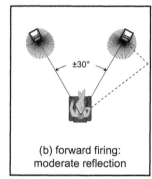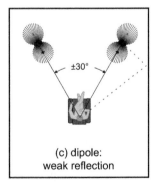

| (a) omnidirectional: strong reflection | (b) forward firing: moderate reflection | (c) dipole: weak reflection |

FIGURE 7.4   *A simplistic look at how some different loudspeaker directivities result in very different room interactions. Here only a single side wall reflection is shown as an example. The same logic can be applied to all of the reflections shown in Figure 7.1.*

can imagine how each of these might fit into the circumstances shown in Figure 7.1. They will be very different. We will be examining subjective comparisons of some of these. Obviously, changing the angle of the loudspeaker relative to the listening position makes a difference to the more directional designs.

7.3 THE PHYSICAL VARIABLES: ACOUSTICAL SURFACE TREATMENTS

The remaining variable in this investigation is the acoustical performance of the surface at the point(s) of reflection. Figure 7.5 illustrates some popular treatments. Reflection,

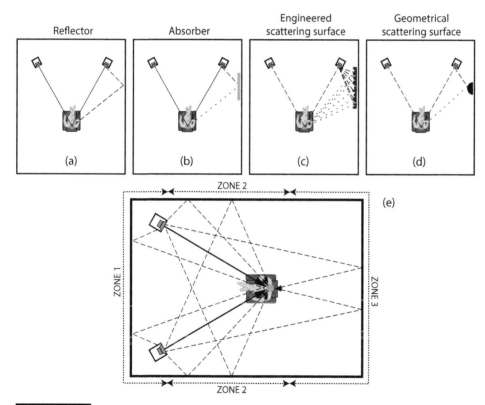

FIGURE 7.5 *shows some popular options for treating the reflection point of a room boundary: (a) reflection, (b) absorption, (c) scattering by engineered surfaces (e.g., Schroeder-type diffusers) and (d) scattering by geometric shapes (e.g., hemi-cylinders). In (c) and (d) only the scattered components that reach the listener are shown; remaining sounds are radiated in many other directions, depending on the design of the scattering device. (e) shows the distinctive zones in a listening room that may require different kinds and amounts of acoustical treatment, or none, depending on the directivity of the loudspeakers radiating the sound, the dimensions of the room, and the expectations of listeners receiving it. The ray pattern shown is only for the first reflections; higher-order reflections will also contribute to the listening experience.*

Figure 7.5a, is straightforward: a hard, massive, flat surface generating a specular reflection attenuated only by the inverse-square law, a function of the distance traveled.

7.3.1 Absorbers

Absorption, Figure 7.5b, is complicated by the fact that the absorption performance of acoustical materials is not specified for specific angles of incidence. For first reflections that matters. Figure 7.6a shows the random-incidence absorption coefficient for 2-inch (50 mm) 6 pcf (100 kg/m³) fiberglass board, as quoted in manufacturer's literature. It shows that this material has essentially no absorption below about 100 Hz and the numbers imply "perfect" absorption above about 500 Hz. This specification is rigorously standardized and requires measurement in a reverberation chamber, with a highly

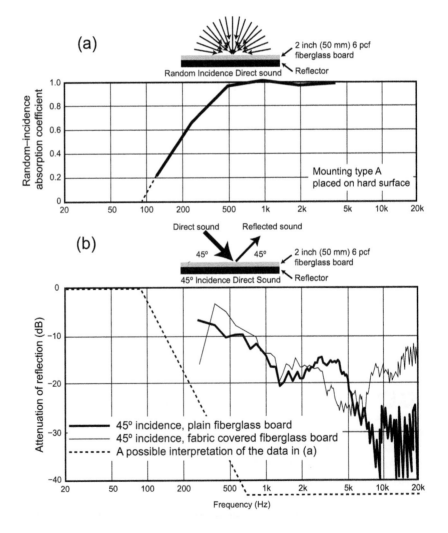

FIGURE 7.6 *(a) shows the published random-incidence absorption coefficient for 2-inch (50 mm) 6 pcf (100 kg/m³) fiberglass board. (b) shows the attenuation of a single reflection from this material. It begins with a simple-minded interpretation of the data in (a): the steeply sloped dashed line that shows no attenuation below 100 Hz, and complete attenuation above about 500 Hz (here the author has decided that 40 dB down qualifies as functionally complete attenuation). Also shown are measurements of what happens with an incident sound arriving at 45° and being reflected by the 2-inch (50 mm) fiberglass board with and without a fabric cover that is widely used in the industry, Guilford of Maine FR701. They are very different.*
Figure (a) data from www. owenscorning.com and Figure (b) data courtesy of Peter D'Antonio.

diffuse—random incidence—sound field. The number is an indication of the relative absorption capabilities of materials, but it is not a percentage. The absorption coefficient often exceeds 1.0, and its measurement is not a precision operation. Its origins lie in the history of controlling reverberation in somewhat diffuse auditoriums for live performances. In the small "dead" spaces used for sound reproduction, from homes to cinemas, there is little diffusion in the sound field. The interest, therefore, is in how the material performs when addressing specific early reflections.

The fabric cover is clearly very acoustically transparent over most of the frequency range, but less so at the highest frequencies. The published random-incidence absorption coefficient for acoustical panels will not show this effect because the measurements typically stop at the 4 kHz octave band, which has a high-frequency limit of 6 kHz, just avoiding the reflected sound shown here. Thicker material will move the curves downward in frequency, thereby absorbing sound over more of the frequency range. Fibrous absorbing panels that are 3 inches (75 mm) or more in thickness should usefully attenuate frequencies down to the transition frequency, below which we enter the room-mode, wave-acoustics domain to be discussed in Chapter 8. Thinner panels will simply roll off the high frequencies, changing but not eliminating the reflection, in effect modifying the off-axis performance of the loudspeaker, thereby possibly degrading it.

There are good reasons to believe that reflected sounds should have spectra that are similar to the direct sound in order for basic auditory processes to function. The precedence effect, which is fundamental to sound image localization is one of them.

In custom listening rooms it is common to cover large wall surfaces with stretched fabric to cover what might be otherwise an unattractive collection of diffusers, absorbers and loudspeakers. This is more than a visual alteration to the room because fabric has flow resistance; if you cannot easily see daylight through a fabric, or blow through it, it will not be acoustically transparent. Such fabrics, with nothing more than empty space behind them, become absorbers, and experience shows that many listening rooms end up being unpleasantly dead because of it. See Figure 7.7 for an example of an ornamental covering becoming a frequency-selective absorber.

7.3.2 Engineered Surfaces and Other Sound Scattering/ Diffusing Devices

Engineered surfaces, with deeply sculptured faces calculated using number theory and other clever concepts (Figure 7.5c), are versatile devices, able to scatter sound in a single plane or multiple planes. Figure 7.7 illustrates the performance of some commercially available diffusers. These devices also absorb some sound, and this is shown in the lower graph. If the installer decides to cover the diffusers with fabric to make them more compatible with the visual décor, the absorption increases substantially and becomes strongly frequency dependent—a bad thing.

These devices are indisputably useful for adding complexity to sound fields in performing-arts venues, but for addressing specific reflections in small sound reproducing spaces they may or may not be appropriate. The issue is that these surfaces are designed so that incoming sounds are scattered in all directions from all parts of

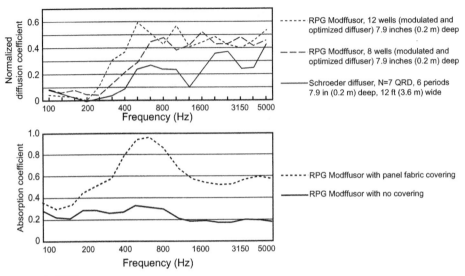

FIGURE 7.7 *A comparison of the normalized diffusion coefficient for a classic Schroeder QRD diffuser with an evolved form, the RPG Modffusor, that has been modulated and optimized for improved performance. It can be seen that the Modffusor improves when used in larger forms. Also shown is the absorption coefficient for the Modffusor with and without a panel fabric covering.*
Data courtesy D'Antonio, 2008, and www.rpginc.com.

the device. Such a surface becomes a distributed source of sound, delivering many "mini-reflections" to the listening position while scattering the rest of the sound to other parts of the room. In rooms for sound reproduction one does not want, nor is it possible to achieve, a diffuse sound field, so when these diffusers are located at points of first reflection, they are functioning as sound reflection attenuators. The improved appearance of ETCs (energy-time curves, see Section 7.7) is often used as proof. Some research indicates that the perceptual sum of several mini-reflections is equivalent to a larger single reflection. This goes back to Cremer and Müller (1982, figure 1.16, shown in Toole (2008) as figure 6.17), reinforced by Angus (1997, 1999) and brought up to date by Robinson et al. (2013). This means that the ETC does not convey reliable information about audibility of reflections—replacing a high spike with a collection of smaller ones may sound less different than is visually implied. It continues to be a topic for more research in the present context. To function down to the transition frequency these devices need to be about 8 inches (200 mm) thick, as seen in Figure 7.7.

In the era during which it was fashionable for the loudspeaker end of a control room (zones 1 and 2 in Figure 7.5) to be acoustically "dead," and the opposite end (zone 3) to be "live," I toured an elaborate recording complex that was nearing completion. I had not experienced one of these new rooms, and was given a demonstration. I bellied up to the center of the console and listened; something was wrong, the center image was strangely unclear at times. Looking around, I saw that the entire back wall (zone 3) was covered with beautiful and expensive wooden diffusers, vertically arranged to spread the sound horizontally. I asked for monophonic pink noise to be played, and from the stereo seat there was simply no center image—the sound spread across much of the space between the loudspeakers. Everybody could hear it; it was not subtle. This was obviously not a good thing. The movers were still on the premises, so we found some moving blankets and hung a couple of them in front of the diffusers. The center image snapped into place where it belonged. The problem: too much uncorrelated sound arriving from behind the listener, scrambling the information in the direct sound from the loudspeakers. Lesson: there can be too much of a good thing. The ETC probably looked just fine, though. Somebody measured but did not listen.

Scattering from **geometric shapes** was the original source of scattered sound in concert venues, and it remains justifiably popular. I can recall a time when some otherwise serious-minded people thought that textured paint made a difference to the sound of a room. If you were a bat, perhaps, but at human audio frequencies this was another audio misconception. Diffusers must manipulate sound waves of significant size, and must

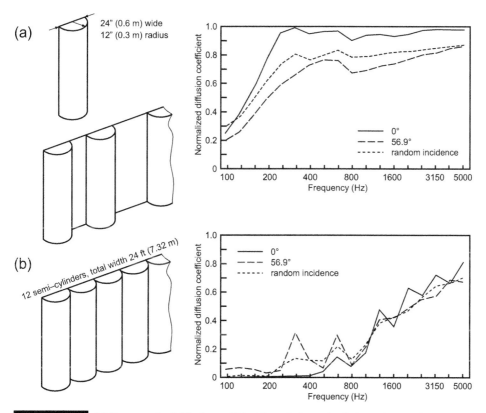

FIGURE 7.8   *(a) The normalized diffusion coefficient for a single semi-cylinder and randomly spaced units. (b) adjacent semi-cylinders covering a surface.*
Data courtesy Peter D'Antonio.

therefore be of comparable size. One of the early popular shapes was the semi-cylinder or, in a slightly flattened form, the polycylindrical absorber/diffuser. Large ones were covered with thin membranes so that they could serve also as absorbers in the upper bass region.

For home listening spaces they are an interesting, and in some custom rooms, a decorative option (Figure 7.5d). Surround loudspeakers can be hidden in them. Of all possible shapes, the hemi-cylinder is desirable for its uniform scattering of incoming sounds from many directions. Other shapes work, including flattened semi-cylinders, but simply not as well (Cox and D'Antonio, 2009). To cover the frequency range down to the transition frequency, it seems that a depth of about 1 ft (0.3 m) is required.

Figure 7.8 shows the normalized diffusion coefficient for a single, and then for multiple semi-cylinders with a 1 ft (0.3 m) radius. The devices work well in isolation and when randomly spaced, and in this thickness it is effective down to a usefully low frequency. However, when placed contiguously the low-frequency diffusion is lost—the geometric regularity is also visible in the cyclical pattern in the curve. As Gilford (1959) noted, other shapes also work well, but sizing and placement are important so that they influence the intended reflections.

In summary at this point, it can be seen that both loudspeaker directivity and acoustical treatments of surfaces can modify the pattern and strength of reflections arriving at the listening position. The next question is: how do we perceive these differences in listening tests?

7.4 SUBJECTIVE EVALUATIONS IN REAL-WORLD SITUATIONS

Whatever the measurements tell us, the final answer must come from blind listening tests, conducted under controlled circumstances and involving several individuals. These are not easy to arrange unless one has access to a laboratory kind of environment. They also consume money and time, which eliminates most enthusiasts. A few rigorous tests are in the literature, and these are reviewed later. In addition there are numerous opinions resulting from less well-controlled tests, by serious-minded enthusiasts, and some of these are discussed. Given human susceptibility and adaptability, it is not possible to judge the merits of these opinions, but as will be seen, many of them are in substantial agreement. Always there is evidence of individual preference, and for this there can be no generalized solution.

7.4.1 Side Wall Treatment: Reflecting or Absorbing— Kishinaga et al. (1979)

Figure 7.9 shows one of the experimental configurations used by Kishinaga et al. (1979) in an elaborate exploration of acoustical treatment options for listening rooms. The rooms were outfitted with reversible panels, one side plywood, the other covered with 25 mm (1-inch) fiberglass along both side walls. Several variations were explored, including some with absorbers behind the loudspeakers. Unfortunately there was no information about the loudspeakers used. As general conclusions the authors state:

1. For the purpose of listening to audio products, absorptive walls are desirable.

2. To enjoy the music, however, reflective walls yield a better effect.

The authors were employed by Nippon Gakki (Yamaha), so "listening to audio products" presumably meant evaluating many forms of electronics and loudspeakers. Adding absorbers behind the loudspeakers appeared to be advantageous for product evaluation, but not for being entertained. The listeners were all "acoustical engineers" who formed their opinions listening to six varied musical selections, and voice.

Included among the detailed technical measurements documenting the tests were interaural cross correlation coefficients (ICCC) obtained using a dummy head microphone. ICCC (also IACC) is known to be a good measure of spatial perceptions, lower values being associated with higher subjective ratings of apparent source width (ASW) and envelopment. Obviously, allowing stronger side wall reflections substantially reduced the ICCC. The adjacent side wall reflection arrived at about 4 ms, and the opposite side wall reflection at 8 ms after the direct sound. Floor and ceiling reflections were unchanged. It seems that the engineers in these tests were pleased by the enhanced spatial cues provided by early side wall reflections when not on duty evaluating products.

A useful perspective on ICCCs can be found in Tohyama and Suzuki (1989), discussed in Section 15.4, illustrated in Figure 15.3. The conclusion was that stereo reproduced in a reflection-free environment is not capable of replicating the spatial impression of a concert hall. However, the Kishinaga et al. results indicate that in a normal room with side wall reflections, the situation is improved.

On a different topic, among the measurements displayed by Kishinaga et al. were time-windowed measurements of the spectra reflected from the side walls. As discussed in Section 7.3.1, the sound absorption of fiberglass or acoustical foam panels varies with both the thickness and the angle of the incident sound. Here we are interested in a small number of discrete early reflections. The author derived the attenuation provided by the 25 mm thick fiberglass absorbing surface by noting the difference between the "reflective" and "absorptive" curves in figure 13 of Kishinaga et al. (1979). The top curve in Figure 7.10 is that smoothed data. For comparison are shown data from Figure 7.6b for 50 mm thick fiberglass board with and without fabric covering (from independent tests).

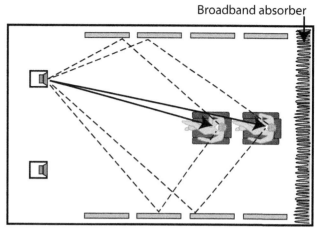

Broadband absorber

Absorbing panels (25 mm fiberglass): ICCC = 0.44
Reflecting plywood panels: ICCC = 0.28

FIGURE 7.9 *One configuration of the listening rooms used in the Kishinaga et al. (1979) tests. Each of the panels on the side walls was reversible—reflecting on one side and absorbing on the other. In other experiments absorbers were added behind the loudspeakers.*

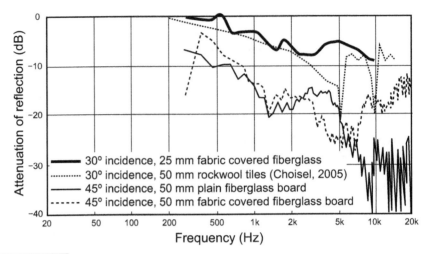

FIGURE 7.10 *The attenuation of a single reflection from flat fiberglass panels. The top curve, for 25 mm thick panels is derived from data in Figure 13 of Kishinaga et al. (1979) at the estimated angle of incidence for the adjacent side wall. The dotted curve pertains to the Choisel experiments described in Section 7.4.3. The 50 mm panel data at the bottom are from Figure 7.6b.*

For comparison, the random-incidence absorption coefficient for 25 mm material is about 1.0 for frequencies above about 1 kHz, and that for 50 mm material is about 1.0 for frequencies above about 500 Hz. Obviously expectations of absorption effectiveness based on these data are wildly optimistic in small listening rooms. It is sufficient to note at this stage that conventional 25 mm material does not "eliminate" a side wall reflection, but merely distorts the spectral balance in the reflected sound field. This may not be beneficial in view of the frequency distribution of audible effects (Figure 6.3). More than 50 mm thickness is required to usefully attenuate reflections above the transition frequency in typical listening rooms (about 300 Hz).

7.4.2 The Effect of Loudspeaker Directivity—Toole (1985)
The author conducted a series of experiments to explore the notion that loudspeaker directivity, and the variations in lateral room reflections that followed from this, were factors in listener opinions (Toole, 1985, 1986).

Figure 7.11 shows the room layout for stereo comparisons of two cone and dome forward-firing loudspeakers and a full-range electrostatic dipole loudspeaker—see Figure 7.4a and c. The room was the prototype IEC 60268–13 (1998) recommended listening room. The program was muted while the turntables were rotated into position. The photograph of the room shows the acoustically transparent screens, with reference markings to help listeners compare image locations and soundstage dimensions, and the two chairs, with the rear one higher than the front one to reduce shadowing effects.

■ The Rega Model 3 was a two-way design, 8-inch (200 mm) woofer and 1-inch (25 mm) tweeter. It shows the expected increasing directivity of the woofer up to

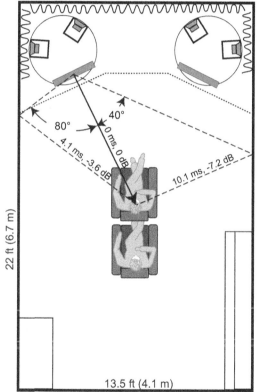

FIGURE 7.11 *The physical arrangement for the listening tests, showing the turntables used to rotate the three loudspeaker pairs into identical positions for listening. An acoustically transparent screen prevented listeners from seeing the loudspeakers. The geometry was such that the adjacent side wall reflections occur at very large angles off-axis (80° and 68° for front- and rear-row listeners respectively), and those from the opposite wall are traceable to moderate angles (40° and 31°). Medium-weight drapes covered the walls behind the loudspeakers. This would have the greatest influence on the bidirectional dipole loudspeaker, the Quad ESL 63, but the product was already equipped with absorbing pads in the rear half of the enclosure to attenuate the output above about 500 Hz. The curtains would further attenuate the rear radiation, and absorb mid-high-frequency sound escaping to the rear from the forward firing designs. The side walls between the loudspeakers and the listeners were hard, flat broadband reflectors. The tests were done in stereo and mono, the latter using only the left loudspeaker.*

the crossover to the tweeter around 3 kHz. This uneven off-axis performance was audible in normal rooms.

■ The KEF 105.2 was a three-way design: 12-inch (300 mm) woofer, 5-inch (110 mm) midrange, and a 2-inch (50 mm) tweeter. This is the loudspeaker used for Figure 5.4 that illustrates the consequences of the uneven off-axis performance. They are also seen here.

■ The Quad ESL 63 was a full-range electrostatic modified dipole, employing a diaphragm subdivided into areas driven in a manner to approximate a spherically expanding wavefront. The center circle, the "tweeter," was about 3 inches (76 mm) in diameter.

The right half of Figure 7.12 shows the author's crude estimate of the sounds arriving at the front listener's ears. Unfortunately, 30 years after the fact, there is no data on the sounds radiated at the specific off-axis angles shown in Figure 7.11, so the 60°–75° data were used as a substitute for the required 80° off-axis adjacent-wall curve, and the 30°–45° data were used to represent the 40° opposite-wall curve. The curves for the specific angles, 40° and 80°, would have been lower than those shown but the comparative levels are probably realistic. These curves were lowered by the amount required

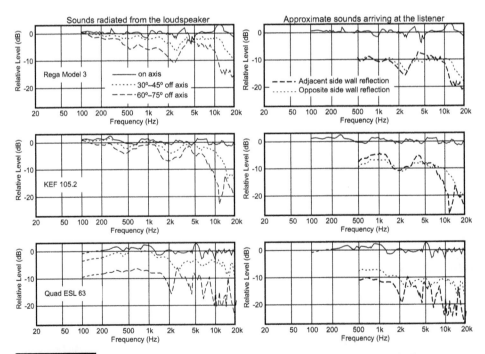

FIGURE 7.12 *The left half of his figure shows anechoic chamber measurements on the three loudspeakers (from Toole (1986), figure 24) with the on-axis, 30° to 45° off-axis, and 60° to 75° off-axis spatially averaged frequency responses shown.*

by the inverse-square law, 3.6 dB for the adjacent-wall reflection and 7.2 dB for the opposite-wall reflection—both compared to the direct-sound arrival, the on-axis curve. The curves are truncated below 500 Hz because at lower frequencies room modes and adjacent-boundary effects progressively dominate.

In Figure 7.12 the higher broadband directivity of the ESL 63 results in lower level first-order lateral reflections than either of the cone and dome designs. Both cone and dome systems exhibit fairly unattractive off-axis performance compared to the electrostatic system. It is worth noting that in this listening arrangement, the combination of loudspeaker directivity and aiming has significantly reduced the amplitude of the reflected sounds, with no acoustical wall treatment.

Ten listeners, audiophiles, audio journalists and recording professionals, participated in the evaluations. These people were selected because they were knowledgeable, experienced and opinionated about audio—prerequisites for such a demanding test. They listened alone or in pairs, being cautioned to avoid verbal or non-verbal communication during the tests and discussion between the randomized tests, which went on for several days. Generous rest periods were built in to the schedule. While listening, they completed a questionnaire interrogating their views on numerous aspects of sound quality and spatial qualities, and at the end requiring a "Gestalt" opinion, a number on a scale of 10 that reflects their overall opinion of fidelity (sound quality), pleasantness, and spatial quality. The music consisted of excerpts of choral, chamber, jazz and popular programs recorded with special care by tonmeister/recording engineering students and staff at McGill University. All were one generation from master-tape recordings done especially for this purpose, employing known microphones and documented processing, which was minimal. The questionnaire is shown in Figure 7.13.

In the results shown in Figure 7.14, the first surprise was that single loudspeakers elicited strong opinions about spatial quality. Those of us who had participated in many single-loudspeaker comparisons were aware that there were differences in the perception of the spatial extent and distance of the single sound source. To us, the most neutral sounding loudspeakers tended to not draw attention to themselves; they almost disappeared, leaving the sense of distance to be conveyed by the recordings. The least neutral loudspeakers were localizable sources; they "sounded like" loudspeakers. In stereo this was especially noticed in recordings with hard-panned left and right images, which are monophonic signals. However, it was still a surprise that this was an impression shared by other listeners not accustomed to this form of critical analysis.

In these results, spatial quality and sound quality ratings were obviously not completely independent—one followed the trends of the other. Is it possible that listeners cannot separate them even though, consciously, most were confident that they could (the author included)?

In monophonic tests, listeners reported large differences in both sound quality and spatial quality. However, in stereo listening most of the differences disappeared in these data that average ratings for all programs. The two highly rated loudspeakers kept their high sound quality ratings, but the loudspeaker with low spatial ratings in mono became

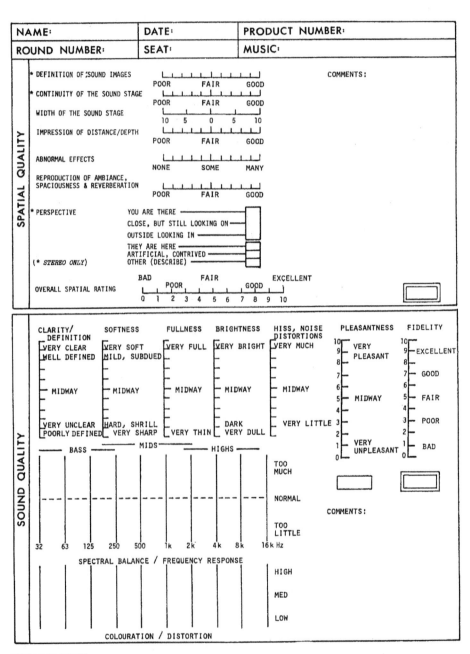

FIGURE 7.13   *The listener questionnaire used in the tests. During the presentations, listeners responded to each of the "perceptual dimensions" and then, at the end, arrived at single number overall ratings of spatial quality, pleasantness and fidelity (sound quality). Comments were encouraged. Loudness levels were carefully matched, and the music selection and product presentation sequences were randomized. The experiment was done in stereo and also in monophonic form, using only the left loudspeaker. Half of the listeners started with the monophonic test and half with the stereo test. Loudspeakers were identified only as "product number."*
Toole (1985), figure 2.

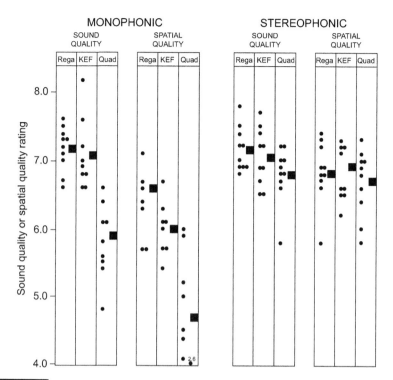

FIGURE 7.14 *Overall sound quality and overall spatial quality ratings for the three loudspeakers when auditioned in mono and in stereo. Each dot is the average of several ratings by a single listener. The ratings are averaged across the four music selections: choral, chamber, jazz and popular. On the vertical scale, 10 represented the best imaginable sound and 0 the worst. The large square symbols indicate the average of all responses for individual loudspeakers.*
Toole (1985), figure 20.

competitive in stereo. This was a puzzle, because it had been assumed that it was stereo that would reveal the relative merits in terms of imaging and space.

Earlier it was speculated, based on current understanding of loudspeaker measurements, that the Quad might have had an advantage in terms of sound quality, or at the very least, not a disadvantage. That speculation was not borne out. For the same reason, the notion that early lateral reflections in the listening room would be audible problems was not reinforced. The reflections in the arrangement in Figure 7.11 arrived at 4.1 ms and 10.1 ms, within the range thought to be problematic, however, as shown in Figure 7.12, due to loudspeaker directivity, the reflections were attenuated by roughly 10 dB when they reached the listener. These were flat, plaster-on-block side walls—perfect specular reflectors. Yet they appeared to contribute something to the listening experience, and it was not all negative. Comb filtering is often thought to be involved in all delayed sounds. However, a delayed sound arriving from a different horizontal angle than the direct sound is perceived as spaciousness, not comb filtering. It adds to the information about the room. It may also do more.

Figure 7.2 shows that for the phantom center image stereo introduces a problem, not an advantage. The massive 2 kHz dip in the direct-sound field is alleviated by room reflected sound: Figure 7.2e, improving not only sound quality, but speech intelligibility. This is a parallel with the seat-dip effect in concert halls, where reflected sounds reduce the audibility of the measured effect.

It is difficult to conceive of a mechanism that would cause the fundamental sound quality character of the loudspeakers to improve when they were used in stereo pairs. The implication, therefore, is that spatial factors were strongly influential, if not the deciding factors, in both tests.

Before addressing that question further, let us examine the raw data. Figure 7.15 shows histograms of judgments in the various categories of spatial quality listed in the questionnaire, Figure 7.13. The spread of these ratings indicates either strong personal differences among listeners or—and this is the more likely reason—opinions that change from program to program. That said, the ratings show a lot of similarity among the loudspeakers; it is difficult to see a clear winner or loser. In the category "abnormal

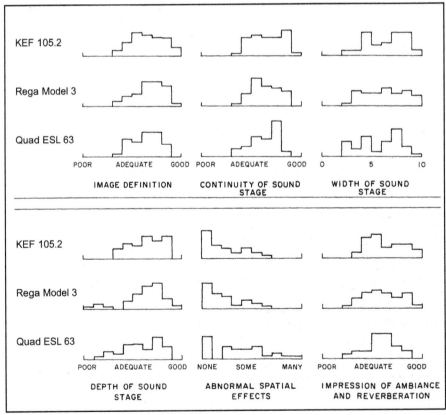

FIGURE 7.15 *Distributions of the analytical judgments in listener responses for all music selections combined.*
Adapted from Toole (1985), figure 22.

spatial effects," it was interesting that the Quad was criticized for creating the impression of sounds being inappropriately close to the listener, and occasionally inside the head—see the box. These are perceptions normally associated with a dominant direct sound field, that is, insufficient reflections to establish a sense of distance. Moulton (1995) noted that "speakers with narrow high-frequency dispersion . . . tend to project the phantom at or in front of the lateral speaker plane."

In the overall description of listening perspective, for the music recorded with a natural perspective (choral, chamber and jazz), the modal listener response was "you are there" for the Rega and the KEF and "close, but still looking on" for the Quad. According to the definitions of those phrases, the Rega and the KEF gave listeners some impression of being enveloped by the ambient sound of the recording environment, with the Quad tending to separate them from the performance (Toole, 1986, p. 342). This all sounds very much like the influence of lateral reflections, ASW, early spatial impression and the associated variations in IACC.

Because these data, and the ratings of Figure 7.14, combined the ratings for all four musical selections, perhaps something more can be learned by separately examining the ratings for each of the musical selections (and the recording techniques embedded therein). See Figure 7.16.

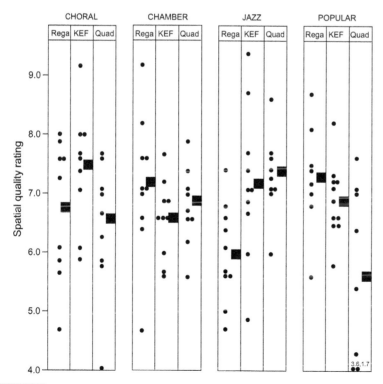

FIGURE 7.16 *Overall spatial ratings of loudspeakers in stereo, for each of the musical selections.*
From Toole (1985), figure 23.

IN-HEAD LOCALIZATION

In-head localization seems like the logical opposite of an enveloping, external and spacious, auditory illusion. Perceptions of sounds originating inside the head, which routinely occur in headphone listening, can also occur in loudspeaker listening when the direct sound is not supported by the right amount and kind of reflected sound. The author and his colleagues have experienced the phenomenon many times when listening to stereo recordings in an anechoic chamber, usually with acoustically "dry" sounds hard-panned to center or, less often, to the sides. It prompted an investigation, Toole (1970), the conclusion of which was that there is a continuum of localization experience from external at a distance through to totally within the head. It is often noted with higher frequencies, and it can happen in a normal room with loudspeakers having high directivity, or in any situation where a strong direct sound is heard without appropriate reflections. In an anechoic chamber it can occur when listening to a single loudspeaker, especially on the frontal axis, in which case front-back reversals are also frequent occurrences. This phenomenon is so strong that it need not be a "blind" situation. I have experienced looking at a loudspeaker in front of me, knowing that it was the only loudspeaker in the anechoic chamber, and hearing the sound in my head or behind me. It is very disconcerting. Interestingly, a demonstration of four-loudspeaker Ambisonic recordings played in an anechoic chamber yielded an auditory impression that was almost totally within the head. This was a great disappointment to the gathered enthusiasts, all of whom anticipated an approximation of perfection. It suggested that, psychoacoustically, something fundamentally important was not being captured, or communicated to the ears. An identical setup in a normally reflective room sounded far more realistic, even though the room reflections were a substantial corruption of the encoded sounds arriving at the ears.

Figure 7.16 shows the spatial quality ratings for each of the music selections. *They are all different.* The choral selection used multiple microphones and the chamber piece employed a Blumlein coincident pair. The listening room reflections are likely to be heavily diluted or masked by the spatial information incorporated into the stereo classical recordings. Nevertheless, there are hints of preferential biases in the distributions, but they are different for the two recordings. The jazz selection revealed lower ratings for the Rega. No explanation for this was found, but there is clearly an interaction with aspects of this specific recording.

In distinct contrast, the pop music selection put the Quad in a position of disfavor. In fact, the subjective ratings in this stereo test are remarkably similar to those seen in monophonic listening (Figure 7.14). Why? Of all the recordings, the pop recording was the only one to have significant amounts of hard-panned, that is, monophonic, sound emerging from the left and right loudspeakers and a hard-panned "double-mono" phantom center image. It is conceivable that listeners reacted to the relative lack of reflected-sound (spatial) accompaniment for these monophonic components of the stereo soundstage. In a stereo mix incorporating substantial reflected sound, as in these classical recordings, listeners would hear more low-correlation sounds (low ICCC) placing the musical instruments in a spatial context. However, close-miked sounds hard panned to left and right, as in many pop/rock and jazz recordings, would be heard from "naked" loudspeakers, with only the listening room to provide a spatial context.

Obviously, recording and mixing methods matter greatly to what is heard. Moulton et al. (1986) had similar experiences.

The principal conclusion is that recording technique can be the prime determinant of spatial impressions perceived in sound reproduction. The directivity of the loudspeakers is a factor, as is the reflectivity of the surfaces involved in the first lateral reflections, especially in recordings incorporating left or right hard-panned sounds. If simplistic "mono" hard-pans had been ameliorated by the addition of spatial cues in the recordings themselves, perhaps the Quad would have received higher ratings. It is interesting to contemplate if the tendency for recording control rooms to be acoustically dead, reflection free, motivates the engineers to incorporate left-right delayed or uncorrelated components of the center image into the mixes. It would be an advantage for all, whatever loudspeaker/room combination is used.

Summary and observations. Sound and spatial qualities both contribute to our musical pleasure, but to what extent? Here they are shown to be of comparable importance. In fact, sound quality and spatial quality ratings had a strong tendency to track each other, suggesting that listeners could not completely separate them. In these tests a loudspeaker with narrower dispersion, but with more uniform output off-axis, was given lower ratings than two loudspeakers with wider dispersion, but uneven output off-axis, suggesting that some amount of laterally reflected energy is desirable, even if it is spectrally distorted. Would loudspeakers with wider, more uniform, dispersion have done even better? This was especially true in monophonic sources whether they were hard-panned (isolated) left- or right-channel sounds in pop/rock stereo recordings or loudspeakers auditioned monophonically.

In movies, the center channel is a monophonic source carrying almost all the dialog and all on-screen sounds much of the time. It and the front left and right channel loudspeakers must therefore perform well in isolation. In "middle-of-the-band" multichannel music recordings, all loudspeakers must do well as "solo" performers. Wide dispersion is of no value if the reflections cannot be heard, so this implies that some amount of room reflectivity should be present. The questions are: how much and where?

In stereo listening, all of the variables in this test were influential, but the nature of the recordings themselves proved to be the overriding factor. The essential ingredients of "imaging" are in the recordings. Early lateral room reflections, to the extent to which they were manipulated by loudspeaker directivity, seemed either be neutral-to-beneficial factors in pop/rock style recordings and in monophonic listening. In classical recordings with substantial uncorrelated left-right information, the effects are not clear.

The acoustical crosstalk associated with the phantom center image is a factor, as discussed in Section 7.1.1. The coloration of the featured artist caused by the 2 kHz dip cannot be ignored in a situation where the direct sound is strong, as it would be for very directional loudspeakers or in rooms lacking early reflections. Again this phenomenon will be most evident in recordings with a "hard" phantom center image—typically pop/rock and jazz. Early reflections from different directions tend to fill in the interference dip, making the spectrum more neutral (Figure 7.2e). This is a spectral reason to prefer wide dispersion loudspeakers and to encourage reflections. Even with room reflections to moderate the interference dip, speech intelligibility is degraded (Shirley et al., 2007) and there are significant effects on instrumental and vocal timbre. This provides a motivation for recording engineers to incorporate some real or simulated lateral reflections of the center image into their mixes. The "sweetening" is both measurable and audible.

Finally, evaluate loudspeakers in monophonic comparisons (to find out what you really have), and demonstrate them in stereo or multichannel (to impress everybody). Choose the recordings carefully—they are a significant factor. Subsequent stereo vs. mono tests in the intervening 30+ years have not changed these conclusions.

Evans et al. (2009) reviewed several studies examining the effects of loudspeaker directivity, concluding that there were several fundamental questions remaining to be properly investigated. They did note that the author's work described here "appears to be the most relevant to date, with regard to loudspeaker directivity effects."

7.4.3 Loudspeaker Directivity and Wall Treatment Together—Choisel (2005)

This study examined stereo image localization using two very different loudspeakers, one (B&O Beolab 5) with unusually wide, uniform frontal dispersion, and the other (B&W 801N) with less than ideal forward-firing directivity (see Figure 5.11b for the similar 802N). The first side wall reflections were either perfectly reflected, or attenuated by "Rockfon absorbing tiles (120x60x5 cm)." Differences in loudspeaker directivity and the way they were aimed yielded reflection-angle sounds being 2 to 3.5 dB lower than the direct sound for the Beolab 5, and 5 to 14 dB lower for the B&W 801N. The test was to judge the influence of first lateral reflected sound on the precision of stereo image localization. Stereo images were positioned by interchannel amplitude or delay panning. Listeners indicated the perceived location with a laser pointer on the acoustically transparent screen that hid the loudspeakers. Stimuli were 1/3-octave bands of noise at 1, 2, 4 and 8 kHz, rhythmic hand-claps (anechoic) and female voice (anechoic). One would expect these signals to elicit well-defined images.

Differences were observed between amplitude- and delay-panned image locations, but "it can be concluded that the direction of panned sources was not affected by the loudspeaker condition (Nautilus 801, Beolab 5 with absorbers on the side walls, or Beolab 5 with reflectors) with the selected stimuli." In other words, large differences in the magnitude of first laterally reflected sound arriving 9.5 ms after the direct sound was found not to have a significant effect on image localization across the stereo soundstage. Bear in mind that this was the *only* perceptual dimension being interrogated. Other spatial perceptions were not reported on. However, a significant fact is that the test room was about 24 ft (7.3 m) wide, with the loudspeakers placed 7 ft (2.15 m) from the side walls (and 9.8 ft (3 m) apart). The first reflection from the adjacent side wall arrived 9.5 ms after the direct sound and that from the opposite wall, about 23 ms (my scaling from a drawing that seems not to be exactly to scale). This is a large room, and earlier-arriving, louder reflections in a smaller venue would have provided more persuasive proof.

The measured data included spectra of the time-windowed reflections showing the reflected sound and the absorbed sound for the Beolab 5 that arrived at the listening location. The difference between these curves is shown in Figure 7.10. Even though the material is stated to be 50 mm deep, the attenuation is not greatly different from the 25 mm material in the Kommamura experiments, and much lower than that from the D'Antonio data. These attenuation curves continue to confirm the fact that single reflections from normal acoustical materials are treated very differently from expectations based on random-incidence absorption coefficients. Reflections are spectrally modified, not eliminated.

7.4.4 The Nature of the Sound Field—Klippel (1990)

A serious examination of listener reactions to complex sound fields in stereo reproduction was undertaken by Klippel (1990a, 1990b). The investigation attempted to relate listener descriptions of what they heard, the perceptual dimensions, to measured quantities. He summarized the process as follows: "All dimensions perceived in the performed listening tests correspond with features extracted from the sound pressure response of the diffuse and direct field at the listener's position. There was no hint of a relation to phase response or to nonlinear distortion." According to this the relevant loudspeaker measurements, therefore, are the anechoic on-axis frequency response and the sound power response—or at least a sufficient collection of off-axis measurements to describe the reflected sounds arriving at the listening location.

Of special interest was his finding that what he called "feeling of space" figured prominently in listener responses. In evaluating listener perceptions Klippel assessed the listener responses against what was judged to be the ideal quantity of each perceptual parameter. Thus, the perceived quality is evaluated as a defect:

$$\text{Defect} = |\text{basic measure} - \text{ideal value}|$$

A defect can therefore indicate that there is too much or too little of a perceived dimension; listeners responded according to what seemed to them to be appropriate. The responses were solicited for two broad categories, "naturalness" and "pleasantness," one relating to realism and accuracy, and the other to general satisfaction or preference, without regard to realism.

When he analyzed the factors that contributed to the perception of naturalness, it was found that:

- 30% was related to inappropriate discoloration (sound quality);

- 20% was related to inappropriate brightness, which was explained as a 70% excess of treble and 30% lack of low frequencies (sound quality);

- 50% to defects in the "feeling of space."

For the second general measure of quality, "pleasantness," the factor weighting was:

- 30% inappropriate discoloration (sound quality);

- 70% defects in the "feeling of space."

Thus sensations of sound quality and spaciousness contribute equally to impressions of "naturalness," and spaciousness dominated the impression of "pleasantness." Therefore, whether one is a picky purist or a relaxed recreational listener, the impression of space is a significant factor.

Perceptions of space are substantially related to laterally reflected sounds. In small rooms, it is improbable for natural reflections to initiate impressions of envelopment; those cues must be in the recordings. However, sensations of ASW (apparent source width), image broadening and early spatial impression are very real and, it seems, desirable. Klippel chose as his objective measure of "feeling of space" (R) the difference between the sound levels of the multidirectional reflected sounds and the direct sound at the listening location.

Figure 7.17 shows that there is an optimum amount of reflected sound; there can be too much and too little, depending on the nature of the program. This is in line with the evidence in my own research, reported earlier, that the recording itself is a major factor. The smallest amount is required to provide a satisfying setting for speech, more is required for music, and music has many varieties. The optimum difference between the direct and reflected sound fields is about 3 dB for speech, 4 dB for a mixed program and 5 dB for music. There is no frequency dependence considered in these numbers, and we know that most loudspeakers do not exhibit constant directional behavior at all frequencies.

A good loudspeaker for this purpose would therefore be one that has two qualities: wide dispersion, thereby promoting some amount of reflected sound, and a relatively constant directivity index, so that the direct sound and reflected sounds have similar spectra. The essence of good design in this respect would be to deliver the optimum proportion of reflected sound for the program being auditioned. An associated requirement of considerable importance is that at least some of the off-axis sounds be allowed to reflect, and any that are acoustically modified by absorption or diffusion are treated in a spectrally neutral manner. In other words, any reflected or scattered sounds should convey the spectral balance of the incident sounds, only at a lower sound level.

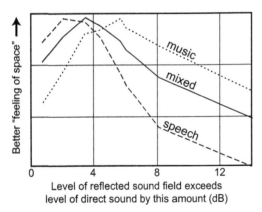

FIGURE 7.17 *The relationship between the subjective evaluation of "feeling of space" on the vertical axis, and the objective measure R (the estimated difference in sound level between the direct and reflected sounds at the listening position) on the horizontal axis. Moving up the vertical scale corresponds to increased subjective satisfaction; moving to the right corresponds to increased reflected sound in proportion to the direct sound. The maximum point of each curve describes the optimum level of diffuse sound compared to the direct sound, for each of the three programs. This is an adaptation of figure 5 in Klippel (1990a).*

7.4.5 Observations of an Audio Enthusiast—Linkwitz (2007)

This is a sample of work by a long-term friend, Siegfried Linkwitz, of Linkwitz-Riley crossover fame. He is a forthright stereo enthusiast, and a competent engineer capable of measuring both loudspeakers and in-room data. He has attempted to improve the stereo listening experience by personal experimentation over many years. This regards a test reported in Linkwitz (2007) wherein loudspeakers having the directivities shown in Figure 7.4b, an approximately omnidirectional loudspeaker, are contrasted with a cone and dome dipole system, as shown in Figure 7.4c. Both were designed by Linkwitz, and were equalized to have flat on-axis frequency responses. As expected, acoustical measurements showed significant differences in the reflected sound fields generated by the two loudspeakers at the listening location. The listening room had an RT of about 500 ms (normally regarded as the upper limit for a listening space), both side walls were hard, reflective glass, and the stereo listening setup was precisely symmetrical between them.

The subjective comments were not the result of double-blind tests, but they appeared to lack obvious bias. It was noted that "in terms of timbre or perceived frequency response, both loudspeakers sound almost identical on program material. The main differences are in their spatial presentation" (Linkwitz, 2007, p. 4). He goes on to speculate that "The different room reflection patterns due to the different polar responses become perceptually fused with the direct sound of the loudspeakers, if the reflections are sufficiently delayed and if their spectral content is coherent with the direct sound" (Linkwitz, 2007, p. 15). The delays for the adjacent side wall reflection from the dipole speaker was 5 ms, and 7.3 ms for the omnidirectional loudspeaker; both for the forward listening location. These delays are similar to those in the earlier experiments discussed, and are not uncommon in typical homes. The requirement for loudspeaker design, therefore, seems to focus on flat on-axis frequency response and constant directivity, so that direct and reflected sounds have similar spectra—assuming that the surface of the room boundary does not change the reflected spectrum.

His comment about the reflected sounds being "perceptually fused with the direct sound" is consistent with the classic precedence effect, which works best when the direct and delayed sounds have similar spectra (Section 7.6.4), and also of the now recognized ability of listeners to adapt to the acoustics of a listening environment. We can significantly separate the room from live sound sources and loudspeakers within them when given time to adapt (Section 7.6.2).

Linkwitz provides a good summary of his thinking:

The type and performance of loudspeakers used, their setup, the listening distance and the room's acoustic properties determine the quality of the rendered aural scene at the consumer's end of the acoustical chain. The aural scene is ultimately limited by the recording.

The type and configuration of microphones used, their setup, the venue, the monitor loudspeakers for the mix, their setup, the listening distance, the studio's acoustic properties and the mix determine the quality of the recording at the producer's end of the acoustical chain between performer and consumer.

Monitor and consumer loudspeakers must exhibit close to constant directivity at all frequencies in order to obtain optimal results for recording and rendering.

(www.linkwitzlab.com, "Stereo Recording and Rendering 101")

I would change this quote only to say at the end of the first paragraph: the aural scene is ultimately *determined* by the recordings, and they differ widely. The contributions of the loudspeakers and room are secondary and they are usually fixed, so a combined good result is a hit and miss affair. One is left therefore to optimize a home listening situation that best suits the kind of program most listened to, or to build into it some acoustical flexibility.

7.4.6 Observations of an Audio Enthusiast—Toole (2016)

I have spent my professional life as a research scientist, and as such I try to ensure that what I write about is based on documented, accurate measurements and observations of subjective reactions in double-blind circumstances. Those are the basic tenets of scientific investigation and reporting. It all started because I was an audiophile, and I still am. People are naturally curious to know what my personal opinions are. Having had the opportunity to technically and subjectively evaluate a very large number of loudspeakers over four decades has influenced those opinions. So, it will be no surprise that the loudspeakers I favor have been those that acquitted themselves well in those evaluations. The following is therefore shameless, biased, personal opinion.

So, what is it that I like to listen to, and what kind of room do I listen in? The answer is that there is no simple answer. I have very eclectic tastes in music, having enjoyed folk, pop/rock, jazz and classical through my life. I go to about a dozen L.A. Philharmonic concerts a year in Disney Hall to remind me of what real sound and real envelopment is like, and a few small-venue popular and jazz events. I have always set a goal of having good sound in my home. Both my wife and I have an affinity for art; paintings and sculpture have been important in our lives, and the house is a backdrop for that art. Traditional acoustical treatments would compromise that goal, so one of my challenges has been to identify what can make loudspeakers more tolerant of different, non-optimum, acoustical environments. After all, voices and musical instruments do not need equalization for different performance venues. Another goal has been to find ways to deliver good bass in rooms that want to "boom," without having to fill them with bass traps. So, in a very real sense my personal needs and desires have motivated much of my research. And, gratifyingly, our research has rewarded me with superb sound in what, to the eyes, seem like ordinary rooms.

In 1975 we built a custom home on 5 scenic acres on the Rideau River outside of Ottawa, Canada. In it were two rooms for listening; one was a family/entertainment room outfitted with forward-firing loudspeakers. In 1988 it became my first-generation multichannel room using a Shure HTS upmixer for Dolby Surround encoded movie sound tracks. Later a Lexicon CP-1 added 7-channel flexibility for upmixing movies and adding tasteful ambiance to music. Appearances mattered, so there were some DIY custom cabinets. I had the skills and wood shop to make it happen. Figure 7.18 shows the diagonal arrangement that allowed for panoramic views of the property and river. This was deliberate. I think many people are not aware of the advantages of a diagonal arrangement. There are essentially no side wall reflections; those from the windows that reach listeners are so far off-axis as to be irrelevant (see Figure 5.7). Although side wall reflections are approved of by many for recreational stereo listening, in a multichannel system they truly are optional.

Initially, the overall sound quality was fine, but the bass was offensively boomy. I wished to avoid conspicuous bass traps, so it took some analysis and thought to come up with the dual-subwoofer mode-attenuating scheme discussed in Section 8.2.5. The

FIGURE 7.18 *The author's first multichannel entertainment room, ca. 1988. At the time this was an excellent CRT video monitor. The large audio panorama made it seem much larger. "Top Gun" was a favorite demo.*

bass was then smooth, deep and resonance free. This was the origin of the active mode canceling research. It was a good sounding system, and provided quality entertainment for family and friends.

The other listening room was much larger, conceived as my "classical" listening room. I had concluded that two stereo loudspeakers needed help in delivering a credible enveloping concert experience, so I built the largest concert hall that I could afford at the time. The room had a high sloping ceiling reaching to a clerestory window at 32 ft, and was otherwise deliberately irregularly shaped. Only two significant parallel surfaces remained, and those posed the only, easily solved, problem (Sections 8.2.2 and 8.2.4). The volume was 7,800 ft^3 (220 m^3). It served as a very open living/dining room in the "Danish modern" style of sparse furnishings, some art, and again no *obvious* acoustical treatment—there was a dense thick clipped-pile carpet on 1-inch (25 mm) felt underlay (requiring a recessed subfloor), and a large wall-hung Danish Rya rug over an air space. Both were efficient broadband absorbers. The abundant scattering made the absorbing material work hard and resulted in a mid-frequency RT of about 0.5 s. Large areas of gypsum wallboard on wood studs provided significant low-frequency absorption, helping to attenuate bass modes (Section 8.2.1). Double-layer glass windows, of which there were a lot, exhibit absorption coefficients similar to the wallboard—these had 5/8-inch air spaces for good thermal performance. As can be seen in Figure 7.19 it was visually interesting, open and spacious, with 180° windows to enjoy the views. Acoustically it was a pleasant space to be in.

FIGURE 7.19 *The author's "classical" stereo listening room, ca. 1990. The stereo seat was just to the left of the camera location. The black "monoliths" are bipolar Mirage M1s.*

Over the years, a parade of loudspeakers went through that room, and all disappointed. The room was an unforgiving critic of loudspeakers in which the direct and reflected sounds exhibited different spectra, and conventional forward-firing loudspeakers drew attention to themselves. With many common, frankly primitive, recording techniques, entire sections of the orchestra inappropriately emerged from a single loudspeaker. Some blurring and expansion of the soundstage in those cases was needed. Apparent source width (ASW) is an important ingredient of the design of a good hall. It was also a large room, so significant acoustical output was required to generate rewarding sound levels. That eliminated many otherwise good consumer products. Then, in 1989, a new loudspeaker came on the scene: the almost omnidirectional, bidirectional-in-phase "bipolar" Mirage M1. They performed well in double-blind listening tests in the small NRCC room, and also in this large one. They simply "became" the orchestra. Figure 7.20a and b show that the direct sound and the sound power were similar, with a slightly lumpy but low DI.

A loudspeaker that approaches omnidirectional radiation should show good agreement between sound power and steady-state room curves. In (b) the steady-state room curve was measured at a single location, the prime listening position. No spatial or spectral averaging was used, so the curve is not smooth. One parametric filter was used, to tame a room resonance at 42 Hz (see Figure 8.11). Figure 7.20c shows the sound power compared to a measurement made at a different place and time, and because of

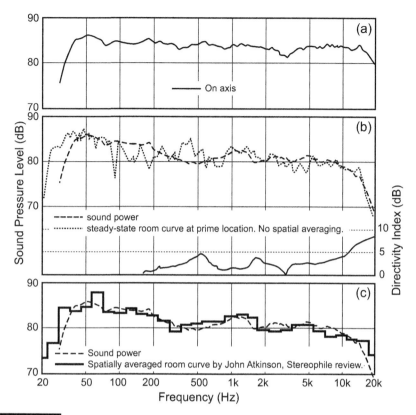

FIGURE 7.20 *(a) the anechoic on-axis curve for the Mirage M1. (b) the calculated sound power from anechoic data compared with the high-resolution room curve measured at the prime listening location in the room pictured in Figure 7.19 (no spatial averaging). (c) the calculated sound power compared to a spatial averaged, 1/3-octave resolution, room curve measured by John Atkinson for a review published in* Stereophile *magazine, June 1989 (used with permission).*

the spatial and spectral averaging the agreement is seen to be good. The in-room measurement has been substantially predicted by anechoic data.

I had experienced other omnidirectional and multidirectional loudspeakers in the past, and they could do some spatially interesting things, but not with the timbral neutrality of this one.

It was interesting to find that, in this situation, these nearly omnidirectional loudspeakers tolerated being quite close ($\approx$ 0.6 to 1 m) to room boundaries, with different spacing to the side and rear walls (see adjacent boundary issues, Chapter 9). I liked them enough to buy a pair and eventually sold them with the house when we moved to California. A key factor was that all of the many reflected sounds closely matched the timbre of the direct sound. The acceptable listening area was quite large at the listening distance of 22 ft (6.7 m). The human precedence-effect processor must have been especially pleased because the soundstage and imaging were surprisingly intact, and the room became a seamless extension of it, embellishing the envelopment illusion that served the intended function—classical music—very well. Such a venue was a luxury, but it turned out to be not only for the classical repertoire. Many popular recordings with high production values employ spatial enhancement. Late at night, alone in the dark, I have been known to play Dire Straits' *Brothers in Arms* at high level—a personal concert performance. I miss it.

I was motivated to explore the extent to which such an experience can be imitated in the small-room multichannel system in Figure 7.18. The Lexicon CP-1 provided considerable freedom to explore spatial enhancements in seven channels, using standard stereo signals as a starting point. Although there were many ways to make a good stereo recording sound contrived and bad, it was also true that some of the modes, tastefully used with the right recordings, sounded very good. It could be a letdown to switch back to stereo. Most of the success can be attributed to the designer Dr. David Griesinger, who has spent much of his life delving into concert hall acoustics and the simulation of those and other acoustical spaces in the Lexicon reverberation and spatial enhancement processors that have become standards of the recording industry. This and his subsequent Logic-7 upmixers tried to respect the stereo soundstage while adding missing spatial effects.

That exploration continued in my new listening room in California, a very different space (Figure 7.21).

Not having the luxury of a custom home, this room required major renovation. A 7.1-channel system is shown. Revel Salon2s are in the L & R locations, a Revel Voice 2 is the center channel below the screen, and 4 Revel Gem2s provide surround information. Three Mark Levinson No. 536 amplifiers drive the front channels. Timbre

FIGURE 7.21 *The author's entertainment/family room in California. The large L & R Revel Salon2s with bases removed are inverted so that the sound is perceived to originate at the bottom of the units. The physical locations of the loudspeakers and the ventriloquism effect provide excellent sound and picture correlation. There is a 65-inch flat screen TV and a motorized 10-foot front-projection screen shown by the dotted lines. The panoramic photograph distorts perspective; the true floor plan is in Figure 8.22.*

matching of all channels is excellent. The subwoofer system comprises four 1 kW closed box units in a sound-field-managed scheme as described in Figure 8.22. Resonance-free bass extends to below 20 Hz. Side wall reflections don't exist because there is no wall on one side, as can be seen, and for balance the opposite side is covered with a motorized heavy velour drape, which covers the large view windows and doors. The wall behind the front loudspeakers is irregular, with sound scattering sculpture niches, and the wall behind the listeners is mostly absorbing. Other surfaces are deliberately sound scattering or absorbing.

A motorized 10 ft (3 m) screen drops down in front of a 65-inch (1.65 m) direct view flat screen TV that uses the same audio system. Motorized drapes and blinds darken the room for front-projector viewing at any time of day. The lightweight chairs are easily moved for movie viewing by six persons. This suits our lifestyle, and it gives me special satisfaction that it has been possible to create impressive audio/video experiences in a room that does not look like a "home theater."

Stereo reproduction is very satisfying, but I still employ tasteful upmixing for many recordings to embellish the sense of space. The system will soon be elaborated to include immersive options, possibly a 9.4.5 configuration. Everything will be controlled by a flexible JBL Synthesis SDP-75 processor that will provide opportunities for more investigation.

7.4.7 Floor Reflections: A Special Case?

Very early in my explorations of loudspeaker/room interactions I took note of the measurable effect of the floor reflection in steady-state room curves. It seemed like a problem that needed attention, so I devoted some time to modifying loudspeakers to minimize it, and followed through with subjective evaluations. I cannot claim to have been exhaustive, but I soon became frustrated when the curves looked better but the sound seemed not to have changed very much. I moved on to other problems that were clearly audible and never returned.

Siegfried Linkwitz also decided that "the floor reflection, which is mostly unavoidable, and which is readily seen in the steady-state on-axis frequency response as dips and peaks, is not necessarily audible on program material" (Linkwitz, 2009).

The Fraunhofer Institute in Germany constructed an elaborate listening room in which different room surfaces could be changed (Silzle et al., 2009). "Regarding the floor reflection, the audible influence by removing this with absorbers around the listener is negative—unnatural sounding. No normal room has an absorbent floor. The human brain seems to be used to this."

Some of this is anecdotal, not the result of thorough scientific investigation. But there is an intriguing logic to the notion that humans evolved while standing on something, and most of what we hear includes reflections from what is under our feet. Wherever we are, reflections from what is under our feet have been useful in gauging distance, among other things. Floor reflections are a part of symphonic performances, jazz club performances, conversations at home or on the street. Is that a problem, a virtue or just a fact of life?

7.5 PROFESSIONAL LISTENING VS. RECREATIONAL LISTENING

Go into almost any recording control room and there is a high probability that early lateral reflections have been eliminated or attenuated by appropriately angled reflecting surfaces or massive absorbers, or both. It seems that recording engineers want to be in a predominantly direct sound field, as I found when I designed my first studios in the 1970s. Reflected sound had become unacceptable in the context of mixing recordings. This led to the "reflection-free zone" concept (D'Antonio and Konnert, 1984), the "live-end-dead-end" (LEDE) style of control room (Davis and Davis, 1980), and countless variations on those themes, including the extreme "non-environment" room, in which virtually all boundary reflections have been substantially attenuated (explained in Newell, 2003).

Having followed this for decades, it is clear that fashion and folklore play roles in this situation. Some of this has become "hand-me-down" acoustical theory, with some misinterpretations of psychoacoustics propping it up. If there is merit to the approach, it should be possible to demonstrate it in a way that does not deteriorate to a strongly asserted personal opinion.

I have not done an exhaustive search of the literature, but it has not been difficult to find several thoughtful investigations of what might be optimum listening and working conditions with respect to early reflections. One of these was shown in Section 7.4.1, where professional listeners opined that for recreational listening some lateral reflections were advantageous, but for examining audio products it was better to attenuate them. Kuhl and Plantz (1978) looked into the directional properties of loudspeakers that would be most suitable for (stereo) control room monitoring. Using only professional sound engineers as listeners, they found that narrow-dispersion loudspeakers were required for good reproduction of voices in radio dramas; dance and popular music was also desirably "aggressive" with "highly directed" loudspeakers. The majority of these same listeners, however, preferred wide-dispersion loudspeakers for the reproduction of symphonic music at home. In the control room, though, only about half of them felt that they could produce recordings with wide-dispersion loudspeakers. So, most of these professionals liked room reflections for recreational listening and about half of them thought that they could mix with wide-dispersion loudspeakers and the lateral reflections that resulted from them. Clearly there are individual differences.

Voelker (1985) had 90 people with varied professional audio backgrounds evaluate a control room that was set up to be (a) relatively reflective, (b) live-end, dead-end and (c) all surfaces damped. He concluded that the reverberant control room was preferred for chamber music and church organ. The LEDE™ room received the most votes for drum solo and disco music, followed by the non-reverberant room. It is concluded that a compromise is necessary in the acoustic design of control rooms when they are to be used with many types of music. This compromise is found in many existing control rooms where short-term (early) reflections from windows, doors, equipment and other furnishings give rise to a sense of reverberance.

Augspurger (1990), a well-known designer of loudspeakers, recording and listening rooms examined the similarities and differences between control rooms and home listening rooms. As noted earlier, he was very aware of the 2 kHz acoustical crosstalk dip in the phantom center image—a characteristic of all stereo systems. Control rooms are configured so that "the recording engineer hears roughly equal parts of direct sound and generally reflected sound at midrange frequencies." In his home stereo experiments he concluded that he preferred "hard, untreated wall surfaces. To my ears the more spacious stereo image more than offset the negative side effects. Other listeners, including many recording engineers, would have preferred the flatter, more tightly focused sound picture." As he says, "any study of real world stereo reproduction involves a strong element of subjective bias."

David Moulton (2003, 2011), an experienced recording engineer and educator in the field, has thought and written about the matter of monitoring rooms and loudspeakers for years. He and his recording engineer colleague LaCarrubba (1999) concluded that loudspeakers should have flat and very wide uniform horizontal dispersion over the front hemisphere combined with control-room acoustical treatment that would leave early lateral reflections intact, while attenuating later reflected sound. He collaborated in the design of a high-end consumer loudspeaker, the B&O Beolab 5, aspiring to those requirements. According to this approach the same enriched early-reflection sound field would exist both at the creation of recordings and their playback.

The Producers' and Engineers' Wing of the National Academy of Recording Arts and Sciences, an assembly of prominent sound recording professionals, produced a report, "Recommendations for Surround Sound Production" (NARAS, 2004). In it, they recommend, for acoustical treatment, "To as great a degree as possible, early reflections should be suppressed." "In addition there should be as much diffusion as a budget will allow." "To summarize: the more uniform (diffuse) the ambience in the professional mixing environment, the more site-independent the resultant mixes will be." There is a fundamental contradiction in this recommendation: suppressing the early reflections also suppresses the ones that would have followed them, and therefore the diffusion of the sound field is impeded before it has a chance to develop (see Figure 10.4). If the early reflections are indeed absorbed, any "diffuse" sound field would be from areas illuminated by later reflections and would be at a low sound level. There was no reference to any research supporting this approach.

Inspired by comments in the earlier edition of this book relating to the relative importance of the direct and early reflected sound, and the seeming ability of listeners to adapt to reflections in a room, researchers in the graduate program in sound recording, Schulich School of Music, McGill University, embarked on an elaborate investigation (King et al., 2012). "The study focused specifically on the working audio professional and the audio production environment." It evaluated changes in mix settings when laterally reflected sound was altered by surfaces that were reflecting, absorbing or diffusing. The tests were blind, and the 26 subjects were professional recording and mixing engineers with over 10 years of musical training, and an average of approximately 10 years of production experience. As it was set up, the dominant acoustical reflections were from the test surface opposite to the active loudspeaker, which is different from the majority of tests that involved first reflections from the adjacent boundary. The fundamental result was that "no significant main effect was found for acoustical treatment." When asked which acoustical treatment created the best listening condition for mixing, 8 subjects voted for diffusion, 7 for absorption, and 11 for reflection. From this it would seem that these professionals quickly adapted to each of the lateral sound conditions and simply got on with the job. More studies of this kind would be worthwhile, employing more aggressive differences in the reflected sound field.

In 2014, Tervq et al. tested the preferences of 15 sound engineers who listened in nine different environments. They found that the preference depended on the task of the engineer, mixing or mastering, and on the specific song. In general it was found that mixing engineers preferred the clarity provided by a lack of reflections, while mastering engineers preferred more reverberant environments. The latter is probably good in that it is closer to how customers listen. Again, they found evidence of significant adaptability.

So, if we are looking to professional sound engineers for guidance in loudspeakers and acoustical treatment, we find that they are not all in agreement—except, perhaps, about what they prefer to listen to when they relax.

7.5.1 Hearing Loss Is a Major Concern

If the ears are not functioning normally, what we hear is not normal. Aspects of hearing that are important to our appreciation of music, movies and life in general are not evaluated by conventional audiometric examinations. Chapter 17 explores some of the details and, frankly, it is a discouraging picture. This is a topic that every person needs to be informed about at an early age so that the necessary precautions can be taken to preserve this essential ability. In the present context, the summary information seems to be that those with temporarily or permanently deteriorated hearing not only hear less sound, but they are able to extract less information from sounds that they hear. The natural instinct is to reduce the complexity of the listening situation: that is, eliminate reflections. In the audio industry hearing performance is a factor that is not controlled, yet it undoubtedly contributes to differences in opinions.

7.5.2 Discussion

It is not surprising that audio professionals and audio enthusiasts have differing opinions about what an optimum combination of loudspeakers and room may be. In both cases they cover almost the entire range of possibilities. Numerous arguments exist that delayed sounds degrade sound quality, imaging, soundstage, clarity, speech intelligibility and so on. For some people this is true. But, for others it is not.

There is evidence that some professionals are able to mix in a variety of different acoustical circumstances, indicating that adaptation is possible. As pointed out in Section 7.1.1, a dominant direct sound field is where the acoustical crosstalk dip in the phantom center image is most audible. Some control rooms put diffusers on the wall behind the mix position, adding uncorrelated sound that would somewhat alleviate the problem. It obviously should not be taken to the extreme of the example given in Section 7.3.2, where excessive use of diffusers degraded the center image. However, if a recording engineer is in a situation where the interference dip is audible, it may be motivation to add some uncorrelated, delayed sound to the stereo mix itself, thereby lessening the problem for everybody (e.g., Vickers, 2009).

Missing entirely is any proof that the personal comfort of the mixer yields recordings that are audibly superior when auditioned by the customer who, like many professionals at home, will be listening in significantly reflective sound fields. There are tales of how well some mixes "translate" better from some studios to other venues. That is a good thing. However, having heard some of those "translations" it seems that a "literal" one was not always a requirement. True, the musical message may get through, but the timbral essence may not.

Delayed sounds are an essential part of live music performances. Without them they become timbrally and spatially deprived. The irony in this is that, for the most part, the recordings that are being constructed are two-channel stereo—itself a directionally and spatially deprived format.

Some promoters of acoustical materials are vociferous in their assertion that because many professionals listen in a certain fashion, that all serious playback facilities, even home listening rooms, should follow the lead of the pros. However, people listening for pleasure, even professionals, have shown a preference for some amount of room-reflected sound. As I stated at the end of the earlier edition of this book, the treatment of early reflection boundary areas is "optional": reflect, diffuse or absorb, as the customer prefers.

I recall my very early experiments at the National Research Council of Canada (NRCC), where in exploring the basics of perception I installed heavy sound-absorbing drapes on a track extending down the side walls and behind the loudspeakers. In uncontrolled experiments with audiophile friends, we decided that absorbing side wall reflections seemed to flatter some recordings (mostly pop/rock) while leaving the walls reflective flattered others (mostly classical and jazz). One of them actually set up a similar arrangement in his home and used it as a "spatial" control.

Conclusion: one size does not fit all. Personal taste, music and the reason for listening are all significant variables. And, deteriorated hearing, in its many forms, does none of us any favors; it makes us distinctive in ways that we may never know.

7.6 PERCEPTUAL EFFECTS OF ROOM REFLECTIONS

7.6.1 Adaptation and Perceptual Streaming

Benade (1984) sums up the situation:

The physicist says that the signal path in a music room is the cause of great confusion, whereas the musician and his audience find that without the room, only music of the most elementary sort is possible! Clearly we have a paradox to resolve as we look for the features of the musical sound that gives it sufficient robustness to survive its strenuous voyage to its listeners and as we seek the features of the transmission process itself that permit a cleverly designed auditory system to deduce the nature of the source that produced the original sound.

In auditory perceptions, adaptation is at work, allowing us to "normalize" the acoustical environment within which we listen. Imagine a scenario in which you and a colleague are conversing while walking down the street (a "dead" space), you enter a building foyer (a large reverberant space), take the elevator (a tiny space), and walk down a corridor (a distinctive reverberant space) into an open-plan office (a relatively dead, semi-reflective space). While you are conversing, your colleague's voice is a constant factor in spite of the enormous changes in the acoustical environment. It is the same voice in different acoustical spaces. The same is true of live musical performances in different venues. A Steinway is a Steinway, a Strad is a Strad, but in different rooms. The only relatively constant factor in all of this is the direct sound.

Out of the complexity of reflected sounds we extract useful information about the listening space, and apply it to sounds we will hear in the future. We are able, it seems, to separate acoustical aspects of a reproduced musical or theatrical performance from those of the room within which the reproduction takes place. This appears to be achieved at the cognitive level of perception—the result of data acquisition, processing and decision making, involving notions of what is or is not plausible. All of it indicates a long-standing human familiarity with listening in reflective spaces and a natural predisposition to adjusting to the changing patterns of reflections we live in and with. It is well described by Bregman (1999) as perceptual streaming, in which the acoustical space is separated from the sound source, the result of "auditory scene analysis," which also serves as the title of his book. It can be said that humans have an ability to "listen through" rooms.

Added to that is growing evidence that listeners who have had brief prior exposure to the sound of a room exhibit higher levels of speech comprehension than those who experience it for the first time (Brandewie and Zahorik, 2010; Srinivasan and Zahorik, 2012). Further, the magnitude of the effect varies with the RT of the room, being most evident in rooms having moderate RTs: 0.4 to 1.0 s—the kinds of rooms we do a lot of listening in (Zahorik and Brandewie, 2016).

The inevitable conclusion is that all aspects of room acoustics are not targets for "treatment." It would seem to be a case of identifying those aspects that we can, even should, leave alone, and focusing our attention on those aspects that most directly interact with important aspects of sound reproduction—reducing unwanted interference on the one hand or, on the other hand, enhancing desirable aspects of the spatial and timbral panoramas.

7.6.2 The Effects of Rooms on Loudspeaker Sound Quality

Olive et al. (1995) published results of an elaborate test in which three loudspeakers were subjectively evaluated in four different rooms. Figure 7.22 shows the rooms and the arrangements within them. The rest of this section is based on the account in Toole (2006).

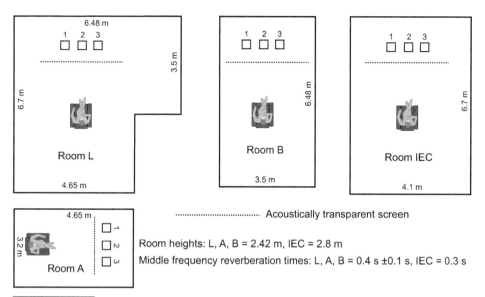

Room heights: L, A, B = 2.42 m, IEC = 2.8 m
Middle frequency reverberation times: L, A, B = 0.4 s ±0.1 s, IEC = 0.3 s

FIGURE 7.22 *Four listening rooms, showing arrangements in which listeners auditioned three loudspeakers. The loudspeakers were evaluated in each of the locations, using three different programs, by 20 listeners. Binaural recordings were made for subsequent headphone reproduction.*
Based on Olive et al. (1995), figure 3.

The loudspeakers were all forward firing cone/dome configurations, having similar directivities and similarly good performance, so it was not a simple task to distinguish the sounds. It was also a situation in which differences might have been masked by differences in room dimensions, loudspeaker location or placement of acoustical materials. In the first experiment, called the "live" test, listeners completed the evaluation of the three loudspeakers in one room before moving to the next one.

Binaural recordings were made of each loudspeaker in each location in each room, and the tests were repeated, but this time with listeners hearing all of the sounds through calibrated headphones. All tests were double-blind. In each room, three loudspeakers were evaluated in three locations for each of three programs. The whole process was repeated, resulting in 54 ratings for each of the 20 listeners. The result from a statistical perspective was that:

- "Loudspeaker" was highly significant: $p \leq 0.05$

- "Room" was not a significant factor

- Results of the live and binaural tests were essentially the same.

A possible interpretation is that the listeners became familiar with—adapted to—the room they were in and, this done, were able to accurately judge the relative merits of the loudspeakers. Because they were given the opportunity to become familiar with each of the four rooms, they were able to arrive at four very similar ratings of the relative qualities of the loudspeakers. Obviously, part of this adaptation, if that is the right description for what is happening, is an accommodation for the different loudspeaker locations. Different rooms and different positions within those rooms have not confused listeners to the point that they were not able to differentiate among and similarly rate the loudspeakers.

Then, using the same binaural recordings that so faithfully replicated the results of the live listening tests, another experiment was conducted. In this, the loudspeakers were compared with themselves and each other when located in each of the loudspeaker positions in each of the four rooms. Thus, in this experiment, the sound of the room was combined with the sound of the loudspeaker in randomized presentations that did not permit listeners to adapt. The result was that:

- "Room" became the highly significant variable: $p \leq 0.001$

- "Loudspeaker" was not a significant factor.

It appears, therefore, that we can acclimatize to our listening environment to such an extent that we are able to listen through it to appreciate qualities intrinsic to the sound sources themselves. It is as if we can separate the sound of a spectrum that is changing (the sounds from the different loudspeakers) from that which is fixed (the colorations added by the room itself for the specific listener and loudspeaker locations within it). This appears to be related to the spectral compensation effect noted by Watkins (1991, 1999, 2005) and Watkins and Makin (1996).

It is well not to over dramatize these results because, while the overall results were as stated, it does not mean that there were *no* interactions between individual loudspeakers and individual rooms. There were, and almost all seemed to be related to low-frequency performance. The encouraging part of this conclusion is that, as will be seen in Chapter 8, there are ways to control what happens at low frequencies.

Pike et al. (2013) were able to confirm these basic findings of timbral adaptation.

7.6.3 The Effect of Rooms on Speech Intelligibility

A sound reproduction system could have no greater fault than to have impaired speech intelligibility. Lyrics in songs lose meaning; movie plots get confused, and the evening news—well. . . . In parts of the audio community, it is popular to claim that reflected sounds within small listening rooms contribute to degraded dialog intelligibility. The concept has an instinctive "rightness" to it. However, as with several perceptual phenomena, when they are rigorously examined the results are not quite as expected. This is another such case.

In the field of architectural acoustics it has long been recognized that early reflections *improve* speech intelligibility. For this to happen, they must arrive within an "integration interval" within which there is an effective amplification of the direct sound—it is perceived to be louder. For speech, single reflections at the same level as the direct sound contribute usefully to the effective sound level, and thereby the intelligibility, up to about a 30 ms delay. For delayed sounds 5 dB lower than the direct sound, the integration interval is about 40 ms. These intervals embrace all consequential early reflections in domestic rooms or control rooms, although in larger rooms longer delays can be problematic (Lochner and Burger, 1958).

More recent investigations confirmed these findings, and found that intelligibility progressively improves as the delay of a single reflection is reduced, although the subjective effect is less than would be predicted by a perfect energy summation of direct and reflected sounds (Soulodre et al., 1989).

Lochner and Burger (1958), Soulodre et al. (1989) and Bradley et al. (2003) found that multiple reflections also contribute to improved speech intelligibility. The most elaborate of these experiments used an array of eight loudspeakers in an anechoic chamber to simulate early reflections and a reverberant decay for several different rooms (Bradley et al., 2003). The smallest was similar in size to a very large home theater or a screening room (13,773 ft^3, 390 m^3). The result was that early reflections (< 50 ms) had the same desirable effect on speech intelligibility as increasing the level of the direct sound. The authors go on to point out that late reflections (including reverberation) are undesirable, but controlling them should not be the first priority, which is to maximize the total energy in the direct and early-reflected speech sounds. Remarkably, even attenuating the direct sound had little effect on intelligibility in a sound field with sufficient early reflections.

Isolating reverberation time as a factor in school classrooms, it was found that optimum speech communication occurred for RT in the range 0.2 to 0.5 s (Sato and Bradley, 2008). This is conveniently in the range of normally furnished domestic rooms and recording control rooms.

Brandewie and Zahorik (2010) found that speech intelligibility improved by about 18% when listeners were given prior exposure to reverberant rooms, leading them to conclude that "the physical effects of acoustic reflections may be suppressed via high-level perceptual processes that require adaptation time to the particular reflective listening environment in order to be effective." The authors noted the similarity to the experiments described in Section 7.6.1, described in Toole, 2006. In subsequent experiments even more demanding tests confirmed this result, noting that "the processes underlying this effect are relatively rapid—on the order of seconds—and do not show longer term improvement" (Srinivasan and Zahorik, 2012).

All of these investigations point to the conclusion that normal listening rooms and control rooms do not impair speech intelligibility, rather they are more likely to improve it. Adaptation to the rooms is effective and quick, sometimes around 1 s. It is easy to imagine that this kind of adaptation applies to the interpretation of non-speech, musical sounds as well.

7.6.4 Sound Localization in Reflective Spaces— The Precedence (Haas) Effect

The precedence effect is fundamental to most of what we hear. In audio it has come to be known as the Haas effect, but others refer to it as the law of the first wavefront. Haas conducted his experiments as part of his PhD studies. Figure 7.23 describes the

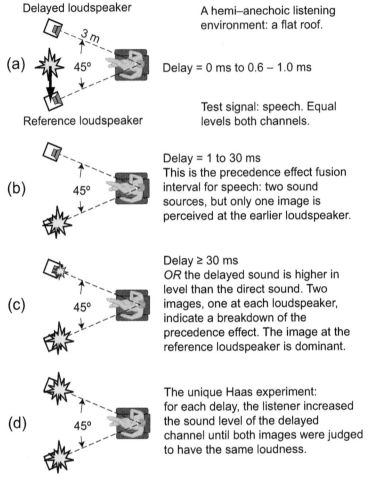

FIGURE 7.23 *A progression of localization effects observed in the experimental setup used by Haas, including stereo (summing) localization (a), the precedence effect (b), the breakdown of the precedence effect (c), and the equal-loudness experiment. Because the experiments were done on a flat roof, to minimize the effect of the roof reflection Haas began by placing the loudspeakers directly on the roof, aimed upwards toward the listener's ears. He found, though, that there was no significant difference if the loudspeakers were elevated to ear level, and that is the configuration used for the experiments.*

experiments, which involve only two sounds: a direct sound from one direction followed by a delayed sound from another direction—both in the horizontal plane. It is important to understand this process.

Whatever it is called, it describes the well-known phenomenon wherein the first arrived sound, normally the direct sound from a source, dominates our impression of where sound is coming from. Within a time interval often called the "fusion zone" we are not aware of reflected sounds arriving from other directions as separate spatial events. All of the sound appears to come from the direction of the first arrival. Sounds arriving later than the fusion interval may be perceived as spatially separated auditory images, coexisting with the direct sound, but the direct sound is still perceptually dominant. At very long delays the secondary images are perceived as echoes, separated in time and direction. The literature is not consistent in language, with the word "echo" often being used to describe a delayed sound that is not perceived as being separate in either direction or time.

Haas was not the first person to observe the primacy of the first arrived sound as far as localization in rooms is concerned (Gardner, 1968, 1969, 1973), but work done for his PhD thesis in 1949—translated from German to English in Haas (1972)—has become one of the standard references in the audio field. Sadly, his conclusions are often misconstrued. Let us review his core experiment.

Figure 7.23 shows the essence of the experiment. On the hemi-anechoic space provided by the flat roof of a laboratory building, a listener faced loudspeakers placed 45° apart. The Haas (1972) translation describes the setup as "at an angle of 45° to the left and right side of the observer" (p. 150). This could be construed in two ways. However, Gardner (1968) in a translation of a different Haas document reports "loudspeakers . . . at an angle of 45°, half to the right and half to the left of him." When Lochner and Burger (1958) repeated the Haas experiment they used loudspeakers separated by 90°. So, there is ambiguity about the angular separation. I have arbitrarily elected to show a 45° separation.

A recording of running speech was sent to both loudspeakers, and a delay could be introduced into the signal fed to one of them. In all situations except that for Figure 7.23d, both signals were radiated with the same sound level.

Figure 7.23a: summing localization. When there is no delay, the perceived result was a phantom (stereo) image floating midway between the loudspeakers. When delay was introduced, the center image moved toward the loudspeaker radiating the earlier sound, reaching that location at delays of about 0.7–1.0 ms. This is called "summing localization," and it is the basis for the phantom images that can be positioned between the left and right loudspeakers in stereo recordings, assuming a listener in the "sweet spot" (Blauert, 1996).

Figure 7.23b: the precedence effect. For delays in excess of 1 ms, it is found that the single image remains at the reference loudspeaker up to about 30 ms. This is the precedence effect: when there are two (or more) sound sources and only one sound image is perceived. It is important to note that the 30 ms interval is only for speech, and only for equal level direct and delayed sounds.

Figure 7.23c: multiple images—the breakdown of precedence. With delays greater than 30 ms, but certainly by 40 ms, the listener becomes aware of a second sound image at the location of the delayed loudspeaker. The precedence effect has broken down because there are two images, but the second image is a subordinate one; the dominant (louder) localization cue still comes from the loudspeaker radiating the earlier sound.

Figure 7.23d: the Haas equal-loudness experiment. In all illustrations to this point the first and delayed sounds had identical amplitudes. Obviously, this is artificial, because if the delayed sound were a reflection it would be attenuated by having traveled a greater distance and also by being reflected by a somewhat absorptive surface. But Haas moved even further from passive acoustical realities and deliberately amplified the delayed sound, as would happen in a public address situation. His interest was to determine how much higher in sound level the delayed sound could be before it became the dominant localization cue—that is, subjectively louder. To do this, he asked his listeners to adjust the sound level of the delayed loudspeaker until both of the perceived images appeared to be equally loud. This is the balance point, beyond which the sound image associated with the delayed loudspeaker would be perceived as being dominant. The objective was to prevent an audience from seeing a person speaking in one direction, and being distracted by a louder voice coming from a different direction. As shown in Figure 7.24, over a wide range of delays the later loudspeaker can be as much as 10 dB higher in level before it is perceived to be equally loud, and therefore a major distraction to the audience. Naturally, this would depend on where the audience member is seated relative to the symmetrical axis of the two sound sources.

Haas described this as an "echo suppression effect." Some people have mistakenly taken this to mean that the delayed sound is masked—it isn't. Within the precedence effect fusion interval, there is no masking; all of the reflected (delayed) sounds are audible, making their contributions to timbre and loudness, but the early reflections simply are not heard as spatially separate events. They are perceived as coming from the direction of the first sound—this, and only this, is the essence of the "fusion." The widely held belief that there is a "Haas fusion zone," approximately the first 20 ms after the direct sound, within which everything gets innocently combined, is simply untrue.

Haas noted audible effects having nothing to do with localization. First, the addition of a second sound source increased loudness. There were some changes to sound quality "liveliness" and "body" (Haas, 1972, p. 150), and a "pleasant broadening of the primary sound source" (Haas, 1972, p. 159). Increased loudness was a benefit to speech reinforcement, and the other effects would be of concern only if they affected intelligibility.

Benade (1985) contributed an insightful summary under the title "Generalized Precedence Effect," stating:

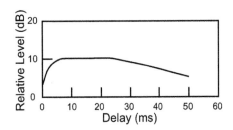

FIGURE 7.24 *In the situation depicted in Figure 7.23d, the sound level of the delayed sound, relative to that of the first arrival, at which listeners judged the two sound images to be equal in loudness.*

1) The human auditory system combines the information contained in a set of reduplicated sound sequences and hears them as though they were a single entity, provided a) that these sequences are reasonably similar in their spectral and temporal patterns and b) that most of them arrive within a time interval of about 40 ms following the arrival of the first member of the set.

2) The singly perceived composite entity represents the accumulated information about the acoustical features shared by the set of signals (tone color, articulation, etc.). It is heard as though all the later arrivals were piled upon the first one without any delay, that is, the perceived time of arrival of the entire set is the physical instant at which the earliest member arrived.

3) The loudness of the perceived sound is augmented above that of the first arrival by the accumulated contributions from the later arrivals. This is true even in the case when one or more of the later signals is stronger than the first one to arrive, that is, a strong later pulse does not start a new sequence of its own.

4) The apparent position of the source of the composite sound coincides with the position of the source of the first-arriving member of the set, regardless of the physical directions from which the later arrivals may be coming.

5) If there are any arrivals of sounds from the original acceptably similar set that come in after a delay of 100–200 ms they will not be accepted for processing with their fellows. On the contrary, they will be taken as a source of confusion and will damage the clarity and certainty of the previously established percept. These "middle-delay" signals that dog the footsteps of their betters may or may not be heard as separate events.

6) If for some reason a reasonably strong member of the original set should come in with a delay of something more than 250–300 ms, it will be distinctly heard as a separate echo. This late reflection will be so heard even if it is superposed on a welter of other (for example, reverberant) sounds.

It is important to notice that these very strongly worked categorical statements all emphasize that there is an *accumulation of information* from the various members of the sequence. It is quite incorrect to assume that the precedence effect is some sort of masking phenomenon that, by blocking out the later arrivals of the signal, prevents the auditory system from being confused. Quite to the contrary, those arrivals that come in within a reasonable time after the first one actively contribute to our knowledge of the source. Furthermore members of the set that are delayed somewhat too long actually disrupt and confuse our perceptions even when they may not be consciously recognized. If the arrivals are later yet, they are heard as separate events (echoes) and are treated as a nuisance. In neither case are the late arrivals masked out.

What was discussed by Hass in his 1949 thesis, his contemporaries, and those that preceded him (well summarized in Gardner, 1968, 1969) was just a beginning. Recent research (e.g., Blauert, 1996; Litovsky et al., 1999; Blauert and Divenyi, 1988; Djelani and Blauert, 2000, 2001) suggests that the precedence effect is cognitive, meaning that it occurs at a high level in the brain, not at a peripheral auditory level. Its purpose appears to be to allow us to localize sound sources in reflective environments where the sound field is so complicated by multiple reflections that sounds at the ears cannot be continuously relied upon for accurate directional information. This leads to the concept of "plausibility" wherein we accumulate data we can trust—both auditory and visual—and persist in localizing sounds to those locations at times when the auditory cues at our ears are contradictory (Rakerd and Hartmann, 1985). Among localization phenomena encountered in audio/video entertainment systems are bimodal (hearing and seeing) interactions, including what we know as the ventriloquism effect, wherein sounds are perceived to come from directions other than their true directions. This happens routinely in movies, where most of the time all on-screen sounds emerge from the center loudspeaker; what we see dominates the localization.

When we put these notions into a three-dimensional context, the elaborated precedence effect appears as follows. At the onset of a sound accompanied by reflections in an *unfamiliar* setting, it appears that we hear everything. Then, after a brief build-up interval, the precedence effect causes our attention to focus on the first arrival, and we simply are not aware of the reflections as spatially separate events. Remember that this is not *masking*; in all other respects, the reflections are present, contributing to loudness, timbre and so forth. This suppression of the directional identities of later sounds seems to persist for at least 9 s following cessation of the sound, allowing the adaptation to be effective in situations where sound is not continuous (Djelani and Blauert, 2000, 2001). Figure 7.25 shows an exaggerated view of how we might localize a sound source in a small room in (a) the first impression and (b) after build-up.

However, it would be wrong to think that this is a static situation. A change in the pattern of reflections, in number, direction, timing or spectrum, can cause the initiation of a new build-up, without eliminating the old one. We seem to be able to remember several of these "scenes." All of this build-up and decay of the precedence effect needs to be considered in the design of experiments where spatial/localization effects are being investigated; namely, are the reported perceptions before or after precedence-effect build-up? In any situation where listener adjustments of acoustical parameters are permitted, adaptation may not occur at all. One has to think that this is a factor in what recording personnel hear in control rooms—and it will be different from what is heard during playback at home. One unknown factor is the role played by spatial cues in the recordings themselves, which are being altered in real time during the mixing operation. Nothing could be more relevant in this context than the recording and subsequent playback of high channel count immersive audio.

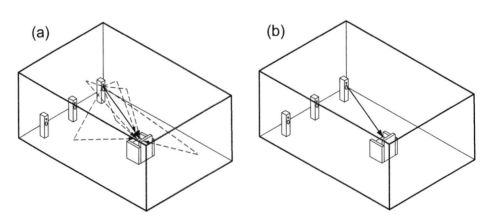

Localization: first impression ....after adaptation

FIGURE 7.25 *A simplistic illustration of how the precedence effect allows us to focus on the true direction of a sound source when listening in a reflective space. It takes time to develop, and it will fade from memory if the reflection pattern is changed or not reinforced from time to time.*

Perrot et al. (1989) and Saberi and Perrot (1990) observed that, with the practice that inevitably comes from prolonged exposure in experiments, listeners appear to be able to learn to ignore the precedence effect, detecting the delayed sounds almost as if they existed in isolation. A recording engineer can adjust the level of a sound component upwards to the point where it is clearly audible, reduce it, and at any time turn it on or off. Under these circumstances, where the component can be aurally "tracked," it is highly probable that it can be heard at levels below those at which it is likely to be audible when listening normally to the completed mix. Thus, sounds that may be gratifying to the mixing or mastering engineer may be insufficient to reward a normal listener, or worse, simply not heard at all.

Important for localization, and very interesting from the perspective of sound reproduction, is the observation that the precedence effect appears to be most effective when the spectra of the direct and reflected sounds are similar (Benade, 1985; Litovsky et al., 1999; Blauert and Divenyi, 1988). If the reflected sound has a "sufficiently" (not well defined at this stage) different spectrum from the direct sound, there is a greater likelihood that it will be separately localized, and not merged with the direct sound as far as localization is concerned. In the extreme, if all early reflections had sufficient spectral differences, it would seem that the illustration in Figure 7.25a would not transition into Figure 7.25b. Because the precedence effect is more effective for broadband sounds than for narrow-band sounds, this seems to be a matter of importance (Braasch et al., 2003).

7.6.5 Bringing the Precedence Effect into the Real Acoustical World

Most of the investigations of the precedence effect used speech as a signal. Speech is important, but obviously not the only sound that matters to sound reproduction. Looking further into the situation, one quickly realizes that some of the generalized rules many of us have been using are wrong if we are listening to music or sound effects, not speech.

Figure 7.26 looks at the precedence effect from a realistic perspective. Most data come from laboratory experiments in which the first arrival and the delayed sounds are equal in level. This cannot happen in real spaces, so this illustration shows the changes in audible effects at different sound levels and delays.

All of the data points and lines are thresholds—the sound levels at which listeners detected a change in their perceptions. One of the audio rules of thumb is that the "fusion interval" is about 30 ms. This number comes from experiments, like the Haas experiments, in which the direct and delayed speech sounds are equal in level. Figure 7.26b shows that in the real world, where the delayed sounds are lower in level, the fusion interval increases to much higher values. It also shows that the precedence effect is not disrupted by early reflections in normal listening rooms.

Some of the perceived changes are beneficial in that, up to a point, listeners find that levels well above threshold provide greater pleasure. For example, the perception described at threshold as "image shift or spreading" may seem like a negative attribute, but when translated into what is heard in rooms it becomes "image broadening" or apparent source width (ASW), widely liked qualities in sound reproduction as they are in live performances. Even "second image" thresholds can be exceeded with certain

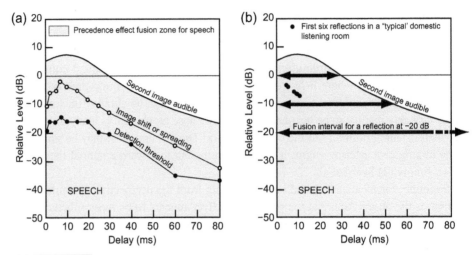

FIGURE 7.26 *(a) The shaded area represents the precedence-effect fusion zone for speech. This is the range of amplitudes and delays within which a reflected sound will not be identified as a separately localizable event. From Toole (2006). Within that zone the detection threshold and the image-shift or spreading thresholds are shown. (b) The precedence-effect fusion intervals for delayed sounds at three sound levels. The classic experiments much quoted from psychoacoustic literature generally used equal level direct and delayed sounds—this is the highest large arrow at 0 dB, showing an interval of about 30 ms. In rooms delayed sounds are attenuated by propagation loss, typically –6 dB/double-distance, and sound absorption at the reflecting surfaces. As the delayed sound is reduced in level, the fusion zone increases rapidly. The set of black dots shows the delays and amplitudes for the first six reflections in a typical listening room or home theater, indicating that in such rooms the precedence effect is solidly in control of the localization of speech sounds.*

kinds of sounds, expanding the size of an orchestra beyond its visible extent in a concert hall, or extending the stereo soundstage beyond the spread of the loudspeakers. In reproduced sound the picture is more confused because there are techniques in the recording process that can achieve similar perceptions. Because all of these factors are influenced by how the recordings are made and how they are reproduced, these comments are observations, not judgments of relative merit. Some evidence suggests that even these small effects might be diminished by experience during listening within a given room (Shinn-Cunningham, 2003)—another in the growing list of perceptual phenomena that we can adapt to.

7.6.5.1 Ceiling vs. Wall Reflections

While the horizontal plane is important, because that is where our ears are located, it is important to know how differently one perceives sounds reflected from the ceiling, or even delayed sounds arriving along the same axis as the direct sound.

Figure 7.27 shows that the thresholds for the side wall and the ceiling reflections are almost identical. This is counterintuitive, because one would expect a lateral reflection to be much more strongly identified by the binaural discrimination mechanism because of the large signal differences at the two ears. For sounds differing only in elevation, we

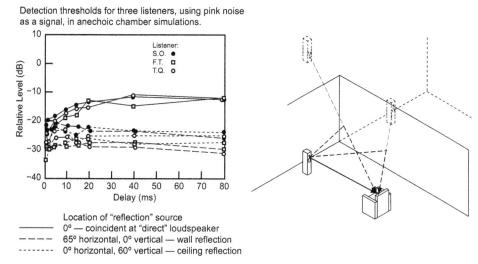

Detection thresholds for three listeners, using pink noise as a signal, in anechoic chamber simulations.

Location of "reflection" source
—————— 0° — coincident at "direct" loudspeaker
− − − − 65° horizontal, 0° vertical — wall reflection
· · · · · · 0° horizontal, 60° vertical — ceiling reflection

FIGURE 7.27 *The detection thresholds for delayed sounds simulating a wall reflection, a ceiling reflection and one arriving from the same direction as the direct sound. The test signal was pink noise. Adapted from Olive and Toole (1989a).*

have only the spectral cues provided by the external ears and the torso (HRTFs, see Section 15.12.1). Although the threshold levels might be surprising, intuition is rewarded in that the dominant audible effect of the lateral reflection was spaciousness (the result of interaural differences), and that of the vertical reflection was timbre change (the result of spectral differences). The broadband pink noise used in these tests would be very good at revealing colorations, especially those associated with HRTF differences at high frequencies. On the other hand, continuous noise lacks the strong temporal patterns of some other sounds, like speech.

This makes the findings of Rakerd et al. (2000) especially interesting. These authors examined what happened with sources arranged in a horizontal plane, and vertically, on the front-back (median sagittal) plane. Using speech as a test sound, they found no significant differences in masked thresholds and echo thresholds sources in the horizontal and vertical planes. In explanation, they agreed with other referenced researchers, that there may be an "echo suppression mechanism mediated by higher auditory centers where binaural and spectral cues to location are combined" (Rakerd et al., 2000). This is another example of humans being very well adapted to listening in reflective environments.

Another surprise in Figure 7.27 is that delayed sounds originating in the same loudspeaker are more difficult to hear—the threshold here is consistently higher than for sounds arriving from the side or above, slightly for short delays, and much higher (10+ dB) at long delays. Burgtorf (1961) agrees, finding thresholds for coincident delayed sounds to be 5–10 dB higher than those separated by 40–80 degrees. Seraphim (1961), used a delayed source that was positioned just above the direct-sound source (~ 5° elevation difference) and found that, with speech, the threshold was elevated by about 5 dB compared to one at a 30° horizontal separation. The relative insensitivity to coincident sounds appears to be real, and the explanation seems to be that it is the result of spectral similarities between the direct sound and the delayed sound. These sounds take on progressively greater timbral differences as they are changed in elevation relative to the direct sound. For those readers who have been wondering about the phenomenon of "comb filtering," this evidence tells us that the situation of maximum comb filtering, when the direct and delayed sounds emerge from the same loudspeaker, is the one for which we are least sensitive. Encouraging news.

All this said, it still seems remarkable that a vertically displaced reflection, with no apparent binaural (between the ears) differences, can be detected as well as a reflection arriving from the side, generating large binaural differences. Not only are the auditory effects at threshold different—timbre vs. spaciousness—but the perceptual mechanisms required for their detection are also different.

7.6.5.2 Real vs. Phantom Images

A phantom image is a perceptual illusion resulting from summing localization when the same sound is radiated by two loudspeakers. It is natural to think that these directional illusions may be more fragile than those created by a single loudspeaker at the same

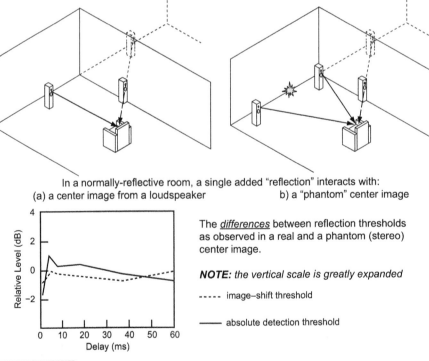

In a normally-reflective room, a single added "reflection" interacts with:
(a) a center image from a loudspeaker   b) a "phantom" center image

The _differences_ between reflection thresholds as observed in a real and a phantom (stereo) center image.

NOTE: the vertical scale is greatly expanded

- - - - - image–shift threshold

———— absolute detection threshold

FIGURE 7.28 An examination of how a real and a phantom center image respond to a single lateral reflection simulated by a loudspeaker located at the right side wall. The room was the "live" version of the IEC listening room used in other experiments. Note that the vertical scale has been greatly expanded to emphasize the lack of any consequential effect. The signal was speech.

location. The evidence shown here applies to the simple case of a single lateral reflection, simulated in a normally reflective room with a loudspeaker positioned along a side wall, as shown in Figure 7.28. When detection threshold and image-shift threshold determinations were done first with real and then with phantom center images, in the presence of an asymmetrical single lateral reflection, the differences were insignificantly small. It appears that concerns about the fragility of a phantom center image are misplaced.

Examining the phantom image in transition from front to surround loudspeakers (±30° to ±110°), Corey and Woszczyk (2002) concluded that adding simulated reflections of each of the individual loudspeakers did not significantly change image position or blur, but it did slightly reduce the confidence that listeners expressed in the judgment.

7.6.5.3 Speech vs. Various Musical Sounds

Figure 7.29 shows detection thresholds for sounds chosen to exemplify different degrees of "continuity," starting with continuous pink noise, and moving through Mozart, speech, castanets with reverberation, and "anechoic" clicks (brief electronically generated pulses sent to the loudspeakers). The result is that increasing "continuity" produces progressive flattening. The perceptual effect is similar if the "continuity" or "prolongation" is due to variations in the structure of the signal itself, or due to reflective repetitions added in the listening environment. In any event, pink noise generated an almost horizontal flat line, Mozart was only slightly different over the 80 ms delay range examined, speech produced a moderate tilt, castanets (clicks) with some recorded reverberation were even more tilted, and isolated clicks generated a very compact, steeply tilted threshold curve.

Figure 7.30 puts these data into a practical context, where it is seen that in small listening rooms the first reflections are all above the threshold of detection.

Based on the stable shapes of the patterns as one moves from the detection threshold, through the image shift threshold to the second-image threshold, Figure 7.30 shows how three very different sounds compare to reflections in typical small listening rooms. It is seen that all of the early reflections are above the detection thresholds, so there will be audible effects. The most destructive effect, the breakdown of the precedence effect, is barely avoided in the absence of any other reflections. However, in normally reflective rooms the curves would be flattened by the "continuity" added by later arriving reflections.

Looking at the 0 dB relative level line—where the direct and reflected sounds are identical in level—it can be seen that the precedence effect (fusion) interval for clicks appears to be just under 10 ms. According to Litovsky et al. (1999) this is consistent with other determinations (<10 ms), and the approximately 30 ms for speech is also in the right range (<50 ms). They offer no fusion interval data for Mozart, but it is

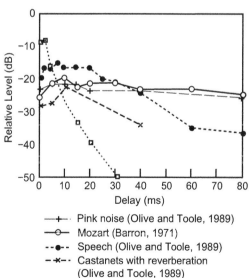

— + — Pink noise (Olive and Toole, 1989)
— o — Mozart (Barron, 1971)
-- • -- Speech (Olive and Toole, 1989)
- × - Castanets with reverberation (Olive and Toole, 1989)
··· □ ··· Clicks (Olive and Toole, 1989)

FIGURE 7.29 *Detection thresholds for a single lateral reflection, determined in an anechoic chamber for several sounds exhibiting different degrees of "continuity" or temporal extension.*
Toole (2006), figure 16.

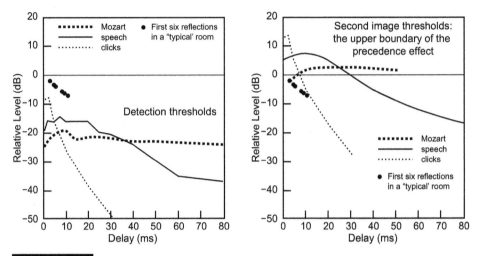

FIGURE 7.30   *Using data from Figure 7.26 and 7.29, this is a comparative estimate of the detection thresholds and the second image thresholds (i.e., the boundary of the precedence effect) for clicks, speech and Mozart. The "typical room reflections" suggest that in the absence of any other reflections, the clicks are approaching the point of being detected as a second image. However, normal room reflections would be expected to prevent this from happening because the threshold curve would be flattened by the sustained room reflections.*

reasonable to speculate based on the Barron data that it might be substantially longer then 50 ms.

The short fusion interval for clicks suggests that sounds like close-miked percussion instruments might, in an acoustically dead room, elicit second images if there were appropriately located reflecting surfaces.

7.7 MEANINGFUL MEASUREMENTS OF REFLECTION AMPLITUDES

It seems obvious to look at reflections in the time domain, in a "reflectogram" or impulse response, a simple oscilloscope-like display of events as a function of time or, a popular alternative, the ETC (energy-time curve). In such displays, the strength of the reflection would be represented by the height of the spike. However, the height of a spike is affected by the frequency content of the reflection and time domain displays are "blind" to spectrum. The measurement has no information about the frequency content of the sound it represents. Only if the spectra of the sounds represented by two spikes are identical can they legitimately be compared.

Let us take an example. In a very common room acoustic situation, suppose a time-domain measurement reveals a reflection that it is believed needs attenuation. Following a common procedure, a large panel of fiberglass is placed at the reflection point. It is respectably thick at 2 inches (50 mm), so it attenuates sounds above about 500 Hz.

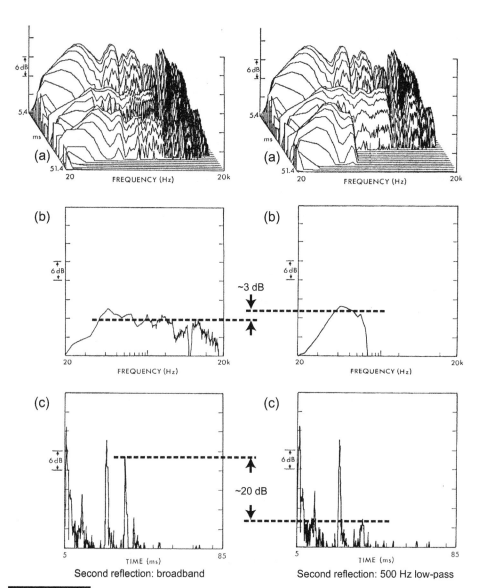

Second reflection: broadband   Second reflection: 500 Hz low-pass

FIGURE 7.31   *The left column of data shows results when the second of a series of reflections was adjusted to the threshold of detection when it was broadband, the right column shows threshold data when the reflection was low-pass filtered at 500 Hz. (a) shows the waterfall diagram, (b) the spectrum of the second reflection extracted from the waterfall, and (c) shows the ETC measured with a Techron 12 in its default condition (Hamming windowing). The signal was speech. The horizontal dotted lines are "eyeball" estimates of reflection levels.*
From Olive and Toole (1989a), figures 18 and 19.

A new measurement is made and, behold, the spike has gone down. Success, right? Maybe not.

In a controlled situation, Olive and Toole (1989a) performed a test intended to show how different measurements portrayed reflections that, subjectively, were adjusted to be at the threshold of detection. So, from the listener's perspective, the two reflections that are about to be discussed are the same—just at the point of audibility or inaudibility.

The results are shown in Figure 7.31. At the top, the (a) graphs are waterfall diagrams displaying events in three dimensions. At the rear is the direct sound, the next event in time is an intermediate reflection, and at the front is the second reflection, the one that we are interested in. It can be seen that the second reflection is broadband in the left-hand diagram and that frequencies above 500 Hz have been eliminated in the right-hand version. When that particular "layer" of the waterfall is isolated, as in the (b) displays, the differences in frequency content are obvious. The amplitudes are rather similar, although the low-pass filtered version is a little higher, which seems to make sense considering that slightly over five octaves of the audible spectrum have been removed from the signal. Recall that these signals have been adjusted to produce the same subjective effect—a threshold detection—and it would be logical for a reduced-bandwidth signal to be higher in level.

In contrast, the (c) displays, showing the ETC measurements, were telling us that there might be a difference of about 20 dB in the opposite direction; the narrow-band sound is shown to be lower in level. Obviously, this particular form of the measurement is not a good correlate with the audible effect in this test.

The message is that we need to know the spectrum level of reflections in order to be able to gauge their relative audible effects. This can be done using time-domain representations, like ETC or impulse responses, but it must be done using a method that equates the spectra in all of the spikes in the display (e.g., by octave-band filtering).

Examining the "slices" of a waterfall, as done here, is an option, as is performing FFTs on individual reflections isolated by time windowing of an impulse response. Such processes need to be done with care because of the trade-off between time and frequency resolution, as explained in Section 4.7. It is quite possible to generate misleading data.

All of this is especially relevant in room acoustics because acoustical materials, absorbers and diffusers routinely modify the spectra of reflected sounds. Whenever the direct and reflected sounds have different spectra, simple broadband ETCs or impulse responses are not trustworthy indicators of audible effects. Spectrum information is necessary, and modern measurement equipment and algorithms provide the means to get it.

CHAPTER 8

Below the Transition Frequency: Acoustical Events and Perceptions

"Small rooms" are notorious for bad bass. Sustained bass notes fluctuate in level as they change frequency. Bass transients "boom," and lack the tightness that can be experienced in live performances. All of these audible characteristics come and go as one moves around the room, or sits in different seats. They exist in recording control rooms used in the creation of music, movie and television sound, and the multitudes of domestic spaces devoted to stereo listening and home theaters. It is a major problem.

In assessing the factors contributing to subjective judgments of sound quality, discussed in Section 5.7, it was shown that about 30% of the overall rating is contributed by factors related to low-frequency performance (Olive, 2004a, 2004b). All of the listening tests in that study were done in the same room, which was equipped with apparatus to move the active loudspeakers to the same locations, shown in Figures 3.13 and 3.14, and where listeners had ample time to adapt to the physical circumstances. All of this helped to neutralize—not to eliminate—the room and loudspeaker location as factors in the evaluations; they were constants, not variables.

Our idealized objective is to achieve high subjective ratings for loudspeakers, and to do it in different rooms. Up to a point, it happens now, as discussed in Section 7.6.2, but even so there was evidence of variations in judgment due to irregularities in bass performance. Humans can adapt to a certain amount of bass misbehavior, but the more extreme the problem, the more difficult it is to completely ignore.

A strategy is needed that can ensure the delivery of similarly good bass to all listeners in all rooms. As discussed in Section 1.4, the "circle of confusion" argues that achieving such consistency is a necessary objective for the entire audio industry. Ideally we want recording professionals and consumers to hear the same quantity and quality of bass. Let us see how far this idea can be taken.

8.1 THE BASICS OF ROOM RESONANCES AND STANDING WAVES

All of the bass we hear, no matter how excellent the woofers/subwoofers may be, is filtered through room resonances. They affect what we hear at home and in our cars, adding to the problem of communicating the recorded art as it was intended, and leaving listeners with a conundrum: if one likes or dislikes something that is heard, who or what does one praise or blame?

This topic has generated countless words in audio forums, with widely differing opinions about what works and what does not. Some of the differing opinions appear to be simply the result of inadequate data and explanations based on ill-informed subjective impressions of what really is happening. Discussions of the effectiveness of "room equalization" or "room calibration" are confounded by the fact that commercial algorithms differ widely in the resolution of the measurements they do, the resolution of the equalization employed, the algorithmic rules by which the corrective filters are created and, last but not least, the target performance being strived for. This topic will be discussed in subsequent chapters.

Room resonances are distinctive because:

- The mechanism creating the resonances involves propagating sound waves.

- The resonances exist as a result of constructive interference (synchronized waveform addition) between the outgoing sound from the loudspeaker and the incoming sound from reflections within the room, as illustrated in Figure 4.21. At resonance frequencies, the incoming sound is delayed by multiples of half-wavelengths, summation occurs, and standing waves are created—a pattern of higher and lower sound pressure levels within the room. Seat-to-seat variations in bass are an issue to be dealt with.

- The frequencies are low, periods and wavelengths are long, and the duration of ringing might be long enough for it to be audible as a prolongation of a brief sound.

Figure 8.1 shows for a rectangular room: (a) the orientations of the three axial modes, (b) the three tangential modes, and (c) one of many possible oblique modes. Because some energy is lost at every boundary interaction, modes that complete their "cycle" of the room with the fewest reflections are the most energetic. The axial modes are therefore the most energetic, followed by the tangential modes and the oblique modes. It is rare for an oblique mode to be identifiable as a problem in a room. Tangential modes can be found in rooms having significantly reflective boundaries, or when multiple sources are appropriately located. Axial modes are omnipresent, and they are the usual culprits in bass problems in small rooms. Fortunately, they are easily calculated, as described in the box.

Calculating all of the modes in a rectangular room is more difficult, as the equation in Figure 8.2 shows. The method of mode identification is helpful if one is trying to

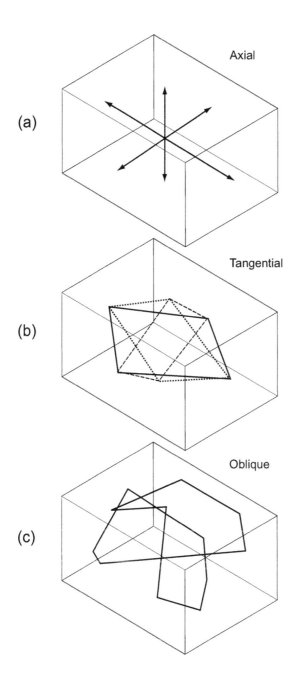

Axial

(a)

Tangential

(b)

Oblique

(c)

FIGURE 8.1 *The three classes of room modes: (a) axial: length, width and height; (b) tangential: each of the three involving two pairs of parallel boundaries and ignoring the third pair. In (c) is shown one of the many possibilities for oblique modes that involve all surfaces.*

track down the origin of an irritating bass boom and from that to decide on a remedy (see Figure 6.2).

Figure 8.3 shows all of the modes for a small rectangular room. This is the kind of information yielded by any of several computer programs that run the preceding equation. Not all of these modes will be problematic in practical circumstances, but they

$$f_{n_x n_y n_z} = \frac{c}{2} \sqrt{\left(\frac{n_x}{l_x}\right)^2 + \left(\frac{n_y}{l_y}\right)^2 + \left(\frac{n_z}{l_z}\right)^2}$$

$f_{n_x n_y n_z}$ = the frequency of the mode defined by the integers applied to dimensions x, y and z (length, width and height).

n_x, n_y, n_z = integers from 0 to ∞ applied to each dimension: x, y, z.

l = dimension of the room in ft (m)
c = speed of sound: 1131 ft/s (345 m/s)

Examples of mode identification:
$f_{1,0,0}$ is the first–order length mode (x dimension)
$f_{0,2,0}$ is the second–order width mode (y dimension)
$f_{0,0,4}$ is the fourth–order height mode (z dimension)
$f_{1,2,0}$ is a tangential mode (involving two dimensions)
$f_{1,3,2}$ is an oblique mode (involving all three dimensions)

FIGURE 8.2 *The general equation for calculating all possible modes in a rectangular room.*

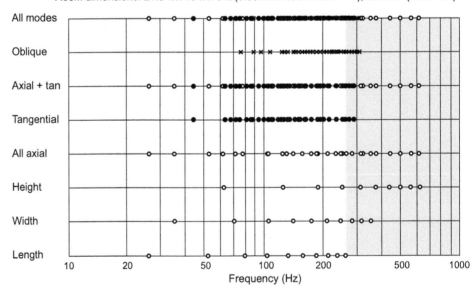

Room dimensions: 21.5 ft x 16 ft x 9 ft (6.55 m x 4.88 m x 2.74 m), 3096 ft³ (87.67 m³)

FIGURE 8.3 *The low-order modes for a small rectangular room. In the shaded area all of the modes have not been calculated. This is based on the presentation of data by a Microsoft Excel program, which is available for download at www.harman.com. An Internet search for "room mode calculator" will reveal many more, along with some opinions about their value that may differ from those expressed here.*

exist, and knowing the frequencies at which they occur can be helpful in analyzing specific installations. It has been traditional to use calculations of this kind to optimize room dimensions and dimensional ratios; the dimensions determine the frequencies of resonances, and the dimensional ratios determine the distribution of those frequencies.

"BACK OF THE ENVELOPE" ACOUSTICAL CALCULATIONS

To compute the frequencies at which axial standing waves occur, simply measure the distance between the walls (this is one-half wavelength of the lowest resonance frequency), multiply it by two (to get the wavelength) and divide that number into the speed of sound in whatever units the measurements were done (1,131 ft/s, 345 m/s). The result is the frequency of the first-order mode along that dimension.

All higher-order modes are the result of multiplying this frequency by 2, 3, 4, 5 and so on. For example: a room is 22 ft long. The first-order length resonance occurs at 1,131 / 44 = 25.7 Hz. Higher-order resonances exist at: 51.4 Hz, 77.1 Hz, 102.8 Hz, and so on. Repeat this for the width and height of the room and all of the axial modes will have been calculated.

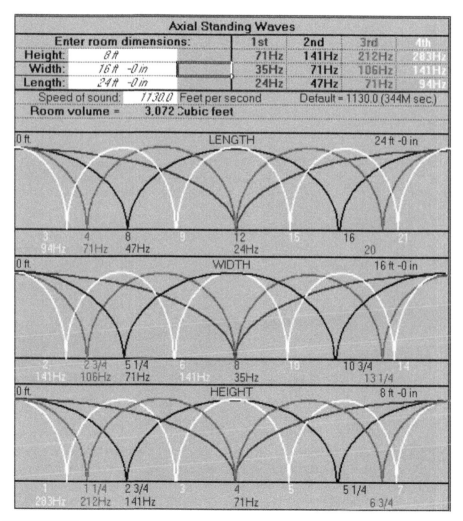

Axial Standing Waves						
Enter room dimensions:			1st	2nd	3rd	4th
Height:	8 ft		71Hz	141Hz	212Hz	283Hz
Width:	16 ft -0 in		35Hz	71Hz	106Hz	141Hz
Length:	24 ft -0 in		24Hz	47Hz	71Hz	94Hz
Speed of sound:	1130.0	Feet per second		Default = 1130.0 (344M sec.)		
Room volume =	3,072 Cubic feet					

FIGURE 8.4 *The standing wave patterns along the length, width and height for a rectangular room. From the same mode calculator referenced in Figure 8.3.*

Figure 8.4, the companion diagram in the Harman calculator, shows the standing wave patterns at four orders of axial modes. Both of these sets of data can be useful in identifying problem room modes.

8.1.1 Optimizing Room Dimensions—Does an "Ideal" Room Exist?

One of the legends of room acoustics is that some room dimensional ratios, height-to-width-to-length, as 1:1.5:2.5, are better than others. The argument goes that these ratios determine the frequency distribution of room resonances, and it is desirable to have these frequencies distributed uniformly in the frequency domain. For example, it is obviously not good to have length and width resonances occurring at the same or very similar frequencies. If a room were to have a tendency to boom or coloration because of resonances, it would be worse if several were to occur at the same frequencies.

Over the years, several noted acousticians lent their support to certain ratios, for example: Louden: 1:1.4:1.9; Sabine: 1:1.5:2.5; Knudsen, Olsen: 1:1.25:1.6; Volkmann: 1:1.6:2.5; and Bolt, whose "blob" became well known, see Figure 8.5.

What is not well known is that the serious deliberations leading to these recommendations had to do with reverberation chambers for the purpose of measuring the sound power output of noisy mechanical devices.

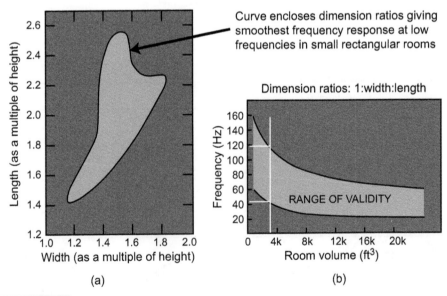

(a)   (b)

FIGURE 8.5 *(a) shows the Bolt "blob," a specification of room ratios, which are interpreted here as length and width, that yield the smoothest frequency responses at low frequencies in small rectangular rooms. (b) shows the frequency range over which the relationship has validity. For a 3,000 ft³ (85 m³) room, the optimum ratios are effective from about 40 to 120 Hz, as shown by the white lines on the graph. Adapted from Bolt (1946). Reprinted with permission from Bolt, R. H. "Note on Normal Frequency Statistics for Rectangular Rooms," J. Acoust. Soc. Am., 18, pp. 130–133. Copyright 1946, Acoustical Society of America.*

Nevertheless, these concepts migrated into the audio field, and certain room dimensional ratios have been promoted as having especially desirable characteristics for listening. In normal rooms, the benefits apply only to low frequencies. Bolt (1946) makes this clear in the rarely seen "range of validity" graph that accompanies his "blob," Figure 8.5. This shows that, in an 85 m^3 (3,000 ft^3) room, the optimum ratios are effective from about 40 to 120 Hz. This is similar to the room shown in Figure 8.3, which shows that this frequency range embraces six or seven axial resonances. This is consistent with the common experience that above the low-bass region the regularity of standing-wave patterns is upset by furniture, openings and protrusions in the wall surface and so forth, so that predictions of standing wave activity outside the bass region (i.e., above the transition frequency) are unreliable. In fact, even within the low-bass region wall flexure can introduce phase shift in reflected sound sufficient to make the "acoustic" dimension at a modal frequency substantially different from the physical dimension (see Figure 8.8).

Nevertheless, efforts to solve the riddle continued, with Sepmeyer (1965), Rettinger (1968) and Louden (1971) all making suggestions for superior dimensional ratios. Others, for example Bonello (1981), proposed superior metrics by which to evaluate the distribution of modes in the frequency domain. The latter was examined by Welti (2009) and found not to improve things significantly. Walker (1993) proposed some generous room ratio guidelines. All of them differ, at least subtly, in their guidance. Driven by the apparently irresistible logic of the arguments, information from these studies has been incorporated into international standards for listening rooms, and continues to be cited by numerous acoustical consultants as an important starting point in listening room design.

However, more recent examinations have given less reason for optimism. Linkwitz (1998) thought that the process of optimizing room dimensional ratios was "highly questionable." Cox et al. (2004) found good agreement between modeled and real room frequency responses of a stereo pair of loudspeakers below about 125 Hz, but ended their investigation by concluding that "there does not appear to be one set of magical dimensions or positions that significantly surpass all others in performance." Fazenda et al. (2005) investigated subjective ratings and technical metrics, finding:

It follows that descriptions of room quality according to metrics relying on modal distribution or magnitude pressure response are seriously undermined by their lack of generality, and the fact that they do not correlate with a subjective percept on any kind of continuous scale.

These people seem to be saying that the acoustical performance of rooms cannot be generalized on the basis of their dimensional ratios, and that reliably hearing superiority of a "good" one may not be possible.

The simple explanation is that there are problems with the basic assumptions underlying determinations of "optimum" room dimensions for domestic listening rooms or control rooms. The normal assumptions are:

- All of the calculated room modes are assumed to be simultaneously excited, and by a similar amount. This requires that the sound source be located at the intersection of three room boundaries, that is, on the floor or at the ceiling, in a corner. *Any* departure from this location will result in some modes being more strongly energized than others.

- It is assumed that the listener can hear all of the modes—equally. This requires that the head be located in another, preferably opposite, three-boundary corner. *Any* departure from this listening position means that all of the modes will not be equally loud or even audible.

- It is assumed that the room is perfectly rectangular, with perfectly flat, highly reflecting (rigid and massive) walls, floor and ceiling.

These assumptions do not describe the circumstances or rooms in which we live and listen. In real rooms, irregularities in geometry and variations in surface reflectivity affect the relative strength and frequencies of modes, even if they had the potential of being uniformly energized. Loudspeaker and listener locations are major factors.

It is difficult to understand how this concept of an optimum room got so much traction in the field of listening room acoustics, and why it has endured. Figure 8.6 illustrates the principles. In (a) it is shown that, even with the greatest of determination, a listener is not likely to put ears in the ideal location, and practical loudspeakers do not radiate all of their sound into a corner. This means that with a loudspeaker and a listener in typical practical locations, all of the calculated modes will not be equally audible, and any of the predictions of modal distribution will fail. We insist on listening to at least two loudspeakers, if not five or more, so the calculations underlying the ideal dimensions come to naught. In (c) is shown an idea that might work. Because we must deal with wave effects below the transition frequency, let us employ a separate sound system that is optimized for this purpose.

In conclusion, it is not that the idea of optimum room ratios is wrong, it is simply that, as originally conceived, it is irrelevant in our business of sound reproduction. This will be revisited in discussions of multiple subwoofer solutions, but employing very different criteria.

8.1.2 Are Non-rectangular Rooms the Answer?

A recurring fantasy about rooms is that if one avoids parallel surfaces, room modes cannot exist. Sadly, it is incorrect. Among the few studies of this topic, Geddes (1982) provides some of the most useful insights. He found that "room shape has no significant effect on the spatial variations of the pressure response" . . . "the spatial standard deviations of the p^2 response is very nearly uniform for all the data cases [the five room shapes evaluated in the computer model]." Source location was a factor in the behavior of the modes of course, as was the distribution of absorption. "Distribution of absorption was far more important in the more symmetrical shapes—[a non-rectangular] shape did help to distribute the damping evenly among the modes" (Geddes, 2005).

(a) The listening arrangement required to hear the benefits of a room having "perfectly optimized" dimensions. All modes are equally energized and all modes are equally audible. Problem: we don't do this.

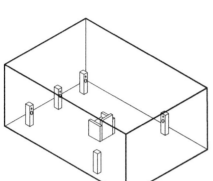

(b) The realities of listener location(s) and the number and locations of loudspeakers, stereo or multichannel, mean that the simple predictions do not apply. Every case will be different. There is no "ideal" room.

(c) It is greatly advantageous to deal with the low bass separately, using one or more subwoofers in a bass managed scheme.

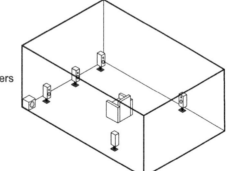

FIGURE 8.6 *(a) shows source and receiver locations necessary for all modes to be energized by the loudspeaker and audible to a listener—being practical and keeping them both at floor level. (b) shows how far a real situation is from the simplistic ideal of (a). (c) shows that separating the reproduction of low frequencies is a good way to address the room resonance problems in a manner that is independent of the number of reproduction channels.*

It seemed that the most effective modification to a rectangular room was to angle a single wall.

The behavior of sound in non-rectangular rooms is difficult to predict, requiring either scale models or powerful computer programs. Figure 8.7 shows two-dimensional estimated pressure distributions for modes in rectangular and non-rectangular spaces. It is clear that both shapes exhibit regions of high sound level, and nodal lines where sound levels are very low. The real difference is that in rectangular rooms the patterns can be predicted using simple calculations. Consequently, most acousticians favor the simplicity of rectangular shapes.

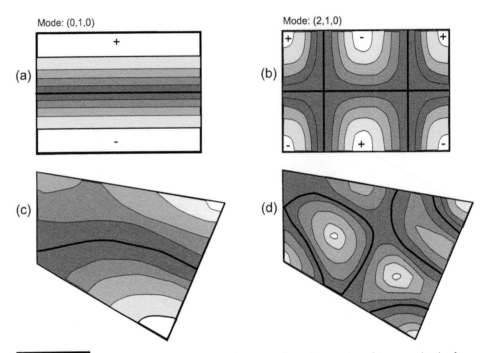

Mode: (0,1,0)   Mode: (2,1,0)

(a)   (b)   (c)   (d)

FIGURE 8.7 *Computed pressure distributions in rectangular and non-rectangular rooms, showing for each a simple mode and a more complex modal pattern. Pressure minima, nulls, are shown by heavy lines. Lighter shades indicate increasing sound pressure levels. Note that the instantaneous polarity of the pressure reverses at each nodal line (pressure minimum).*

8.2 SOLUTIONS FOR THE REAL WORLD

All of the topics to be discussed are in the peer-reviewed literature, some of it for a decade or more. There is nothing new here, but new generations of audiophiles need to become aware of it and learn how to use it to their advantage. Internet forums continue to have debates over issues long settled, and those with business interests continue to agitate to their advantage. The old expression: "If all you have is a hammer, everything looks like a nail" has an audio parallel: "if you sell acoustical materials, all listening rooms need extensive acoustical treatment." Maybe, maybe not. Acoustical materials are important ingredients in the solutions, but where historically they were the only solutions, there are now some interesting options.

There are three audible problems related to modes:

1. Frequency: The amplification or attenuation of specific sustained bass notes (e.g., bass guitar, synthesizer, organ) at different frequencies. Pitch shifting is another subtle possibility, as a rapidly decaying mode at one frequency is replaced by a mode with a slower decay at an adjacent frequency.

2. Time: The prolongation of brief bass sounds, like a kick drum, by ringing that can occur at one or more resonance frequencies (bass "booms").

3. Space: The variations in (1) and (2) that occur at different seats in the room. No two listeners hear precisely the same bass.

4. Beating: A lesser-known effect that occurs when two modes are very close in frequency. It can cause uneven decay of sound near the resonance frequency.

The audible degradation depends on:

- The amount of energy transferred from the woofer/subwoofer(s) to the resonances. This is determined by the locations of bass energy sources in the standing wave pattern, how many there are, and the specific nature of the signals driving them. Those factors determine which room resonances are energized and the extent to which they are energized.

- How audible individual resonances are to listeners. This is determined by where the listener sits. The standing waves associated with the resonant modes cause large seat-to-seat variations. As you move around a room, substantial variations in bass level and quality can be heard—different modes are audible at different locations. At very low frequencies, loudness changes by a factor of two with a mere 5 dB sound level change (see Section 4.4), compared to about 10 dB at middle and high frequencies. Therefore, small changes in bass level seriously influence impressions of spectral balance; a deficiency of bass is often perceived as an excess of treble, and vice versa.

- Source material.

For decades the only way to improve bass sound in small rooms has been to employ low-frequency absorbers or bass traps to damp the resonances. These are necessarily large, altering the visual décor, which may not be an issue in professional recording studios or control rooms, or even in dedicated stereo listening rooms or home theaters. However, high-quality reproduced sound is also desired in rooms that serve as normal living spaces, where visual appearances, even style, are a high priority. Consequently, as will be seen, the author has been motivated to find alternatives to the traditional remedies.

The decision of which method to employ depends on your circumstances, balancing the factors of appearance, performance, and cost. The good news is that there are choices. Here are some of the popular options to addressing the "room boom" problem:

1. Deliver energy to the modes and dissipate some of that energy with absorbers.

2. Deliver energy to the modes and reduce the coupling of that energy to the listener by optimizing the listening location.

3. Reduce the energy delivered to a bothersome mode by optimizing the subwoofer location.

4. Reduce the energy delivered to a bothersome mode by using high-resolution measurements and parametric equalization (matching frequency and Q) to attenuate the mode.

5. Reduce the energy delivered to bothersome modes by using mode-manipulation techniques: Strategic location of multiple subwoofers in the standing wave pattern to achieve destructive acoustical interference (mode canceling).

6. Computer-optimized multi-subwoofer solutions including signal processing of the subwoofer signals that allows for optimized mode manipulation.

In the following sections, each of these options is discussed in detail. It is said that "necessity is the mother of invention," and in the following it will be seen that my personal needs and perspectives drove the development of my own knowledge about sound fields in rooms and some unconventional solutions to common problems.

8.2.1 Deliver Energy to the Modes and Dissipate Some of That Energy with Absorbers

This is the time-honored method, and it works. It is virtually a requirement in recording studios—spaces that become part of the microphone-captured "original" sound and must be relatively neutral. It is often used in control rooms and in dedicated home listening rooms, but can be an aesthetic challenge in normal living spaces. To be maximally effective, the large low-frequency absorbers, often called "bass traps," must be located in the high-pressure regions of the standing waves of bothersome modes, meaning that they may be difficult to hide or disguise. The energy that is "trapped" is lost, of course; the total quantity of bass energy is reduced as a sacrifice to getting better quality bass. When one feels a wall, floor or membrane absorber vibrating, the mechanical energy in the movement has been removed from the sound field energy. Addressing specific bothersome modes without attenuating benevolent bass frequencies requires membrane absorbers tuned to those frequencies and/or absorbers placed in specific locations. Some thought is required.

In normal wood- or metal-frame buildings, that is, many homes, the single layer of drywall is a moderately effective absorber at low frequencies. The existing room boundaries "trap" some bass energy. Figure 8.8 shows measurements of sound level versus distance along the length of a listening room, at frequencies representing the first- and second-order axial modes along the length of the room. The room was perfectly rectangular, constructed of 2 × 6 inch (50 × 150 mm) studs with two layers of 5/8-inch (15.9 mm) gypsum board on the interior surface—it was designed with sound isolation as a consideration. Structurally, these walls are more massive and stiffer than those found in typical North American homes. The only opening was a heavy solid-core door located in the middle of the end wall on the right of Figure 7.2.

Several important observations can be made about these diagrams.

- In Figure 8.8b it can be seen that the standing-wave ratio of the first-order mode is only 9 dB. This indicates that the walls are absorbing substantial energy at this frequency: calculated to be 23.6 Hz for the 24 ft (7.32 m) long room. Energy lost in the boundaries causes the standing wave (constructive interference) peaks to be lower and the (destructive interference) minima to be higher. Further confirmation is in the shape of the sound pressure distribution, smoothly undulating, without

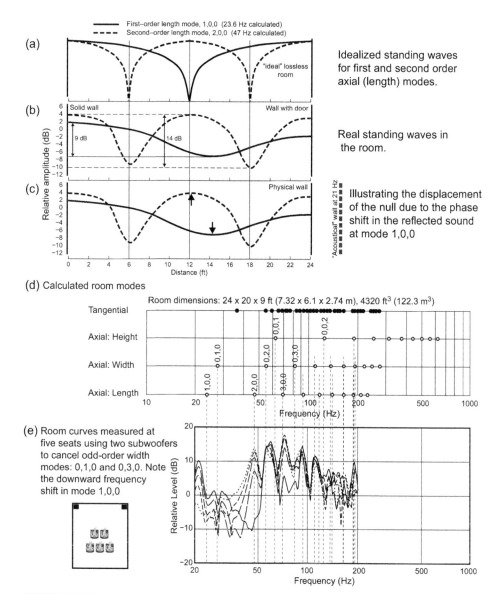

(a) Idealized standing waves for first and second order axial (length) modes.

(b) Real standing waves in the room.

(c) Illustrating the displacement of the null due to the phase shift in the reflected sound at mode 1,0,0

(d) Calculated room modes

Room dimensions: 24 x 20 x 9 ft (7.32 x 6.1 x 2.74 m), 4320 ft³ (122.3 m³)

(e) Room curves measured at five seats using two subwoofers to cancel odd-order width modes: 0,1,0 and 0,3,0. Note the downward frequency shift in mode 1,0,0

FIGURE 8.8   *(a) The stylized standing waves that might occur in a lossless, perfectly reflective room. (b) and (c) show measurements of sound level as a function of distance along the length of a real room, measured at the frequencies of the first- and second-order modes (1,0,0 and 2,0,0). (b) shows the difference between the maximum sound level and the minimum sound level (the standing-wave ratio) for each of the modes, indicating that there is more acoustical absorption (damping) at the lower frequency. (c) shows the predicted locations for maxima and minima for the modes (the fine vertical lines at quarter- and half-wavelength distances from the boundaries). The measured pressure maxima and minima are displaced from their anticipated locations. The discrepancy is large for the first-order mode and small for the second-order mode. See text for explanation. Adapted from Figure 13.7 in Toole, 2008. (d) and (e) show the calculated room modes and room curve measurements confirming their presence, and the absence of the deliberately canceled modes. The frequency shift of mode 1,0,0 is clearly seen. Adapted from Figure 13.8 in Toole, 2008.*

a sharp null as shown in Figure 8.8a. The spatial Q of this resonance has been reduced compared to the second-order mode.

■ The second-order mode exhibits a larger max/min spread of 14 dB, indicating that the walls are more reflective at this frequency. The curves also exhibit sharper, albeit not sharp, dips. This mode has a higher Q than the first-order mode.

■ The minimum for the first-order mode is not where it should be—at the halfway point down the length of the room. It is shifted by about 2 ft (0.6 m) toward the wall having a door. The phase-shifted (delayed) reflection from that wall makes the room "acoustically" longer than the physical dimension, and that the extension is at the end where the door is located. A boundary that moves is a membrane absorber, absorbing a portion of the energy falling on it, and reflecting the remainder with a phase shift. In this case, strong vibrations were felt in the wall at 21 Hz. This is true for all membrane, or diaphragmatic, absorbers and it was a feature that I deliberately used in designing membrane absorbers for the National Research Council of Canada (NRCC) listening room many years ago, simultaneously damping modes and moving bothersome nulls away from the prime listening locations (Toole, 1982, appendix). In this case, at this particular frequency, the phase shift has the same effect as moving the wall by some distance beyond the physical location. If this is so, there should be a corresponding lowering of the frequency at which the resonance is observed. This is confirmed in Figure 8.8e, which shows that the first modal peak (1,0,0) in frequency-response measurements is substantially lower than the predicted frequency shown in Figure 8.8d.

■ The second-order mode (2,0,0) nulls and maximum are all very slightly displaced toward the wall with the door. However, the movement is small, indicating that at this higher frequency (47 Hz), the wall with the door is substantially less absorbing than it is at the lower first-order mode frequency. Figure 8.8e confirms that the shift in resonant frequency is not evident.

■ The odd-order modes that were acoustically canceled by the two-subwoofer configuration (0,1,0 and 0,3,0) are not visible as peaks in Figure 8.8e. In fact, mode 0,1,0 appears as a pronounced narrow acoustical interference dip. In contrast, the even-order mode 0,2,0 is amplified, showing up prominently. This is explained in Section 8.2.5.

The simple explanation for all of this is that in spite of the solid construction of the room, the stiffness of one end wall of this room was reduced by the presence of a door in the center of it. The wall became more flexible at very low frequencies. The performance at higher frequencies—above about 50 Hz—was unchanged. Does this matter? Not really. Low-frequency absorption is a good thing, damping the mode. The small frequency shift is normally of no consequence. It is of interest to note that, in the spatial domain, the peaks are broad and the dips are narrow. Attenuating a peak is beneficial to more listeners than elevating a dip, although damping achieves both.

The energy lost to the room-boundary membrane-absorption process can considerable, resulting in 2 to 3 dB reduction in amplitude over a significant frequency range as can be seen in Figure 8.9a. Obviously the thick, heavy wall constructions necessary for good sound isolation aggravate low-frequency resonances, as do room walls and/or floors constructed of heavy masonry or concrete. Figure 8.9b shows how much more effective optimally designed and located low-frequency absorbers can be. Figure 8.9c shows an interesting comparison of active and passive low-frequency absorbers. As can be seen, these curves are substantially smoothed compared to the high-resolution

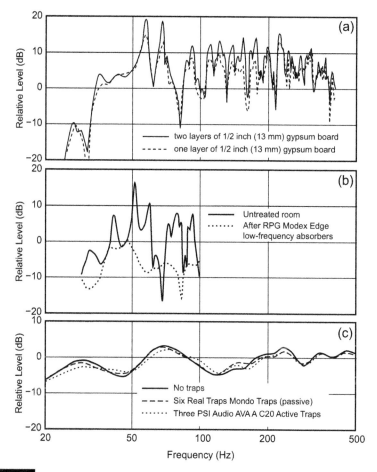

FIGURE 8.9 *(a) Measurements made with a subwoofer placed in a front corner, and a microphone located at the prime listening location in a 3,000 ft³ (85 m³) listening room. The solid curve shows results with two layers of gypsum board and the dashed curve shows results for a single layer of gypsum board mounted normally on 2 x 6 inch (50 x 150 mm)) wooden studs. (b) An example of damping provided by some commercial low-frequency absorbers. Courtesy RPGinc.com. (c) A comparison of active and passive low-frequency absorbers in a music mastering room. Katz (2016), figure 2, reproduced with permission. These data are spectrally smoothed so individual modes are not visible.*

data in Figure 8.9a and b. Information about individual modes is lost, meaning that it is not possible to observe the extent to which different modes are damped. This would be associated with the locations of the absorbers and could lead to better optimized arrangements. But the overall trends in absorption can be seen.

With energy disappearing into "traps," the loudspeakers will work harder to generate sound levels, but the perceived sound will be less resonant—there is a trade-off involved. Narrow-band, tuned membrane absorbers are a logical solution when only a few modes are problematic, placed so that they address the offending modes (use standing-wave patterns like those in Figure 8.4 to find the high-pressure regions). There are several suppliers of low-frequency absorbers, most of which are broadband, and some of them are not effective at very low frequencies, so look for realistic specifications. Broadband absorption is useful in some situations, just as narrow band is in others.

Alternatively, active bass absorbers are available. These resemble powered subwoofers but they have microphones and control electronics that turn them into absorbers of bass, not producers of bass. With electronics involved they can be very flexible in what they absorb.

Passive or active, all absorbers work, especially if optimally located.

Were the resonant modes eliminated? No, but they were acoustically damped to reduce the height of the peaks and the depth of the dips, reducing "booms" and seat-to-seat differences. Broadband bass traps remove energy at frequencies other than those involved with the resonant modes; the room will be less reflective overall, which may or may not be desirable. Narrow-band, tuned, membrane absorbers can address the problem modes in relative isolation, as can some of the following solutions. In general, though, low-frequency absorption is desirable, as it will benefit any additional remedies that may be selected.

8.2.2 Deliver Energy to the Modes and Reduce the Coupling of That Energy to the Listener by Optimizing the Listening Location—"Positional" Equalization

Moving closer to a null in the standing wave of a bothersome mode reduces the sound pressure at the ears. This is a solution for a solitary listener, or in some situations, a small number of listeners in the right locations. See Figure 8.10 for a real example with frequency and time-domain data. This was in my "classical" music listening room seen in Figure 7.19. The deliberately irregular shape of the room was intended to inhibit the development of strong room modes, and whether it did or not it would cause the damping to be better distributed among the modes as discussed in Section 8.1.2. In this case the energy in resonances was concentrated in a second-order length mode between the only two parallel walls of significant area. This was a room with one problem resonance, but it was a monster. Bass transients boomed outrageously and certain organ pedal notes were overwhelming.

Numerous large view windows and my aversion to unattractive objects caused me to think of alternatives to low-frequency absorbers. This situation is described in Figure 8.10,

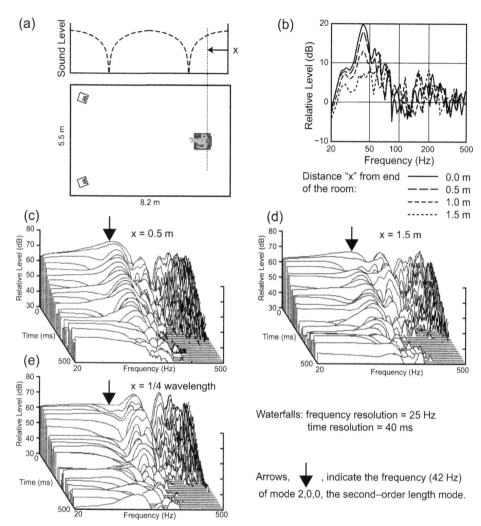

(a)

(b)

Distance "x" from end of the room:
— 0.0 m
– – – 0.5 m
– – – – 1.0 m
· · · · · · 1.5 m

(c)

(d)

(e)

Waterfalls: frequency resolution = 25 Hz
time resolution = 40 ms

Arrows, ↓ , indicate the frequency (42 Hz)
of mode 2,0,0, the second–order length mode.

FIGURE 8.10 *Measurements made in a large (7,770 ft³/220 m³) living/listening room. (a) the floor plan showing the second-order length mode (2,0,0) with 1/4-wavelength minima at 2 m from the end walls. (b) High resolution frequency responses at different distances from the right end wall showing the reduction of the resonant peak as the minimum is approached. (c), (d) and (e) show waterfall diagrams at two of the locations in (b) and at the 1/4-wavelength minimum. The waterfalls have frequency resolution of 25 Hz, which corresponds to a time resolution of 40 ms.*
Adapted from Toole (1990), figure 13.20 in Toole (2008).

where I illustrate what I call "positional equalization"—finding a seating location where the ears were not strongly coupled to the offending standing wave. I show that when it sounded right, the frequency response and waterfalls looked right. The "boom" went away. Note that the optimum seating location was not the 1/4-wavelength null—there the resonance was notably absent, but bass was audibly deficient. The evidence that the boom was gone is in the steady-state frequency response: the resonance peak is attenuated. Room resonances behave as minimum-phase systems.

In Figure 8.10c it can be seen that the reduced frequency resolution has smoothed the 42 Hz resonant peak seen clearly in Figure 8.10b. However, the reward is that we are able to see the prolonged ringing as time moves forward in the waterfall. A sharp resonant peak in the steady-state room curve correlates with significant ringing. The bass was intolerably boomy. In Figure 8.10d the listening/measurement point has been moved toward the null and, as seen in Figure 8.10b the resonant peak seen in the back curve is attenuated. This is accompanied by a time domain response that drops very quickly by about 12 dB and then decays. The bass boom was greatly reduced, and the bass sounded substantial and tight.

Going to an extreme, Figure 8.10e shows behavior in the 1/4-wavelength null. There is no evidence of the resonance at all in either the frequency or time domains but, subjectively, there was insufficient bass. So, the ideal solution is to attenuate the resonance by an optimum amount, but not to eliminate it. The best sounding condition was $x = 1.5$ m, a slight bass rise, as shown in (b).

This was a totally passive solution—no modifications except that my chair ended up in a silly location 5 ft out from the back wall. I put up with this for a while, but when I started to see tracks in the carpet from dragging the chair out for my listening sessions I decided that a better solution was needed.

> Was the mode eliminated? No, but the excessive energy in it was not coupled to the ears of the listener in the prime location. Others in the room got no benefit. There was no change to the overall reflective characteristics of the room, which, in this case, was not a problem.

8.2.3 Reduce the Energy Delivered to a Bothersome Mode by Optimizing the Loudspeaker/Subwoofer Location

Moving the sound source away from the other end of the room and closer to a null reduces the energy transfer to the problematic mode. This is the inverse of Section 8.2.2, with the difference that the attenuation of the room mode is experienced everywhere in the room. The optimization can work for a single listener, or a few, but others take their chances—seat-to-seat variations still exist. In my situation this was not a practical solution in this room, so the following remedy was adopted.

> Would the mode be eliminated? No, but the energy in the mode could be reduced to a useful degree for the listener in the prime location and possibly for other locations. But there would be substantial seat-to-seat variations. The room acoustics were unchanged.

8.2.4 Reduce the Energy Delivered to a Bothersome Mode by Using Parametric Equalization

Here we introduce electronic manipulation of a physical acoustical phenomenon. It is possible because low-frequency room resonances behave as minimum-phase systems,

and therefore respond to corrective equalization of the frequency response peak by attenuating the time-domain ringing. To be clear, *this does not eliminate the resonance when
there is no equalized audio signal present*. Pause the playback, have an interlude of live
acoustical music and the room resonances will have their effects. However, press the
"play" button and music reproduced through loudspeakers supplied with the appropriately equalized signal will be heard with the problematic room resonance(s) attenuated.
This works best for one listener, where the mike is located, or possibly for a few in a row
or column of listeners oriented with respect to simple axial modes. Others in the room
take their chances—seat-to-seat variations can be substantial, as they were before the
equalization.

Figure 8.11 shows data comparing approach (2) with the equalization method. On
the left, (a) shows the frequency response and (c) the waterfall for "positional" equalization, with the listener at 1.5 m from the back wall (from Figure 8.10b and d). Figure 8.11b and d on the right show the comparable results for the listener moved back
to 0.5 m from the wall and matched parametric equalization engaged. The main point
of the figure is to show how very similar the waterfall plots look. Most of the small
differences in both the frequency responses and the waterfalls are the result of making
the measurements at different locations, with the consequent different interactions with

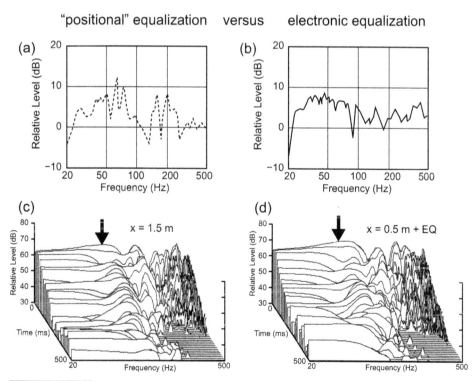

FIGURE 8.11 *A comparison of steady-state frequency responses (a) and (b) and waterfall diagrams (c)
and (d) for two methods of addressing a problematic room resonance.
From Toole (2008), figure 13.21.*

other standing waves in the room, none of which was as consequential as the "monster" mode at 42 Hz.

Looking similar is one thing, but how did they sound? Over several months, colleagues, audio journalists and interested social visitors were subjected to simple A vs. B comparisons of the two conditions. The overwhelming conclusion was that they sounded remarkably similar in every respect and very much better than the original condition. The most dramatic demonstration sound was a well-recorded kick drum that, before treatment, was amusingly fat and flabby and, after treatment, became an abrupt "slam"—as it should be. It was audibly obvious that the time domain problem had been repaired.

Equalization had the huge advantage of allowing the listener to sit in a sensible location. Acoustically, there were advantages too. With up to a 14 dB amplitude reduction around 42 Hz, the woofers no longer had to work so hard, distortion was lower and they could play louder. There was also *much* less energy everywhere in the room at 42 Hz.

> Was the mode eliminated? No, but audible evidence of it was removed when the sound system was operating. It only works for sounds radiated by the loudspeakers, not from a bass guitar that someone might pick up and play. It also only worked for my seat, and one on each side on a sofa. Everyone else in the room took their chances. The room acoustics were unchanged.

If this solution is chosen, it is important to know what kind of equalizer to use. Figure 8.12 illustrates what happens when measurements on a subwoofer in a room lack the resolution to reveal the problem in sufficient detail, and when the equalizer lacks the resolution to address the resonance. It is evident that high-resolution (1/20-octave in this case) measurements were necessary to reveal the true nature of the resonance and that matching the Q of the resonance with the parametric equalizer had a significant beneficial effect on the reduction of time-domain ringing. Excessive smoothing of room curves disguises the true nature of the resonance and prevents a maximally effective equalization.

8.2.5 Reduce the Energy Delivered to a Bothersome Mode by Using Simple Mode-Manipulation Techniques

- Relocating the subwoofer or the listener in the standing wave pattern.

- Use multiple subwoofers situated and driven to achieve destructive acoustical interference (mode canceling).

Before embarking on this topic, it is necessary to understand some properties of standing waves and their interactions with woofers and subwoofers. Figure 8.13 shows the basic methods of manipulating the supply of bass energy to low-frequency modes.

Perceptive readers will have noticed that conventional arrangements of stereo loudspeakers will result in the cancellation of the first-order width mode because most bass in stereo recordings is essentially monophonic. The attenuation works, thereby liberating the symmetrical axis of the room for the stereo seat, which is no longer in a profound null. Taking it further, those listeners who are able to place their loudspeakers at the 25% positions will find three width modes attenuated. I doubt that many people ever

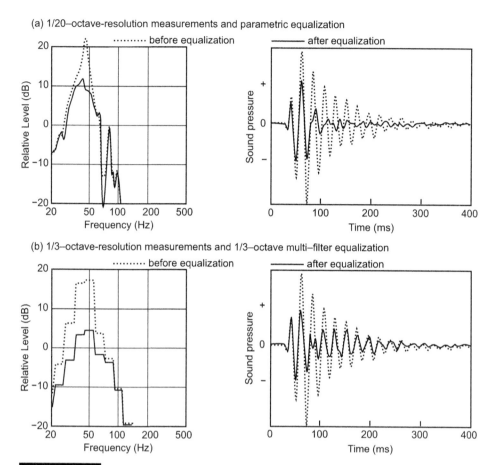

(a) 1/20–octave-resolution measurements and parametric equalization

(b) 1/3–octave-resolution measurements and 1/3–octave multi–filter equalization

FIGURE 8.12   *Before and after measurements in the frequency and time domains for two different measurement and equalization methods. This is figure 13.24 in Toole (2008).*

thought about basic stereo as a room-mode-manipulating scheme. It has been working for us for decades, and does so very well because most bass in stereo recordings, and all bass in LPs, is monophonic.

As luck would have it, my other listening room, the home theater shown in Figure 7.18, also suffered from an enormous bass boom. It was quickly traced to a combination of a first-order width mode and a second-order length mode having almost identical frequencies, both being well energized by a single subwoofer in a corner. Because of the large view windows, large bass traps were not an option. I took it as a challenge to solve the problem without degrading the appearance of the room. The floor plan is shown in Figure 8.14a. Employing the logic of Figure 8.13b. I replaced the single sub in the corner (position 1) with two identical subs located in positions 2 and 3. The subs, driven with identical signals, were in opposite polarity lobes of the width mode and also of the length mode, so both of the problem modes were destructively energized and substantially attenuated. The resonant peak was attenuated, and one of

(a)

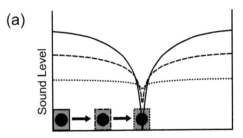

Using a first-order width mode as an example, it is seen that the subwoofer maximally energises the mode when it is against the wall, and progressively decouples from the mode as it it moved toward the null. A sound *pressure* source located at a pressure minimum - a *particle-velocity* maximum - does not couple energy efficiently and that mode is less energised.

(b)

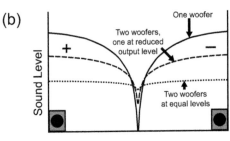

The sounds on opposite sides of a null have opposite polarities at any instant in time. Therefore placing identically polarized, in phase, subwoofers in each of the lobes results in destructive acoustical interference. The mode is attenuated by an amount that can be controlled by varying the signal level to one of the two subwoofers. Inverting the polarity of one of the subwoofers will amplify the mode, and that too is adjustable.

(c)

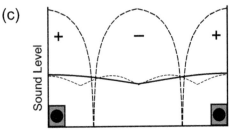

Looking at the second- and third-order modes, the third has also been attenuated because the lobes containing the subwoofers have opposite polarity. The second-order mode has been amplified because the subwoofers are in lobes having the same polarity.

——— first order
---- second order
------- third order

(d)

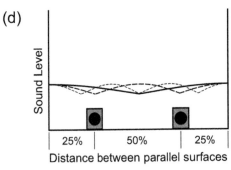

25%   50%   25%
Distance between parallel surfaces

Moving the subwoofers to the null locations for the second-order mode reduces the efficiency of energy transfer to that mode, attenuating it. The other modes remain attenuated because the subwoofers continue to be located in lobes having opposite polarity. These locations are the 25% points across the width of the room. Stereo setups in this configuration are a mode attenuating scheme that minimizes the problem of the listener being in a position of left–right symmetry at low frequencies.

FIGURE 8.13 *A simplified, exaggerated illustration of how mode manipulation works. (a) Subwoofer location is used as a means of controlling the amount of energy supplied to a bothersome mode. (b) Taking advantage of polarity reversals in the adjacent lobes of standing waves, two subwoofers are used to vary the amount of attenuation of a specific mode, potentially achieving cancellation. (c) presents a more realistic view by including three orders of width modes. The odd-order (first and third) width modes are attenuated because the subwoofers are in the opposite-polarity lobes of these modes, but the second-order mode is amplified because the subwoofers are in lobes exhibiting the same polarity. (d) shows a combination solution, in which approach (a) addresses the second-order mode and approach (b) addresses the first- and third-order modes.*

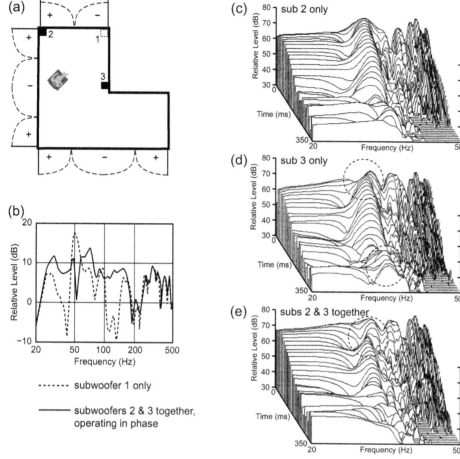

FIGURE 8.14   *A family room attached to a dining area and kitchen in an L-shaped space. (a) shows that there were three axial modes at similar frequencies, all of which were strongly energized by a single subwoofer located in the corner behind the video display—subwoofer 1. (b) shows (dotted curve) that there was a powerful resonance around 50 Hz, and large holes in the bass response above and below that frequency. Inspection of the polarities in the standing wave lobes suggested that a pair of subwoofers, one in a lobe of each polarity, should alleviate the problem. This was conveniently possible, at locations 2 and 3, and the solid curve in (b) was the result, with no equalization. The prominent resonant peak has been transformed into a narrow acoustical interference dip. (c) shows the waterfall at the prime listening position for subwoofer 2 only, (d) shows one for subwoofer 3 only, and (e) shows one for both subwoofers. In (c), the energetic ringing is obvious, and it appears that, with time, the ringing skews downward in frequency. In (d) it becomes clear that the reason is that there are at least two closely adjacent modes that are revealed because they have different decay rates (amounts of damping). See what appears to be a quite discrete downward frequency shift in the top circle, from one rapidly decaying mode to a more energetic lower-frequency one, leaving a single resonant tail in the bottom circle. In (e) it is clear that, with both subwoofers operating, the aggressive ringing is substantially attenuated. This is seen as a rapid 12 dB drop in amplitude in the very early decay before low-level ringing disappears into room noise. Evidence of the two modes can still be seen early in the decay. (a) and (b) are adapted from Toole (1990), figures 14 and 15. This is figure 13.22 in Toole (2008).*

the modes was truly canceled, as is demonstrated by the narrow destructive-interference dip in Figure 18.14b. The other improvements were fortuitous, and never analyzed, following the "leave well enough alone" principle. The good behavior can also be seen in the waterfall diagrams, as can evidence of energy sharing between the two closely spaced, overlapping resonances. In Figure 8.14d it is clear that the modes are merged, and with both subs operating the higher frequency one is seen to decay very quickly, and the lower frequency one is greatly attenuated before decaying more slowly. The latter could have resulted in some pitch shifting in the very early part of the decay, but no listeners appeared to hear it; the bass sounded satisfyingly smooth and tight. There was no equalization. Again, the trustworthy evidence of good behavior is in the steady-state room curve; the waterfall diagrams provide decorative confirmation.

I presented this data at an AES Conference (Toole, 1990). It may have been the first application of deliberate mode canceling in an audio room-acoustics application.

Figure 8.8e shows another example of simple mode canceling.

> Were the modes eliminated? In effect, yes, but only when the sound system was operating, and only for the sounds radiated by it. The benefits were clearly audible in all of the seats in the room.

8.2.6 Selective Mode Activation in Rectangular Rooms Using Passive Multiple-Subwoofer Mode Manipulation

Not long after the preceding events took place, I moved to California to work for Harman International as corporate vice president of acoustical engineering. I carried with me the notion of having substantial control of room resonances without the visual and real estate commitments to large bass traps. A room could be an ordinary room, visually and acoustically, with ordinary acoustical properties and problems, until the "play" button was pressed—then it could become extraordinary. The goal would be to provide similar bass for multiple listeners in the room, reducing seat-to-seat variations. Then equalization can be useful to more than just the listener in the sweet spot.

I set up a research group, and one of my first employees was Todd Welti, an acoustician skilled in acoustical modeling. We agreed that mode manipulation should be able to work in a more generalized sense, not just for one mode in a specific custom room installation. He first tackled the case of simple rectangular rooms, and after substantial analysis, involving simulations using from one to 5,000 subwoofers (acoustic wallpaper?), concluded that two or four subwoofers—identical, and identically driven—created a central zone in which seat-to-seat variations were usefully reduced (Welti, 2002a, 2002b). This means that equalization can benefit several listeners.

Figure 8.15 illustrates the underlying logic of the process. Figure 8.15a shows that a single subwoofer in a corner energizes all of the horizontal-plane modes in the room producing null (anti-node) lines as indicated in the floor plan. A metric, Mean Spatial Variation (MSV), was calculated for a square area in the center of the room, indicating the potential seat-to-seat variations in bass level. With all the modes active, the MSV was high. Moving the single subwoofer from the corner to a mid-wall position did what

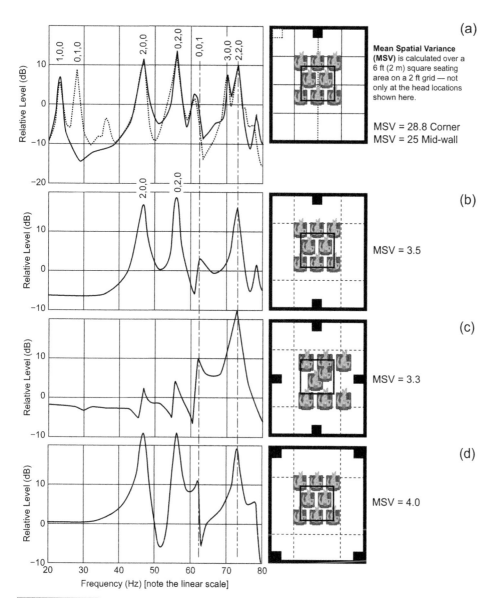

FIGURE 8.15 consists of "average total" frequency response curves taken from Welti (2002a, 2002b). This is the combination of direct sound from the subwoofers and the modal energy within the room. They have been traced and vertically rescaled to conform to the standard used throughout this book. The Mean Spatial Variance (MSV) values are from Welti and Devantier (2006, figure 5(c)) for the same loudspeaker configurations.

was predicted in Figure 8.13a—the first-order width mode (0,1,0) at 28 Hz was attenuated because the subwoofer was located at a pressure minimum. Because the one of the nulls running through the seating area was removed (the dotted line), the MSV was reduced, but only slightly.

In Figure 8.15b, the addition of a second identical in-phase subwoofer at the center of the opposite wall cancels the first-order length mode (1,0,0) at 23.5 Hz (as shown in Figure 8.13b) and the accompanying odd-order mode (3,0,0) near 70 Hz. Eliminating these nulls in the listening area results in a dramatic reduction in the MSV. However, the two woofers now amplify the second-order length mode (2,0,0) at 47 Hz, giving it more amplitude. The gain in the second-order width mode (0,2,0) at 56 Hz appears to be the result of having subs at both ends of the room, symmetrically driving this mode from the central high-pressure region. But, because the pressure nulls of second-order modes are at the 25% and 75% locations between room walls, they miss the designated listening area. The amplitude gains in modes 2,0,0 and 0,2,0 help boost the bass efficiency. However, the MSV is attractively low, and equalization to reduce the resonance peaks in fact gives the system increased headroom at those frequencies, so this is a viable option.

In Figure 8.15c, it is seen that adding two more identical mid-wall subwoofers puts in-phase excitation in all three pressure maxima of the second-order modes along both length and width axes. They are greatly attenuated, significantly reducing the system efficiency. But what is happening around 74 Hz? While we were attending to the lower-frequency axial modes, a tangential mode has been creeping up in amplitude and now dominates the acoustic delivery, thereby compensating somewhat for the loss of the axial modes. This is because we have placed in-phase subwoofers at in-phase regions in the standing-wave pattern for this mode, and it is responding. Figure 8.16a shows

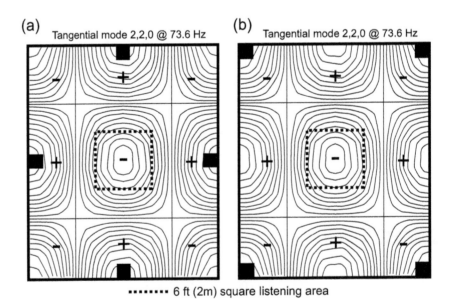

6 ft (2m) square listening area

FIGURE 8.16 *(a) The pressure contours for tangential mode 2,2,0 are shown for the test room. Subwoofers in two configurations are shown to be located in high pressure points that all share the same polarity, resulting in maximal stimulation of this mode. The listening area is shown to be fully within the central high-pressure region, resulting in this mode being strongly visible in the frequency response of Figure 8.15(c)–(d), and in the low seat-to-seat variation metric, MSV.*

what is happening. The seat-to-seat variations are low because this mode includes all listeners within a broad high-pressure region, and this is now, by far, the strongest mode in the room.

Moving to Figure 8.15d, it is seen that placing the subwoofers at the four corners is also a great improvement in reducing seat-to-seat variations, with a considerable increase in overall sound level compared to the four mid-wall locations in (c), but only slightly better than the two mid-wall subwoofer solution in (b).

These methods worked, but there were questions about the tradeoffs between reducing the seat-to-seat variations and the efficiency of the resulting systems. Welti (2012)

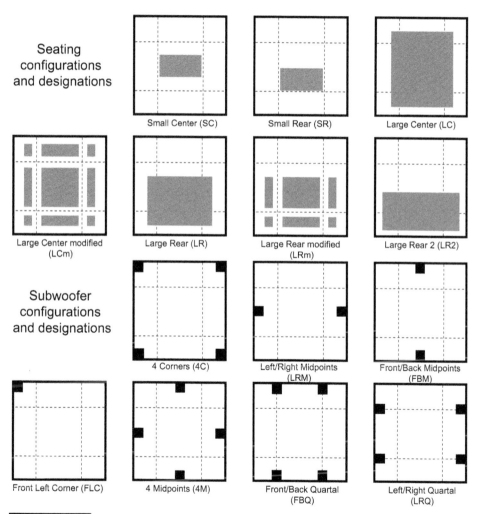

FIGURE 8.17 *The seating and subwoofer configurations used in the Welti (2012) study. As before, all subwoofers are identical, operating with the same polarity and gain. The dashed lines are the null locations for the second-order axial and tangential modes.*

revisited this topic and elaborated the investigation to include more variables. He added more seating arrangements, including larger, more realistic, listening areas, with more close to the rear of the room, as happens often in home theaters. The results in Welti (2012) run to many pages of plots and tables, all of which are available in color online at www.harman.com/innovation. Figure 8.17 shows the variables in his calculations.

The listening areas are different from that used in the preceding examples, so the MSV numbers will not be directly comparable. Given that we have already seen that moving listeners away from null lines is advantageous, it is clear that some of the large seating configurations in the new study will yield higher MSVs than the smaller, more centrally located areas. Those that avoid the null lines would be expected to show improvements over those that do not.

Figure 8.18 shows mean MSV values for these configurations. It can be seen that the small, centrally located listening areas SC and SR are advantageous, along with the Large Rear modified (LRm) arrangement that avoids the second-order null lines. This shows up very clearly in the averaged values along the bottom of the table. If one were to choose the best of the listener arrangements, SC, SR and LRm, the averages along the right side of the table indicate an advantage for the 4M loudspeaker arrangement, but the spread among them is small compared to other numbers in the table.

Figure 8.19 answers the second important question: now that one can identify listener and loudspeaker arrangements that yield superior seat-to-seat consistency, which listener and loudspeaker arrangement(s) generate the highest sound levels from a given set of subwoofers? The MOL metric (Mean Output Level) is a metric giving a relative low frequency efficiency for each configuration (averaged SPL over all seats from 20–40 Hz, normalized to account for the number of subwoofers used).

The answer is the same for the listener arrangements, but with respect to loudspeaker arrangements the outstanding winner for efficiency is 4C—four subwoofers,

	SC	SR	LC	LCm	LR	LRm	LR2	Ave.	
FLC	27	32	41	36	41	36	41	36.3	
4C	12	22	35	28	35	26	37	27.9	20
LRM	12	19	28	23	28	21	31	23.1	17.3
FBM	10	19	28	22	28	21	31	22.7	16.6
4M	11	15	21	21	21	21	22	18.9	15.6
FBQ	10	24	26	24	26	21	30	23.0	18.3
LRQ	16	18	28	22	27	22	30	23.3	18.6
	11.8*	19.5*	27.6	23.3	27.5	22*	30.1		

FIGURE 8.18 *Mean MSV values (dB²) for the configurations shown in Figure 8.17, listener arrangement horizontally and loudspeaker arrangements vertically. In this table, lower values indicate smaller seat-to-seat variations. The best values for each seating arrangement are in bold. The basic table is from Welti (2012), table 2. The additional numbers added outside the table boundaries by the author are, horizontally, the averages of listener-configuration scores, ignoring the poorly performing reference condition: FLC. The vertical numbers on the right are the average loudspeaker-configuration scores for the best listener arrangements SC, SR and LRm. Asterisks indicate the top three scores.*

	SC	SR	LC	LCm	LR	LRm	LR2	Ave.	
FLC	−6	−5	−3	−3	−3	−4	−1	**−3.6**	
4C	−2	−3	−5	−4	−5	−4	−7	**−4.3**	−3
LRM	−10	−9	−10	−9	−10	−9	−9	**−9.4**	−9.3
FBM	−9	−10	−10	−10	−10	−9	−10	**−9.7**	−9.3
4M	−8	−8	−9	−9	−9	−9	−10	**−8.9**	−8.3
FBQ	−4	−6	−7	−7	−7	−6	−9	**−6.6**	−5.3
LRQ	−5	−5	−7	−6	−7	−6	−8	**−6.3**	−5.3
	−6.3*	−6.8*	−8	−7.5	−8	−7.2*	−8.8		

FIGURE 8.19   *Mean MOL values (dB) for the configurations shown in Figure 8.17, listener arrangements horizontally and loudspeaker arrangements vertically. In this table, higher values (the smaller negative numbers) indicate higher subwoofer efficiency over the frequency range 20–40 Hz (higher sound levels for a given signal level). The best values for each seating arrangement are in bold. The basic table is from Welti (2012), table 3. The additional numbers added outside the table boundaries by the author are, horizontally, the averages of listener-configuration scores, ignoring the poorly performing reference condition: FLC. The vertical numbers on the right are the average loudspeaker-configuration scores for the best listener arrangements SC, SR and LRm. Asterisks indicate the top three scores.*

MSV = Mean Spatial Variance
(seat-to-seat variations within
the shaded areas)

MOL = Mean Output Level
(subwoofer output efficiency)

Most preferred listening configurations

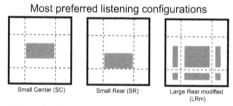

Small Center (SC)   Small Rear (SR)   Large Rear modified (LRm)

Loudspeaker configurations that work best for the most preferred listening configurations

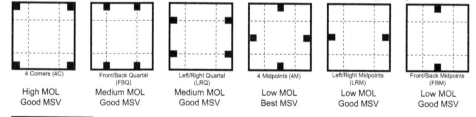

4 Corners (4C)	Front/Back Quartal (FBQ)	Left/Right Quartal (LRQ)	4 Midpoints (4M)	Left/Right Midpoints (LRM)	Front/Back Midpoints (FBM)
High MOL	Medium MOL	Medium MOL	Low MOL	Low MOL	Low MOL
Good MSV	Good MSV	Good MSV	Best MSV	Good MSV	Good MSV

FIGURE 8.20   *The author's personal summary of Welti's investigation, assuming preferred seating configurations used.*

one in each corner. Not far behind are the FBQ and LRQ subwoofer configurations. The mid-wall subwoofer arrangements, which did well in terms of MSV (Figure 8.18), do so at the expense of higher power requirements and/or larger subwoofers.

The lower-performing listener configurations are those that embrace problematic null lines, thereby increasing seat-to-seat variations. However, the prime listening areas SC, SR and LRm are included in those larger configurations. So, if one of the larger areas is necessary to accommodate a large audience, a logical decision might be to employ a subwoofer configuration that performs well for the smaller areas and simply accept the reality that some seats will be better than others. That way at least some number of listeners will have the potential of hearing similarly good bass, and the good seats are identified. The author's simple summary of Welti's very thorough investigation is in Figure 8.20. Of course, low MOL is addressable by using more powerful subwoofers. Low MSV is inherent to the arrangements.

Readers wishing to explore more deeply should go to the original reference (Welti, 2012) and access the supplemental illustrations that are available on the Internet. An attempt was also made to achieve a single metric to quantify MSV and MOL.

8.2.7 Mode Manipulation for Rectangular Rooms Using Multiple Subwoofers and Signal Processing

As has been demonstrated, rectangular rooms have advantages for passive multi-subwoofer strategies. They also are advantageous if one goes to the next level and introduces signal processing of the signals being supplied to multiple subwoofers. The idea seems to have originated with Goertz et al. (2001), who used an arrangement of subwoofers on the front wall, generating an approximation of a plane wave that propagates to the rear of the room where a similar arrangement of subwoofers, appropriately delayed and polarity inverted, canceled the wave when it arrived. Ideally this eliminates standing waves and everyone in the room should hear good bass. It requires a perfectly rectangular room to work well. The scheme was called Double Bass Array (DBA).

Santillán (2001) demonstrated that a system of 16 subwoofers arranged in rows in the walls could establish a large useful listening area in a rectangular room. The installation would only be practical in a dedicated installation, and the implementation is not simple.

Celestinos and Neilsen (2008) developed the Controlled Acoustic Bass System (CABS), a DBA scheme that could be retrofitted into an existing rectangular room. They examined the process in models and real rooms and demonstrated that it works well, yielding reduced variations in bass over much of a room.

Again the requirement for these strategies is a rectangular room; geometrical or structural irregularities would introduce errors. The best performance can be expected when four front-wall and four rear-wall subwoofers are arranged to be 25% of the way from floor, ceiling and side walls, thereby attenuating the width and height modes, leaving only the length modes to be avoided. These identical subwoofers would be mounted in or on the walls.

8.2.8 Mode Manipulation for Any Room Using Multiple Subwoofers and Signal Processing: Sound Field Management (SFM)

There are several reasons why a room may not be optimum for the solutions discussed in the previous sections:

- The room is not rectangular.

- The room is basically rectangular, but
  - There is an opening to another space;
  - There is acoustical asymmetry (e.g., one or more walls with distinctive construction, including brick, stone, expanses of glass);

□ One or more walls is not flat: a large fireplace protrusion, alcoves and so forth. In custom theaters the screen wall is often a culprit; in addition to the screen, there is often custom cabinetry to house loudspeakers and equipment. In highly ornamented rooms, surfaces may be broken up with fake columns, a bar and so forth;

□ Massive leather seats, arranged in rows on a staged floor, can hardly be ignored.

■ Desirable subwoofer locations are not all available, or practical to use.

■ Listener locations are not within the desirable areas.

In practice, such rooms are very common. What then? The pieces of the puzzle exist, and those with the time and good instincts can, by trial and error, find combinations of location and settings that provide improved listening situations in "difficult" rooms. But what is needed is assistance from ubiquitous computers to reduce the labor content in finding the right combinations of settings. While we are at it we could add more to the wish list, including allowing the use of subwoofers that are not required to be identical.

The basis of these solutions is real acoustical measurements—full complex data: amplitude and phase or impulse response—between each subwoofer location and each listener location. Then some form of signal processing is used to modify the signal sent to each subwoofer so that seat-to-seat variations are minimized. Welti and Devantier (2006) discuss some of the options. Harman's Sound Field Management (SFM) will be discussed here.

In SFM, measurement data are processed by an algorithm that combines the sound from all active subwoofers at each listening position (superposition). The algorithm allows for signal manipulation in the path to each subwoofer—gain, delay and equalization—with workable values chosen for the parameters, and limits applied to the permitted ranges of the manipulation. Then, an optimization program is run, systematically varying the parameters of the signal processing and monitoring the predicted frequency response at the listening positions, with the objective of minimizing the seat-to-seat variations. It is a brute-force trial and error system, but it is something that computers are very good at and, with the power and speed of modern laptops, solutions are quickly reached.

To summarize, Sound Field Management has the ability to assist in the selection of optimal subwoofer positions and modify signals fed to individual subwoofers with respect to:

■ Signal level (gain)

■ Delay

■ One parametric filter, with values of center frequency, Q and attenuation (no gain).

The optimization process starts with the operator selecting some number of subwoofer locations and some number of listening locations. Obviously some locations will be better than others, as was found in the rectangular room solutions, but there is no restriction on the subwoofer locations that can be explored. The algorithm will simply do the best it can with the circumstances it is presented with.

In the extreme, one subwoofer can be located in numerous possible locations and measurements made from each location to each of the seats being optimized. Then, limited only by patience and time, the operator can let the computer optimize different combinations of subwoofers, in number and location, to find the solution that works best, or best meets the needs and budget of the situation. The installer has considerable flexibility.

If the same subwoofer is used for the initial measurements, and it is found after optimization that some locations are not driven to full output (the gain is reduced by 6 or 12 dB, for example) this means that significantly smaller, lower power, subwoofers can be used at those locations. A 6 dB reduction is 25% of the 0 dB power, and a 12 dB reduction is 6.3%, so much smaller, less expensive subwoofers can be employed if desired. SFM does not require identical subwoofers, unlike other solutions, but they should be of similar design (e.g., closed box or reflex) and capable of similar bandwidth. If this option is taken, a new optimization must be done after the chosen subwoofers are installed. Figure 8.21 is a good example of the impact of both the choice of subwoofer locations and SFM on seat-to-seat sound consistency and on the power required to deliver a given sound level at low frequencies. Figure 8.21a shows the reference condition in the same room used in computer simulations in Figure 8.15. These are real measurements in the room. The seat-to-seat variability (MSV) is high. Figure 8.21b shows that employing the four mid-wall subwoofer arrangement reveals the sound level reductions attributable to the loss of the axial modes 2,0,0 and 0,2,0 (Figure 8.15c). The MSV, though, is dramatically reduced with SFM implemented. Substantial equalization would be needed to flatten and smooth the frequency response. In Figure 8.21c the higher sound output and smoother frequency response of the four-corner subwoofer arrangement is seen. (MSV is 8.2 compared to 33.42).

Focusing on system efficiency, Figure 8.21c shows the average room curve for configuration (b). If one wanted to roughly match the sound output from both of these configurations, the gain of the four-mid-wall arrangement would need to be turned up by about 9.4 dB, the difference between the MOLs of (b) and (c): −17 dB and −7.6 dB. This corresponds to a power difference of 8.7× for the 20–40 Hz frequency range over which the MOL is calculated. In practical terms, if the subwoofers used in (c) used a total of 100 watts to generate an entertaining level of low bass, condition (b) would require 870 watts. The MSV is marginally better in (b), but not enough to justify the additional power demand.

The finishing touch on all of these systems is to apply global equalization to achieve a smooth, flattish frequency response. In the case of (c), the preferred arrangement, it is a matter of attenuating 4 dB peaks at 55 and 60 Hz, and a 3 dB peak at 73 Hz. The deficiency near 80 Hz would be handled in combination with the high-pass filtering for the satellite loudspeakers in the bass-managed system.

The examples shown so far have been from installations in the listening rooms at Harman. Figure 8.22 is of my entertainment room constructed in 2002. It was the motivation

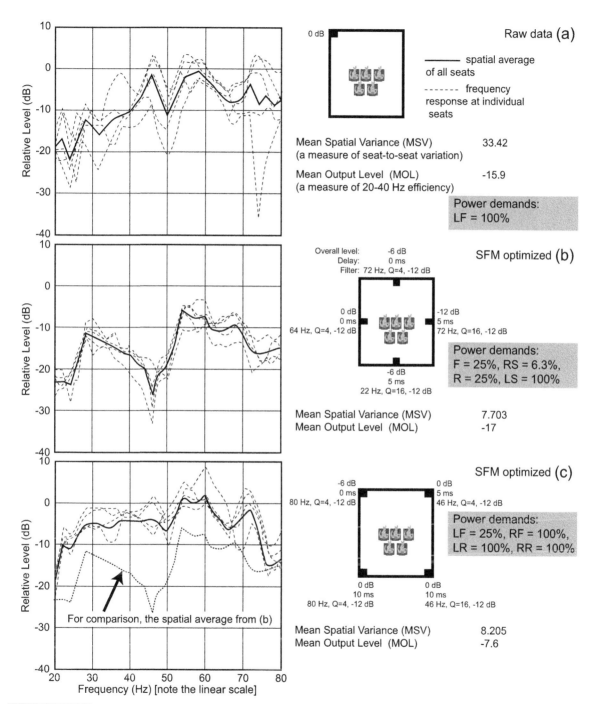

FIGURE 8.21 *Measurements made in the example room used in several earlier figures.*
Data from Welti and Devantier (2006): (a) their figure 20, (b) their figure 24 and (c) their figure 25.

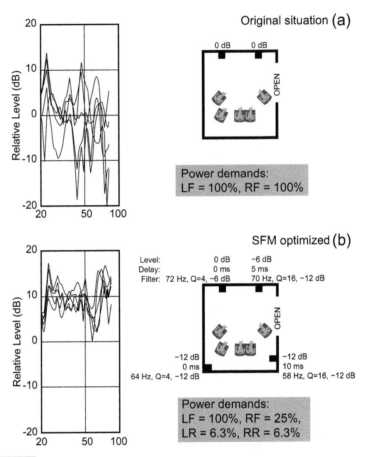

FIGURE 8.22 *Measurements made in the author's entertainment/family room. Subwoofers were located as space and visual considerations permitted, not by any acoustical rules. Listeners were arranged to allow verbal and visual communication with and without video. The room is 22 ft × 20 ft (6.7 m × 6.1 m), with a ceiling that slopes from 8 ft to 12 ft (2.4 m to 3.6 m).*
Data from Welti and Devantier (2003).

for Welti and Devantier, my colleagues in the Harman research group, to go beyond the rectangular room multi-sub-solutions and solve the problem in my "acoustically inconvenient" room. It was a living space, not a dedicated home theater, with one incomplete wall, open to the rest of the house, the opposite wall was mostly glass, the ceiling was sloped, and the listeners were not in rows. Instead they were arranged for conversation, away from the desirable central area, although the seats are easily rearranged for shows. It was intended to be a comfortable multipurpose room that could convert to a superb home theater with a roll-down screen and 7.1 channels. It was a challenge.

The magnitude of the problem can be seen in Figure 8.22a. With the original two front subwoofers, the room curves at the six seats were very disorderly, ranging 15 to 20 dB above and below the estimated average 0 dB level. Confirming audible evidence was abundant; the bass boomed and changed dramatically from seat to seat. Equalization could have improved things for one listener, but nobody else. Massive bass traps were not an option in this rather attractive room, with paintings, sculptures and fine cabinetry on display (see Figure 7.21).

Adding another pair of compact, powered subwoofers and performing SFM optimization resulted in Figure 8.22b. Not only were the seat-to-seat variations greatly reduced, but the need for global equalization was all but eliminated. Overall sound levels were significantly elevated—by about 10 dB if my eyeball averages are right. And this was achieved with *less* total power from the subwoofers. Three of the subwoofers operated at substantially reduced levels, and with planning they could have been scaled down to considerably smaller products. How did it sound? Personally, it sounded superb then, and still does 15 years later. The room, in acoustical terms, essentially disappears when "play" is pressed—no booms, only tight, deep, clean bass. Again SFM optimization resulted in what appears from all perspectives to be desirable performance in what seemed at the outset to be a hostile situation, and all with minimal visual and physical intrusion into the living space.

An interesting side story: the seat closest to the lower-left corner subwoofer suffered from seriously excessive low bass in the original two-subwoofer configuration. In fact, there was a tendency to localize bass to that rear corner of the room, partly because there was so much concentrated energy there that small noises were emitted from vibrating structural elements and windows. There was no loudspeaker at that location. Adding the rear woofers eliminated the problem, and the woofers went unnoticed from a localization perspective.

Different rooms present different circumstances. Figure 8.23 shows measurements in a customer's home theater installation. It is another persuasive illustration of what can be achieved by combining multiple subwoofers with digital processing and optimization. The large seat-to-seat variations have substantially disappeared, leaving an opportunity for uncomplicated global equalization—one medium-Q parametric filter at 50 Hz—to flatten the frequency response for all listeners. The peak will be reduced by about 7 dB, meaning that in that frequency range the subwoofers will be radiating about 1/5 of their original output power. The time domain will then also be well behaved. This example employed no bass traps.

Figure 8.24 shows another successful application of SFM, this time in the listening room of an audio journalist and reviewer. The improvement was easily heard, and a small amount of global equalization to attenuate a peak around 60 Hz is all that is left to do. Obviously, all listeners benefitted from the installation while enjoying movies, but products under review were still heard through the original room, with its massive variations.

The modal control is exercised only on sound reproduced through the engineered system, but that is the objective. The fact that the processing delivers good sound to several locations in the room means that the sound field between and around those locations tends also to be improved. One can play a loop of kick drum and bass guitar and walk around the room with little variation in bass timbre or tightness.

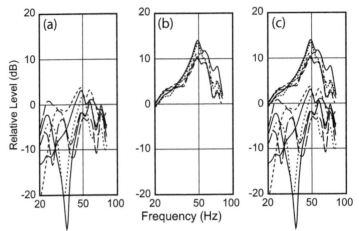

(a) One corner subwoofer, measured at six seats
(b) Four corner subwoofers with SFM, measured at six seats
(c) (a) and (b) combined

FIGURE 8.23 *In a home theater (a) shows steady-state room curves at six seats using a single subwoofer. (b) shows the measurements with four subwoofers with levels, delays and one parametric filter for each subwoofer selected by a computer optimization program—Harman's Sound Field Management. (c) shows a direct comparison. No global equalization in these measurements.*

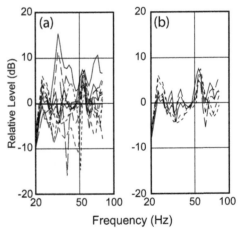

(a) Original single subwoofer
measured at five seats.
(b) Four corner subwoofers with SFM
measured at five seats. Average
sound levels normalized to match (a).

FIGURE 8.24 *The application of Sound Field Management to an audio journalist/reviewer's listening room. There is no global equalization in these measurements. The room was quite irregular in shape, with an opening in one wall.*
Adapted from figure 19 in Toole (2006).

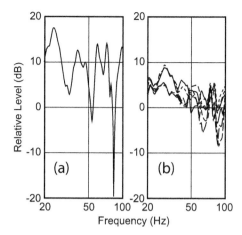

FIGURE 8.25 *(a) the steady-state room curve at the prime seat using the original front loudspeakers. (b) the results at all six seats after trial-and-error placement of additional subwoofers and delay, gain and equalization adjustments using a digital processor. Sound levels in (a) and (b) cannot be compared. Data courtesy Gene DellaSala, Audioholics.com.*

The only unfortunate fact is that, right now, SFM is only available in JBL Synthesis custom installation products. The good news, though, is that using trial and error, good measurements and patience, satisfying results can be achieved by do-it-yourself enthusiasts. Figure 8.25 shows the results obtained in a difficult L-shaped room using five subwoofers and manually adjusted digital processing (DellaSala, 2016). In (b) the bass is very uniform, gently rising and, although not shown, it extends well into subsonic frequencies. Different rooms will obviously require different solutions.

8.2.9 Revisiting Room Resonances in Time and Space

Figure 8.26 looks at the current topic from a slightly different perspective, illustrating how standing waves associated with room resonances are revealed in multiple measurements along the specific trajectory shown in Figure 8.26a. Figure 8.26b shows results using only subwoofer 1. It can be seen that:

- Length mode 1,0,0 is weakly energized because the woofer is in the null of the mode, and is attenuated further because the microphone trajectory runs along the null line across the width of the room.

- Width mode 0,1,0 is strongly energized and shows decaying amplitude as the microphone approaches the null at the halfway point (see also bottom of (a)).

- Tangential mode 1,1,0 is not very strong, and also shows declining amplitude as the center of the room is approached.

- Length mode 2,0,0 is strongly energized, and the amplitude remains steady across the width of the room.

FIGURE 8.26

(a) shows the room layout with an illustration of the width mode being addressed. Unlike the stylized mode illustrations in Figure 8.13, this is a sketch of the sound level along the measurement axis using peak values extracted from the curves in (b). Only subwoofer 1 was used for the measurements in this figure. In that location mode 0,1,0 is maximally energized. (b) shows measurements made at the one-foot intervals starting at the side wall and moving toward the center. The resonant peak is evident in all of the curves, as is the declining sound level as the microphone is moved toward the null in the center of the room. (c) shows the attenuation of the peak by parametric equalization, and confirms that the beneficial effect is seen in all of the measurements. There is no change in the spatial distribution.

Data courtesy Todd Welti, Harman International.

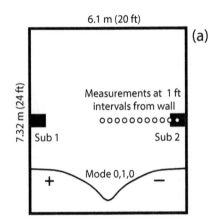

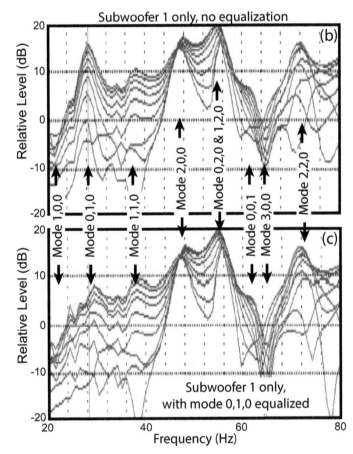

- Width mode 0,2,0 and tangential mode 1,2,0 are very close (56.5 and 54 Hz), and in the modal interaction one can see a slight frequency shift in the dominant peak as the microphone is moved to the center of the room.

- Height mode 0,0,1 is poorly represented because the microphone was close to the vertical null.

- Length mode 3,0,0 is essentially absent because both the woofer and the microphone trajectory are in the null of the mode.

- Tangential mode 2,2,0 is strongly energized and shows an untidy set of curves as the microphone is moved through a null and then back to a maximum, as can be seen in Figure 8.16.

This brief analysis is a good example of just how predictable modal events are in a known rectangular room when the woofer and microphone locations are well defined. This underlines the huge advantage of rectangular rooms—the low-frequency sound fields can be substantially predicted in advance and measurements interpreted after the fact. Knowing which modes cause problems allow solutions to be focused. Obviously, only high-resolution measurements allow for this depth of analysis.

The experiment involved using a parametric filter to attenuate mode 0,1,0. The filter used was not a perfect match for the mode—it had a slightly lower Q—so the attenuation is less than it might have been. Perhaps this makes it even more of a real-world example. It is first worth examining the plot of sound pressure vs. distance at the bottom of (a). This is derived from the peak sound levels of mode 0,1,0 as a function of distance from the wall as shown in (b). The notable feature is the flatness of the pressure peaks compared to the narrower interference dip. The lesson is that resonant peaks are bothersome because they are resonances, with accentuated amplitude and ringing, and also because the peaks occupy most of the floor area of the room. More listeners are annoyed by peaks than dips for this reason and also because the presence of excessive sound is more annoying than an absence—see discussion of peak and dip audibility in Section 4.6.2. The measurements in Figure 8.26c indicate that the mode is attenuated at all microphone locations, which is to be expected, and also that none of the other modes is significantly affected. Substantial bass energy has been removed; perhaps more than is optimum for good listening.

If one were to equalize this system it would probably be reasonable to start by aiming for the 10 dB horizontal line as a target, attenuating mode 0,1,0 by about 6 dB, and introducing other parametric filters to attenuate modes 2,0,0, 0,2,0, and 1,2,0 combined, and 2,2,0 to the 10 dB line. Attempts to fill the dips associated with the 0,0,1 and 3,0,0 modal nulls are not advised. This is why mode cancelling and manipulation techniques are so advantageous—they reduce peaks and elevate nulls while using less energy. After the "high resolution" equalization is complete, broadband spectrum adjustments can be made to bring the result into agreement with the desired overall room curve target, as discussed in Sections 12.2 and 12.3. Automated low-frequency equalization algorithms should be capable of doing both operations simultaneously, but some are overly zealous in trying to achieve a smooth target curve.

Figure 8.27 shows the time domain performance of two methods of attenuating a room resonance: equalization of one sub and mode cancelling using two subs (1 and 2 in Figure 8.26a). These energy-time curves illustrate the amplitude vs. time events at the

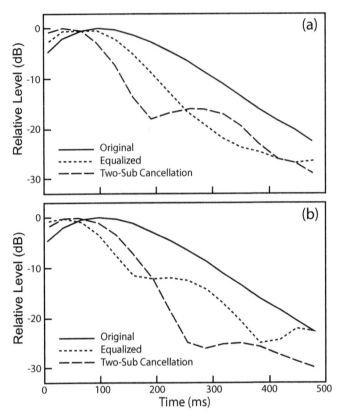

FIGURE 8.27 *Energy-time curves measured at (a) 1 ft from the wall in the floor plan shown in Figure 8.27, and (b) at 10 ft from the wall, the midpoint of the room width. The signal was filtered at the center frequency of mode 0,1,0, so what is shown reveals the time course of sound decaying at the modal frequency in three conditions: the original modal decay, the decay after the mode was attenuated by parametric equalization (as shown in Figure 8.26), and after the mode was attenuated by employing subwoofers 1 and 2 in a "mode cancelling" configuration, driven with identical signals.*
Data from Todd Welti, Harman International.

frequency of resonance 0,1,0, seen earlier. The data are not perfect because the filter used to isolate these events in the frequency domain has a decay time of its own, but the differences between the curves are valid. In this comparison the curves have been adjusted so that they all exhibit the same peak sound levels. In reality the equalized and mode-canceled curves would start at lower levels, thereby adding further assurance of reduced audibility. Figure 8.26b and c show a 9 dB reduction in amplitude due to the equalization, to which the increased decay rate is an added improvement. It is interesting to see the buildup of the resonant energy in the "original" situation, as was seen in the tone-burst waveforms of Figure 4.10. All of the bass energy at this frequency must be communicated through the resonance, and it has a build-up time and a decay time. Both the equalization and cancellation treatments reduce the build-up interval and the ringing, but they behave differently in their interactions with the prevailing sound field. In both cases it is not difficult to hear the improvement in bass uniformity and temporal definition.

Looking back at Figure 8.11d for equalization and Figure 8.14e for cancellation, one can see the waterfalls drop quickly in the early time interval moving forward in the illustrations. They then stabilize to a lower decay rate. This is also seen here, especially in the mode cancellation situation. The initial drop in all cases is about 15 dB or more, which is perceptually very significant. Recall that at these low frequencies perceived loudness halves for each 5 dB or so of amplitude reduction (Section 4.4). When listening to music or movies, the resonance is perceptually insignificant.

8.3 DO WE HEAR THE SPECTRAL BUMP, THE TEMPORAL RINGING OR BOTH?

The preceding sections have enumerated and explained several ways to attenuate bothersome room modes, ways to manage which are active and which are not, and electronically elegant ways to deliver similarly good bass to several listeners. Because low-frequency room resonances generally behave as minimum-phase systems, individual modes can be addressed with parametric-filter equalization to reduce the magnitude of the spectral bump associated with a room resonance. If there is a bump, there will be ringing. If the bump is attenuated by equalization, or in fact any of the alternative schemes, so is the ringing. Waterfall diagrams and impulse responses routinely confirm the improvements in time domain behavior.

In many cases, as shown in the preceding examples, the problems and the improvements are not difficult to measure or to hear. As is seen in Figure 8.27, a resonance, in this case a medium-Q resonance, does not respond to an input instantaneously. There is a build-up interval and a decay interval, the former occurring while a signal is present (see also Figure 4.10). A kick drum may appear to be a sharp transient, but the impulsive "kick" is at frequencies much higher than the fundamental resonance "boom," which can last 200 ms or so. What sounds "tight" is actually a drumhead/volume resonance superimposed on room resonances. During that time the amplitude at the room resonance frequency increases before the ringing decay begins; the higher the Q, the longer the build-up interval, and the longer the ringing. So the perception of such an event is not a simple one. The notion of "fast" subwoofers is misguided, in that the time-domain performance of minimum-phase systems is predictable from the amplitude response. With an 80 Hz bass management crossover frequency, all subwoofers are equally "fast." So, what we confront in the real world is essentially "fast" subwoofers communicating through "slow" rooms.

Several studies have examined what happens in the time and frequency domains when tailored filters address specific modes as distinct from what happens with "typical" room equalization, where for historic reasons the frequency resolution is reduced from what is easily achieved today, to 1/3-octave smoothed curves. Obviously, with less precise data, the equalization is not as well matched to the resonance, and the ringing is less controlled (see Figure 8.12).

But how much precision is necessary? In real rooms several modes can be simultaneously active and stimulated by ever-changing music containing long and short tonal components and various forms of transient stimulation. There can be subtle things like two adjacent modes that share energy, one decaying faster than the other and effectively shifting the pitch of the resonance (Figure 8.14c, d and e). If they are of similar Q, they will beat as they decay. A mode can have high Q, long ringing, but be at a level significantly below the overall spectral level, as in Figure 8.14e and others. These are all situations difficult to subjectively examine. So, investigations that have been done tend to be conducted under very controlled circumstances, most involving computer simulations of both the problems and the solutions.

The following experiments involve comparisons of various forms of equalization, but it is important to note that they address audible performance only for a single listener. Just one test employed a multi-subwoofer mode-manipulation scheme, and it turned out to be more effective than necessary from a mode attenuation perspective, but of course had the unique virtue of creating multiple seats where similarly good bass can be heard.

Toole and Olive (1988) concluded that at frequencies above 200 Hz, the threshold of detection of resonances decreased about 3 dB for each doubling of the Q value. This means that the higher the Q, and therefore the longer the ringing, the less detectable the resonance. This is not an indication that ringing *per se* is not a problem, rather it most likely relates to the probability of a narrow-band resonance being energized by typical music. Consequently, at these frequencies the duration of ringing is not a reliable indicator of audibility (see Section 4.6.2).

Those investigations were in the frequency range above the transition frequency, where reflected sounds were an important factor. Reverberation—reflected sounds—provide masking for high-Q ringing, and perceptual amplification for medium- and low-Q resonances by providing repetitions of the events. Reverberation time measurements at these frequencies are dominated by the decay of reflected sounds: repetitions of the direct sound. At bass frequencies the ringing from strong resonances is likely to be the dominant factor in RT. This is a continuation of the direct sound, not reflected repetitions of it coming from different directions. Binaural hearing is not very effective at low frequencies, so measured RT can have different perceptual consequences at different frequencies. This is often ignored in discussions.

Olive et al. (1997) elaborated the study to include lower frequencies, finding again that increasing Q yielded increasingly high detection thresholds as revealed by the amplitude peak in a room curve. Decreasing frequency also yielded increasing thresholds except for the lowest Q resonances. Q = 1 resonances were detected at low threshold levels whether they were resonances (peaks) or anti-resonances (dips), but it must be noted that such low-Q, wide-bandwidth resonances embrace most of the frequency range of a typical subwoofer, so this result is hardly surprising. It really is changing the overall spectral balance in an unsubtle way. Here it was found that impulsive signals increased the audibility of high-Q resonances, suggesting that ringing was itself a contributor to audibility. None of this is surprising, but more detailed information would be useful. Fortunately other investigators have added insights, but with so many variables simultaneously at play, it is difficult to be absolutely definitive.

Antsalo et al. (2003) compared the audibility of simplistic 1/3-octave smoothed room curve equalization to that equalization combined with modal equalization that addressed specific room modes with higher precision. The waterfall diagrams show that the synthetic modes were usefully addressed by the modal equalization, reducing both the initial amplitudes and the decay times of the ringing. In simulations over headphones, listeners compared the original room with the room curve smoothed by the magnitude equalization, and a combination of that curve and modal equalization achieved by two different methods. The result was that listeners reported little or no improvement over the simple smoothed magnitude equalization, which alone could only reduce the spectral prominence of the resonances, without addressing the decay rate. The inference is that listeners responded to the reduction of the "bump" rather than the reduction of the decay rate, and did so even with 1/3-octave equalization. The tests relate to a single listening location.

Karjalainen et al. (2004), in work shared with the previous authors, varied the duration of ringing in modes that had been amplitude equalized to within ±3 dB of the 1/3-octave smoothed response that left the modal ringing essentially intact. The room had a reverberation time of 0.28 s at mid-frequencies, rising gradually to 0.5 s at 50 Hz. Test signals included speech, 0.5 s noise bursts, drum hits, and rock music. They found that the detection thresholds of ringing decay times were above the RT of the room, rising gradually from about 0.5 s at 800 Hz to about 0.7 s at 100 Hz. Incredibly, below 100 Hz the listeners became unresponsive to modal decay times up to almost 2 s. Noting that the amplitude responses had been equalized, they concluded that "Without such equalization the differences can be noticed much more easily, which indicates the obvious fact that the frequency-domain properties of resonances are prominent over the time domain properties." This is in agreement with the Toole and Olive finding for resonances above 200 Hz. In any event, the conclusion was that modal ringing at frequencies below about 100 Hz is not an issue—but there is more.

Avis et al. (2007) examined the audibility of a room resonance as a function of Q, without specific control in modeling or in listening test programs that favors either decay time or spectral variance. Q is translatable into a corresponding decay time, so there is continuity of intent. A summary result is that "reductions to the Q factor below a value of about 16 will be subjectively imperceptible," which translates into calculated modal decay times of about 1 s at 32 Hz, 0.7 s at 50 Hz and 0.28 s for 125 Hz. In the context of normal listening spaces the high-frequency value will likely be below the measured RT, so there is the matter of masking of the resonant ringing by room reflections. The authors addressed RT as a variable, but the data applied only to frequencies above 250 Hz, leaving the key question about simultaneous masking at the resonance frequencies unanswered. Unless the room is inherently very dead, as happens in a few recording control rooms, these durations of ringing may not be audible. There is also the uncertainty of the height of the resonant peak above the average spectrum level; that is, how much does it perceptually stand out? In any event, these data add to the impression that ringing at very low frequencies are not easy to hear.

Adding further information and complexity to the discussion, Fazenda et al. (2012) subjectively and objectively compared several methods of addressing room modes. Unfortunately, from my perspective, they chose to degrade the measurement prior to equalization to 1/3-octave resolution. They then noted that this had detrimental effects in both the high-resolution frequency responses and in the decay rates of resonances. So, why was it done? The most useful observation was that the active multiple-subwoofer mode-control scheme C.A.B.S., discussed in Section 8.2.7, worked very well, and better than they thought was required by resonance detection criteria they had developed. This is in principle similar to SFM discussed in Section 8.2.8. So, if resonance detection at a single listening location is the only thing of value, the electronically processed mode control methods might be too good, but the additional value they bring is that multiple listeners can share the excellence, which cannot be said of other schemes.

On the practical side, they point out that improvements are possible by relocating loudspeakers and, if that is not possible, by simple magnitude equalization. I would add that simple equalization is advisable even when repositioning is possible. I would add further that it is time to abandon deliberately low-resolution (1/3-octave) measurements, even though they exist in legacy industry standards. All of the data indicate that modal equalization is superior, especially in demanding listening situations.

Fazenda et al. (2015) continued the investigation of the audibility of the time decay component of resonances. They examined resonances in isolation and in simulated room contexts, finding that with music stimuli the thresholds were 0.51 s at 63 Hz, 0.3 s at 125 Hz, and 0.12 s at 250 Hz. In practical terms this means that in the subwoofer frequency range (typically below 80 Hz), these decay times can very likely be achieved with conventional parametric equalization of problem room modes, and almost certainly with any of the multiple subwoofer active control schemes that bring the additional and important advantage of providing several listeners with similarly good bass sound.

All of these studies indicate that waterfall diagrams are not a definitive statement of the audible consequences of resonances. The uncertainties in what one is seeing in the decorative waterfalls adds to the confusion (see Section 4.7).

8.4 STEREO BASS: LITTLE ADO ABOUT EVEN LESS

Apologies to William Shakespeare. This issue relates to the fact that, in order for all of the systems described earlier to function fully, the bass must be monophonic below the subwoofer crossover frequency. Most of the bass in common program material is highly correlated or monophonic to begin with and bass-management systems are commonplace, but some have argued that it is necessary to preserve at least two-channel playback down to some very low frequency. It is alleged that this is necessary to deliver certain aspects of spatial effect.

Experimental evidence thus far has not been encouraging to supporters of this notion (Welti, 2004, and references therein). Audible differences appear to be near or

below the threshold of detection, even when experienced listeners are exposed to isolated low-frequency sounds. The author has participated in a few comparisons, carefully
set up and supervised by proponents of stereo bass, but each time the result has been
inconclusive. With music and film sound tracks, differences in "spaciousness" were in
the small to non-existent category, but differences in "bass" were sometimes obvious,
as the interaction of the two woofers and the room modes changed as they moved in
and out of phase. These were simple frequency-response matters that are rarely compensated for in such evaluations. Even with contrived stereo signals spatial differences were
difficult to tie down. This is not a mass-market concern. In fact, some of the discussion
revolved around the idea that one may need to undergo some training in order to hear
the effects.

Another recent investigation concludes that the audible effects benefitting from channel separation relate to frequencies above about 80 Hz (Martens et al., 2004). In their conclusion, the authors identify a "cutoff-frequency boundary between 50 Hz and 63 Hz,"
these being the center frequencies of the octave bands of noise used as signals. However,
when the upper frequency limits of the bands are taken into account, the numbers change
to about 71 Hz and 89 Hz, the average of which is 80 Hz. This means, in essence, that it is
a "stereo *upper*-bass" issue, and the surround channels (which typically operate down to
80 Hz) are already "stereo" and placed at the sides for maximum benefit.

Of course audiophiles dedicated to LP playback face an impossible dilemma: the
bass is mixed to mono to prevent the stylus from being thrown out of the groove.
Enough said.

8.5 BASS MANAGEMENT MAKES IT ALL POSSIBLE

All of the multiple subwoofer schemes discussed earlier in this chapter assume
that there is a single signal for all bass frequencies, no matter how many satellite
channels there are. Individual channels in the original film sound tracks are broadband, and the main loudspeakers operate usefully down to about 40 Hz—some
lower, others not so low. In cinemas this was adequate until the first "blockbuster"
films wanted more low bass. Existing cinema systems were unable to handle the
load, so a new channel was created, called either "low-frequency enhancement" or
"low-frequency effects," but always LFE. The bandwidth was restricted to frequencies below 120 Hz and the signal level was reduced by 10 dB on the soundtrack
to allow for more headroom, which was a problem with analog soundtracks of the
day. The 10 dB was restored on playback. In cinemas, dedicated subwoofers reproduced the LFE signal, which was used for extra impact in dramatic sound effects.
This additional limited-bandwidth channel was identified as the "0.1" in 5.1 or 7.1
configurations.

The playback system was modified to be more practical in home situations, and bass
management was born. Bass management is a signal processing option in multichannel
receivers and surround processors, with which it is possible to extract the lowest frequencies from the individual channels, combine them, add them to the low-frequency

effects (LFE) channel and deliver the combination signal to a subwoofer output. The normal crossover frequency at which this is done is 80 Hz (this can often be changed), the frequency below which it is difficult or impossible to localize the source. Bass sounds will be stripped from all channels in which the loudspeakers have been identified as "small," and will be reproduced through those identified as "large," as well, of course, as the subwoofer(s). Holman (1998) gives a good history. Figure 8.28 illustrates a common implementation.

Setting up such a system involves accessing the right menu in the receiver or processor and identifying all of the five or seven channel loudspeakers as "small" (thereby introducing high-pass filters), activating the "sub" output, and, if it is an option, choosing a crossover frequency, most often 80 Hz. With the L, C, R and surround loudspeakers relieved of low-frequency duties, they can play louder with less distortion, and the purpose-designed subwoofer(s) are left to their task.

The weakness of the simple bass management used in mass-market consumer equipment is that the acoustical crossover from the subwoofer array to each of the satellites is not controlled. The electrical crossover filters are defined, but what matters is how the

Typical bass management system - crossover frequency 80 Hz

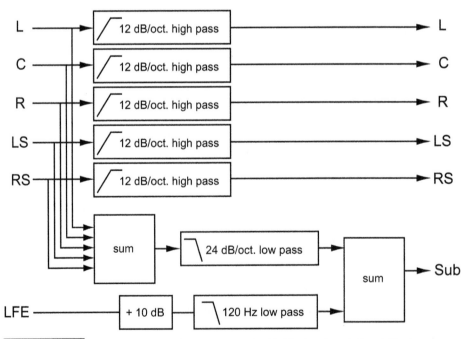

FIGURE 8.28 *A common version of bass management with a basic subwoofer-to-satellite crossover frequency of 80 Hz. In this the LFE, containing frequencies up to 120 Hz, is allowed to pass unmodified to the subwoofer. Some implementations low-pass filter the LFE at 80 Hz before the subwoofer output, which discards some of the LFE signal. This is not encouraged. Neither are implementations that simply ignore the LFE.*

sounds in the room combine at the listening location(s). This requires in-situ acoustical measurements and equalized low- and high-pass filters. This is done in some custom installations, but not in mainstream systems. The result, unfortunately, is that bass management is justifiably criticized for not always sounding as good as it should. The alternative commonly proposed is five or seven full-range loudspeakers, believed to keep the sound sources somehow "pure." Figure 6.2 shows that this is a fantasy—each channel then has its own imperfect, and distinctive, acoustical coupling to the listener, and no practical control is possible.

A further complication arises if a customer wishes to use full-range floor-standing loudspeakers in the L&R locations and smaller satellites elsewhere. The woofers are not likely to be in the best locations for sound field control, and integrating the bass component of these loudspeakers with additional, different subwoofers in the room to achieve better seat-to-seat uniformity will be a matter of trial and error. It is often best to treat them as "small," along with the other multichannel satellites, and to employ a separate arrangement of subwoofers for room mode control. That said, in custom installations, sophisticated digital processors are able to separate the bass component from the line-level signal driving full range L&R loudspeakers, combine it with two other subwoofers, and return the processed signal to the signal feed.

In cinemas, the separate low-frequency effects (LFE) channel, operating to 120 Hz, combines with the main channels that operate down to about 40 Hz, so there is overlap and uncontrolled acoustical interference in the playback space. The existing setup and calibration standards for cinemas do not prevent this. Consequently, in current discussions about future, improved calibration standards, thought is being given to implementing bass management in cinemas, simply because with digital sound tracks a separate LFE is not necessary, and if there is to be an LFE, it needs to be properly integrated into the total playback system (Toole, 2015b).

8.6 SUMMARY AND DISCUSSION

The basic techniques discussed in this chapter are all in the scientific literature, some for decades. Acoustical absorption has been around since the 1930s, when dozens of companies were manufacturing versions of resistive absorbers. The design of bass traps came later, and fit the general pattern of acoustical treatment being synonymous with acoustical absorption. "Dead" acoustics became fashionable. Fortunately we have moved on from those times and, as discussed in Chapter 7, we can contemplate allowing some room sound to be not only a part of the recording control room, but also the playback venue. In the latter instance, the existence of much better loudspeakers makes the task less difficult than in the past. But bass in small rooms remains a challenge.

New generations of audiophiles need to become aware of current knowledge and learn how to use it to their advantage. Unfortunately, Internet forums continue to have protracted debates over issues long settled, good measurements are rare, opinions abound and those with business interests continue to agitate to their advantage. My observations of the professional audio field indicate that the situation is similar.

One of the most common misconceptions is that simply adding another subwoofer or two will automatically improve things. Most often they are not identical subwoofers, and they are not located where they need to be, and they may or may not be driven with the same bass-managed signal. Unpredictable results and debates are almost inevitable.

When measurements are done, it is common for some individuals to try equalizing each subwoofer separately and then to combine them. Again, disappointment is almost guaranteed because when the system is in "play" mode all subwoofers are operating simultaneously, and the result of the acoustical summation is unpredictable. The measurement that matters is the one done with all subwoofers operating simultaneously. For individually measured subwoofer outputs to sum in a predictable manner, the measurements must be of the transfer function—amplitude and phase—not the steady-state amplitude response. This is the basis of the room mode manipulation schemes described in Sections 8.2.7 and 8.2.8, in which after individual subwoofer measurements are done at the seating locations, predictions of room response can be calculated in the computer with high precision, including experimenting with delay, gain and equalization in each of the subwoofer signal paths. Acoustical superposition applies.

There is also the "spatial averaging" confusion. By measuring at several seats and averaging the result, one gets a spectrally smoothed overall impression of what the audience hears, but not of what any individual in the audience hears. Unless one employs a massive bass-trap array or a multiple-subwoofer strategy to reduce seat-to-seat variations, equalization to satisfy a spatial average may end up truly satisfying none of the listeners. It is arguably better in such cases to satisfy the prime listener and let the rest take their chances. The latter approach is a requirement if one wishes to optimize the subwoofer-to-satellite crossovers. With digital processors, one can have both "group" and "just me" equalizer settings.

So, let us summarize this section about bass by saying that with intelligent use of multiple subwoofers there is now an alternative to or an addition to bass traps. Low frequency absorbers work, and will always make things better if one chooses one of the multi-sub options. But the reality that faces most people in homes is that large passive low-frequency absorbers are difficult to accommodate. Multiple subwoofers also present complications. Nevertheless, to the extent that absorption can be installed, starting with a smaller problem makes any additional solution work better. Acoustical absorbers remove energy, multi-sub solutions tend to add efficiency, so there is a trade-off to consider. The benefactors are the listeners.

CHAPTER 9

Adjacent-Boundary and Loudspeaker Mounting Effects

As has been demonstrated, loudspeakers and rooms operate as a system. Understanding that system better means getting close to the individual mechanisms of that interaction. When a loudspeaker is close to a room boundary, the boundary has an effect on the sound radiated into the room. When a loudspeaker is mounted on or in a wall, there are effects, and if there are open cavities close to the loudspeaker, they too influence the radiated sound. This chapter will examine each of these phenomena.

9.1 THE EFFECTS OF SOLID ANGLES ON THE RADIATION OF SOUND BY OMNIDIRECTIONAL SOURCES

At 20 Hz, the wavelength is 56.5 ft (17.25 m). At this and similar frequencies, any practical separation between a loudspeaker and a room boundary is "small." In addition, any practical loudspeaker will be small relative to the wavelength, and therefore it will radiate sound in an essentially omnidirectional manner. For both of these reasons, the conditions are met for the classic set of relationships shown in Figure 9.1.

The technical description of full spherical, omnidirectional radiation is that the sound source "sees" a solid angle of 4π steradians. It is a "free field" with no surfaces to reflect or redirect the radiated sound. Placing the sound source on or in a large plane surface reduces the solid angle into which the sound radiates by half, to 2π steradians. Energy that would have traveled into the rear hemisphere is reflected forward; there is a reflected acoustical "image" of the source. Additional surfaces, positioned at right angles, reduce the solid angle by half, to π steradians, and then to $\pi/2$ steradians. The number of reflected images increases correspondingly.

It is commonly heard in the audio industry that each factor-of-two reduction of the solid angle increases the sound level by 3 dB. The truth is that this may or may not

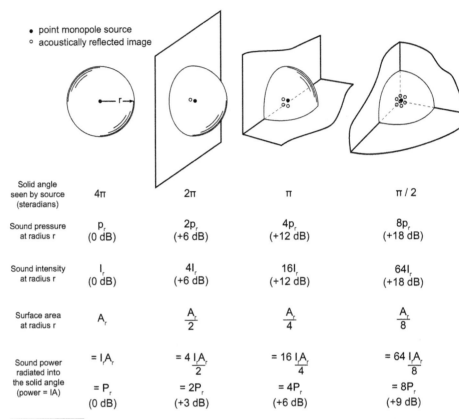

FIGURE 9.1 *Various measurable quantities at a fixed distance from a physically small omnidirectional sound source (a point monopole) in several basic locations closely adjacent to large flat surfaces.*

be the case, depending on circumstances. As seen in Figure 9.1 sound pressure level increases in 6 dB increments, but sound *power* (the total sound energy radiated into and distributed over the solid angle) goes up in increments of 3 dB. It can be seen that the sound pressure level, measured at a constant distance from the sound source *in these otherwise reflection-free circumstances*, goes up by 6 dB for each halving of the solid angle. Strictly this applies only to the direct sound. In a large, reflective venue, the reflected sounds represent the total power radiated at all angles, and in the incoherent summation at a distant point the gain will be closer to the 3 dB per halving of the solid angle. However, in rooms that are small compared to the wavelengths at issue, the gains can be higher within the subwoofer frequency range.

But, there is more to be aware of. What is shown in Figure 9.1 happens only at wavelengths that are long compared to the source size and the separation distance. With respect to where we place loudspeakers or listeners relative to the boundaries of a room, this normally applies only at very low frequencies. How low?

Theory is nice, but practical measurements are the confirmation. Figure 9.2 shows measurements made at the National Research Council of Canada (NRCC) in Ottawa,

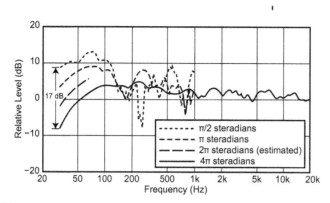

FIGURE 9.2 *A loudspeaker measured when it radiates into different solid angles as approximated by an anechoic chamber (calibrated at low frequencies) for the 4ϖ condition, and by an outdoor parking lot adjacent to a rectangular building. All measurements were made at 2 m.*

in the mid-1980s, in an anechoic chamber, and outdoors, in a parking lot adjacent to an isolated slab-sided building. The only reflecting surfaces were the adjacent boundaries. The loudspeaker was a conventional 12-inch (305 mm) driver in a closed box, about 17 inches (432 mm) on a side. Noise problems prevented the acquisition of data below about 35 Hz, and a technical problem resulted in the corruption of the 2π steradian data, but the trend is very evident. First, at the lowest frequency, the curves are separated by very close to 6 dB, as Figure 9.1 predicts. The fact that there is a tiny discrepancy can easily be accounted for by the fact that the large powerful woofer is not exactly a vanishingly small "point monopole," it is not located "in" the wall/floor, and the building was sheathed in corrugated metal siding, which is not a perfect reflector. In any event, the approximate 6 dB difference begins to diminish immediately and just below 200 Hz all of the curves converge in a null. The reason is that the woofer was facing forward, with the rear of the enclosure close to the wall. This separation allowed the sound to travel backwards from the driver to the wall, and back again toward the driver where it interacted with sound being radiated by the diaphragm. The round-trip distance, about 36 inches (914 mm), is one-half wavelength, at 188 Hz, which is where the destructive acoustical interference occurs. This is evidence of the adjacent-boundary effect—a reduction in sound output from a loudspeaker related to the distance between the loudspeaker and adjacent boundaries.

Figure 9.3 shows evidence that normal listening rooms are different in several important ways. The untidy curves are difficult to interpret, but it is clear that the gains in sound level for each halving of the solid angle are closer to 3 dB than 6 dB. Is this because even in this small room the reflected sounds are summing incoherently at these very low frequencies? That is possible, but it is hard to ignore the disparity between the dimensions of the small room and those of the relevant wavelengths (see Figure 4.19). The boundaries assumed in Figure 9.1 and that existed in Figure 9.2 were large compared to the wavelengths being propagated into a reflection-free environment. In typical listening rooms the boundaries are small, and the space is anything but reflection free.

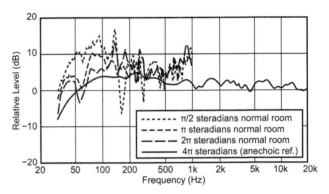

FIGURE 9.3 *The averages of four to six measurements made in the same loudspeaker/microphone relationship used in the previous figure, but moved to different locations within a normal listening room.*

What happens is a combination of incoherent acoustical interference, and, at room resonance frequencies, structured acoustical interference. These are lossy phenomena, clear evidence of which is seen in the irregular curves in Figure 9.3, especially at 62 Hz: the first-order vertical mode. So, energy will indeed be reflected from the adjacent boundaries, and sound levels will be elevated in the direct sound, but the amount by which it is elevated is not as shown in Figure 9.1, and at the measuring (listening) location, there will be substantial destructive acoustical interference to reduce the steady-state sound level. The 188 Hz adjacent boundary interference dip seen in Figure 9.2 is gone; it is swamped by other reflected sounds in the room. There is a lot going on in this space—the kind we live in.

9.2 CLASSIC ADJACENT-BOUNDARY EFFECTS

Figure 9.2 shows an acoustical interference dip at 188 Hz, caused by reflected sound from an adjacent boundary, the wall behind the loudspeaker in this case, destructively interfering with the direct sound radiated by the loudspeaker—the loudspeaker is essentially omnidirectional at these frequencies. It shows up clearly in these outdoor measurements with only adjacent boundaries. In rooms there are multiple boundaries, some adjacent to and none very far away from the loudspeaker, each contributing something to variations in frequency response at listening locations.

The adjacent boundary phenomenon was known to acousticians (Waterhouse, 1958), but it was Allison and Berkovitz (1972) and Allison (1974) who brought it into the consciousness of audio people in papers that described, in measurements, the significant dimensions of the problem. It is sometimes known as the "Allison effect."

Allison (1974, 1975) presents many examples of room curve shapes resulting from different arrangements of loudspeakers and corner boundaries. The curves are different from each other, depending on the distances of the woofers to the floor and each of the nearby walls. An average of 22 of these, in eight rooms, is shown in Figure 9.4.

To get an impression of how this curve relates to anechoic measurements, the figure also shows 2π and 4π anechoic curves. The precise vertical alignment of these curves is uncertain, so focus on the shapes. It seems that the anechoic facilities were not calibrated at very low frequencies.

The broad dip in output between about 80 Hz and 200 Hz is attributable to adjacent boundary effects that are averaged over all measurements in the eight rooms. Each would have had a distinctive signature.

All of this is significantly predictable for a specific situation, and Figure 9.5 shows the result of computer modeling of the interaction between a loudspeaker and adjacent boundaries. In the model, the loudspeaker has a perfectly flat frequency response, so what is seen is due to the loudspeaker/boundary interaction. The dotted curve shows the predicted effect of the adjacent boundaries on in-room frequency responses. To assist in understanding what is happening, the figure shows a superimposition of many steady-state frequency responses calculated for many different locations within the listening area. Each one is different because of standing waves and reflected sounds. Also shown is the average of all of these measurements. This is, therefore, an estimate of the effect of the adjacent boundaries on the radiated sound power, which is not specific to a single listener location.

The effects of room boundary interactions are in all room curve measurements, but in any individual one they may be obscured by other factors, such as standing waves. However, by averaging several room curves, measured at different locations—a spatial average—the effects of the position-dependent variations are reduced, and evidence of the underlying adjacent-boundary effects is more clearly seen. The average of the room curves is obviously similar in shape to the predicted curve.

This tells us that eliminating the adjacent-boundary effect will not eliminate all problems, but it is definitely one of the problems. The adjacent-boundary effect changes the acoustical radiation resistance experienced by the loudspeaker, and as a result the sound power radiated by the loudspeaker at different frequencies is altered. Figure 9.10 adds perspective. How can such a problem be addressed?

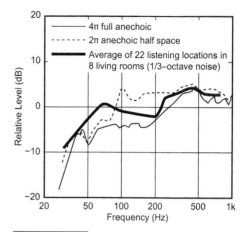

FIGURE 9.4 *A comparison of 2ϖ and 4ϖ anechoic measurements on an Acoustic Research AR-3a compared with the average of 1/3-octave measurements made at 22 listening locations in eight living rooms. Anechoic data from Allison and Berkovitz (1972), figures 4 and 9. The room data was also in this paper, but was more conveniently presented in Allison (1974, 1975).*

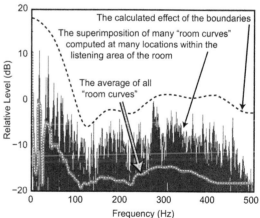

FIGURE 9.5 *A computed simulation of an omnidirectional loudspeaker situated in a "normal" relationship to the floor, ceiling and wall boundaries, in a "normal" listening room.*
Simulation by Todd Welti, Harman International.

9.2.1 Alleviating Adjacent-Boundary Effects

The approach offered by Allison (1974, 1975) and Ballagh (1983) involves choosing the position of the loudspeaker with respect to the boundaries in a manner that minimizes the variations in frequency response at the listening locations. Obviously, the goal is to arrange for the woofer-to-boundary distances, floor and walls, to be as different as possible, while not being multiples of each other. The more different these distances are, the less will be the effect. Finding an optimum result may involve some trial-and-error in practical listening. It may also result in visually unappealing, asymmetrical, or incorrect (in terms of stereo or multichannel imaging) locations for the loudspeakers.

Absorbing the boundary reflections is an option, but to be effective at these low frequencies, one is looking at large, thick devices beside and behind the loudspeakers, and treating the floor would not be practical. Equalization is another option. The attraction of equalization is that it allows the loudspeakers to be located according to other criteria, and then the performance is electronically optimized.

Figure 9.5 shows that averaging several measurements within the listening area can reveal the underlying shape of the room-boundary effects, and thereby provide a basis for correcting the frequency response to meet whatever target curve is decided on. However, there is another method, described by Pedersen (2003), in which a clever device measures the acoustic power output of the loudspeaker in situ, and makes the appropriate equalization correction to the frequency response to achieve a more uniform output.

Figure 9.6 shows a comparison of the two methods in the same room, one the result of making frequency response measurements at nine very different listening locations, and the other the result of a measurement of the sound power radiated by the loudspeaker. They are remarkably similar and, if any amount of spectral smoothing were incorporated, they would be even closer than shown. Obviously this means that an in-room measurement of radiated sound power can identify the adjacent-boundary problems, and so can a spatial average of in-room measurements. Both methods allow us to separate out the adjacent-boundary problems, but the solution is distributed uniformly throughout the volume of the room, which may or may not be what is needed for a single listening location. The effects of room resonances are added to this, so there will be seat-to-seat variations. If the equalization is performed at a single listening location, it will obviously be correct for that seat, and it will be an amalgam of adjacent boundary and room resonance problems.

Angus (2010) offers suggestions for using loudspeaker bass alignments to compensate for the large-scale bass effects of nearby boundaries. Small undulations and standing waves were not included.

Chapter 8 was a convincing argument for equalization of the low-bass subwoofer frequencies, whatever method is adopted to diminish seat-to-seat variations. Here we see that in the frequencies above the subwoofer range, through the transition frequency range, equalization again is beneficial.

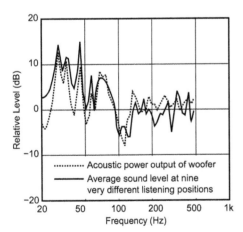

FIGURE 9.6 *A comparison of two methods of evaluating adjacent-boundary effects: a measure of the sound power radiated by the loudspeaker and the average of several room curves measured at different locations in the listening area.*
From Pedersen (2003).

9.3 LOUDSPEAKER MOUNTING OPTIONS AND EFFECTS

In general, conventional loudspeakers should be used "free-standing," away from walls. So-called bookshelf loudspeakers, presumably so named because they can fit a (sometimes large) bookshelf, are as often as not used on stands to bring the tweeters to seated ear level. The engineers designing these products cannot anticipate the details of any other manner of use, and so such loudspeakers are generally optimized to sound good in a free-standing mode. There are exceptions of course: some loudspeakers are optimized to be mounted on walls or ceilings, and others are optimized for in-wall or in-ceiling mounting. These products are clearly identifiable. It is interesting to take a standard small bookshelf loudspeaker and examine what happens to the sound it produces if we violate the rules and use it as it was not intended. Such a test took some preparation.

The chamber used for the following measurements was chosen because it could accommodate the apparatus, an 8-foot (2.44 m) square section of "domestic frame wall" on and into which loudspeakers were mounted. This chamber was not perfectly anechoic at the lowest frequencies and, although it had been calibrated for the measurement of loudspeakers in isolation, the presence of the massive wall structure introduced errors. Consequently, no claim is made for absolute accuracy at low frequencies in the measurements that follow, but they should be reliable in a comparative sense.

The tests began with an Infinity Primus 160 in the free field, an anechoic chamber, shown in Figure 9.7a. The acoustical measurements shown are components of the spinorama described in Section 5.3: the conventional on-axis frequency response, measured at 2 m, the sound power output, and the directivity index, the difference between the two. Directivity index (DI) is 0 for an omnidirectional source, and here we see that this little 6.5-inch (165 mm) woofer approximates that quite well up to about 150 Hz. One can see the directionality progressively increase to about 7 dB at 2 kHz, above which the crossover network progressively attenuates the woofer and the tweeter takes over. Being small, the tweeter exhibits much better dispersion, about 4.5 dB around 4 kHz, and then it too becomes progressively more directional, reaching a DI around 9 dB at the highest frequencies. Just for perspective, this is extremely good performance for an inexpensive product, listing at $220/pair at the time.

Figure 9.7b shows what happens when it is then mounted in a wall, with its front face flush with the surface. This is the classic 2π, half-space condition, which happens to be met by all in-wall and in-ceiling loudspeakers. What happens?

First, the bass increases, exactly as predicted by Figure 9.1. The sound pressure level goes up by roughly 6 dB (remember the measurements include some errors). Around 100 Hz it can be seen that in (a) the on-axis curve is about 3 dB below the 80 dB line, and in (b) it is about 3 dB above the line, for an increase in sound pressure level of about 6 dB. The sound power increases by about 3 dB in going from the 4π to the 2π conditions, again as predicted in Figure 9.1. The gains in acoustic output drop at higher frequencies because, as seen in (a), the loudspeaker is no longer perfectly

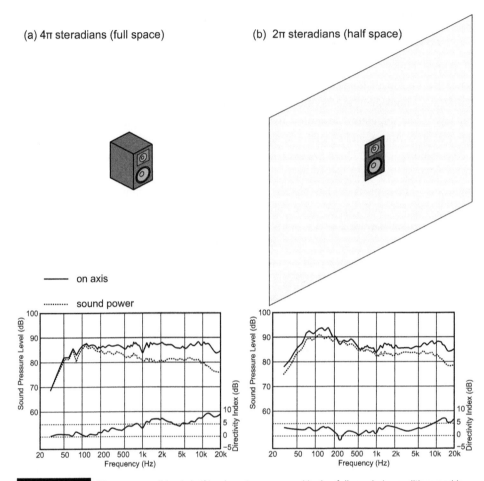

FIGURE 9.7   *The same small bookshelf loudspeaker measured in 4ϖ, full anechoic conditions and in 2ϖ, half-space conditions. The enclosure was carefully flush mounted.*

omnidirectional; more of the sound is being radiated forward and not being reflected by the boundary. The DI at low frequencies is, as theory would predict, 3 dB for a half space.

The overall conclusion is that mounting this excellent little loudspeaker in a wall has left its overall performance substantially intact, but the bass output has been greatly increased, making it sound fat, thick and tubby. After all, it was not designed to be used in this manner. The solution in this case is to turn the bass down. Any competent equalizer can do it, or the old-fashioned bass control may just be optimum if the "hinge" frequency is around 500 Hz.

The proper solution, if a loudspeaker is to be used in this manner, is to design it from the outset so that it has a flat axial frequency response when it is mounted in a wall. All in-wall and in-ceiling loudspeakers *should* be designed in this manner, but not all are.

Moving on, Figure 9.8b shows what happens when the loudspeaker is simply mounted on the surface of a wall. For comparison purposes, the half-space data from Figure 9.7 are repeated in (a). It can be seen that there is a strong acoustical interference dip at about 220 Hz. We saw this before, back in Figure 9.2, but at a lower frequency. This loudspeaker is smaller, and the round-trip distance from the woofer to the wall and back is shorter: about 31 inches (787 mm), which is one-half wavelength at about 220 Hz, the destructive-interference condition representing the first "tooth" in a comb filter. There is a hint of a second tooth, a partial cancellation, in the on-axis curve at 660 Hz, but one can presume that increasing source directivity at higher frequencies eliminates any higher-order cancellations. But, why is there no corresponding "hole" in the sound power curve? It turns out that there is, but it is not so easy to see as the dramatic event in the on-axis curve. This will be explained in the next section.

We have all seen it: loudspeakers sitting in otherwise empty cavities in bookcases, entertainment furniture and, embarrassingly, in expensive custom installations. Figure 9.9

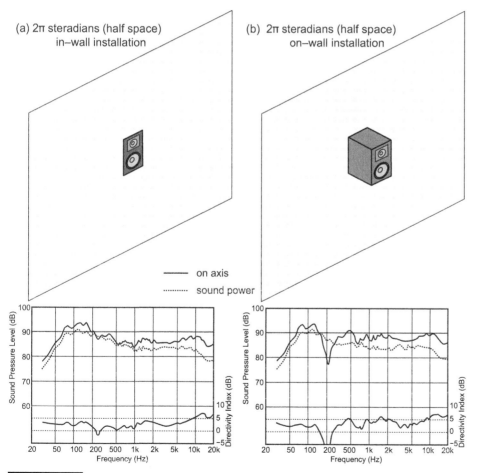

FIGURE 9.8 *A comparison of in-wall and on-wall placement of the bookshelf loudspeaker.*

shows what can result: in this case a perfectly good little loudspeaker has been seriously compromised by the installation. There is evidence of high-Q cavity resonances and diffraction effects created by the edges of the cavity. The everyday remedy of filling the cavity with absorbing material helps, but the root problem is still in evidence. Figures 9.7b and 9.8a show the much-improved performance in a mounting where the cavity openings have simply been closed off with hard material.

Reflecting on what has just been discussed, it can be concluded that there are really only two locations in which a loudspeaker has the potential of performing at its best: free standing, or flush mounted in a wall (or ceiling). All other options involve compromises.

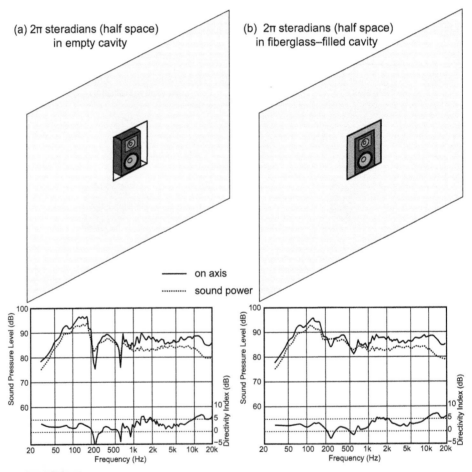

FIGURE 9.9 *A comparison of acoustical performance when the same small bookshelf loudspeaker is placed in a cavity—a bookshelf perhaps—as often happens in "entertainment system" furniture. In (a) the cavity is empty. In (b) the cavity has been filled with fiberglass.*

9.3.1 An Example of Adjacent-Boundary Interference

Figure 9.8b showed an obvious acoustical cancellation dip in the on-axis frequency response, but not in the sound power. Figure 9.10 explains why.

The ray diagram in Figure 9.10 illustrates that the direct and reflected path length difference is maximum at 0° (on-axis), trending to zero at 90° off-axis. Therefore the frequency at which the first destructive acoustical interference occurs rises from the 200 Hz dip seen in Figure 9.8b to ever-increasing frequencies as the measurement point moves away from the forward axis. The declining off-axis radiation from the loud-speaker is a major factor in the progressive reduction in the cancellation dip. In the sound power spatial summation, this results in a shallow depression in acoustic output, compared to the in-wall installation where there is no interference.

If the loudspeaker were positioned at a distance from the wall, as when using a mounting bracket, the dip would begin at a lower frequency. The greater the separation, the lower the frequency at which the depression begins. The more directional the loud-speaker, the narrower the frequency range over which it exists. This is the mechanism of the classic adjacent-boundary effect.

What we need are loudspeakers designed with knowledge of how they are to be mounted, where they are to be placed. The idea of a universally applicable, one-type-does-all, loudspeaker is an obsolete concept, but it is the basis of most of today's designs: bookshelf loudspeakers that don't work in bookshelves, free-standing loudspeakers that for practical reasons cannot avoid at least some adjacent-boundary problems. It would seem that there is an opportunity for something different.

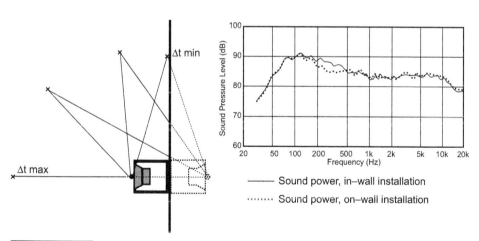

FIGURE 9.10 *A direct comparison of sound power measurements made with the same loudspeaker mounted in and on a wall, as shown in Figure 9.8. The ray diagram applies to the on-wall situation, showing the direct sound paths from the loudspeaker and the reflected sound paths from its acoustical mirror image behind the wall.*

9.4 "BOUNDARY-FRIENDLY" LOUDSPEAKER DESIGNS

Having expounded on the problems created by adjacent-boundary effects, it was no surprise that Allison (1974) proposed a loudspeaker design that minimized the effects. Figure 9.11 shows the configuration of the loudspeaker and how it was intended to be placed in a room.

First, the woofers were located close to both the floor and the side walls, eliminating issues with those boundaries and taking full advantage of solid angle gains. The effect of the side wall was minimized by placing the loudspeaker some distance away from it. Around 350 Hz the woofers crossed over to the midrange-tweeter array situated at the top of the enclosure, at ear level. The drivers radiating the lower frequencies were located close to the wall to minimize that boundary problem, and two sets of drivers at 90° to each other were intended to approximate a hemispherical radiation pattern. It was a very thoughtful design but, unfortunately, I have no measurement data on it.

The Acoustic Research AR-9 incorporated some of Allison's adjacent-boundary compensation ideas. Some anechoic data are shown in Figure 18.3h, repeated in Figure 9.12 along with a spatially averaged room curve. Controlling adjacent boundary effects should improve the uniformity of sound radiated into a room, but it does not exercise control over the large effects of standing waves at the listening locations; these would be strongly room dependent. The room curve shown was the result of

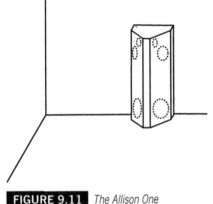

FIGURE 9.11 *The Allison One loudspeaker as described in Allison (1974), figure 16.*

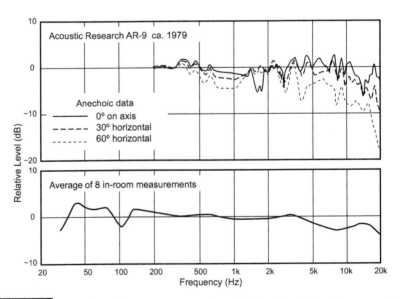

FIGURE 9.12 *Anechoic and in-room measurements on a loudspeaker designed to compensate for adjacent boundary effects at low frequencies. The anechoic curves are truncated at 200 Hz because of chamber errors in measurements of that period.*

several combinations of loudspeaker and microphone locations, attempting to reveal the performance of the loudspeaker-and-adjacent-boundary combination. The result confirms that the design was successful, even with contributions of standing waves included.

Loudspeakers designed to function well when close to boundaries offer an advantage given the present popularity of on-wall flat-screen video displays. Low profile loudspeakers situated on or close to walls are clearly desirable and not all loudspeakers are designed with this in mind, as illustrated in Figure 9.8b.

Good ideas don't go away; they just morph, evolve or get reinvented. In this case the surround loudspeaker shown in Figure 9.13a is an example of a class of products developed to cater to home theaters. It was designed with on-wall mounting in mind, and considerable thought went into the physical layout and the crossover network to control the acoustical interactions of the drivers with each other and with the wall behind. It can be switched among three radiating patterns (Figure 15.11), but here we look at the most favorable one, called "bipole," meaning that both sets of drivers are radiating in phase with each other. This was the configuration it was optimized for and performance shown in (b) is excellent for a loudspeaker aimed at the mainstream market.

The measured curves in (b) are all very similar. In spite of an appearance suggesting bi-directional radiation, over most of the frequency range this loudspeaker behaves as a hemispherically omnidirectional radiator. For surround-channel loudspeakers mounted

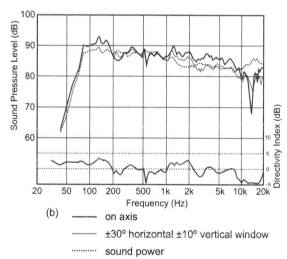

(a) A bidirectional in–phase
(aka "bipole") surround loudspeaker
designed for on-wall mounting

(b) ——— on axis

——— ±30° horizontal ±10° vertical window

·········· sound power

FIGURE 9.13 *(a) An on-wall surround loudspeaker with switchable directivity, an Infinity Beta ES250. (b) Frequency response measurements on axis and averaged over the listening window ±30° horizontal and ±10° vertical (9 curves) compared with the total sound power. The directivity index was computed using the listening window as the reference. For the measurements the loudspeaker was used in the "bipole" mode: both sets of drivers were radiating in phase with each other, and it was mounted on a large section of wall in an anechoic chamber.*

on side and rear walls this ensures delivery of similar direct sound to all parts of the audience. These bipole (bidirectional in-phase) designs are also appropriate for the current immersive systems; the dipole versions are not.

It is worth a reminder that, in spite of what one sometimes reads, such loudspeakers are not "diffuse" sound sources; they are wide-dispersion sound sources. Diffuse sound fields simply do not occur in home theaters, and diffusers are something entirely different.

Obviously, to avoid adjacent-boundary problems, loudspeaker drivers must be less than half a wavelength separation from large reflecting surfaces. In-wall flush mounting is excellent, but with good design, on-wall configurations work very well and, as shown in this example, they allow for nearly hemispherical radiation. Many surround loudspeakers are designed in this fashion, a welcome trend. Ironically, it is the front loudspeakers, the most important ones, which routinely are designed with little or no regard for the adjacent-boundary settings into which they will be placed.

In control rooms, it has been common practice for decades to mount the main front monitor loudspeakers in some form of half-space mounting. Eargle (1973) and Makivirta and Anet (2001) are examples of advocates of this form of installation. Some rooms, because of the location of the viewing window into the studio, force the loudspeakers to be installed in an overhanging structure placed against the ceiling and with the loudspeaker surface some distance forward of the window and remainder of the wall. In the audio industry this is called soffit mounting, although the word normally has a different meaning. It is a corruption of half-space mounting; consequently, non-optimal boundary interactions may be anticipated.

9.5 ARRAY LOUDSPEAKERS—OTHER WAYS TO MANIPULATE BOUNDARY INTERACTIONS

Traditional loudspeakers employ multiple transducers, each operating over a fraction of the audible frequency range. They individually interact with reflecting surfaces around them, and they exhibit varying directivity with frequency. In tower loudspeakers it is common to see multiple woofers and more, and this has led to them being called "line sources" or "line arrays." They consist of transducers arranged in a line, but they are absolutely *not* either line sources or arrays. They are just tall loudspeakers functioning very similarly to their shorter versions.

However, a continuous narrow strip radiator, or a line array of closely spaced small loudspeakers (e.g., Figure 10.10), is a very different device with its own distinctive properties. Computer simulation technology now permits us to explore domains with precision previously denied us, and loudspeaker and electronics technology allow us to turn some of it into physical reality. In sound reinforcement such designs have been used for decades to control sound radiation patterns in large venues.

However, in small rooms it is a significant complication that adjacent reflecting surfaces must be considered. This means that the far field (see Section 10.5) for the combination of a conventional loudspeaker and one or more adjacent boundary reflections (now part of the "source") can be a long distance away. What we hear and measure at normal listening distances in small rooms is subject to variations. These include the low-frequency adjacent-boundary effects discussed earlier, but go further, with acoustical interference affecting much of the audible frequency range.

Figure 9.14 is a small excerpt from Keele and Button (2005)—a collection of thought-provoking predictions and measurements. In it they examine the performance of a theoretical point source compared to several variations of truncated lines: straight and curved, "shaded" (drive power reduced toward the end) and unshaded (all transducers driven equally), all standing on a reflecting surface. For comparison I have added a floor-to-ceiling true line source at the top that embraces the floor and ceiling reflections as extensions of itself.

Figure 9.14a shows the orderly behavior of a true line source in its near field: −3 dB per double distance at all frequencies. In Figure 9.14b, a single reflecting surface, the floor, disrupts the radiation pattern of a point source—an omnidirectional, infinitely small, theoretical starting point for discussing acoustics. Instead of tidy, expanding circular contour plots, we see an example of gross acoustical interference with alternating lobes of high and low sound levels. The constant directivity of the source, indicated on the right, means that this problem exists at all frequencies, but the patterns will be different because of differing wavelengths. Additional boundaries, ceiling and side walls add more of the same, of course.

Typical forward-firing loudspeakers are omnidirectional only at low frequencies, becoming more directional at higher frequencies, which helps in that the highest frequencies arrive at listeners mainly as direct sound only, negligible reflections (e.g., Figure 5.4). The sounds arriving at a listening location in a room are therefore complicated, but if the reflected sounds have a spectrum that is similar to the direct sound, the merged combination usually ends up being perceptually more satisfactory than this single-frequency, single-dimensional perspective suggests. Figure 12.4b shows examples of well-designed loudspeakers in rooms, proving the point.

Section 9.4 showed examples of conventional cone/dome loudspeakers designed to minimize boundary interface problems. Figure 9.14c and those that follow show what happens when considering the floor boundary as an integral part of vertical array loudspeaker designs. In Figure 9.14c, a simple truncated line seems to be an improvement over the elevated point source, but uniform directivity has been sacrificed. The directivity index has a sharply rising character, indicating high frequency beaming. Figure 9.14d shows that shading the output, reducing the drive delivered to the transducers closer to the top of the line according to a Hann contour, greatly simplifies the pattern, yielding stable sound levels at ear level over the length of the room. However, it still beams dramatically at high frequencies. We are not there yet.

Curving the line, as shown in Figure 9.14e, is a step in the right direction. The contour lines are not yet smooth, but there is an underlying desirable order to them. The constancy of the directivity index tells us that it applies over a wide bandwidth.

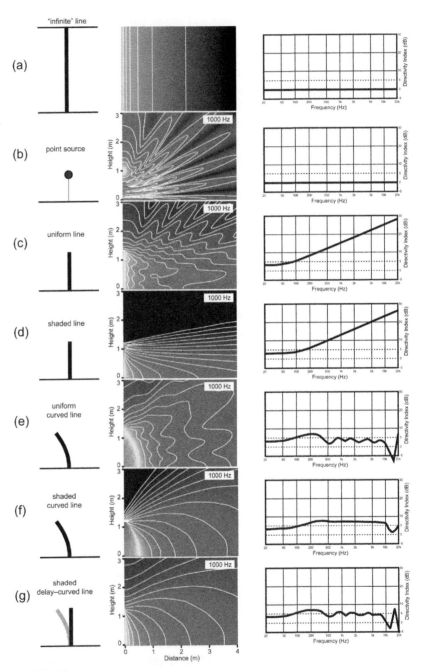

FIGURE 9.14 *Illustrations of the near sound fields generated above a ground plane by several sound sources. The author has added an idealized line source at the top as a contrast to the truncated lines shown below it. The shading in the middle illustrations gets darker as sound levels drop; adjacent contour lines represent sound levels that differ by 3 dB. The original paper displays results for several frequencies, all of those shown are for 1 kHz. The words and graphics on the left explain the sources. On the right are far-field directivity indexes. Data from Keele and Button (2005).*

In Figure 9.14f, we see that shading the curved line using the Legendre contour yields a set of plots that have two desirable characteristics: a stable sound level at ear level out to 3 m or more from the loudspeaker, and a relatively constant directivity, meaning that it applies over a wide frequency range. This is a good result, and it has been called the Constant Beamwidth Transducer (CBT). This is not merely coping with the floor as an adjacent boundary, a low-frequency concern; it is using the floor reflection to create a sound field that has desirably uniform broadband sound level over a wide range of listening distances.

Curved loudspeakers may not be to everyone's liking, so applying the appropriate delays to the individual drivers can, in effect, contour a straight line. When Legendre shading is applied to the delay-curved line, the resulting CBT is very similar, as seen in Figure 9.14g.

A comparable effect can be created without a floor, but the array will need to be differently optimized. JBLPro has a series of innovative free-standing CBT products, designed by Doug Button, that have found use in sound reinforcement and high-end cinema and home theater surround applications. The stability of both sound quality and sound level as a function of listening distance is striking, which makes such designs generally attractive and especially for use as side surrounds (see Figure 14.4).

9.6 LISTENERS ALSO HAVE BOUNDARIES

A lot of effort and talk goes into getting the right acoustical setting for the loudspeakers, but little, it seems, for the listener. Just today I was perusing a high-end audio magazine, and saw a photo of a reviewer's listening room. My first impression was that it was probably not a bad room: the loudspeakers were well away from the boundaries, the walls were geometrically irregular with many large shelving units full of books and recordings, and there was what looked like a good carpet. But, the solitary chair was very close to the back wall. It is an unfortunate fact of many home listening rooms, and I have experienced it in some hotel room audio-show demonstrations. The ideal solution is to move the listener away from the wall, but with limited space this may not be possible. Figure 9.15 shows what happens; a whole new set of reflections is created, and the ones immediately following the direct sounds from the loudspeakers are very early and very strong. There will be ceiling reflections as well.

Multiple reflections from the rear add poorly correlated sounds to those arriving from the front. So, in addition to the timbral colorations added by the acoustical interference, as described in previous sections, there will be degraded image and soundstage quality.

When confronted with such a situation, a persuasive demonstration is to have the listener play a stereo recording, asking him or her to focus on imaging and soundstage. Then hold a thick fabric-upholstered cushion or pillow filled with fibrous stuffing, not foam (seating foam is not the same as acoustical foam), or the equivalent behind the head. It has been my experience that most listeners notice an improved clarity of sound

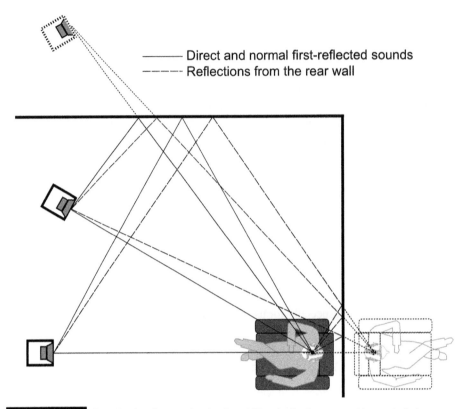

Direct and normal first-reflected sounds
Reflections from the rear wall

FIGURE 9.15 *A ray-tracing diagram showing the additional reflections caused by proximity to a reflecting rear wall. Not only loudspeakers have acoustical mirror images.*

and less confused imaging. Obviously the permanent solution to this problem is to attenuate the rear-wall reflections. Place a panel of thick—at least 6 inches (150 mm) if possible—fibrous absorbing material covering a significant area behind, beside, and above the listener's head. I have seen this hidden behind a visually attractive, acoustically transparent, wall hanging. It won't be perfect, but it will be an improvement.

CHAPTER 10

The Sound Fields in Sound Reproduction Spaces

Intuition might tell us that there is little in common among home listening rooms, home theaters, recording control rooms, movie dubbing stages and cinemas, but there is. They are all points on a continuum of acoustical and perceptual dimensions (Figure 1.4). Some of the measures we use to describe circumstances and events in these spaces have their origins in concert hall acoustical investigations, a long way from the present context. But when we look closely at the sounds heard by listeners in most sound reproduction venues, there are clear parallels. There is much to discuss.

10.1 REVERBERATION

In large performance venues, concert halls and auditoriums, reverberation is real and necessary. It consists of sounds from voices and musical instruments that are reflected many times from many surfaces, penetrating to all parts of the room, gradually decaying in amplitude with time until they are inaudible. It is measured as reverberation time (RT)—the time for a 60 dB reduction in amplitude. The hall is part of the performance. In live, unamplified performances, halls are essential for the entertainment of large audiences. Voices and musical instruments have limited sound power output, so halls are as reflective as possible to keep loudness high, without interfering with the "intelligibility" of the music. It is a difficult compromise. Typical values for RT in performance venues are in the range 1–2 s, optimized for the music most commonly performed, and occasionally slightly variable.

In sound reproduction, the essential hall reverberation is captured in the recording, so nothing additional is required. Typical reverberation times for domestic listening rooms and recording control rooms are 0.2–0.4 s. The reverberation in classical recordings completely overpowers that of the listening room. Figure 10.1a shows how

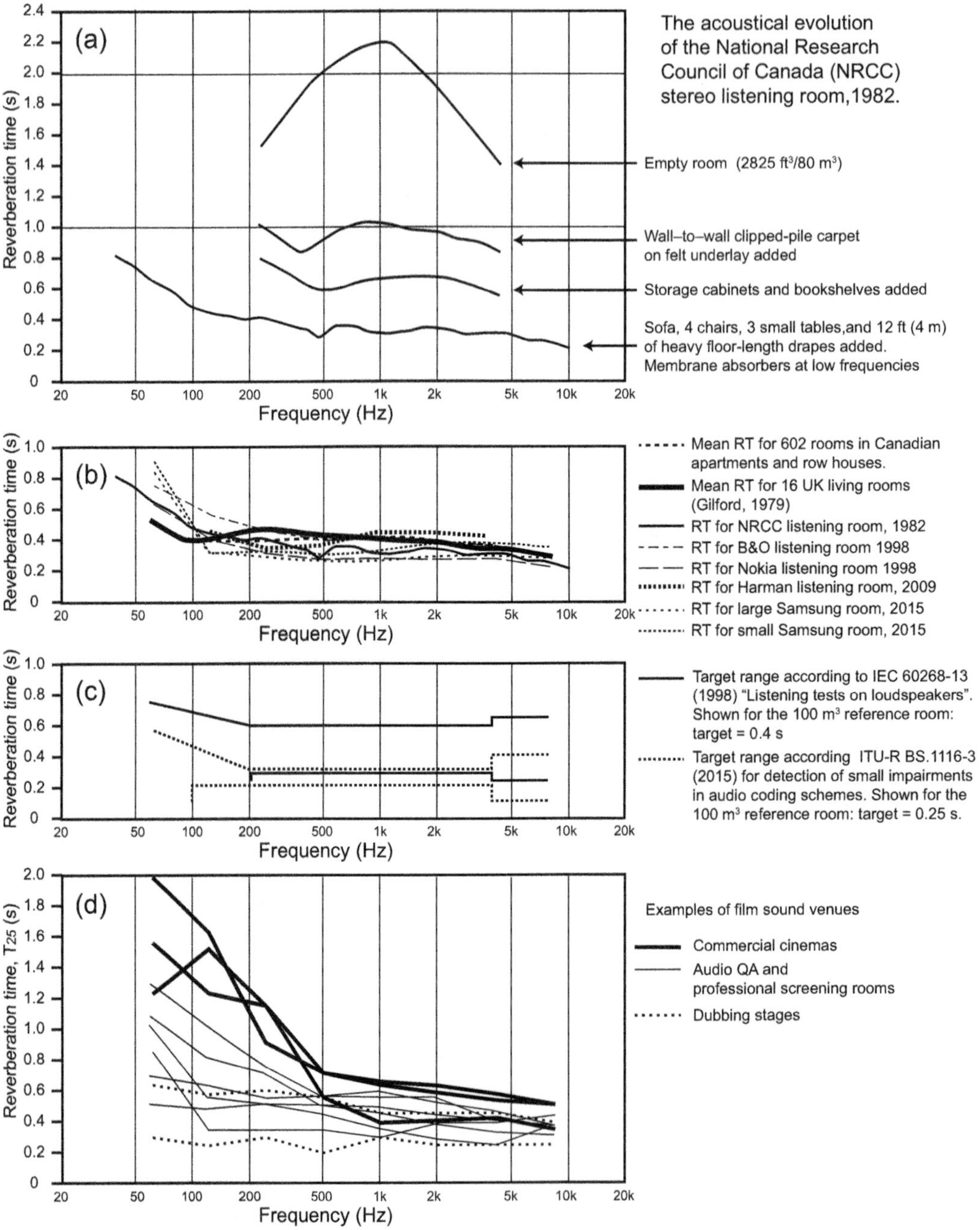

FIGURE 10.1 *(a) An illustration of how common home furnishing items can result in a good listening space. (b) RT averages for surveys in Canadian and UK homes, and example performances for several professionally constructed listening rooms. (c) The recommended ranges for RT from the noted IEC and ITU documents. Reproduced with permission. (d) RT data for typical cinemas, a reference cinema and two dubbing stages. From Toole (2015b), figure 3, and SMPTE TC-25CSS (2014) reproduced with permission. Screening room data courtesy of Linda Gedemer.*

progressive room treatments and furnishings modify RT in a typical domestic room as exemplified by the original National Research Council of Canada (NRCC) listening room, the prototype for the original IEC 60268–13 (1998) recommended stereo listening room for loudspeaker evaluations. It is seen that adding an acoustically absorbent carpet has a major effect. Adding scattering from cabinets causes that absorbent material to work harder. Finally, adding the absorption and scattering of normal furnishings brings the RT into the desirable range. Low frequency membrane absorbers were designed and built because the original room was a laboratory space with concrete floor and plaster on block walls (to contain any mad-scientist experiments that might go wrong). The resulting RT is consistent with data on typical domestic rooms. From this it is clear that a well-furnished domestic room can be a good basis for a listening environment, without additional acoustical devices.

Figure 10.1b shows RT data on a selection of rooms, most of which were constructed for listening evaluations on loudspeakers and other audio products. Data for those rooms mingle with average data for Canadian and UK living spaces, which suggests that the results of tests done in those environments may translate well into the world of consumers.

Figure 10.1c shows target ranges for RT from two well-known international recommendations. The IEC 60268–13 (1998) document is intended to provide guidance for conducting listening tests on loudspeakers. The author was one of the writers of the original document that was published in 1985. The ITU-R BS.1116–3 document is aimed at a very different task; that of subjectively detecting the typically very subtle degradations introduced by perceptual encoders/decoders (codecs) used to reduce the digital data content of audio signals. The introduction to the document says:

This Recommendation is intended for use in the assessment of systems which introduce impairments so small as to be undetectable without rigorous control of the experimental conditions and appropriate statistical analysis. If used for systems that introduce relatively large and easily detectable impairments, it leads to excessive expenditure of time and effort and may also lead to less reliable results than a simpler test.

In other words, this is *not* what is needed for evaluating loudspeakers. Nevertheless the document often appears as one of the performance requirements for rooms used for general audio evaluations. Fortunately, the curves in Figure 10.1b indicate that most of the rooms fall into the IEC target range.

Figure 10.1d shows measurements from several venues where movie sound tracks are created, inspected for excellence or experienced by customers. It is not surprising that budgetary constraints compromise the acoustical performance of mass-market cinemas, with low-frequency reverberation times clearly elevated. However, several of the audio quality assessment and screening rooms and the dubbing stages where the sound tracks are mixed fall in the range of the quality listening rooms in Figure 10.1b and would be considered acceptable by IEC recommended RT values in Figure 10.1c.

If one is looking for a concurrence of objectives for movie sound and home theater sound, this would seem to be a good beginning. Just note that the auditory experiences in the local cinema multiplex may be less impressive than in a good home theater.

Cinemas exhibit mid-frequency RTs of 0.3–0.6 seconds so that the acoustics of the venue displayed on the screen will can be convincingly delivered. In terms of speech intelligibility, anything below 0.5 s is considered to be safe, but the elevated RTs seen at low frequencies can cause coloration in voices and possibly degraded intelligibility.

The reality, though, is that rooms as acoustically "dead" as these don't have a traditional diffuse reverberant sound field. It cannot exist. What is measured is a decaying sequence of relatively discrete reflections, not a dense, diffuse sound field. This is discussed in Toole (2015b), Section 3.1 and in Section 10.1.3, where it is suggested that cumulative energy time—the time required for sound level to build up, rather than decay—might be a better metric. It is a topic deserving of some attention.

10.1.1 Measuring Reverberation Time

For professional acousticians measuring the RT of performance venues there are highly ritualized, internationally standardized, procedures. In general, they involve a sound source that radiates omnidirectionally—a special loudspeaker, a starter's pistol or a popping balloon—that is located where the orchestra or band plays. Measurements are made at several locations in the audience area using an omnidirectional microphone, and the results considered in the context of how the venue reverberation is likely to affect the musical performances.

For sound reproduction, the process needs to be modified because the sound sources are the reproduction loudspeakers, and the venue reverberation is not expected to be a significant factor in the delivery of the program, whether it is music or movies. In fact, the principal consideration is well summed up in the requirements for good speech intelligibility. In a home theater or cinema situation, the logical loudspeaker to provide the stimulus is the center channel, where virtually all the dialog originates, and the desirable RT is less than about 0.5 s at middle and high frequencies. This is an achievable objective, as is seen in Figure 10.1.

As will be seen, the reflectivity of a room at low frequencies, which is well described by RT in large rooms, is a factor in how much bass will be generated by sustained sounds. The reason is that all practical sound sources, loudspeakers, voices and musical instruments are essentially omnidirectional at low frequencies.

10.1.2 Calculating Reverberation Time

In large highly reflective rooms the reverberation time is often well predicted by the original Sabine formula:

$$RT = .049 \ V/A$$

where V is the total volume in ft^3 and A is the total absorption in the room in sabins. The total absorption, A, is calculated by adding up all of the piecemeal areas (carpet, drapes, walls, etc.) of the boundaries multiplied by their individual absorption coefficients:

$$A = (S_1\alpha_1 + S_2\alpha_2 + S_3\alpha \ldots)$$

where S is the area in square feet and α is the absorption coefficient for the material covering that area. Absorption coefficient is ideally a measure of the proportion of randomly incident sound power that is absorbed by a material. The product of S and α is a number with the unit sabins. The absorption of some items, such as people or chairs, is sometimes quoted directly in sabins.

The metric equivalent of the Sabine formula is $RT = 0.161 V/A$, where the volume is in m^3, areas are in m^2 and A is in metric sabins.

As rooms get more absorptive and smaller, and as the materials on the room boundaries become less randomly distributed (e.g., wall-to-wall carpet on the floor), this equation becomes progressively less reliable. Over the past 100 years, several increasingly more complex equations have been developed in order to accommodate asymmetry in rooms and the fact that the sound field is not diffuse, among them Fitzroy (1959) and Arau-Puchades (1988). However, all of them, in order to be practical, make assumptions. Dalenbäck (2000) says, "these two formulas give a better estimate than the classical formulas [Sabine and Eyring] in *some cases* but here a central question is: *how can one be sure they are better in a particular case?* So far no equation with universal applicability has been shown" [his emphasis]. Fortunately, as will be seen, in small rooms for sound reproduction, high precision is not required. The simple Sabine formula provides estimates that are adequate for our purposes.

10.1.3 Is There a More Useful Metric for Our Purposes?

As useful as RT is for evaluating highly reflective venues, there may be a better metric for our purposes: cumulative energy time—the time taken after the arrival of the direct sound for the sound field to rise to the steady-state level. This is a much shorter time than the corresponding RT, and seems to more directly address the perceptual processes when brief sounds are involved. For example, Figure 10.2 shows that in a cinema with a 2.5 s (2,500 ms) RT at 50 Hz, a level within 2 dB of steady state is reached in 90 ms. At 500 Hz, RT is 800 ms, and the cumulative energy time is 25 ms. These are enormous differences, and although RT is a related parameter, it is far removed from the temporal events that are likely to matter. Events very early in the RT decay data could be relevant, but a new form of interpretation would be required.

Germane to this discussion are the findings discussed in Section 7.6.3, indicating that for speech intelligibility—a crucial consideration for movies—it is the early reflections that are the main contributors. Early reflection energy arriving within about the first 50 ms following the direct sound has the same effect on speech intelligibility scores as an equal increase in the direct sound energy. This was true for both normal and hearing-impaired listeners. Bradley et al. (2003) say: "Although it is important to avoid excessive reverberant sound, adding large amounts of absorption to achieve very short reverberation times may degrade intelligibility due to reduced early reflection levels."

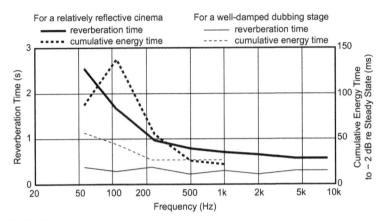

FIGURE 10.2 *A comparison of reverberation time and cumulative energy time for two very different movie playback venues.*
Data from Toole (2015b), figures 8 and 9, based on SMPTE TC-25CSS (2014).

They suggest a ratio of the energy within the first 50 ms of an impulse response to the energy associated with the direct sound as a new figure of merit for a room, called the early reflection benefit (ERB).

On a related topic, it is worth noting that early reflections also reduce the thresholds for detecting resonances in loudspeakers (Toole and Olive, 1988), making flawed loudspeakers more noticeable and revealing more timbral subtleties in music. For stereo recording and listening, early reflections can reduce the timbral degradation and speech intelligibility loss in the phantom center image. It helps to fill the large spectral dip around 2 kHz, created by stereo/interaural crosstalk (Figure 7.2).

Leembruggen (2015a) showed several examples in which cumulative energy data appeared to relate to details in frequency-domain measurements in ways that RT did not, and thereby, possibly to perceptions. In the context of concert halls, Bradley et al. (1997) found that the perceived strength of bass sounds was related to early and late reflection arrivals, and was not significantly related to low-frequency reverberation time.

Multiple factors are at play in this situation, but there are several persuasive reasons to pay special attention to early reflections, and to consider alternatives to RT (and its familiar derivatives, such as EDT, early decay time) as a criterion of acoustical performance in sound reproduction venues. In any event, both RT and cumulative energy measures are frequency dependent—confirmation that direct and steady-state sound fields have different spectra in most rooms.

10.2 DIFFUSION

A perfectly diffuse sound field is isotropic: at any point within the sound field, sounds may be expected to arrive from all directions with equal probability. It is also homogeneous: it is the same everywhere in the space. Small listening and control rooms, and cinemas do not have diffuse sound fields at middle and high frequencies. In fact, true diffusion exists only as an academic ideal. Reverberation chambers used to measure the absorption of acoustical materials are designed to be diffuse and can come close, but as soon as a test sample of absorbing material is introduced into the space, it ceases to be—hence the errors and variations in measured sound absorption coefficients.

Diffusion can be improved by using sound scattering or dispersing objects, irregular, curved and angled surfaces, and specially designed devices, often called diffusers (note the spelling: the word "diffusor" applies to products from RPG). Perceptually, a diffuse sound field in a concert hall sounds spacious and enveloping. However, a diffuse sound field is not a requirement for the perception of spaciousness and envelopment. Much simpler sound fields work too, especially in multichannel sound reproduction where it is possible to deliver sounds to the ears that are perceived to have those qualities—with or without a reflective room; see Section 15.7.

When people talk about diffusion in small listening rooms, they misspeak. One hears of "diffusing" loudspeakers, when what is meant is a widely dispersing loud-speaker that radiates its sound in many directions. We talk of "diffusers," but these devices really scatter or disperse incoming sounds in many directions. Whether either of these contribute to additional diffusion in the sound field depends on the acoustics of the rest of the room. It is impossible to change this word misusage at this stage, but let us at least understand that a diffuse sound field is not desirable in a sound reproduction venue, and, fortunately, because of the amount of absorption in these spaces, it cannot exist. Section 7.3.2 discusses these diffusing/scattering/dispersing devices.

With modern computer modeling techniques, it is easier to predict a sound field than it is to examine an existing one, but Gover et al. (2004) provided important measured evidence of what is going on in the sound fields in some small rooms. Using a novel spherical steerable-array microphone, the authors explored, in three dimensions, the decaying sound field in several small rooms. None of the rooms exhibited isotropic distributions at the measurement locations. Strong directional features were associated with early reflections. Small meeting rooms and a videoconferencing room with rever-beration times of 0.36–0.4s, in the range of typical listening rooms, had anisotropy indi-ces and directional diffusion measures that fell roughly halfway between anechoic and reverberant conditions. Moreover, the values changed with time. Later sound showed increased anisotropy and even changed orientation in the room according to the surfaces that were more reflective (Figure 10.3).

First, it is clear from Figure 10.3 that the energetic acoustic events happen in the first 50 to 100 ms in a room with a measured RT (really reflection-decay time) of 400 ms. These are early reflections, not reverberation. In the earliest time interval, the direct sound and the opposite and side wall reflections dominate. All of these are within about 7 dB of the direct sound. Once these early sounds pass, the next collection of reflections ($t > 50$ ms) in the example room are 12 to 17 dB down and the reflection pattern has shifted 90° to align with reflections between the parallel side walls. After 100 ms the lateral directional bias remains and levels have dropped to about −20 to −27 dB. At no time is the sound field diffuse, that is, the pattern is not circular.

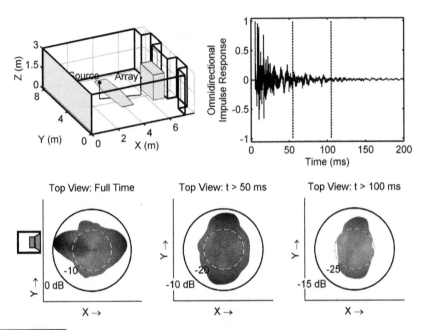

FIGURE 10.3 *Diffusivity measurements made in a videoconferencing room (7.23 m × 8.33 m × 3.01 m) with a mid-frequency RT of 0.4 s. The omnidirectional source and the measurement microphone array were 2.03 m apart. The shapes across the bottom of the figure are the horizontal plane diffusivity patterns. The loudspeaker symbol shows the orientation of the direct sound. A perfectly diffuse sound field would show a circular pattern. The pattern on the left is for the entire time record, shown in the upper right. It shows prominent lobes for the direct sound, first-order lateral reflections, and a rear wall reflection. The middle and right patterns represent diffusivity of the later portions of the impulse response: the diffusivity rotates to a side-to-side orientation, the result of reflections between the side walls.*
Reprinted with permission from Gover et al. (2004). Copyright 2004, Acoustical Society of America.

If this were a home theater with the audience aligned with the long axis of the room, one could imagine that these side-to-side early reflections might assist the surround channels in creating a sense of space and envelopment, which is the result of low inter-aural cross-correlation (the sounds are different at the two ears). I know of no tests that prove this, but it is a logical conjecture.

This example plainly illustrates how the disposition of smooth and irregular surfaces, and absorbing material, in a room influences the structure of the decaying sound field. It is a variable that smart acousticians can use to customize the pattern of early-reflected sounds to avoid potential problems, or to enhance some desirable perceptual effects.

Another example of this kind is shown in Figure 10.4, a much earlier study by Erwin Meyer (1954), showing that the total metric "diffusivity" is dramatically reduced when first reflections are absorbed.

These tests relate to the diffusivity as measured at the microphone location, not throughout the room volume. This makes the results directly relatable to what might be

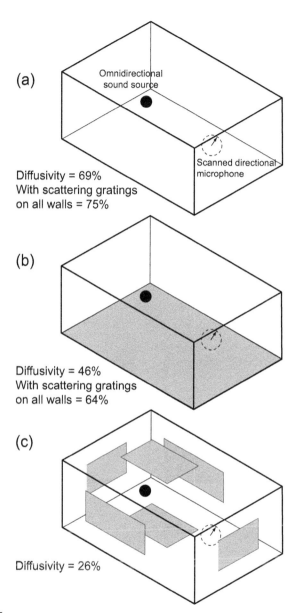

(a)

Omnidirectional
sound source

Scanned directional
microphone

Diffusivity = 69%
With scattering gratings
on all walls = 75%

(b)

Diffusivity = 46%
With scattering gratings
on all walls = 64%

(c)

Diffusivity = 26%

FIGURE 10.4 *Meyer (1954) scale model experiments, in which the diffusivity of an omnidirectional sound source was measured using a rotating directional microphone with 10° as the half-energy value. The top figure shows an empty room. In the middle figure, the floor has been covered by absorbing material. In the bottom figure the same absorbing material has been cut up and located so that it would attenuate the first reflections between the source and the microphone. He claimed that the reverberation time was relatively unchanged in the two configurations of absorbing material, although the diffusivity changed dramatically. He also added scattering gratings (diffusers) to the walls in some configurations. Unfortunately, Meyer published no drawings of the setups, so these are how this author imagined them to be from his verbal descriptions. The diffusivity number is a calculation from his measurements; 100% would describe a situation in which sounds arrived at the listening location equally from all directions—perfect diffusivity.*

heard by a listener. Meyer did several tests with and without sound scattering gratings on various surfaces, with the following results:

1. Bare room with smooth walls: diffusivity = 69% (Figure 10.3a)

2. Adding scattering gratings to all walls: diffusivity = 75%

3. Bare room, but floor totally absorbent: diffusivity = 46% (Figure 10.3b)

4. As above with scattering gratings on all other walls: diffusivity = 64%

5. Bare room, using absorbent that was on floor, divided in pieces to suppress the first reflections between the source and the microphone: diffusivity = 26% (Figure 10.3c).

It is evident that absorbing the first reflections has a powerful effect on the diffusivity, the IACC, and thereby the perceived spaciousness of sound in a room. Adding large areas of sound scattering devices increased the diffusivity, as would be expected.

Views from the center channel loudspeaker in a room showing:

(a) what the eye sees

(b) what the loudspeaker sees at frequencies above about 300–500 Hz - leather furniture

(c) what the loudspeaker sees at frequencies above about 300–500 Hz - fabric upholstered furniture

FIGURE 10.5 *A view of the author's listening space as "seen" by the center channel loudspeaker. It shows that much of the direct sound from the loudspeaker is absorbed at the first encounter with a surface, thereafter being unavailable to contribute to later reflections. In (b) the leather-covered furniture is shown as being translucent, because there will be some mid-high-frequency reflection from the surface. With fabric upholstery (c), even this is gone. There is little reflecting or scattering surface area left, although a wider-angle lens would have shown that there is no side wall on the left side—it is an opening to the rest of the house— and on the right this lost energy is balanced by velour drapes over a window wall.*

Absorbing first reflections not only eliminates those specific components of sound, but significantly alters all subsequent acoustical events. Listeners would be in a strong direct sound field if those reflections were the only ones absorbed.

It is important to note that the sound source used here was truly omnidirectional—not the horizontal omnidirectionality we accept in audio loudspeakers with that claim. If the sound source had the directivity of conventional forward-facing cone/dome or cone/horn loudspeakers, the diffusivity numbers would have been much lower and, of course, frequency dependent. Figure 10.5 illustrates this situation for a home listening room employing conventional forward-firing loudspeakers.

It is often instructive to put yourself at the loudspeaker location and imagine what happens to the radiated sound. Although every room will be different, it is a safe generalization that a high percentage of the mid-high-frequency energy is absorbed at the first surface it encounters. This is also true in cinemas, where the loudspeakers are significantly directional, being designed to deliver sound to the (highly absorptive) audience area and not to the ceiling and side walls. This helps to explain why the mid-high-frequency RT of cinemas is comparable with that of domestic venues, as seen in Figure 10.1.

Therefore a diffuse sound field is not a possibility in normal sound reproduction venues, nor would we want it if it were. The information required to generate perceptions of such sound fields is in the multichannel audio system, where it can be manipulated as required to be appropriate to the musical content or dramatic needs of a plot.

10.3 DIRECT SOUND AND EARLY REFLECTIONS

With insubstantial reverberation and little diffusion it is evident that there are few "late" reflections at middle and high frequencies. We are left with the direct sound and some number of "early" reflections. These highly consequential acoustical events have been discussed from a physical perspective in Chapter 5 and from a perceptual perspective in great detail in Chapter 7. In fact, above the bass-frequency range it is the direct sound combined with some number of early-reflected sounds that dominate what is measured and heard in sound reproducing systems. Figure 10.6a shows a simplistic concept of the dominant sound fields arriving at listening locations in typical domestic rooms and cinemas. The principal difference is that in cinemas the transitions occur at lower frequencies due mainly to the higher directivities of the loudspeakers and the acoustical absorption on the walls and that provided by the audience and seating. Illustration (a) is from Figure 5.4, where it is explained.

Figure 10.6b shows what happens in a typical cinema based on information from SMPTE TC-25CSS (2014). From about 200 Hz to around 600 to 1000 Hz, the energetic sound events occur within about the first 50 ms—listeners are exposed to the direct sound and a few early reflections. Above this, for the top three octaves or more, the direct sound is the dominant factor. There is no need to be concerned about the omnidirectionality of the measurement microphone at very high frequencies; aim the flat axis at the source.

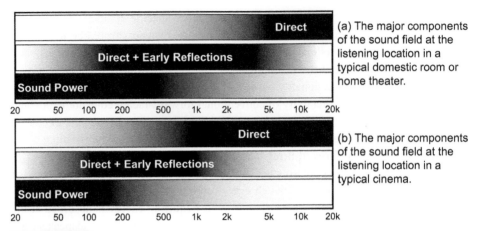

(a) The major components of the sound field at the listening location in a typical domestic room or home theater.

(b) The major components of the sound field at the listening location in a typical cinema.

FIGURE 10.6 *Concept diagrams showing the dominant sound fields in (a) typical domestic rooms and (b) in cinemas.*

These patterns are the result of combining the directivity of the loudspeakers with the reflectivity of the rooms. This matters to some extent at all frequencies, but mostly at low frequencies, where all common loudspeakers are omnidirectional (see Section 4.10.1).

10.4 NEAR AND FAR FIELDS OF ROOMS— SOUND LEVEL VS. DISTANCE

Classical acoustics often begins with explanations of events in reverberant spaces, a concert hall being the most thoroughly studied of all. These spaces have the difficult task of preserving the finite amount of energy radiated by musical instruments and voices (meaning as little absorbing material as possible), delivering it to all parts of a large audience (meaning large, carefully angled and shaped reflecting surfaces), while adding the right amount of reverberation for the kind of music being performed (meaning high ceilings, large volumes, to allow the sound to reflect multiple times). It is not simple.

Figure 10.7 shows the textbook explanation of a concert hall performance, where the direct sound is seen to fall according to the inverse-square law (−6 dB per double distance) until it meets the reverberant sound field. The distance at which they are equal is called the critical distance, or the reverberation radius or distance. It is the point where a draw-away measurement of sound level transitions to remaining relatively constant.

Common wisdom is that beyond the critical distance, speech intelligibility and perception of some kinds of musical details progressively degrade.

As the total absorption in the hall is increased, the "steady-state" reverberation is lowered in level, and the critical distance moves farther from the source. As the

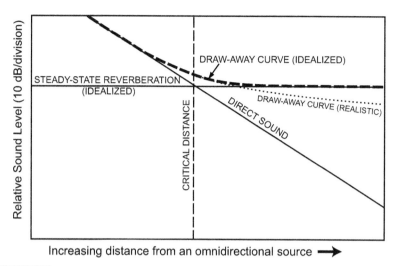

FIGURE 10.7   *The structure of the steady-state sound field in an idealized large performance space. The dotted draw-away curve indicates the declining sound level that, in some form or other, is seen in real halls.*
Based on Schultz (1983), figure 10.

directivity of the sound source is increased, less energy is delivered to the reverberant sound field, and again the critical distance moves further from the source. In real terms, this means that during a concert the critical distance varies, being longer for directional horns than for multidirectional violins.

In some publications the region ahead of the critical distance is called the "near field" and that on the distant side is called the "far field." This is an unfortunate use of the terms because, as will be seen in the next section, there is another use with a longer history. More accurate, terms could be "direct-sound dominant" and "reflected-sound dominant," describing circumstances in the performance venue. In Section 10.5 the near/far field terms relate to the sound source.

The situation is very different in the smaller, more absorptive, rooms used for sound reproduction, whether it is a cinema, a recording control room, a home theater, or a stereo listening room. Loudspeakers have significant directivity, gradually increasing toward high frequencies (Figure 5.9). This means that the critical distance would be frequency dependent. In addition, the reverberation times are so short that the concept of a "steady-state" background reverberation is dubious, and that is the basis for the critical distance.

Chapter 4 in the earlier edition of this book explained a progression of acoustical changes starting with concert halls, moving through industrial and commercial spaces, and ending with those that pertain in sound reproduction venues. The intermediate phase is also shown in figure 3 of Toole, 2006. Part of the difference is attributable to the amount of absorbing material and where it is deployed. Some of the difference is related to the low ceilings and the presence of surfaces and proportionately large objects—including humans—that reflect some sound back toward the source. Figure 10.5 exemplifies some of these points. The result is summarized in Figure 10.8.

Figure 10.8 shows the manner in which the direct sound and the direct sound combined with reflected sounds behave in typical domestic listening spaces. From the perspective of sound localization and a reference direct sound to initiate the precedence effect, it is clear that the complications increase with distance from the source—the reflected sound field becomes progressively dominant. In these small, low reverberation rooms, there will be no truly diffuse "reverberation." The reflected sounds contributing to the elevated steady-state sounds are best described as early reflections, some of which are reflected back toward the source by objects and surfaces in the room. Obviously the less reflective the listening environment, the steeper will be the draw-away curve, ultimately uniting with the −6 dB/dd curve for a completely non-reflective, anechoic environment. Close to the loudspeaker, and certainly at 1 ft (0.3 m), the microphone is in the acoustical near field of the loudspeaker and spectral variations are likely to occur (Section 10.5).

Of separate interest are the data points from the Zacharov et al. (1998) study, which were determined not by measurement but by subjective loudness matching. They fall on the −2.5 dB/dd line, very likely decided by the specific combination of loudspeaker directivity and room acoustics used in the experiments. This is reassuring, in that we now have confirmation that these graphic data can be used to anticipate the perceived loudness of steady-state acoustical events in typical rooms.

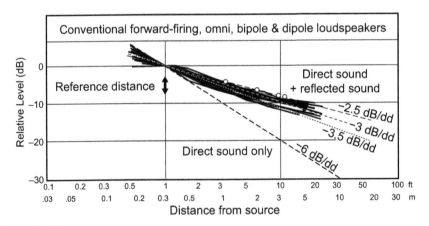

FIGURE 10.8 *The cluster of curves consists of draw-away measurements of steady-state sound in four living rooms using four approximately omnidirectional sound sources (Schultz, 1983). dd = double distance. These are combined with measurements by the author using five loudspeakers with different directivities (omnidirectional, dipole, bipole and forward-firing) in two domestic listening rooms. All curves were measured using A-weighting of broadband sounds and were normalized to an arbitrary reference distance of 1 ft (0.3 m). Through the cluster are drawn three lines having different slopes, approximating what might happen in different room/loudspeaker combinations. The four white dots are data from Part II of the experiments discussed in Zacharov et al. (1998), in which listeners did subjective loudness matching of loudspeakers at different distances. At the bottom is the inverse-square law attenuation of the direct sound with distance.*

The monotonic decline in sound level shown in all of the draw-away curves indicates a source-to-sink energy flow at increasing distance from the source. Variations in the curves at short distances are probably near-field effects caused by being so close to the sources, some of which (the electrostatic panel dipole loudspeakers especially) were quite large. At the far end of the curves, some of the measurements were made close to the back wall of the listening space where boundary effects may be expected. There may be rooms, unusually live or dead, or loudspeakers of sufficient directivity that could result in draw-away curves that fall outside this range, but that is precisely what would be expected in the real world. In the cases shown here, the surprise is that the curves exhibit such similarity, in spite of real differences in source directivity and rooms.

Considering the distances at which we listen in entertainment spaces and control rooms, it is clear that we are in the transitional region, where the direct and early-reflected sounds dominate, and late-reflected sounds are subdued, and progressively attenuated with distance. The sound field is not diffuse, statistical acoustics do not apply, and there is *no critical distance*, as classically defined. This, of course has been convincingly shown in Section 5.6, where steady-state room curves are well predicted by a combination of direct and early-reflected sounds.

10.5 NEAR AND FAR FIELDS OF SOUND SOURCES

In audio, the "inverse-square law" is widely known. It says that for every doubling of distance from a source of sound, the sound level falls by 6 dB. Two things need to be noted about that statement:

1. It applies only to the direct sound arriving from the source. No reflections are included. As shown in the previous section, in normally reflective rooms the steady-state sound level falls at a rate close to −3 dB/double distance.

2. It applies only when one is in the far field of the sound source. What is that?

10.5.1 Point Sources and Real Loudspeakers

Figure 10.9 shows the situation for point sources and combinations of "point-like" sources: loudspeaker drivers in a box.

Figure 10.9a shows an ideal point source radiating sound equally in all directions in free space. The sound energy is distributed uniformly over a spherical surface that, as a function of distance, experiences a rapid increase in area over which the sound energy is distributed. The sound energy per unit area (called sound *intensity*) is inversely proportional to the square of the distance from the source, so this relationship has come to be called the "inverse-square law." The sound level correspondingly falls rapidly, at a rate of −6 dB/double-distance).

With an ideal point source this relation holds for any distance. However, practical sound sources are not infinitely small points. When sound radiates from a complex source like a loudspeaker or large musical instruments, what is measured and heard at

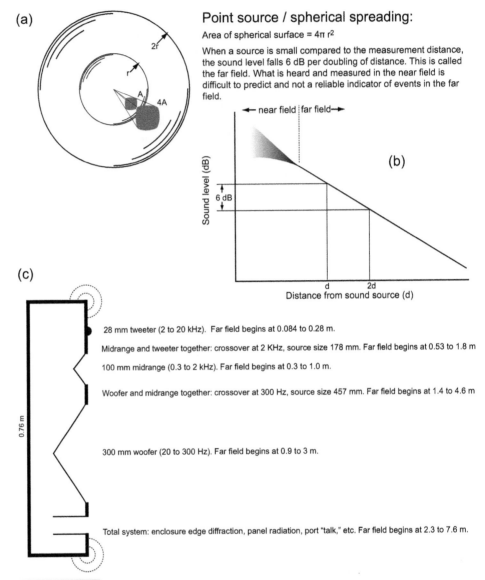

(a)

Point source / spherical spreading:

Area of spherical surface = $4\pi r^2$

When a source is small compared to the measurement distance, the sound level falls 6 dB per doubling of distance. This is called the far field. What is heard and measured in the near field is difficult to predict and not a reliable indicator of events in the far field.

←— near field ┊ far field —→

Sound level (dB)

6 dB

(b)

d   2d

Distance from sound source (d)

(c)

0.76 m

28 mm tweeter (2 to 20 kHz). Far field begins at 0.084 to 0.28 m.

Midrange and tweeter together: crossover at 2 KHz, source size 178 mm. Far field begins at 0.53 to 1.8 m

100 mm midrange (0.3 to 2 kHz). Far field begins at 0.3 to 1.0 m.

Woofer and midrange together: crossover at 300 Hz, source size 457 mm. Far field begins at 1.4 to 4.6 m

300 mm woofer (20 to 300 Hz). Far field begins at 0.9 to 3 m.

Total system: enclosure edge diffraction, panel radiation, port "talk," etc. Far field begins at 2.3 to 7.6 m.

FIGURE 10.9 *(a) The classic illustration of spherical spreading, originating with a point source. In the far field the sound level falls at a rate of –6 dB per double-distance. (b) A graphic illustration showing the disorderly near field and the predictable far field behavior of a source. (c) Estimates of the distances at which far field conditions are established for a three-way loudspeaker system and for its components, singly and in combination.*

different frequencies is different at short distances and long distances. In the near field, as shown in Figure 10.9b, the sound level at any frequency is uncertain. Figure 10.9c shows estimated distances at which far-field conditions should prevail for a loud-speaker system and for its components. This would be the minimum distance at which

NEAR-FIELD MONITORS

In recording control rooms it is common to place small loudspeakers on the meter bridge of the recording console. These are called near-field or close-field monitors because they are not far from the listeners. As shown in Figure 13.1c, the near field of a small two-way loudspeaker (the midrange and tweeter of the example system) extends to somewhere in the range 21 in. to almost 6 ft (0.53 to 1.8 m). Including the reflection from the console under the loudspeaker greatly extends that distance. There is no doubt, then, that the recording engineer is listening in the acoustical near field, and that what is heard will depend on where the ears are located in distance, laterally and in height. The propagating wavefront has not stabilized, and as a result this is not a desirable sound field in which to do precision listening. Mid-field locations, behind the console and arranged to minimize reflections, are preferred.

a microphone should be placed for measurements, and at which listeners should sit in order to have a predictable experience.

Beranek (1986) suggests that the far field begins at a distance of 3 to 10 times the largest dimension of the sound source. At this distance the source is small compared to the distance, and a second criterion is normally satisfied: distance^2 = wavelength^2 / 36.

Diffusers behave as secondary sources of sound, and they can cover significant areas of room surfaces. Cox and D'Antonio (2004b, p. 37) point out that listeners should be placed as far from scattering surfaces as possible, at least three wavelengths away. This is to avoid hearing colorations from repeating features of the devices. For devices that are effective down to 300–500 Hz, this is a minimum distance of about 10 ft (3 m). As they realistically point out, "In some situations this distance may have to be compromised."

10.5.2 Line Sources

While on this topic, it is interesting to look at the behavior of true line sources. I say "true" because they are rare. Tall loudspeakers are common, but these at best are truncated line sources and they behave very differently, and much worse (see Figure 9.14). Figure 10.10 shows the orderly cylindrical spreading from an infinite line source.

Practical line sources have finite lengths, so the critical issue becomes one of keeping listeners within the near field of the line, where the desirable −3 dB/double distance relationship holds, and out of the far field where even line sources revert to −6 dB/double-distance. Obviously the distance at which the near-/far-field transition occurs is a function of frequency and the length of the line. So, interestingly enough, for conventional loudspeakers we may wish to be in the acoustical far field, but for true line sources we need to be in the acoustical near field to fully appreciate the benefits.

Figure 10.10 shows a stereo pair of full-height lines, taking advantage of the ceiling and floor reflected images to make them appear to be even longer. A portion of one line has been expanded to show that it is a two-way system using conventional cone and/or dome loudspeaker drivers, densely packed (ideally spaced by less than about

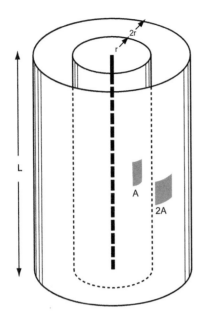

Line source / cylindrical spreading:

Area of cylindrical surface = 2πrL

When a source is long compared to the measurement distance, the direct sound falls 3 dB per doubling of distance. For a line loudspeaker this requires that it run from floor to ceiling, using "image" reflections from those surfaces to extend the effective length of the line. Most practical line loudspeakers are truncated (shortened) lines and they behave differently.

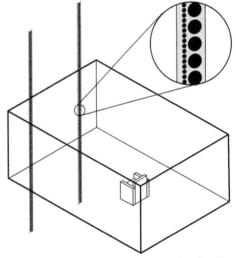

A stereo pair of line sources in a room, showing "images".

FIGURE 10.10 *An illustration of a theoretical infinite line source, and of a practical approximation.*

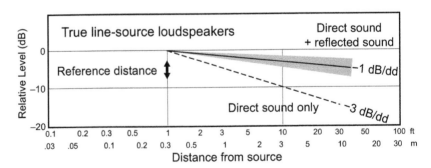

FIGURE 10.11 *Predicted draw-away curves for a full-height line source.*

1/2 wavelength of the highest reproduced frequency) in order to simulate a continuous sound source. Obviously, planar transducers can also be used, but they too have modular repetitions.

It is possible to use less than a full-height floor-to-ceiling array if one understands the variables and how they can be traded off. Lipshitz and Vanderkooy (1986) provide a thorough theoretical background to the behavior of "finite length" (not full height), truncated, line sources and they point out a number of problems, ultimately concluding that "there is little to recommend the use of line sources as acoustic radiators." They did grant that full-height lines had potential if the −3 dB/octave tilt in the frequency response is corrected.

Griffin (2003) gives a comprehensive and comprehensible presentation of what is involved in designing practical line sources that approach the performance of full-height lines using less hardware. Smith (1997) describes a commercial realization.

Figure 10.11 shows the anticipated draw-away curves for a full-height line source. The upper steady-state curve is based on very limited data, but seems reasonable.

10.6 AIR ABSORPTION AT HIGH FREQUENCIES

The explanations involve some serious physics, but for our purposes it is sufficient to understand that as sound propagates some of the energy is lost. This is different from the amplitude reduction as a function of distance—the inverse-square law—in which the direct sound is reduced by 6 dB per double distance because the sound energy is distributed over a spherical surface of ever-increasing area. That energy is not lost; it is merely distributed more thinly, and the effect is not frequency dependent.

Air attenuation is very frequency dependent, affecting only the highest frequencies, and it also is dependent on the prevailing air pressure, temperature and relative humidity. In large concert venues these change as audiences heat the air and exhale humidity. Figure 10.12 shows data that explains what happens in venues and conditions likely to relate to sound reproduction circumstances.

This is a phenomenon that we experience daily: sounds at a distance are less loud and sound duller than sounds close up. It is sometimes hypothesized that humans compensate for these effects in the spectra of direct sound components in a sound field. Note that it applies to a single "ray" of sound, such as the direct sound, or a reflected sound. The effect it has on a steady-state sound field depends on the complex summation of the direct and reflected sounds. However, as Figure 10.6 shows, in most sound reproduction

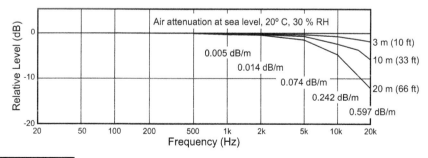

FIGURE 10.12 *High-frequency air losses as a function of distance, and the basic data from which they were calculated. Several Internet sources provide calculators for these numbers: for example, search for "air attenuation of sound."*

venues, at frequencies where air absorption is significant we tend to be in dominantly direct sound fields.

10.7 SCREEN LOSS IN HOME THEATERS AND CINEMAS

When loudspeakers are placed behind perforated or woven screens, there is loss of energy because some of the sound is reflected back toward the loudspeaker. The amount of this reflected energy depends on the thickness and open area of the screen. The losses increase with increasing frequency.

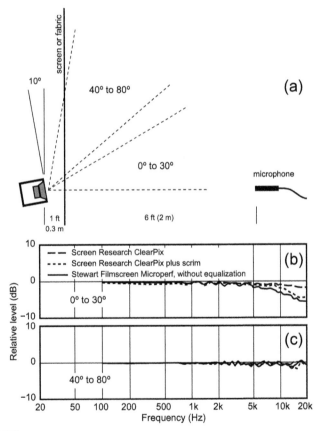

FIGURE 10.13 *(a) shows the measurement setup in the anechoic chamber. The loudspeaker was angled to minimize comb filtering, the angling can be in any direction. The screens are slightly reflective, the front panel of the loudspeaker is greatly reflective, and reflections can occur between them. The more transparent the screen, the less the need for angling the loudspeaker. In (b) are shown the transmission losses over the 0° to 30° angular range of three screen configurations, a woven screen alone (Screen Research ClearPix), the same screen with a dark scrim behind it to minimize optical reflections from objects behind, and the Stewart Filmscreen Microperf. In (c) are shown the transmission losses averaged over the 40° to 80° range. Harman data.*

Figure 10.13 shows measurements on some screen materials that exemplify what is available for use in home theaters. It is recommended to request data from the manufacturers for specific screens of interest. It is clear that there is an acoustical advantage to woven screens, but the trade-off, at present at least, is reduced optical reflectivity. The screen loss is known in advance and can be compensated for by a fixed filter available from the manufacturer, or accommodated in the overall system equalization.

The physical size of cinema screens is a challenge for woven materials, and 3D pictures need high optical performance, so the common screen material is perforated vinyl. Figure 10.14 shows the on-axis loss of some of these screens.

The behavior off-axis has been a topic of some discussion because there is evidence indicating that the dispersion of the sound radiated by horns is increased at frequencies above about 4 kHz. Some measurements have indicated that it is not an issue (Long et al., 2012, and Newell et al., 2013), while others have shown reduced off-axis high-frequency attenuation (Eargle et al., 1985, Benjamin, 2004, in an unpublished PowerPoint presentation). The studies that showed little or no change employed small loudspeakers placed at 6 or 12 inches (150 or 300 mm) from the screen. One study that showed directional effects employed large horns (31.3 in. (800 mm) square in Eargle et al., 1985) placed 2 inches (50 mm) from the screen in a cinema installation. Much later, Eargle, in a 1998 internal JBL Professional memo stated:

When we look at the resulting polars through the screen, what appears to be happening is multiple reflections between the screen and the horn's internal boundaries, with sound at 4 kHz and above eventually exiting through the screen over a wider angle than if the screen were not present.

This would be much more obvious with large cinema-scale horns close to the screen than with smaller horns farther from the screen. Even Figure 10.13 shows a small effect in these home-theater simulations. It appears to be a real effect under some, but not all, circumstances. The consequences to listeners off-axis are most likely to be beneficial, especially in cinemas.

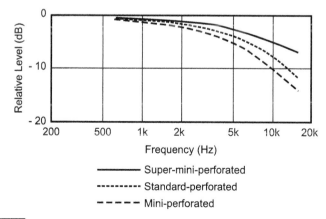

FIGURE 10.14 *Measurements of on-axis losses in some popular cinema screen materials.*
Data from www.harkness-screens.com/publications.html, "Choosing the Right Cinema Screen," 2008.

10.8 THE DIRECTIVITIES OF COMMON SOUND SOURCES

The audible spectrum, 20 Hz to 20 kHz, has wavelengths ranging from about 57 ft (17 m) to 0.6 in (17 mm). As a result, the dispersion of radiated sounds changes with frequency depending on the size of the sound-radiating surface. It matters not whether we are considering voices, musical instruments or loudspeakers. Low frequencies from most sources radiate essentially omnidirectionally because the wavelengths are long compared to the size of the source—the directivity index (DI) is near zero. As shown in Figure 5.9, the DI can be interpreted as the difference in dB between the on-axis or listening window curve and the total radiated sound power. In a room this translates into the difference between the direct sound and the reflected sound field. As frequency increases, so does the directivity of most sources. The higher the DI, the higher is the level of direct sound relative to later arriving reflections. Therefore, as a rule, humans are exposed to more energetic reflected sound fields at low frequencies than at high frequencies, whether we are at a live concert performance, carrying on a conversation in a corridor or listening to loudspeakers in a room. Figure 10.15 shows examples of some musical instruments, voice and loudspeakers.

Some musical instruments, especially stringed instruments, exhibit very complicated radiation patterns, so these are simplifications; J. Meyer (2009) shows much more detailed data. It is important to note that they fall into a similar range of directivities as the loudspeakers. Very clearly shown in the loudspeaker data is the importance of the

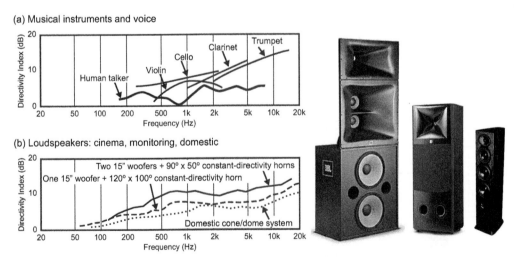

FIGURE 10.15 *(a) The simplified directivity indices (DI) for some musical instruments (J. Meyer, 2009). The voice data are from Toole (2008), Figure 10.3. (b) Directivity indices for a large cinema loudspeaker system with a double-woofer bass unit and 90° × 50° horns (JBL Pro 5732), a single woofer studio monitor system with a 120° × 100° horn (JBL Pro M2), and a three-way domestic cone/dome system (Revel F206), which are shown approximately to scale.*
Based on Toole (2015b), figure 2.

size of the low frequency energy source, with the directivity index curve progressively flattening as the radiating area—the size and number of woofers—shrinks. There is also a reduction in DI as the angular dispersion of the high frequency horn expands, reducing further with the small dome tweeter in the domestic loudspeaker.

Whatever the shape of the spectrum of the direct sound from musical instruments, or loudspeakers reproducing recordings of those musical instruments, the steady-state sound in a normal room will exhibit a version of that spectrum that rises at lower frequencies. Highly reflective concert halls attempt to preserve the limited sound output of musical instruments and voices in reflections, delivering as much of it as possible to listeners, while not masking temporal details in the music, and still creating a pleasant sense of envelopment. It is a difficult acoustical balancing act. Sound reproduction spaces are much less reflective: the principal cues to space and envelopment are in the multichannel recordings, and a volume control changes loudness at will. Loudspeakers designed with flat on-axis frequency responses, so as to accurately reproduce the initial timbral signature of the recorded sounds, will therefore exhibit a rising sound power output at lower frequencies. The exceptions would be arrays designed to maintain high DI at low frequencies.

In a measured steady-state room curve, therefore, one can anticipate that the bass will be elevated compared to the direct-sound, on-axis, curve. The amount of the bass rise will depend on the reflectivity of the room at low frequencies, which is, in turn, related to the RT of the room. Looking back at Figure 10.1d, one can imagine that bass levels in cinemas could be very different from those in reference cinemas, dubbing stages and home theaters.

CHAPTER 11

Sound in Cinemas

11.1 THE CLOSED LOOP OF CINEMA SOUND

Movies are art. They tempt us to "suspend disbelief," drawing us into plots that generate all conceivable emotional reactions. As emphasized in Chapter 1, audiences should get to see or hear art as it was created. If we experience distorted renderings, we can never be certain who or what was responsible for what we liked or disliked in it. In movies, the picture and the sound combine to deliver the message. If it is a good movie we suspend disbelief, and from that point on we probably look less critically at picture quality and listen less critically to sound quality. In professional parlance, the artistic experience "translates" from the point of creation to the point of delivery even though the physical delivery may not be perfect. Absolute accuracy in the replica is not a requirement because humans routinely adapt to some amount of variation in color balance in pictures and some changes in sound quality. The same principle applies to music, where we find pleasure in reproductions that are significantly flawed. But the essences must be there, and distractions that break the mood are to be avoided. That said, the highest possible "fidelity" in sound and image should be the objective, at least to the extent that it is practical.

The movie industry has long been proud that it was standardized. Pictures should be similar from venue to venue and soundtracks created in calibrated dubbing stages should sound similar in cinemas where they are reproduced. It was conceived as a closed loop—no "circle of confusion" (Figure 1.7). Two similar documents spell out the audio requirements: ISO 2969 (1987) and SMPTE ST 202 (2010). Teams of technicians the world over visit cinemas and dubbing stages to measure their performance and align them to meet the target, called the X-curve. The steady-state target curve and its tolerances are shown in Figure 11.1.

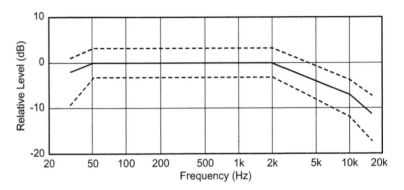

FIGURE 11.1 *The X-curve and its tolerances.*
From SMPTE ST 202 (2010).

To many in the audio industry such a target curve is odd, not fitting with what would be considered standard practice. Most steady-state room curves tilt slightly downward over most of the frequency range—an abrupt transition such as that at 2 kHz is unfamiliar. Normally such a transition, if needed, would be achieved gradually. As it is, this discontinuity, this "knee" is likely to be perceived as coloration. Nevertheless, within a closed loop embracing soundtrack recording in dubbing stages and reproduction in cinemas, an idiosyncratic target curve can work. Calibrations need only ensure that the resulting sound is consistent to the standard. Recording engineers are expected to adjust their mixes to sound good in the dubbing stage, and everything being equal, customers in cinemas should hear what they created. The assumption was that meeting the measured steady-state in-room performance target guarantees consistent sound quality. No specific knowledge of the loudspeakers or room was required.

But the reality is that when the X-curve was decided on in the 1970s it was not thought of as being idiosyncratic. It was believed to be the steady-state room curve necessary for mixers in dubbing stages and audiences in cinemas to hear a neutral, flattish, direct sound. Subsequent events have shown that this is not the case.

The X-curve has remained a feature of movie sound production and reproduction venues for decades, although it has not been adopted by any other area of the audio industry. Over the years the movie business has changed. Now, more than ever, compatibility with the outside world is important because movies are probably more often viewed outside of cinemas than in them. If soundtracks are mixed to sound good through monitoring systems calibrated to the X-curve, there is the question of how they would sound through the systems in our homes and elsewhere that do not employ the X-curve target. Some amount of frequency response adjustment is therefore required in repurposing sound tracks for delivery to consumers outside of cinemas. Reverse compatibility also matters because cinemas exhibit music, opera and sports programs created outside the movie-sound domain.

It is an important topic with a long history. First it is useful to understand the sound fields that exist in typical cinemas and dubbing stages.

11.2 SOUND FIELDS IN CINEMAS

Cinemas are relatively well-damped acoustical spaces, although some facilities apparently have mid-to-high-frequency reverberation times (RT) that average 1 s, rising to close to 2 s at low frequencies (Allen, 2006). These RT values are relatively high, and such venues are very likely to exhibit speech intelligibility problems, especially in complex multichannel sound tracks. At the very least, conveying a sense of spatial "intimacy" in the soundtrack would be impossible.

Reverberation times of many modern cinemas through middle and high frequencies are similar to those found in homes and professional listening rooms, namely 0.3–0.6 s (Figure 10.1). However, below about 500 Hz RT rises, especially below about 200 Hz, where reflectivity significantly increases in many mass-market venues. Low-frequency absorption is expensive. When that is combined with the progressively widening dispersion of loudspeakers at lower frequencies the bass can be expected to rise significantly following the direct sound. Toole (2015b) explains this in detail.

11.2.1 A Loudspeaker in a Cinema

Let us begin with a generic cinema loudspeaker radiating directly into a modern cinema space—no screen. In Figure 11.2a the directivity index curve for the two-woofer cinema system from Figure 10.15b is inverted, thereby giving us an estimate of total radiated sound power when the loudspeaker is equalized to radiate a flat direct sound—flat anechoic on-axis and/or listening window responses.

Figure 11.2b shows a prediction based on the data in Figure 11.2a with the flat direct sound normalized at 0 dB. High-frequency air attenuation progressively rolls off the direct sound at increasing listening distances (see Figure 10.12). At low frequencies, rooms with different reflective properties could yield curves that fall anywhere in the shaded space. At very low frequencies, woofers are essentially omnidirectional, and the more of that sound that is reflected by the room boundaries, the higher will be the steady-state sound level. The maximum bass rise is about 10 dB in a highly reflective venue. According to Figure 10.2, the cumulative energy time (the time taken for the steady-state sound level to be reached after the direct sound arrives) is less than 150 ms in a very reverberant cinema. The question arises: is this bass rise audible in transient sounds?

Figure 11.2c shows two cinema loudspeaker systems that radiated flattish direct sound, one from Snow (1961) and the other from Eargle et al. (1985). Snow showed a 2π direct sound measurement, and this was used to compensate the room curve to what it would have been for a flattish direct sound. The more recent Eargle data were compensated for screen loss, which he published. Both systems showed evidence of mid-frequency crossover dips, notably the 1961 system. In both, the bass rise is evident, as is the flattish trend through the midrange and lower treble regions.

Although the loudspeaker radiates flat frequency response, the high-frequency direct sound arriving at the listening location is not flat because of air attenuation. Here the high-frequency rolloff is a function of distance only, having nothing to do with the

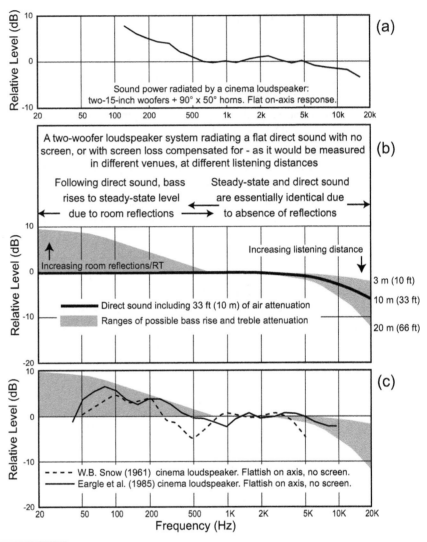

Sound power radiated by a cinema loudspeaker: two-15-inch woofers + 90° x 50° horns. Flat on-axis response.

A two-woofer loudspeaker system radiating a flat direct sound with no screen, or with screen loss compensated for - as it would be measured in different venues, at different listening distances

FIGURE 11.2 *(a) the estimated sound power radiated by the two-woofer loudspeaker system described in Figure 10.15b assuming a flat on-axis amplitude response. Figure 11.2b Anticipated steady-state frequency responses for a two-woofer cinema system radiating a flat direct sound on the audience side of the screen. The shaded area shows the range of opportunities for bass rise in modern cinema venues having different reflectivity/RT at low frequencies. High-frequency air attenuation is shown at different listening distances. Figure 11.2c shows measured room curves compared to the prediction in (b).*

room acoustics. In fact, above 500 to 1000 Hz there are no consequential reflections in modern cinemas because of loudspeaker directivity and room absorptivity (SMPTE TC-25CSS (2014) and Toole (2015b)). The direct and steady state sound fields are therefore identical. There is no significant reverberant sound field at high frequencies. Measurement microphones are not required to be omnidirectional at these frequencies, but they must be flat on the axis of direct sound arriving from the loudspeaker being measured. The notation at the top of Figure 11.2b shows the *approximate* frequency regions over which the described behavior occurs.

Data in SMPTE TC-25CSS (2014) indicated that above about 200 Hz the most energetic acoustical events are likely to happen in the first 80 ms. Clearly these were the direct sound followed by a small number of early reflections from room boundaries between the listener and the loudspeaker. At lower frequencies longer measurement time windows reveal more accurate and useful data—partly as a result of increased frequency resolution. The higher resolution allows for better interpretation of, and compensation for, standing wave and adjacent boundary effects.

This is the basic structure of sound fields in cinemas.

11.2.2 Adding a Screen and Applying the X-curve

Figure 11.3 shows a progression of acoustical events in a cinema. In Figure 11.3a a loudspeaker radiating a flat on-axis/listening window sound is placed behind a standard-perforation cinema screen. At mid- to high frequencies, it is seen that the screen loss is substantial, dropping the curve very close to the rolloff required by the X-curve. In addition there is air attenuation, 10 m of which is shown (see Figure 10.12). At low frequencies the widely dispersing woofers generate reflections from room boundaries and the steady-state sound level rises. Although it is not shown, the LFE would be subject to the same venue-dependent bass rise.

Figure 11.3b shows that the high-frequency sound delivered to the listening position in Figure 11.3a is close to the X-curve target up to about 7 kHz. The rolloff is smoother than the 2 kHz knee requires, but common sense says that the smooth curve is a better choice. Beyond this some high-frequency boost will be required if the target is to be hit, although the performance is still within the tolerance limits.

Figure 11.3c shows the predicted situation after full bandwidth X-curve equalization. The difference between the direct and steady-state sound fields at low frequencies has not changed, but with the steady-state required to be flat, the direct sound is now depressed by an amount that depends on room reflectivity. Only in anechoic space can the direct and steady-state sounds be the same.

Figure 11.3d shows an important set of measurements from Fielder (2012), in which all of the previous predictions are shown to be true. In the measurements of 50 X-curve-calibrated front loudspeakers in 18 cinemas, he showed that measurements of direct sound basically filled the shaded area in Figure 11.3c. Real-world variations in cinema acoustics have a profound effect on the direct sound at lower frequencies when the steady-state sound is equalized to be flat. However, because there is negligible reflected sound above about 1 kHz, there is no difference between the direct sound and the steady-state sound.

Summarizing, in calibrated dubbing stages and cinemas the direct sound is rolled off toward the bass by an amount that depends on the acoustics of the venue, and the direct and steady-state sound are both rolled off at high frequencies by an amount dictated by the X-curve target.

(a) shows predictions of the acoustical events that follow the direct sound radiated by a loudspeaker having a flat on-axis frequency response. (b) shows the SMPTE ST 202 X-curve compared to the predicted sound at a listener location 10 m from the screen. (c) shows a prediction of the situation after the steady-state bass level has been attenuated to hit the X-curve target. (d) shows measurements of both direct sound and steady-state sound made using 50 front loudspeakers in 18 cinemas.
The author has reformatted this data from Fielder (2012).

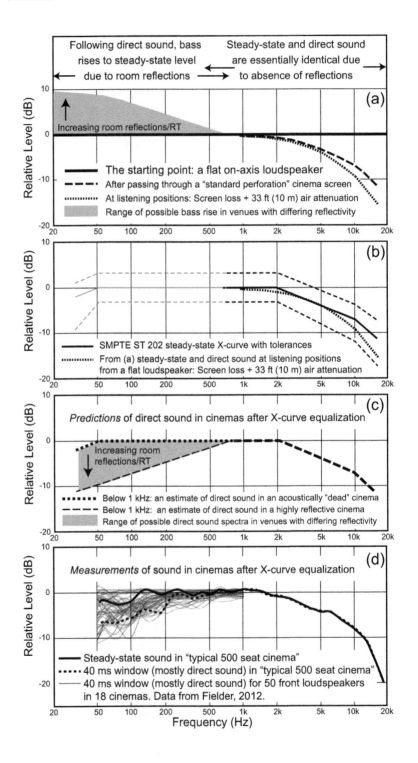

This is not "neutral" sound reproduction. In everyday life, including live musical performances, we hear unadulterated direct sounds followed by some amount of bass rise that is auditory confirmation of the venue that the eyes see.

But nature is subverted when the low-frequency portion of the steady-state room curve is forced to be flat. Then the direct sound is variable, depending on the reflectivity of the venues, which can be quite different from one another, as can be seen in Figure 10.1d. The fact that dubbing stages tend to be more absorptive at low frequencies than many cinemas may be a cause of audible mismatches.

In order for cinema audiences to experience sound that is naturally balanced, neutral, the mixers must compensate for the frequency response peculiarities. The auditory impact of the sharp discontinuity at 2 kHz is a particular issue; a more gradual transition would have been preferred. In discussions of this topic it has been the high-frequency rolloff that gets the attention, but as can be seen, the bass-rise issue is significant, especially as our perception of loudness grows much more rapidly in the bass frequency range (Figure 4.5). If the bass is allowed to rise, the perceived high-frequency excess is diminished, and the 2 kHz knee is less obvious.

Perceptually, manipulating either the bass or treble levels modifies the perception of spectral balance, as can be experienced with traditional tone controls. We are very sensitive to these spectral tilting effects (Section 4.6.1). Turning up the bass has a similar effect to turning down the treble, and vice versa. In this situation the mixer is presented with a sound spectrum that is likely to be perceived as deficient in both bass and treble, prompting adjustments to the spectral balance during the mix. It is totally unknown how much the frequency response is manipulated to deliver a satisfactory balance during the mix (and ultimately in cinemas). The problem is that the X-curve forces mixers to make such decisions.

In practice, the subjective pre-equalization delivered by mixers turns out to be highly variable (Gedemer, 2013, and several anecdotal communications to the author). It is a topic of some sensitivity for the industry, as it suggests that some mixers have adapted to the imposed spectrum and make little or no correction, others make corrections, but not necessarily full corrections and not for all components of the mix, and so on. See the box.

HEARING LOSS: AN "ELEPHANT IN THE ROOM" ISSUE

Hearing loss is an occupational hazard in many professions, unfortunately including professional audio. At the very least, mixers exposed to periods of very loud sound incur temporary hearing loss. Repetitions of such exposures result in permanent hearing loss. As illustrated in Section 3.2 and discussed in detail in Chapter 17, both of these seriously affect what is heard in mixes, and it is not consistent. Judgments of sound quality are impaired. Paradoxically, the people who do the mixes may not be the best judges of the results. If they think their work has "translated" to a different venue, it is very possible that others with more normal hearing may have different opinions not only about how successful the translation was, but also about the original work.

Tragically, if one seeks an evaluation of hearing performance, what an audiologist considers to be "normal" hearing is not an adequate criterion. Audiometry is focused on speech intelligibility, not the audible attributes and intricacies of high-quality broadband, spatially complex music and movies. A threshold determination is merely a starting point. There is much more to the perception timbre and of sounds in spaces than merely detecting that they are present. With gaping 4 kHz notches in hearing thresholds attributable to noise-induced loss, and the eroded high-frequency hearing limit, the ability of listeners has been more than slightly compromised.

All of this is a gross intellectual and emotional insult to audio professionals, but it is an unfortunate reality. It is not possible to learn your way around these problems after the "microphones" are broken.

Applying occupational hearing conservation guidelines (OSHA and the like) does not prevent hearing loss. Their goal is to preserve enough hearing at the end of a career to be able to carry on a (flawed) conversation at a distance of 1 m. These are guidelines created for factory workers, not audio engineers. Musicians' earplugs are essential equipment for prudent audio engineers. Choose ones with uniform broadband attenuation that deliver good sound quality at a reduced volume (e.g., www.etymotic.com/consumer/hearing-protection/erme.html). Custom earmolds are important for function and comfort. Insert them *before* exposure to loud sound and temporary hearing loss has set in.

If the industry and unions are not active in preserving the professional capabilities of their key employees, it is up to individuals to protect their own hearing. Again, see Chapter 17.

Another reality that seriously undermines the intentions of the standards is the fact that some facilities find the calibrated spectral balance to be unacceptable and modify the equalization to suit their preferences. All of this is motivated by the monitoring systems that are inherently not neutral sounding.

There is a fundamental compatibility problem arising when the soundtracks are taken outside the closed dubbing-stage-record/cinema-playback loop, as happens when movies are delivered to television sets, home theaters and portable devices by cable, satellite, discs, and streaming services. None of these playback systems are calibrated to the X curve. The soundtracks are often, but not always, repurposed for these media. The changes frequently involve reducing the dynamic range of blockbuster films, acknowledging the sound output limitations of most domestic systems, and possibly rebalancing the dialog and other components for playback at lower than cinema sound levels.

Anecdotal reports indicate that the spectrum may or may not be altered for non-cinema delivery, which is interesting. There are two possible interpretations. Either the recorded soundtrack sounds adequately good through the presumably "flattish" loudspeakers in the repurposing venue, or it doesn't sound good and they don't care. The latter is unlikely, so the former must be considered as a possibility.

Because of progress in the audio industry, even inexpensive domestic audio systems can deliver quite neutral sound quality. See Chapters 5, 12 and 18. Those who have spent the time and money to set up state-of-the-art home audio systems deserve appropriate program material.

11.3 THE ORIGINS OF THE X-CURVE

This is not a definitive history—not a lot of documentation exists from the period, at least in the public domain. Allen (2006) provides a good summary, but detailed measurements and analytical results of the subjective tests are absent. There is no doubt that the intentions were good: finding an acceptable steady-state room curve target for purposes of standardization across the movie industry. At the time, and still to a large extent, there is a belief that the output from an omnidirectional microphone processed by a 1/3-octave analyzer is a definitive measurement—equivalent to the complex analytical capabilities of two ears and a brain, as shown in Figure 2.2. They are related, but certainly not equivalent. In the 1960s and 1970s acoustical measurements were limited almost exclusively to steady-state measurements. This means that if one wanted accurate data on a loudspeaker, a large anechoic chamber or the great outdoors were required. Such data were not common, and, as will be seen, this was a problem. Audio "motherhood" requires that frequency responses be flat. All electronics are flat and loudspeakers are designed with a flattish on-axis performance target, thereby delivering a neutral direct sound to listeners. In cinemas, perforated screen loss would need to be compensated for to restore a neutral direct sound.

The problem is that the notion that flat is ideal was applied to steady-state in-room curves, which could easily be measured. As seen in Figure 11.2 if the direct sound is flat, the steady-state frequency response in a normally reflective room cannot be flat; the bass must rise, and the high frequencies must roll off as a function of listening distance (air attenuation). In the context of cinema sound this relationship was not universally understood.

Ljungberg (1969) provides good insight into the international debate as it unfolded during the critical period. Participants in standards writing at that time appeared to follow the popular assumption that a flat steady-state room curve was desirable, up to some frequency at which a downward tilt was introduced, otherwise it sounded too bright. Not appreciated was the fact that if the bass were allowed to rise in a natural fashion, the brightness would be a non-issue.

Also at that time in cinema sound, high-frequency rolloffs were employed to attenuate noise and distortion in optical soundtracks. Part of the debate was about the nature of the rolloff, and how much of it should be in the upstream electronics, the A-chain, and how much should be in the playback system, the B-chain, consisting of the power amplifiers, loudspeakers and the room. Interestingly, Allen (2006) credits Ljungberg with splitting cinema sound systems into A and B chains, and also with introducing the idea of using broadband noise as a test signal.

Different countries had different approaches to the problem. Some of the proposals employed smoothly curved rolloffs, while others used straight-line segments. American practice at the time had rolloffs in both A and B chains, and reportedly resulted in a combined A and B-chain curve that yielded about 30 dB of attenuation at 8 kHz with the steady-state target being flat up to about 2 kHz.

The core of Ljungberg's paper was a series of experiments attempting to correlate steady-state measured room curves with subjective impressions. For these he employed a wide range of program material, including domestic and foreign films, magnetic dubbing masters, full-frequency range dialog recordings on magnetic tape, master recordings of music on tape and film, and commercial pressings. Ljungberg's (1969) opinion was that the B-chain

should be designed for a sufficiently good wide-range reproduction to do justice to a high-quality magnetic program (or LP disc music) while on the other hand the total frequency response for the optical sound must be restricted in the usual way owing to noise and distortion. The A part [chain], therefore is the logical place for all equalizations that are peculiar to the recording-reproduction method.

In other words, compensate for soundtrack limitations in the A-chain electronics, where changes are easy to make as technology advances, and let the B-chains deliver the best, most neutral sound possible. But, that is not what happened.

Ljungberg's experimental results indicated that listeners, using the method of adjustment of a multi-band equalizer, preferred steady-state room curves that exhibited a gentle downward tilt over most of the frequency range. Figure 11.4a shows the curves his listeners created compared to the anticipated room curves from a loudspeaker radiating a flat direct sound (from Figure 11.2). Considering the 48 years that have passed since these tests were done, the fit to today's prediction is persuasive. Listeners responded favorably to a steady-state sound field generated by loudspeakers radiating a flattish direct sound.

Following the then common practice, Ljungberg proposed a target that was flat to 250 Hz, falling at 1.5 dB/octave above that, with a tolerance window of 4 dB. With octave-band resolution, no sharp discontinuities were possible in the resulting sound field. An argument could be made that his transition frequency could be much lower and still accommodate his curves. He correctly said: "It is not expected to be valid for all types of loudspeakers in all types of rooms." He would be astonished at how generalizable his first proposed tolerance area is—August 19, 1967 was the date.

He noted during his experiments "that characteristics that went flat as far up as a couple of thousand Hz always sounded over accentuated in the middle-high register." That describes the X-curve; Figure 11.4b shows the comparison. It is apparent that the X-curve allows no bass rise. The knee at 2 kHz is perceived as a low-Q spectral feature. I have heard such coloration in theaters, including high profile "reference" rooms. It is made worse when calibrators take pride in replicating the knee, not smoothing it. In some places one pays extra for this attention to detail. Suitable equalization during mixing could alleviate these issues, but this does not always happen. It is not, as has been commonly thought, a matter of boosting the high frequencies. That alone restores the original problem: excessive brightness. Significant sound spectrum rebalancing is needed, including boosting the bass and smoothing or eliminating the 2 kHz knee.

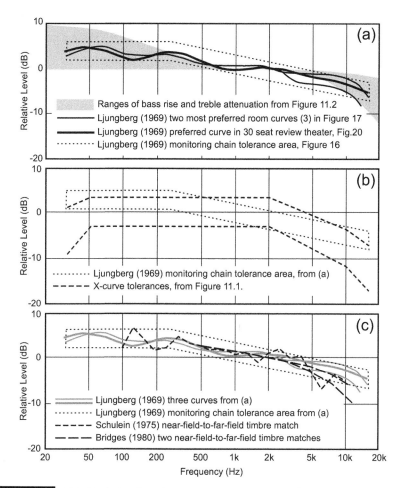

FIGURE 11.4 *(a) Subjectively adjusted room curves from Ljungberg, 1969 compared to anticipated steady-state room responses from a loudspeaker radiating a flat direct sound. His suggested tolerance is shown. (b) The proposed tolerance range for monitoring chains compared to the X-curve tolerances shown in Figure 11.1. (c) Because the X-curve was selected based on near-field-to-far-field listening tests, Figure 11.4c shows the results of two other such tests conducted in large venues. The results of Ljungberg's original tests in (a) are included for comparison.*

As described by Allen (2006), the X-curve high frequency rolloff was decided as a result of adjustments to the spectrum of the screen channels 40 ft away to subjectively match the sound of hi-fi loudspeakers located 6 ft away (his Figure 10). The small loudspeakers, KEFs, had flattish on-axis frequency response, and at that distance it is probable that this flattish direct sound was the dominant sound field heard by the listeners. It was believed that when the X-curve was measured at the listening location "the first-arrival signal will be closer to flat than the 3 dB per octave seen on an analyzer would suggest" (Allen, 2006). As seen in Figure 11.3d and other evidence presented in

SMPTE TC-25CSS (2014) and Toole (2015b), we now know that this is not true: the direct sound and the steady-state sounds are the same above 1 kHz, but can be very different at lower frequencies.

These experiments were conducted in 1971–1972. In 1975 Schulein published results of similar tests with a small loudspeaker at 4 ft and a large loudspeaker at 50 ft. Bridges (1980) also did tests of that kind with a small loudspeaker at 2.5 ft and two different large loudspeakers at 60 ft. Figure 11.4c shows these results, along with Ljungberg's 1969 results (which were done by allowing listeners to adjust the spectrum to suit their preferences; they did not use timbre matching). All of these are superimposed on Ljungberg's suggested target window because it seems clear that this is a good fit; all are well described as relatively linear, downward tilting curves over most of the frequency range.

I can think of no satisfactory explanation for the apparently anomalous results of the Elstree studio experiment described by Allen (2006). Allen himself admits that the "HF [high-frequency] droop is not very easy to explain." He speculates that there could be three contributing factors:

1. Some psychoacoustic phenomena involving faraway sound and picture.

2. Some distortion components in the loudspeaker, making more HF objectionable.

3. The result of reverberation buildup.

He then goes on to explain the concept of broadband reverberation buildup of the steady-state sound that results in the HF droop. He was correct in that the issue is bass rise, but the discussion centered on high-frequency droop for which there is no acoustical mechanism other than air attenuation. According to Allen's (2006) figure 12, the typical medium- to large-size theaters had 1 s reverberation times from about 250 Hz to 6 kHz, rising to 1.8 s at 63 Hz. These are approaching opera-house values; much higher than those in modern cinemas (Figure 10.1d). Still, there is nothing to suggest a motivation for a moderately steep high-frequency rolloff abruptly beginning at 2 kHz. Coincidentally, screen loss is similar to the required X-curve rolloff, as seen in Figure 11.3a and b. But both the air attenuation and screen loss are included in the X-curve target, measured at the 2/3 distance along the length of the cinema.

My thoughts on these three options are (1) highly unlikely; (2) possible, but would have been easy to prove at the time; and (3) impossible in modern cinemas. There will be bass rise/buildup related to low loudspeaker directivity and longer venue RTs at low frequencies, but the X-curve does not accommodate that. It eliminates that.

In fact, I am inclined to speculate that the fixation on flat response in the steady-state curves appears to have been the determining factor. All of the other investigations found preferences for rising bass responses, and natural acoustics and everyday listening experiences incorporate it. The critical experiments at Elstree appeared to have prevented that from happening because the final target curve is ruler flat to the 2 kHz knee, something that appears deliberate. See Figure 11.4b for a comparison. If the natural steady-state bass rise is prevented, the direct sound at lower frequencies is correspondingly attenu-ated, and the overall spectrum will be perceived to be excessively thin and bright—requiring treble attenuation to restore something resembling a normal spectral balance.

As discussed in the context of Figure 10.15, it is normal for sounds from both live sources and loudspeakers to exhibit bass rise following the arrival of the direct sound. It seems that the peculiarities of the X-curve were predestined by a decision that the steady-state room response through the low and middle frequencies must be flat. This is conjecture, but something about those specific tests yielded a result that does not fit the expected pattern.

Above a few hundred hertz in modern cinemas there is no consequential reverbera-tion when the excitation comes from the directional screen loudspeakers. In fact, above 500 to 1,000 Hz listeners are in a dominant direct sound field, the direct and steady-state sound fields are the same. How the specific knee frequency of 2 kHz came to be is not known, except that it shows up in some of the early proposals in which avoiding the noise and distortion of optical sound tracks was a consideration (Ljungberg, 1969; Allen, 2006).

Ljungberg (1969) also concluded: "the overall listening impression is very well mapped out with octave noise measurements." Narrower bandwidths were needed for loudspeaker design, but "for assessing listening character in an auditorium, third-octaves usually give a more ragged curve than is in agreement with listening impressions." A final comment exhibits a firm grasp of the elements: "A third-octave measurement in more or less anechoic conditions, to check speaker details, plus octave measurement in the auditorium, to check listening balance, is a good combination, although not always feasible" (Ljungberg, 1969).

This is consistent with findings discussed throughout this book, indicating the supremacy of loudspeaker performance in the description of sound quality heard in rooms above the transition/Schroeder frequency. Measurements in rooms can show details that are not perceptual problems and miss details that are. Wider bandwidth *steady-state* in-room analysis can prevent unnecessary and possibly counterproduc-tive equalization while revealing useful spectral balance trends. In contrast, only narrow-bandwidth descriptions of the *direct* sound radiated on the audience side of the screen can provide a definitive description of potential sound quality at frequencies above a few hundred hertz. The important fact is that Ljungberg recognized the impor-tance of having data on the loudspeakers before venturing into a room to finalize the system (B-chain) design.

So, from the earliest stages there were serious reasons to consider options to the X-curve in Figure 11.1. Even though there was international agreement on a target curve that eventually emerged as the X-curve, Ljungberg said of an earlier curve of similar basic form: "It was not a B-curve for all-purpose monitoring, and several coun-tries reported the disgusting practice of using different B curves for different sorts of work." According to Gedemer (2013) and the author's discussions with persons well established in the movie sound world, it is not unusual to find non-X-curve dubbing stages and those with multiple EQ settings, including one for "calibration" and another for daily use. In the rest of audio, the X-curve does not exist.

I found the Ljungberg paper to be a remarkable example of clear thinking about topics that were not then, or even now, well understood by many people. Too bad it was not taken more seriously at the time.

11.4 A RECENT STUDY ADDS CONFIRMATION AND CLARITY

The preceding experimental evidence is interesting and persuasive, but it pertains to sound sources and playback apparatus that are 37 to 48 years out of date. Things have changed. As a PhD thesis project at the University of Salford, Manchester, UK, Linda Gedemer undertook a re-examination of the cinema sound situation, employing a modern loudspeaker, modern measurement methods and controlled double-blind listening tests. Harman International loaned some equipment and provided advice on the making of physical measurements and the running of the Binaural Room Scanning (BRS) record/replay headphone system that was used for listening tests. In truth, it was a massive undertaking for a thesis project.

There were two parts to the investigation. The first phase involved making measurements in six screening rooms ranging from 24 to 516 seats. The second phase involved subjective evaluations of several room curve targets in some of those venues.

The loudspeaker used was a JBL Professional M2, a large studio monitor. It was powerful enough to generate useful sound levels in the cinemas, it was reasonably portable, and it had wider dispersion than typical cinema loudspeakers—$120° \times 100°$ vs. $90° \times 50°$—so it would be a more demanding test in that more reflected sounds would be generated by it. All of the in-room measurements in the SMPTE TC-25CSS (2014) document were done with no prior measurements on the loudspeakers or screens. The Gedemer tests were distinctive in that the loudspeaker had been thoroughly documented anechoically: Figure 5.12. It was placed in front of the screen at the center channel location so there were no screen complications.

Ideally such tests would be done in real time, using some number of cinemas. However, liberating these money-earning facilities for days on end was simply not practical with the available budget. Fortunately technology has provided BRS methods for capturing (using a dummy head), and binaurally reproducing through headphones, listening experiences in different places at different times. During playback, head movements are tracked, and translated into changes in the binaural signals, so that the sound sources and room do not move. It can do a remarkable job of replicating the experience of listening to loudspeakers in rooms, without being in the rooms. It is widely used in psychoacoustical investigations. This is a precision operation, incorporating calibrated headphones and accurate tracking of head movements to help externalize the sound. Once the BRS data file exists, different programs can be played back through the rooms, and listened to at any seat where the BRS files were recorded. These tests employed this method, as described in Gedemer (2015b), with results shown in Gedemer (2016). The process is explained here in Section 3.5.1.3.

The purpose was to evaluate subjective reactions to sound recordings made under different circumstances when they are auditioned in three professional screening rooms (60, 161 and 516 seats) in the Hollywood area, employing five different room equalization target curves, including the X-curve.

In theory, a program mixed in a control room or dubbing stage calibrated to the X-curve should be preferred if the playback also employed the X-curve. There were 10 trained listeners and four professional cinema sound designers and editors. All had normal hearing. Programs included orchestral film scores recorded in X-curve and non-X-curve calibrated rooms, and two music CD selections. Music was used because voice is not very revealing of broadband timbral characteristics (Figure 3.15), and sound effects are substantially artificial. All originals were carefully repurposed for monophonic reproduction in a postproduction facility.

Figure 11.5 shows the combined results of the listening tests. The equalization target curve is by far the dominant experimental variable. It is clear that the listeners strongly disapproved of curves A and B, and had much higher ratings for curves C, D and E, with curve C receiving the highest ratings.

Figure 11.6 shows the five room equalization targets as they were heard by the test listeners through the binaural headphone BRS system placed at the 2/3 distance reference seat in the 516-seat screening room, responding to sound delivered by the JBL M2 loudspeaker placed in front of the screen. These curves include no spatial averaging, and therefore show acoustical interference effects occurring at the reference seat. They also include the (small) calibration errors in the BRS headphone system, which would have been a constant factor. In Figure 11.6c the electronic signal path version of the target curve is shown.

Figure 11.6a shows the three target curves that received the high ratings by listeners, compared to the SMPTE ST 202 X-curve tolerances. Figure 11.6b shows the

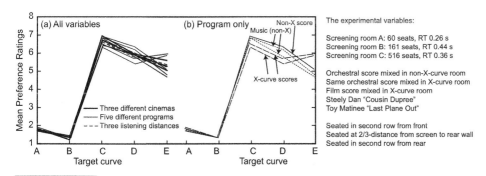

FIGURE 11.5 *A summary of listening test results for all of the variables, including details of the program results.*

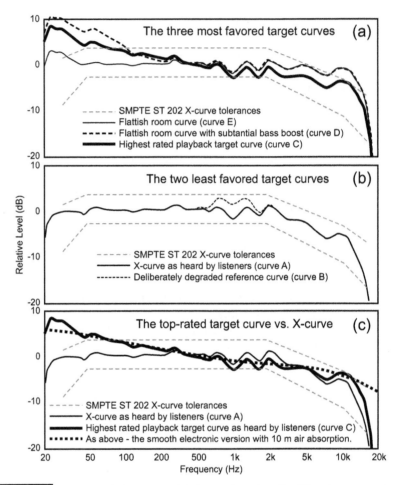

FIGURE 11.6 *shows the anticipated sounds at listeners' ears from five different room curve target shapes, as measured at the 2/3-distance location in the 516-seat screening room, and delivered through calibrated headphones in a binaural-room-scanning (BRS) system. (a) shows the three steady-state room target curves that received high scores in the listening tests. (b) shows the two room curve targets that received low ratings by listeners. (c) shows a comparison of the most favored room curve target and the SMPTE X-curve target, both as heard by listeners. Note that the detailed ripples in these curves are attributable to the specific seat in the venue in which they were measured, and also include the (smaller) residual imperfections of the calibrated headphones. Also shown is the smooth electrical target curve including 10 m of air absorption. See Gedemer (2016) for more information.*

two target curves that received low ratings. It is clear that listeners preferred elevated levels of bass or treble or both, compared to the X-curve. Figure 11.6c shows a comparison between the most preferred target curve and the X-curve. It is interesting that above about 70 Hz the preferred target curve fits within the X-curve tolerances, yet the subjective impressions of it and a true X-curve target were very different. It would appear that there are three significant causes for audible differences:

1. Low frequencies progressively rise in the manner anticipated by Figure 11.2.

2. A spectral prominence associated with the "knee" at 2 kHz in the X-curve has been lowered and smoothed.

3. The very high frequencies are slightly elevated compared to the central X-curve but still within the tolerance limits.

As seen in Figure 11.5, the ratings for the different rooms were very similar, as were those for different seating distances in those rooms. The latter is of special interest because listeners had no visual information that might allow them to apply perceptual compensation for the high frequency air attenuation. The program material elicited some variations, but not enough to alter the overall preference for target curve C. The score and music mixed in non-X-curve calibrated venues received the highest ratings when auditioned using the non-X-curve targets.

However, the observation that the X-curve monitored mixes were greatly preferred with the non-X-curve playback targets needs thought. "Flat" room curve E provides treble boost, which was approved of. Curve C provides this and some bass boost, which was given even higher ratings. Curve D got lower ratings than Curve E for the X-curve monitored scores (speculation: X-curve mixers already boosted the bass and Curve D made it truly excessive). Curve D got higher ratings for the non-X-curve monitored material (speculation: mixers had no need to boost the bass and the boost in D was perceived as being within reason). Curve C, the happy compromise, was highly rated for all programs. Summarizing and simplifying: a slight treble boost is good; bass boost is good; the right amount of both combined with the elimination of the 2 kHz knee was judged to be best for this selection of programs. All of the preferred target curves avoid the knee at 2 kHz.

11.5 FLAT, DIRECT SOUND IS AN ENDURING FAVORITE

Starting with my earliest listening tests in the late 1960s, through a prolific research period in the 1980s (see Figure 5.2), up to the present (see Chapter 12), it has been a monotonous truth that in double-blind listening tests, the highest rated loudspeakers had the flattest, smoothest on-axis and listening-window frequency responses. Listeners liked neutral, uncolored, direct sound. Beyond that, loudspeakers that exhibited similarly good behavior off-axis achieved even higher scores—reflected sounds would then have similar timbral signatures. These findings have remained valid in many different rooms over the years. These were small rooms: stereo listening rooms, home theaters and recording control rooms. As has been discussed earlier, listeners have a significant ability to separate the sound of the source from the sound of the room

(Figure 5.16). The two sets of information appear to be perceptually streamed, with the result that loudspeakers retain their relative sound quality ratings in different rooms (Section 7.6.2).

The direct sound is associated with numerous perceptual processes that are triggered by or referenced to the first arriving sound. Some are timbral, others directional and spatial. In live, unamplified performances the direct sound delivers the timbral essence of the voice or instrument, which is embellished and enriched by reflected sounds arriving later. Reflected sounds also contain information about the listening space, putting the experience into an acoustical context. It seems sensible to deliver a high-quality, neutral, direct sound.

Queen (1973) concluded:

The results thus far tend to confirm the presupposition that spectrum identification depends principally upon the direct sound of the sound source. It would suggest that any equalization for the reverberant field [steady-state] which is detrimental to the desired response in the direct field will reduce the naturalness of the system or will be detrimental to any other desired characteristic designed for. It suggests that most necessary equalization for speech reinforcement systems may be accomplished by proper design, choice, and fixed equalization of the transducers.

This means that the loudspeakers need to be designed and equalized as necessary to deliver a neutral direct sound. In cinemas this means on the audience side of the screen. Steady state in room measurements are necessary to evaluate low-frequency room reflection effects and to allow for the identification and treatment of those problems.

In movie sound, the goal is to give the listener the impression of being in the space portrayed on the screen. Consequently, the acoustics of the cinema should not dominate the artistic space portrayed in the soundtrack, whatever it is. A lover whispering sweet nothings into the ear of his mate loses credibility when the sound is reverberating in a large reflective room. For this reason, cinemas in general need to be, and usually are, relatively well damped acoustically, demonstrating reverberation times that are similar to homes at least over the mid- to high-frequency range (Figure 10.1). In addition the loudspeakers are very directional, focusing the sound on the audience, not the surroundings. The dominance of the direct sound in the mid- and high-frequency ranges is a huge advantage. A sense of acoustical intimacy is possible, and multichannel movie sound systems can do a good job of replicating the impression of being in almost any space.

However, it is a generalizable fact that at low frequencies the reflectivity of normal rooms increases. The exceptions are special-purpose rooms that have been treated with large areas of low-frequency absorbers. Large cinemas are costly to acoustically treat at low frequencies, and it is not often done (Figure 10.1d). As a result, bass rises following the direct sound. It tends to happen quickly, in less than about 200 ms (Figure 10.2), so it is a spectral balance issue, not protracted reverberation in the concert hall or cathedral sense. As seen in Figure 11.2, it can rise as much at 10 dB, which is not a trivial amount. Of course, if there are powerful room resonances the bass can "boom"; this is when acoustical consultants earn their pay.

At high frequencies listeners are in a predominantly direct sound field, so the main factor affecting sounds arriving at listening locations is propagation distance and the associated air attenuation.

Figure 11.7a shows the results of several subjective tests. There are two basic kinds: preference tests and timbre-matching tests.

Preference tests:

- In the Ljungberg tests, listeners adjusted the spectrum of the sound with a multi-filter equalizer until it sounded "right" for a large variety of program samples.

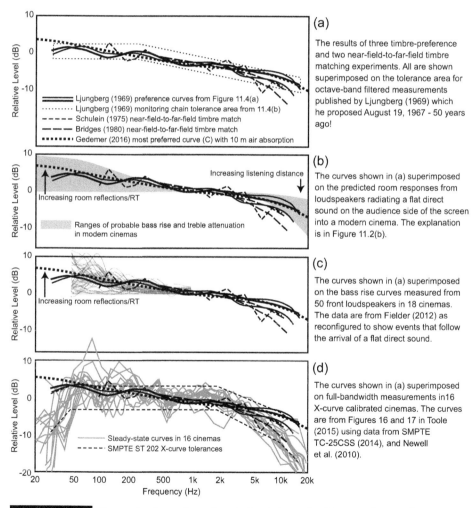

(a)

The results of three timbre-preference and two near-field-to-far-field timbre matching experiments. All are shown superimposed on the tolerance area for octave-band filtered measurements published by Ljungberg (1969) which he proposed August 19, 1967 - 50 years ago!

Ljungberg (1969) preference curves from Figure 11.4(a)
Ljungberg (1969) monitoring chain tolerance area from 11.4(b)
Schulein (1975) near-field-to-far-field timbre match
Bridges (1980) near-field-to-far-field timbre match
Gedemer (2016) most preferred curve (C) with 10 m air absorption

(b)

The curves shown in (a) superimposed on the predicted room responses from loudspeakers radiating a flat direct sound on the audience side of the screen into a modern cinema. The explanation is in Figure 11.2(b).

Increasing listening distance

Increasing room reflections/RT

Ranges of probable bass rise and treble attenuation in modern cinemas

(c)

The curves shown in (a) superimposed on the bass rise curves measured from 50 front loudspeakers in 18 cinemas. The data are from Fielder (2012) as reconfigured to show events that follow the arrival of a flat direct sound.

Increasing room reflections/RT

(d)

The curves shown in (a) superimposed on full-bandwidth measurements in16 X-curve calibrated cinemas. The curves are from Figures 16 and 17 in Toole (2015) using data from SMPTE TC-25CSS (2014), and Newell et al. (2010).

Steady-state curves in 16 cinemas
SMPTE ST 202 X-curve tolerances

FIGURE 11.7 *The results of several experiments aimed at revealing listener preferences for sound systems in large venues. The explanations are in the figure.*

■ In the Gedemer tests listeners were presented with several frequency-response options and they scored each of them on a preference scale while listening to a selection of programs.

Near-field to far-field timbre matching tests:

■ The underlying hypothesis is that a small, high-quality loudspeaker placed close to the listener is the "perfect" reference sound source, and the task is to adjust the spectrum of the distant loudspeaker until it sounds the same. The close loudspeakers would deliver predominantly direct sound over most of the frequency range. At the short distances used in these tests, 2.5 ft and 4 ft, there are questions about the near-field (see Section 10.5) artifacts of the loudspeakers. The distant loudspeakers would deliver a combination of direct and reflected sounds, the ratio of which would change with frequency, depending on the directivity of the loudspeakers and the reflectivity of the venue. Bass rise would be included in the delivered sound. There would be spatial/temporal differences and spectral differences, but only the spectrum is being adjusted. It is a compromised concept for a test, but interesting to examine. Schulein (1975) and Bridges (1980) provide examples.

Of the two methods, the preference tests are more definitive. Examples in this book indicate that for about 50 years there has been continuous confirmation that human listeners with normal hearing are remarkably stable and agreeable in their ratings of sound quality if the tests are blind. They respond to small differences in spectral balance and resonant colorations, and they do this in different rooms, with different programs. A smooth flat forward radiated (on-axis) sound is the fundamental requirement.

Preference tests make no assumptions about the "perfection" of a reference sound, and do not question the influence of spatial, timbral and temporal effects of reflections in large spaces. These ingredients are all present at all times, as they are when listening to movies or music for mixing or entertainment in the venue. Differences would arise because of personal preferences, or personal hearing capabilities, and, of course, interactions with the program material because the circle of confusion applies.

The concern is that not all programs that are entertaining are reliably informative about sound quality. In movies dialog is obviously important, but voice is not the most revealing sound when it comes to judging sound quality (see Section 3.5.1.7). Sound effects arouse emotions and are exciting, but usually it is impossible to say what they should have sounded like. In the end, music, including film scores, is the basis for the most meaningful listening tests, and even then some music is better than others (Figure 3.15). "Translation" of the artistic cinema experience is not the issue; reproduction of the timbral essence, the sound quality, is. If a sound system can reproduce a variety of music with credible accuracy, voices and sound effects are certain to be gratifying. The mixer would as usual individually manipulate program passages in which speech intelligibility is challenged.

Figure 11.7a is extraordinary in that seven curves arrived at by four experimenters over a 47-year time span cluster closely around each other over most of the frequency range. Over most of the frequency range, they all fall within the 4 dB tolerance window proposed by Ljungberg in 1967 (published in 1969). The tolerance is for curves measured with octave-band resolution. Given the differences in the equipment, programs and experimental methodologies used in arriving at these curves, the agreement is impressive. Clearly listeners' tastes have not changed and carefully conducted experiments can be trustworthy sources of data.

Although great importance is placed on steady-state room curves, nothing is known about the performance of the loudspeakers themselves. Unfortunately only the Gedemer experiments include comprehensive anechoic data on the loudspeaker used. What we know of some of the older loudspeakers indicates that bandwidth limitations and high-frequency misbehavior were highly probable and these would influence subjective impressions. See Figure 18.5 for examples of some professional monitor loudspeakers of that period.

Figure 11.7b shows the seven subjectively determined curves superimposed on the predicted room response for a loudspeaker delivering a flat direct sound on the audience side of the screen (as explained in Figure 11.2b). All of them closely track the prediction below about 3 kHz. Above that, three of the four preference test results closely follow the predicted results, with one falling slightly below. The three near- to far-field timbre-matching results all fall below the shaded area, possibly signaling a peculiarity of this kind of test methodology, or of the loudspeaker/room combinations used. The recent Gedemer result falls comfortably within the predicted range.

Figure 11.7c moves closer to reality by showing the bass rises measured using 50 front loudspeakers in 18 cinemas (Fielder, 2012). In this display the Fielder data have been reconfigured to show bass rise starting from a flat direct sound. This confirms the validity of the predicted shaded area in Figure 11.7b, and showing that the subjectively preferred bass rises are comfortably within the results measured in contemporary cinemas.

Figure 11.7d shows an important comparison. First, the background data in gray show steady-state room curves for screen channels in 16 cinemas in both Europe and the US, and the X-curve tolerances. The data show that calibrating technicians in many of these 16 cinemas allowed the bass level to stray toward and beyond the upper limit of the X-curve tolerances. Was this because it sounded better? If so they agree with the listeners in all of the tests being discussed here. All of the black curves from tests discussed here fall within the X-curve tolerances above about 60 Hz. The prolonged flat bass to midrange response and 2 kHz knee that Ljungberg objected to in his 1969 paper are clearly absent from these seven curves.

But there is a mystery. The calibration technicians in many of these 16 cinemas let the very high frequencies drift downward toward and beyond the lower limits of the tolerance range. Why? The errors are not insignificant, and the trend is not subtle. Listeners in these tests preferred more very-high-frequency energy for all programs, especially those in the preference tests. The Gedemer experiments also included mixes done in X-curve calibrated facilities. Current speculation involves the limitations of some high-frequency compression drivers and the consequent distortion at high sound levels. Lacking a technical, and therefore verifiable, justification, the calibration errors remain a mystery.

The difference in the curves around 2 kHz is significant. In all studies of the audibility of spectral variations, it has been the low-Q, wider bandwidth bumps that have been the most audible. Section 4.6.2 discusses this in detail, showing that, as they are represented in frequency-response measurements, low-Q phenomena are detectable at amplitude deviations in the range 1 to 3 dB. Addressing the low-Q spectral problems in a loudspeaker often yields the greatest reward (see Figure 4.11 and discussion). That is what may be happening here because the subjectively preferred curves have no knee at 2 kHz. If a soft transition was the goal, the target X-curve should have been drawn with a French curve, not a straightedge.

Directly related to this is the benefit that lowering the target curve in this frequency region significantly reduces the power demands on compression drivers because much of the spectral energy in program is concentrated there.

11.6 ALTERNATIVE TARGETS—IS IT TIME TO MOVE ON?

It may be premature to say that the subjective data provided by Ljungberg, Schulein, Bridges and Gedemer are absolutely definitive, but the evidence is persuasive. The curves are consistent with acoustical physics of cinema sound fields and the preference for a flat, neutral, direct sound agrees with countless subjective evaluations in small rooms. A flattish direct sound target is employed elsewhere in professional audio, and is the default situation in domestic audio and most recording-studio monitoring. This has been explained in Chapter 5, will be revisited in Chapter 12 and is historically confirmed in Chapter 18.

Although steady-state room curves cannot be decisive descriptors of sound quality, if there is confidence in the loudspeakers and some consistency in the room acoustics, they are useful indicators. In the present cases it is interesting to contemplate what such targets might look like.

Figure 11.8a shows a segmented-line approximation to the set of gently undulating curves being discussed. It follows the guidance of Figure 11.3a that is based on the physical sound field relevant to listeners in cinemas. There is a short horizontal section in the middle frequencies that is in the predictions and that seems to have a subtle presence in the curves. The transitions (knees) have been smoothed. If that is not persuasive, or simplicity is desired, the simple tilted line in Figure 11.8b is a plausible option. In both cases the high-frequency end of the lower limit has been shown with a possible option that accommodates measurements made at large listening distances—greater air attenuation at 20 m or more. All of this is obviously approximate, but equally obviously, it is quite orderly.

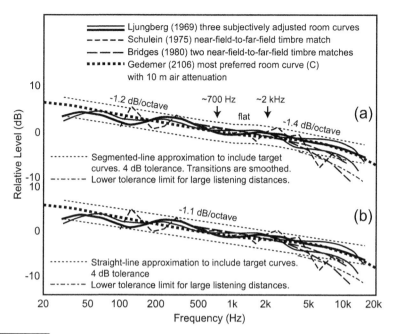

FIGURE 11.8   *(a) The seven subjectively adjusted room curves fitted into a smoothed segmented-line 4 dB tolerance window. (b) the same as (a) but with a simple tilted spectrum as the target. Additional air attenuation is shown for large listening distances.*

Looking farther afield in old documents one finds that the desirability of a downward tilting room curve can be seen in other data. Boner and Boner (1965) were interested in ways to eliminate feedback in sound-reinforcement systems in large (37,000 ft³ to 2,000,000 ft³) highly reverberant (RT = 1.0–1.5 s at 500 Hz) venues that seated several thousand people. However, they also paid attention to audience preferences, relying on steady-state room curves averaged over the seating area for evidence of system performance. The loudspeakers included "theater-type" horn loaded systems and distributed ceiling systems. Figure 11.9 shows three of these (unspecified) systems that pleased audiences. The upper and lower branches of the "with bass" curves were not explained, although the upper one closely follows the trend of the "somewhat preferred" curve. The lower one maintains the same tilt, but is simply displaced downward at a 600 Hz (woofer-to-horn crossover?) discontinuity. The simplified tolerance curve from Figure 11.8 is superimposed and it includes two of the "somewhat preferred" curves.

Boner and Boner noted that audiences did not like rising treble at all. They rejected rolled-off bass, liking "big bass" if room resonances, "bull notes," were attenuated by filtering. These systems in large auditoriums and ballrooms were far removed from modern cinema B chains, but the underlying trend of a fundamentally smooth, gently downward tilted steady state room curve is evident. It is probable that the loudspeakers exhibited much the same directivity trends as their modern counterparts, suggesting that the underlying listener preferences were associated with a flattish direct sound. The sagging high frequencies in the top two curves of Figure 11.9a is consistent with air attenuation at the listening distances in these very large venues, although loudspeaker performance and orientation are unknown. The incongruous downward shift of the lower curve has no obvious explanation.

Schulein (1975) speculated about the desired shape of a house curve, saying:

A very logical answer is that the house curve should be flat and, consequently, not emphasize any particular frequency. This conclusion unfortunately does not appear to hold true in practice. It has been the consistent experience of this author, and many other experimenters, that a flat house curve is subjectively too bright, and should be rolled off by as much as 10 dB at 10 kHz.

The amount of the attenuation is about right: 10 dB, which if distributed over a 20 Hz to 10 kHz range (9 octaves) yields a slope of −1.1 dB/octave, the slope of the target curve tolerance in Figure 11.8b. However, the question is: at what frequency should the rolloff begin? The weight of the evidence points to a relatively linear tilt starting at a low frequency. The notion that a frequency response should be flat is, as Schulein said, "logical," but it is the direct sound that should be flat, not the steady-state room curve.

The 4 dB window is a continuation of Ljungberg's suggestion for octave-band resolution measurements. As stated earlier, it is my opinion that above the transition frequency there is no advantage to basing room equalization on narrower-band steady-state data; it is likely only to motivate inappropriate adjustments. If anyone wishes to pursue this line of thought for future consideration by the movie sound industry, the tolerance window would be a matter for discussion, taking into account the acoustical realities of loudspeakers, venues and the expertise of the calibration teams.

However, a viable alternative may be to simply arrange for the loudspeakers to deliver flat direct sound on the audience side of the screen. This would apply above a few hundred hertz and Section 14.1 discusses optional methods. Steady-state room

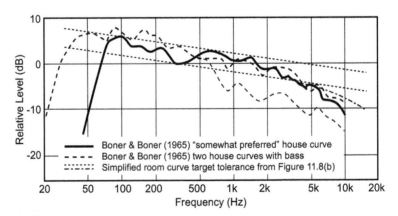

FIGURE 11.9 *Steady-state room curves from Boner and Boner (1965) compared to the simplified target curve tolerance shown in Figure 11.8b.*
Copyright 1965, Acoustical Society of America.

curves will still be needed for evaluating bass performance, and can be useful as a means of periodic testing for system defects. A single in-situ measurement for each channel, compared to stored references, could be used to confirm that all components are functioning. The B-chain components—room, amplifiers, wires and loudspeakers—do not "drift" in performance, but they do occasionally malfunction. Voice-coil rub, buzz and other distortions will need to be evaluated by listening to slow frequency sweeps at various levels. Wearing musicians earplugs to avoid hearing damage while listening is advised.

11.6.1 Compatibility with the Rest of the Audio World

The ideal situation would be one in which the mixing and exhibition of movie sound tracks involved the same sound quality signature as those outside of that domain. We have seen that there is some agreement on a possible revised target for movie sound. The question is how well it matches the performance objectives and practices in other parts of the audio world. Figure 11.10 summarizes the preceding data and discussion and compares it to decades of accumulated experience in the consumer audio world. Literally hundreds of loudspeakers and listeners have participated in double-blind subjective evaluations starting with my early experiments (Toole, 1985, 1986) through to the present day. The answer has been monotonously the same.

A JBL Professional M2 loudspeaker (see Figure 5.12) was used for the Gedemer tests and it is also among the collection of subjectively highly rated loudspeakers used to generate the idealized curve for small rooms. As shown in Figure 12.4, above the small-room resonance frequencies they exhibit very predictable room curves in a variety of rooms. These curves therefore not only represent listener preferences in cinemas

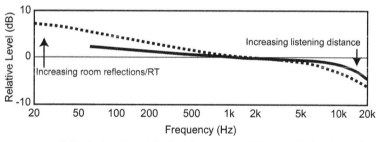

········ Subjectively preferred steady-state room curve in three professional screening rooms: 60, 161 & 516 seats. From Figure 11.5(c) and Gedemer (2016). 10 m of air attenuation is included.

———— Idealized steady-state room curve for subjectively preferred domestic and monitor loudspeakers in typical home listening rooms and home theaters. 3 m of air attenuation is included. From Figure 12.4(d).

FIGURE 11.10 *A comparison of subjectively preferred target curves for cinema sound, using the Gedemer curve to represent the collection of curves shown in Figure 11.8, and the idealized room curve for domestic and monitor loudspeakers from Figure 12.4d.*

and domestic listening conditions, but they represent preferences when the *same* loudspeaker is used.

> In both cases shown in Figure 11.10 the underlying common factor is that these are the steady-state room curves that result when using loudspeakers delivering flattish direct sound.

The difference between the curves at low frequencies is attributable to the greater bass rise resulting from higher low-frequency RTs in cinemas (see Figure 10.1b and 10.1d) and the rapid directivity rise in large woofers (Figure 10.15). At high frequencies the greater air attenuation at the increased listening distances accounts for the difference. In this comparison the results are generalizable to numerous well-designed loudspeakers, but most certainly apply to the one loudspeaker that was common to both curves, the M2.

The conclusion: if the direct sound is flattish and smooth, and if there are no anomalies in the directivity indexes of the loudspeakers, the perceived sound quality should be high and the steady-state room curves should resemble those shown in Figure 11.10.

In practical terms, from the perspective of ensuring a reliable transfer of the auditory experience from dubbing stages to cinemas, the only certain solution is to exercise control over the low-frequency reflectivity/reverberation of the production and reproduction venues.

11.6.2 Compatibility within the Cinema World

There are several reasons to wonder about consistency within the movie world itself. The variations in cinema calibrations seen in Figure 11.7d are enough to raise questions about how precisely the art is transferred from the dubbing stages to the cinemas. Artistic "translation" is important, no doubt, and audiences can experience the drama and emotions of films without absolute precision in the reproductions of visual images and sound. However, it is obviously better if the reproductions are more accurate.

The industry goes through the motions of calibration, yet measurements indicate significant variations in performance. Professionals within the industry are on record with complaints about inconsistencies in sound quality within the production venues used to create soundtracks. Sound mixers are on record with comments indicating different approaches to compensating for the lack of neutrality in X-curve calibrated production facilities. This has led to some production facilities that depart from the X-curve calibration in order to create what is for them a more pleasant working situation. If a music CD does not sound good in a dubbing stage because of the X-curve, there is clearly a problem.

If a new performance target is selected for film sound, there is the important issue of how legacy films will sound. In theory, the playback of legacy soundtracks is not difficult, as that involves only an equalization switch, which can be activated by metadata. X-curve monitored soundtracks ought to sound best through X-curve calibrated playback venues. Yet anecdotal accounts exist of soundtracks being played in large and small venues calibrated for flattish direct sound and apparently sounding just fine. This is the normal experience in domestic home theaters. This is what Gedemer found in cinema venues (Figure 11.5).

But acceptable sound is not necessarily accurate sound reproduction. I have noticed clearly different spectral balances among movies in cinemas and at home. Over the years, things have improved, and the best examples are very good indeed. But if one attempts to install a frequency response correction filter for legacy movies mixed in X-curve calibrated facilities, what form should that filter take? If anyone chooses to examine the situation closely, they may find that one correction curve is not ideal for all movies.

Taking a scientific approach to the problem, two questions need answers:

- How consistent is the spectral content of existing films?

- If there is demonstrable consistency, how do they sound through:
    - ☐ X-curve calibrated facilities?
    - ☐ Facilities calibrated to a new standard?

Given the numerous variables in soundtrack content, such an investigation is a major undertaking.

Any change in a frequency response target raises the issue of whether existing B-chains are able to handle the power demands. In the case of these suggested targets, Figure 11.7d shows that there would be a reduction in the power demands on the compression drivers around 2 kHz and little or no change at the highest frequencies. The required bass rise is already implemented in some number of cinemas, but there is a question about whether less-well-equipped cinemas can handle the additional demands. If bass management were implemented, the LFE subwoofers would share the load all of the time, not just part of the time. Unknown at this time is any notion of how much bass boost is applied during the mixing operation. Logically there would be some.

However, as unambiguously laid out by Newell et al. (2016), a related problem already exists. Present day cinema sound systems are being driven very hard because the undistorted headroom provided by digital delivery formats has been filled with program. Playback sound levels are higher than were possible in analog days. Marginal loudspeaker systems distort or are damaged. Audience members complain of excessive loudness and/or distortion and walk out in sufficient numbers that there are threats of sound level limits being imposed by local authorities. Some of the limits being discussed are based on misunderstandings about hearing loss (occupational vs. recreational exposure) and a lack of sensitivity for preserving the art that is in the soundtracks, especially intelligible dialog. Some sounds need to be loud, but there are suspicions that sustained loudness is sometimes being used as a substitute for a good plot.

All of these thoughts are an important beginning to a serious discussion about the present and future of movie sound. It is unfortunate that in the 4+ decades since cinema sound standards were created, none of this has been done before. Circumstances in all respects have changed, and untold numbers of films have been, and continue to be, created using what now appears to be an inappropriate target curve and B-chain calibration process.

11.7 THE EFFECTS OF ROOM SIZE AND SEATS

Part of the original SMPTE and ISO documents included modifications to the high frequency tilt of the X-curve based on audience size. The smaller the venue, the lower the slope above 2 kHz. The SMPTE TC-25CSS (2014) B-chain report showed that above a few hundred hertz audiences were in a dominant direct sound field, so what is measured and heard is independent of room acoustics, but obviously dependent on listening distance and the associated air attenuation.

An important study by Gedemer (2015a) provides data that differs from most cinema measurements, including those in the SMPTE report, in that the performance of the specific loudspeaker was known beforehand and no room equalization was used. The objective was to compare steady-state cinema curves with the anechoic loudspeaker data, looking for useful relationships. The sound source was a large studio monitor loudspeaker, a JBL Professional M2, with full anechoic documentation shown in Figure 5.12. It was set up in front of the screen, at the center location, in six film sound venues. Measurements were made at many seating locations at ear heights and at differing microphone heights. There was no room equalization.

The curves shown in Figure 11.11a are averages of steady-state measurements made at every second seat in each of the facilities, except for cinema G where 109 locations were used. Above about 1 kHz the curves agree very well, indicating minimal room effects. The curves were normalized to this frequency range. Below about 150 Hz the variations are of a kind expected for differing amounts of reflected energy, adjacent boundary effects and, in the smaller venues, room modes. The irregular behavior in the 150 to 800 Hz range was troubling. Seat dip or other seat interaction effects were suspected (the microphones were at estimated ear level: 40 to 47 in. (1.0 to 1.2 m) above the floor).

Ear level microphones in close proximity to orderly rows of seats, some of which have high backs, generate effects more complex and of wider bandwidth than the well-known seat-dip phenomenon in concert halls. The fact that the cinema sound field is considerably less reflective is a probable factor. It is unlikely that these non-minimum-phase acoustical interference effects can be treated by equalization, if indeed they need correction at all. Two ears and a brain are much more analytical and adaptable than an omnidirectional microphone and analyzer. Once stationary in a chair, listeners usually are unaware of, or attribute innocent spatial descriptors to, measured variations suggesting repugnant interference or comb-filtering phenomena.

Anecdotal information, personal experience, and experiments by Newell et al. (2015) indicate that stand-up/sit-down comparisons yield insignificant changes in sound quality. If so, delivery of a high-quality direct sound can be ascertained using microphones elevated above the seats, thereby avoiding distracting measurement anomalies.

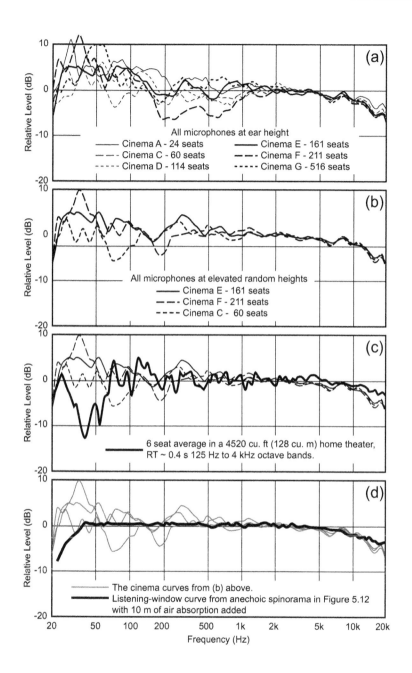

FIGURE 11.11 *A JBL Professional M2 loudspeaker radiates directly into six different film-sound venues from the center channel location in front of the screens. (a) shows the result of microphones at nominal ear height. (b) shows the results when three of the cinemas were re-measured with the microphones at elevated randomized heights. (c) shows the steady-state cinema measurements from (b) with a six-seat average of the same loudspeaker measured in a typical domestic listening room/home theater. All cinema data are shown with 1/6-octave frequency resolution. The home theater curve has 1/12 octave resolution. (d) superimposes the listening-window curve from the spinorama in Figure 5.12 with 10 m of air absorption included. Figure 5.6b illustrates the listening window concept.*

Figure 11.11b shows that when the microphones were elevated to heights random-ized over the range 51 to 79 in. (1.3 to 2 m), in three of the venues the curves congre-gate more closely to each other above about 400 Hz. Newell et al. (2015) suggest a distance of at least 2 ft (60 cm) above the back of the seat to avoid acoustical interfer-ence. It seems reasonable.

Figure 11.11c takes the range of audience size to a low of six seats in a domestic home theater. The superimposed curve shows excellent agreement, but is slightly elevated at high frequencies because of the reduced listening distance (less air absorption). Except for the lowest frequencies, the cinema curves and the home theater curve appear to fluctuate around the same central tendency. Obviously, if the curves were smoothed to the traditional 1/3-octave resolution, the differences would be even smaller.

Figure 11.11d superimposes the listening-window curve from the spinorama in Figure 5.12. Above about 500 Hz listeners in cinemas are in a progressively more dominant direct sound field so this should be a good predictor of what is measured. With the addition of some air attenuation, it is. At low frequencies there is only slight bass rise, as is expected in venues that have been as thoroughly acoustically treated as these professional screening rooms have been (Figure 10.1d). The M2 has a single woofer, and therefore would not exhibit as large a bass rise as more typical two-woofer cinema systems (see Figure 10.15).

Summarizing: In all of the measurements, above about 500 Hz the steady-state room curve would be well predicted from anechoic loudspeaker data, especially if the curves were 1/3-octave smoothed. If a perforated screen is between the loudspeaker and the audience, appropriate equalization could be applied to compensate for screen loss, which is also predictable. Any high-Q irregularities measured at a single microphone location, especially those at ear level, should be ignored—most likely, they are evidence of acoustical interference at that location, a phenomenon not correctable by equalization, but also not needing correction.

Gedemer (2015a) showed that in a 516-seat cinema, there was little difference between the spatial averages of four microphone locations and 109 locations. At low frequencies, standing waves and adjacent boundary effects dominate results, and equalization is a remedy for some of these. Calibrators must avoid boosting narrow interference dips.

Even with a seat count that ranges through 6, 24, 60, 114, 161, 211 and 516, there is no obvious pattern of change in steady-state room curves beyond the small difference in air attenuation in the direct sound. The scaling corrections based on audience size advised in SMPTE ST 202 and ISO 2969 recommendations are incorrect. If the intent was to acknowledge air attenuation as a function of propagation distance, the loss can be easily calculated for any seat in any venue, and it will be different from the straight-line variations suggested in the recommendations (see Figure 10.12).

The large variations are at low frequencies where several factors contribute, and all need to be identified and treated in situ. Some will be stable over the seating area and others will change from seat to seat. Simple equalization is not an automatic corrective measure. The unavoidable standing wave effects in small rooms are distinctive for different rooms, loudspeaker and listener locations. Fortunately, there are effective methods of attenuating their effects, as explained in Chapter 8.

Measurements with microphones at ear level should be discouraged because they introduce acoustical interference irregularities that are tempting to equalize, but that are not audible problems to human listeners. Microphones should be elevated well above the seat backs, preferably at varied heights to average out regular patterns of interference. However, all such measurements are reliable at low frequencies.

11.8 CINEMA SOUND—WHERE TO NEXT?

Perhaps nowhere. It is a huge industry with extensive infrastructure and countless legacy films, and change could be disruptive. How much disruption and what form it would take is a matter for serious consideration. The fact that soundtracks mixed in X-curve facilities sound good when replayed over systems with flattish direct sound, as in the Gedemer experiments and in countless home theater systems, is reason for optimism.

There is evidence that movie sound facilities are not uniformly calibrated, and evidence that mixers have different approaches to establishing a satisfying spectral balance in soundtracks. Both inevitably lead to variations in the artistic product. Any revisions to current practice would have to address the calibration issue.

There are people who think that the existing situation works well enough, and change would impose an unacceptable burden on the industry. It has been said that customers are not complaining. This could also be construed as evidence of the tolerance and adaptability of moviegoers, who get caught up in a plot, and sound quality is not a large factor in their enjoyment. Dramatic dynamics and a lot of bass seem to satisfy many. Now the focus is on many more channels of immersive sound, all of which need to be calibrated. With the elevation localization cues being concentrated in the highest frequencies (Section 15.12.1), special care may be required to ensure their delivery.

Customers walking out of movies have brought another matter to the fore: excessive loudness and/or distortion. In response, cinema operators are turning down the volume, and the cinematic art and speech intelligibility both suffer.

An underlying problem in the industry is that a significant fraction of cinemas have sound systems that were borderline performers in the days of analog soundtracks, and have problems dealing with the expanded dynamics of digital sound tracks. This is aggravated by the practice of compressing more of the sound into higher levels (Allen, 2006; Newell et al., 2016).

If a new target with a significantly altered spectral balance is imposed, the stress on B-chains will change, it needs to be carefully evaluated. Figure 11.7d indicates a reduction in power demand around the 2 kHz knee, a high-energy portion of the spectrum, and an increase in power demand at very high frequencies, a low average energy portion of the spectrum; there are tradeoffs. The bass would also be boosted, but no more than is already done in some cinemas. This requires careful evaluation of program demands and transducer delivery capability, as was done by Leembruggen (2015a). Finally, the effect of the spectral rebalancing on overall loudness needs evaluation for purposes of level calibrations. Section 14.2 provides some guidance.

Here I have attempted to summarize the physical and perceptual evidence in a manner that compares where we are to where we ought to be. If it should happen, for the first time in history soundtracks created in the movie industry would be timbrally compatible with non-cinema professional systems and consumer playback systems everywhere. Mixers will no longer be required to subjectively "pre-equalize" soundtracks compensating for the unnatural spectral balance of an X-curve-calibrated dubbing stage. No repurposing of soundtrack frequency response would be required for distribution outside cinemas. Such a scenario would also allow cinemas to do justice to programs created in the music and television worlds. It seems like a winning situation.

CHAPTER 12

Sound in Home Listening Rooms, Home Theaters and Recording Control Rooms

Rooms of a size appropriate for home theaters, home studios and recording control rooms do not require massive, highly directional loudspeakers. The low-frequency sources are smaller, exhibiting lower directivity indexes. Mid- and high frequencies tend to be radiated by wider-dispersion horns or small cones and domes. These wide dispersion sources generate more reflected sounds over a wider frequency range than is seen in cinemas. Consequently, predictions of acoustical events must be different from those in the previous chapter, but the fundamental principles are the same.

12.1 GOOD SOUND STARTS WITH GOOD LOUDSPEAKERS

Chapter 5 explained how to describe loudspeaker performance in measurements in a way that allows us to predict steady-state room curves above the transition frequency with reasonable precision, and to anticipate sound quality. It turns out that humans have considerable ability to "listen through" rooms, to identify essential characteristics of the sound source. This happens in real-time listening to "live" sounds, and to reproductions of recordings of those sounds.

The anechoic data needed to recognize good loudspeakers was described in Section 5.3, and examples of subjectively highly rated loudspeakers are shown in Figure 12.1. There is a clear bias toward Harman brands because spinorama data and double-blind listening test results were available to the author. The message from this is that, under unbiased listening circumstances, listeners favored loudspeakers that exhibited significant similarities in their anechoic data.

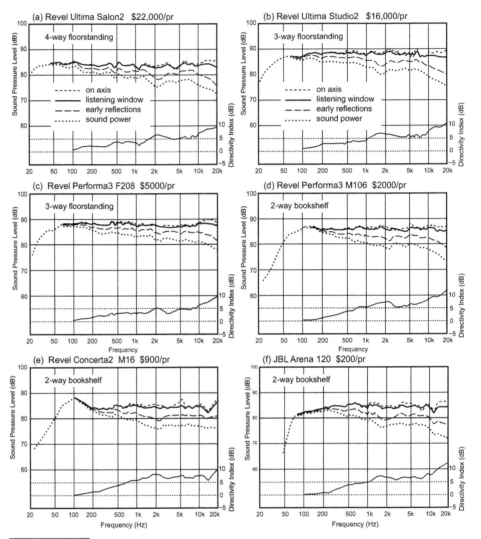

FIGURE 12.1 *Products covering a 110:1 price range, all exhibiting similar timbral signatures. Technically they differ primarily in low-frequency extension and output, distortion and power handling capacity. Visually they range from simple boxes to sculptured art.*

These are fundamentally neutral reproducers even though their prices range over an enormous 110:1 ratio. The differences are in:

■ Bandwidth (primarily how low a frequency can it reproduce);

■ Power-handling capability (how loud it can play);

■ Power compression (how much does the sound quality/timbre change with level);

■ Non-linear distortion (better and multiple drivers sharing the load reduce distortions);

- Resonances in transducers and enclosures;

- Appearance/industrial design (from hand-rubbed exotic woods in sculptured enclosures down to wood grain vinyl–wrapped rectangular boxes);

- Consistency in manufacture (precision end-of-line testing and tweaking of individual drivers and crossover elements is possible only in high priced products).

However, within the limitations of each of these products, the essential timbre of the recording should be well reproduced. (a) is a large, powerful, four-way floor-standing system with useful output down to 20 Hz. (b) and (c) are simplified three-way versions offering quality performance at reduced cost. (d), (e) and (f) are two-way bookshelf systems that can be entertaining on their own, but that benefit from subwoofer assistance for full bandwidth reproduction. As components in a bass-managed system (designated "small" and thereby high-pass filtered) they can play much louder than in their solo versions. Loudspeaker (f) was designed with potential wall mounting in mind, which would provide elevated bass.

Some high-cost products undergo meticulous end-of-line testing and adjustment to ensure similarity to the "golden" prototype. This is part of the price. It is interesting to see how this is revealed in published product reviews. Figure 12.2 shows different samples of such a product measured in different ways in different places. It is evident that all of the measurements in (a) fall well within the widely used ±3 dB tolerance limits. More impressive is the similarity of the anechoic chamber measurements in (b), one

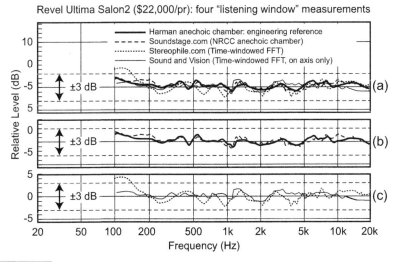

FIGURE 12.2 *Listening window measurements on different samples of a loudspeaker performed in different ways by different reviewers. (a) combines all measurements. (b) compares the two anechoic-chamber measurements, each using a different sample of the loudspeaker, one in Ottawa, Ontario, Canada, the other in Northridge, California, US. (c) compares two time-windowed (quasi-anechoic) FFT measurements.*

from the Harman engineering lab and the other from the lab at the National Research Council of Canada (NRCC). The latter provides measurements to www.soundstage.com. Given the differences in the chambers, measuring equipment, microphone locations, the calculated listening windows and, of course, different samples of the loudspeakers, these are impressively similar curves.

However, audio reviewers normally do not have access to anechoic chambers, so compromises are necessary. To minimize the influences of reflections in the measurement environment two important changes are made:

1. The measurement distance for the *Sound and Vision* data was an adequate 2 m. But *Stereophile* used 1.27 m. This distance puts the microphone in the acoustical near field of large loudspeakers like the Salon2, and errors are expected.

2. Time windowing the FFT measurements to avoid reflections reduces the frequency resolution, which shows up as errors at lower frequencies. *Stereophile* uses 5 ms, which yields a frequency resolution of 200 Hz. Atkinson (1997) notes that measurements below about 1 kHz are compromised, as can be seen in these data. The *Sound and Vision* magazine data were significantly smoothed, meaning that potentially useful high-frequency information was discarded.

Figure 12.2 shows unequivocal data supporting the usefulness of anechoic chambers for measuring loudspeakers. Chambers are very expensive and take up a lot of space, but they yield high-resolution data over the entire audible bandwidth, and are sufficiently quiet that non-linear distortion measurements can be made. Nevertheless, it is understandable why product reviewers and many manufacturers live with compromised quasi-anechoic methods. Such measurements are still usefully revealing of loudspeaker performance. Many reviews contain no measured data at all.

Some argue, defensively, that measurements are not useful. In such reviews the absence of both measurements and double-blind comparison listening tests there is really no verifiable information, measured or subjective—just stylish prose and opinions. For now, www.soundstage.com provides access to NRCC anechoic chamber measurements on many loudspeakers. They are not in spinorama format, but there is enough data to form reliable opinions. There are some surprises.

12.1.1 Typical Loudspeaker Specifications—Part of the Problem

The public has only rarely had access to useful loudspeaker specifications. The average customer may not understand the meaning of the measurements without some guidance, and there are some that simply distrust measurements. Some loudspeaker manufacturers lack the facilities to generate comprehensive and accurate measurements, and many are content not to have to reveal what in some cases would be embarrassing information. The world has come to accept a poor substitute as the norm.

Customers have learned that the only way to select loudspeakers is to listen to them because no matter what the oversimplified published specifications say, the resulting sound quality is not predicted. Listening in reasonable circumstances is almost impossible, as enthusiast audio stores are scarce. Even then, equal-loudness comparisons among several products are not likely to be available, and the tests will be fully sighted with a salesperson adding prejudice to the situation.

The result is a purchasing public making decisions based on imperfect information. Luck is a large factor. Figure 12.3 illustrates the situation.

The published specifications tell us that this loudspeaker falls within a 4 dB window between 39 Hz and 20 kHz. Figure 12.3b shows that the manufacturer met the specification, but undulations in the curve indicate the potential for coloration; it is not a neutral reproducer. In double-blind listening tests that included 268 listeners, and over 300 by the time the test was terminated, Olive (2003) found that this product (identified as "B"

(a) Manufacturer's frequency response specification: 39 - 20 kHz ± 2 dB

(b) What the numbers mean: the solid lines. The reality: the dashed line.

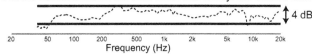

(c) The information that is needed to anticipate sound quality:

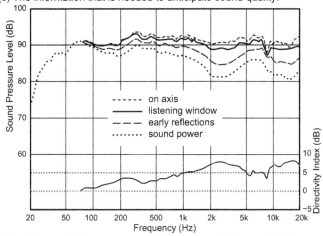

(d) And from which a room curve can be substantially predicted:

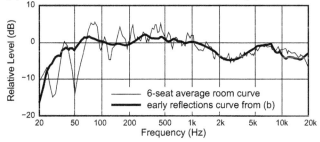

FIGURE 12.3 *The manufacturer's specification and measured data on a B&W 802N.*

in the paper) was rated slightly lower than two loudspeakers that exhibited performance similar to those seen in Figure 12.1. It also appears as "B" in Figure 18.14 in the earlier edition of this book.

Why it sounded as it did cannot be ascertained from the specification. The non-flat on-axis response curve is a clue, but the reasons are not completely known until one looks at the far off-axis performance, seeing directivity fluctuations that affect the reflected sound field. They influence what is heard and dominate the shape of the room curve. Because directivity cannot be changed by equalization, the problem cannot be corrected. The directivity index shows the rising directionality of the twin woofers, crossing over to the large 6-inch (150 mm) midrange driver around 400 Hz that, in turn, becomes increasingly directional until it crosses over to the unbaffled 1-inch (25 mm) tweeter around 4 kHz. The dip in sound power and the rise in directivity around 3 kHz are predictable results of the transducer size disparity. This is behavior consistent with a conventional 6-inch two-way (look ahead to Figure 12.10). This problem can be anticipated by a visual inspection of the loudspeaker. In contrast, Figure 12.1d, e and f show comparable two-way designs that use a shallow waveguide on the tweeter to better match its directivity with that of the woofer at crossover. As Figure 5.12 shows, a well-designed horn also performs well in this respect, delivering very constant directivity in its operational frequency range and closely matching the directivity of the woofer at the 800 Hz crossover.

> When expectations from specifications differ significantly from audible reality it is easy to understand how the old expression "we can't measure what we can hear" came to be. The problem is inadequate measurements, not the supremacy of the human hearing system.

An interesting aside is that this loudspeaker has found its way into some high-profile music recording facilities and scoring stages. Some recording engineers believe it to be especially useful for classical music. The sagging response around 3 kHz is a familiar compensation for excessive brightness in strings—the result of placing microphones above the violins where they project energetic high frequencies that are not heard in that proportion by live audiences (Meyer, 2009). If this is the case, the remedy is to equalize the signal, not to use monitor loudspeakers with a compensating error.

So, why then do we continue with the traditional, uninformative specifications? Familiarity is one argument, often accompanied by statements to the effect that consumers cannot be expected to understand the meaning of graphs. In fact, if they were simply told that flattish and smooth are good indicators of sound quality and that similarity among a family of curves is advantageous, they would be well on their way to being expert diagnosticians. Surely it is better than the uninformative specs that are currently divulged. Only when they are specified to small tolerances do they convey significant meaning, and that is rare. A common tolerance of ±3 dB is not small; ±1.5 dB is. Smoothness is a necessity and uniformity of directivity is a virtue. In the end, curves convey more meaning.

As purchasing of most things is moving to the Internet, it is time for the loudspeaker industry to adopt meaningful specifications so that customers can make intelligent choices in the absence of listening evaluations. After my many years of listening intently to loudspeakers, I have reached the point where I would confidently choose one on the basis of a *truthful* spinorama or other comprehensive anechoic data. Much of the auditory experience is dictated by the room at low frequencies (see Chapter 8), and much of the remainder by what is in the recordings. These are not consistent, so fastidious listeners will still find tone controls to be useful. The starting point, though, should be a loudspeaker with essentially neutral, uncolored sound.

If you don't see useful data in a spec sheet, ask the manufacturer for it. The spinorama shown in Figure 12.3(c) is part of ANSI/CTA-2034A (2015) "Standard Method of Measurement for In-Home Loudspeakers," so this is not a personal or a single manufacturer's opinion. If it or a reasonable facsimile is not published or available on request, draw your own conclusions.

12.2 LOUDSPEAKERS IN SMALL ROOMS: THE MEANING OF ROOM CURVES

Section 5.6 showed persuasive similarities between anechoic measurements on loudspeakers and steady-state measurements in typical listening rooms. Above the transition frequency, quite precise predictions were possible. Figure 10.6a shows that in these situations the direct sound is the dominant factor in measurements only at the highest frequencies. Over most of the frequency range early reflected sounds contribute substantially, and indeed the early-reflections curve from a spinorama provides a good estimate of a steady-state room curve. As the story evolves, it will be clear that the "right" room curve results from a good sounding loudspeaker, but equalizing an inferior loudspeaker to achieve the "right" room curve does not guarantee good sound. Figure 12.4 shows the progression.

The sequence begins with Figure 12.4a showing anechoic data for six loudspeakers that consistently got high ratings in double-blind subjective evaluations. The author has drawn a subjectively judged average curve through the anechoic early-reflections data. The dip around 2 kHz is caused by off-axis directivity effects in the midrange-to-tweeter crossovers. The smoothness of the curves indicates an absence of resonances.

In (b) the predicted room curves in (a) are compared to seven spatially averaged steady-state room curves measured in five different rooms by four different people and measurement systems using the Revel Salon2, F208, M106, the JBL Professional M2 and the Infinity Prelude MTS. The measurements include sample variations, room variations and measuring system variations yet it is evident that the predictions represent a good central tendency for the fluctuating room curves. Look ahead to Figure 13.1 to see that 30 years have not changed listener preferences or the essence of good loudspeaker performance.

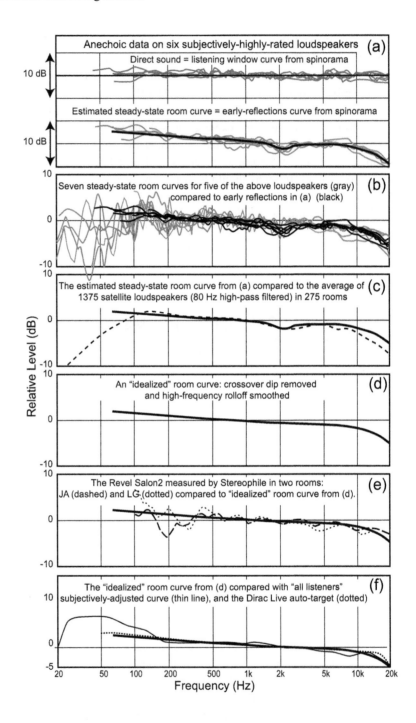

FIGURE 12.4

*Comparisons of anechoic
measurements to steady-
state room curves.
Explanations are in the
illustrations. The six
loudspeakers in (a) are the
four Revel loudspeakers
in Figure 12.1a–d, the
JBL Professional M2 in
Figure 5.12, and the Infinity
Prelude MTS in Figure 5.6.
The room curves in (b)
are for the Revel Salon2,
F208 and M106, Infinity
Prelude MTS and JBL
Professional M2, measured
in five different rooms by
four different people using
different measuring systems.
The survey data in (c) are
from Green and Holman,
2010. (d) shows an idealized
version of the average room
curve prediction shown in
(a). The room curves in
(e) are from* Stereophile
*magazine, March 2009 (vol.
32, no. 3). The subjectively
adjusted room curve in (f)
is from Figure 14 in Toole,
2015b, using data from Olive
et al., 2013. The Dirac Live
default "auto-target" is from
the online manual, 2016.
Discussion of these data are
in the text.*

The residual fluctuations in the steady-state room curves are acoustical interference effects that the microphone responds to, but human listeners largely or completely ignore. We know from the anechoic data that the loudspeakers do not radiate them. These variations cannot be corrected by "room" equalization. Room curves are often

1/3-octave band smoothed, which would eliminate most of the detailed fluctuations seen here, thereby reducing the temptation to equalize them.

When one sees a *very* smooth high-resolution steady-state room curve after equalization, there is a high possibility that something inappropriate has been done, and the sound quality may have been degraded. Yet this is frequently shown in promotions for room equalization/correction products addressing both consumer and professional customers. This is marketing, not science. The room curve is not the definitive statement of sound quality.

Figure 12.4c compares this estimated room curve to data from a very large survey of home-theater satellite loudspeakers (Green and Holman, 2010). The match is excellent: within 1 dB from 100 Hz to about 10 kHz. The crossover dip is evident here as well. These loudspeakers were "as found," without documentation or pedigrees. The rolloff at high frequencies suggests that some of the loudspeakers had rolled-off high frequencies or may have had their tweeters aimed away from the listening/measuring area as in many on/in-wall surround loudspeaker designs.

In Figure 12.4d an attempt is made to arrive at a room curve that might result from a "perfect" loudspeaker; one without a crossover dip and one that generated a smoother high-frequency rolloff. The result is not a perfectly straight line, but a straight line with a slope in the range of −0.4 to −0.5 dB/octave is a good descriptor.

In (e) this curve is compared to room curves measured by John Atkinson of *Stereophile* in his own and another room. The imperfections below about 500 Hz are attributable to the rooms. The small deviations observed in LG's room at high frequencies are puzzling. At these frequencies the direct sound should be a dominant factor. The deviations do not originate in the loudspeaker, so suspicions mount that there might be some acoustical interactions close to the microphone. People are often careless about keeping the microphone away from chairs and even microphone clamps and stands that alter measurements. Overall, though, the comparison confirms that if one begins with a well-designed loudspeaker, the room curves will be close to predictions. Given that anechoic data are definitive, and room curves are subject to variations, the confirmation of excellence is reassuring.

Figure 12.4f compares the idealized curve to one created by listeners with access to bass and treble tone controls, free to adjust them to produce the most pleasing effect. Above about 200 Hz they compare favorably. The bass boost will be discussed in Section 12.3. Also included in this comparison is the default "auto-target" curve for Dirac Live, a well-known room measurement and equalization system. The curves are almost identical above about 200Hz. There was no collusion between the author, Harman and Dirac on the generation of these curves, which makes the similarity impressive.

That said, it is important to remember that the room curves do not unambiguously describe good sounding loudspeakers. The highly rated loudspeakers in (a) and Figure 12.1 generate room curves that look much like (f), but equalizing a flawed loudspeaker to have a room curve that looks like (f) guarantees nothing. The necessary information is in comprehensive anechoic data. Nevertheless a target room curve is a useful guide to establishing a baseline for overall spectral balance when dealing with low-frequency room problems.

12.2.1 The Effect of Loudspeaker Directional Configuration

Examples in this book have emphasized loudspeakers of the most common configuration: forward firing cone/dome or cone/horn. It needs to be asked: What about other designs, like dipoles, bipoles, and omnidirectional? Do the rules change? From what has been learned it seems that a flat direct sound followed by similar sounding reflected sounds (i.e., smoothly changing or constant directivity) produces sound that is highly rated by listeners. There is no stated requirement for any specific directivity, only that it be relatively constant across the frequency range above the transition frequency.

Unfortunately I do not have data to provide a comprehensive answer, but there is a piece of relevant data. As described in Section 7.4.6, the Mirage M1 was a bipole (bidirectional-in-phase) design that was very flat on-axis and, over much of its frequency range, approached horizontal omnidirectionality. In the NRCC double-blind listening tests it was very highly rated. Figure 12.5 shows the essential measurements.

As was found for forward-firing loudspeakers, a flat direct sound and relatively uniform directivity is sufficient to create a steady-state room curve that follows the trend of the idealized room curve. These data show that this conclusion applies also to a bipole loudspeaker measured in two rooms by two people. The downward slope of the room curve is therefore significantly attributable to rising room reflectivity at low frequencies and air attenuation at high frequencies.

As illustrated in Figure 7.4, multidirectional loudspeakers would experience more and different interaction with rooms over a wider bandwidth than conventional

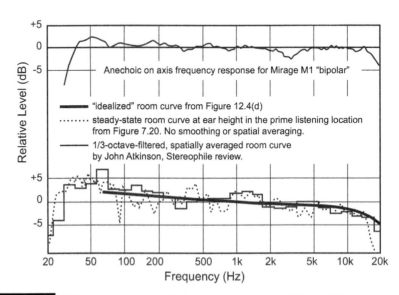

FIGURE 12.5 *1/20-octave anechoic and in-room measurements on a Mirage M1 loudspeaker. The steady-state room curve was measured at the head location in the stereo seat in my listening room. The 1/3-octave spatially averaged room curve is from the* Stereophile *review, June 1989. It is compared to the idealized room curve from Figure 12.4d. Other data are from Figure 7.20.*

forward-firing designs thereby incurring greater variations, especially in the middle- to upper-frequency range. All of these configurations have yielded good sounding loudspeakers over the years, but, in my experience, the best sounding ones are the ones that had the best looking set of anechoic curves. Two ears and a brain manage to perceptually separate loudspeakers from rooms, just as they do multidirectional musical instruments from performance spaces.

12.2.2 Looking Back 42 Years: the Møller/Brüel and Kjaer Experiments

Over the years a few investigators have attempted to identify advantageous room curve targets for small rooms. The studies that the author is aware of have been compromised by a lack of adequate loudspeaker measurements and/or information about the room acoustics. A few of the older studies employed loudspeakers that would not be acceptable by today's standards, thereby biasing the results. Listening tests, if any, were often not well controlled. However, those conducted by Møller (1974), in collaboration with Brüel and Kjaer, the respected Danish manufacturer of precision acoustical measurement apparatus, were different. The erratic room curves that are shown agree with my recollections of loudspeakers of that period.

However, Møller was able to generate a target curve representing a crude average. Even now, this curve appears in discussions of what might be useful target curves for room equalization.

As can be seen in Figure 12.6, it is similar to the idealized curve developed here, except that it falls away more rapidly at high frequencies. Looking at the data for the subjectively highest rated product in his tests (H1) as measured in the three rooms, the curves fluctuate around both of the target curves but at the high frequencies they trend

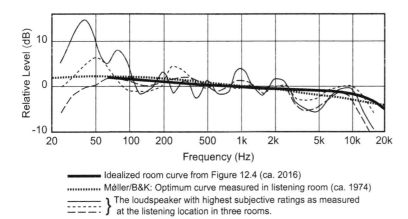

——— Idealized room curve from Figure 12.4 (ca. 2016)
··········· Møller/B&K: Optimum curve measured in listening room (ca. 1974)
——— } The loudspeaker with highest subjective ratings as measured
– – –: } at the listening location in three rooms.

FIGURE 12.6 *Data from a 1974 study of room curves compared to current data. Measurements were made with 1/3-octave fixed-frequency filters responding to pink noise and curves were hand-drawn through the histogram data.*

closer to the 2016 target than to the 1974 target. It would seem that loudspeaker technology has improved, but listener tastes have not changed: a downward tilted steady state room curve is an indicator of a good sounding loudspeaker.

12.2.3 Room Curves and Equalization

Data accumulated thus far indicate that listeners award the highest ratings to loudspeakers radiating flattish direct sound (on axis and/or listening window). No additional information from room curves is necessary. Olive's listeners (Section 5.7.3) were clearly attending to the direct sound in forming their impressions of timbre. Repeating the quote from Queen (1973):

> The results thus far tend to confirm the presupposition that spectrum identification depends principally upon the direct sound of the sound source. It would suggest that any equalization for the reverberant field [steady-state] which is detrimental to the desired response in the direct field will reduce the naturalness of the system or will be detrimental to any other desired characteristic designed for. It suggests that most necessary equalization for speech reinforcement systems may be accomplished by proper design, choice, and fixed equalization of the transducers.

This means that if the loudspeaker is properly designed, there is little need to look at room curves except at low frequencies.

So, what does it mean when a room curve closely matches the idealized target in Figure 12.4d? It means that the customer has probably selected excellent loudspeakers, and that, above the transition frequency of 200–400 Hz, no equalization is required; leave it alone. Deal with the low-frequency problems as discussed in Chapters 8 and 9, and enjoy high-quality listening experiences. Depending on variations in the program, periodic tone-control adjustments may be needed to satisfy fussy listeners.

If the measured curve deviates from the target, does this mean that applying equalization to make it match the target will ensure satisfaction? Unfortunately not. This is the basis for the user-friendly adjustable target curve features on "room correction" devices. Many customers do not have state-of-the-art loudspeakers, so curves deviate from the ideal. Changes are then made to the target curve until the sound is more satisfactory.

Various explanations are offered, but in general they are well summarized by this instruction from one room correction provider: *Installers can customize the target curves and finely tune the target sound to resolve the acoustical problems of each individual room.* The manual for another room correction algorithm states: *Care should be taken to create a target curve that works well with your speakers and room, and suits your personal preferences.* These are open admissions that the room curve is not a reliable indicator of audible performance—so why is it being measured? In spite of mathematical manipulations, and possible allusions to "phase" and "transients," in the end they admit that there is no universally applicable equalization target. So, they provide convenient user interface tools to modify curves and the best curve is the one that sounds best.

When equalizer settings are determined subjectively they will include peculiarities of the programs being auditioned (the circle of confusion), and may require changes if the program changes. There can be no certainty about what is "right," and who or what is at fault if subsequently something does not sound "right." Traditional tone controls remain useful devices for fussy listeners.

This need to search for the "right-sounding" room curve is a farce. The installer and customer are involved in a trial-and-error search for a curve that seems to improve a situation that was most likely caused by an unfortunate choice of loudspeakers. Electronic equalization does not change the room; it is a physical object. If genuine room acoustical problems exist, they will need to be corrected using genuine room acoustical treatments, not equalization.

As discussed in Chapters 8 and 9, equalization is a component in schemes for treating small room bass problems, but over most of the frequency range audible problems are more likely to be associated with the loudspeaker than the room. If equalization is to be involved, it will be most effective if applied to the loudspeaker itself, as illustrated clearly in Figure 4.13. This is not the normal strategy.

Loudspeaker directivity is one common failing in loudspeakers that alters how the sound is communicated through the room, but it cannot be corrected by equalization, as can be seen in the example in Figure 12.3. An improvement might be possible, but a complete remedy is improbable. If the problem with the loudspeaker is resonances, the algorithm may or may not be able to detect and correct them, as illustrated in Figure 4.13. Because in-room measurements include acoustical interference artifacts that may be automatically and inappropriately "corrected," there is a very real possibility that a good loudspeaker may be degraded.

In summary, at frequencies above the transition region, a room curve that is close to that shown in Figure 12.4d can offer reassurance that excellent loudspeakers have been purchased. This fact would have been easily recognized if complete anechoic data had been available on the loudspeakers. If the measured curve significantly deviates from the target curve, there is no certainty that equalizing the system to better match the target will improve the sound quality. If the loudspeakers have relatively constant or smoothly changing directivity, and the problem is a poor frequency response, there is a chance of improvement. As shown in Figure 4.12, above the transition frequency the most reliable basis for equalization is anechoic data on the loudspeakers.

If one is fortunate enough to start with good loudspeakers, only one problem remains, and that is shared by all: bass in small rooms. This is one problem where equalization can be a useful component of the solution, as discussed in Chapter 8.

12.3 SUBJECTIVE PREFERENCES FOR SOUND SPECTRA IN LISTENING ROOMS

Comparing loudspeakers in double-blind subjective evaluations has provided us with clear indications of what constitutes a good-sounding loudspeaker. Chapter 5 describes it thoroughly. However, all listening includes characteristics of the program and the playback apparatus and listening room, so there is always room for "adjustments."

In loudspeaker design, this is the final "voicing" process where, ideally, one begins with a system designed to be as free from colorations as possible. Then small adjustments to the spectrum are made in hopes that it will flatter the inherent spectral trends in the recordings that might be favored by the imagined customers for the product. It needs to be mentioned that this subjective process includes a room, in which bass problems may or may not have been addressed. This, again, is the "circle of confusion" at work. Over the years, using a wide variety of recorded material, it has been found that loudspeakers with a flattish on-axis frequency response have been favored by most listeners in the double-blind, multiple-comparison tests.

But what would happen if listeners were given tone controls to adjust the spectral balance to something they might individually prefer?

Research by Olive et al. (2013) was distinctive in that the loudspeaker used was anechoically characterized, the room described and steady-state room curves measured. In the double-blind tests, listeners made bass and treble balance adjustments to a loudspeaker that had been equalized to a flat, smooth room curve. The loudspeaker had previously received high ratings in independent double-blind comparison tests, without equalization. Three tests were done, with the bass or treble adjusted separately with the other parameter randomly fixed, and a test in which both controls were available, starting from random settings. It was a classic method-of-adjustment experiment. For each program selection, listeners made adjustments to yield the most preferred result.

Figure 12.7 shows the result of evaluations by a mixture of trained and untrained listeners. This is compared to the idealized room curve from Figure 12.4. The "all listeners" average curve is close to the predicted target, except at low frequencies where it is apparent that the preferences of inexperienced listeners significantly elevated the curve. In fact, the target variations at both ends of the spectrum are substantial, with untrained listeners simply choosing "more of everything." An unanswered question is whether this was related to overall loudness or a spectral balance preference—more research is needed. However, many of us have seen evidence of such listener preferences in the "as found" tone control settings in numerous rental and loaner cars.

More data would be enlightening, but this amount is sufficient to indicate that a single target curve is not likely to satisfy all listeners all the time. Because these preferred curves were subjectively determined, they include the "circle of confusion" issues. Just as they changed with listener experience and training, they are likely to change with different programs. It is proof that we need easily accessible bass, treble and/or tilt tone controls in playback equipment. The first task for such controls would be to allow users to optimize the spectral balance of their loudspeakers in their rooms, and, on an ongoing basis, to compensate for spectral imbalances as they appear in movies and music. This is not a new concept. In the 1980s, Peter Walker of Quad introduced a tilt control (linked bass and treble controls) in the Model 34 preamplifier. The Lexicon MC-12 digital implementation, ca. 2000, a very linear tilt, combined with tone controls, added to my own listening pleasure. There may be others I'm not aware of.

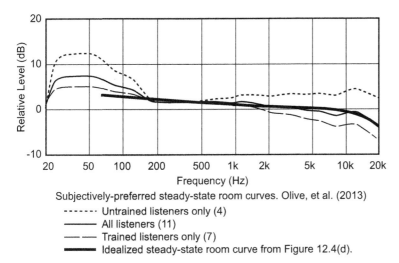

Subjectively-preferred steady-state room curves. Olive, et al. (2013)

------ Untrained listeners only (4)
——— All listeners (11)
— — Trained listeners only (7)
▬▬▬ Idealized steady-state room curve from Figure 12.4(d).

FIGURE 12.7   *Subjectively preferred steady-state room curve targets in a typical domestic listening room (described in Olive, 2009), from Olive et al. (2013). The prediction of Figure 12.1 is superimposed.*

The good thing about such spectral trends is that humans seem to adapt to them unless they are extreme, or suddenly change, as when listening to a varied musical program, or as happens in multiple-comparison blind listening tests. This makes it especially interesting that the same experienced listeners who here deliberately tilted the high frequencies down in these method-of-adjustment tests, voted for loudspeakers close to the predicted curve in the multiple-comparison tests. Is this a consequence of the different experimental methods or program material? There is an opportunity for more research.

The significant observation is that all of the results indicate a preference for a steady-state spectrum that rises toward the low frequencies. In previous experiments, Olive et al. (2008) compared five different "room correction products, finding that the most preferred product followed a curve similar to the "all listeners" curve shown here—those having less bass were not rated as highly.

Why are listeners, even trained listeners as in Figure 12.7, so attracted to boosted bass? There is a possible explanation. Movies and music are commonly mixed while listening at very high sound levels—much higher than is desirable, or sometimes even can be accomplished, in home playback. If the playback sound level is lower, the equal-loudness contours (Section 4.4) explain that the low bass is disproportionately reduced in perceived loudness. Some of the lowest frequencies vanish below the hearing threshold. Boosting the bass restores some of it—see Section 4.4.1, where Figure 4.6e explains what might be happening.

12.4 DIALOG INTELLIGIBILITY IN HOME THEATERS

Nobody disputes the importance of dialog in movies or television. But movies are distinctive in that they employ multiple channels to deliver music, atmospheric sounds and sound effects: surround and immersive sound. That means that human binaural hearing is needed to separate a voice from one direction from music or sound effects arriving from other directions. Two normally functioning ears and brain do a remarkable job of adding clarity, but unfortunately age and hearing loss degrade the function, as some of us realize in noisy restaurants. Some of the problem is simple signal-to-noise ratio, and some of it is degraded binaural discrimination (Chapter 17).

To achieve high scores in speech intelligibility, a signal-to-noise ratio of 5 dB is good, and 15–20 dB nearly perfect. Noise, in this context, is everything other than the speech we want to hear. When several people talk at the same time, the noise is speech itself. In music it is the sound of the band competing with a vocalist. In movies it is everything else in a soundtrack occurring at the same time as the dialogue. For long passages in films and television programming, this is atmosphere-inducing music. When the action starts, all caution is abandoned and things can get very raucous.

The biggest problems are intrinsic to the programs themselves. Obviously, the sound mixers pay attention to this, but they have advantages over the rest of us. They get to hear each section of a film many times as they develop the audio design. They may know the dialogue before they even hear it and understand it even if they don't pay attention.

The experience of a great many consumers is the ability to rewind and play, and the options of having subtitles or closed captions, exist for a reason. The reason has nothing to do with inferior loudspeakers or room acoustics. As we get older, we expect to have occasional difficulties, but one hears the same complaints from people who have no hearing problems.

Why is it that, within the same room and playback system, one can switch from highly intelligible "talking head" television programs to watch a movie, and dialogue is not always perfectly understood? Part of it is the dramatic effect of mumbles and whispers; movies have a significant dynamic range, whereas TV news and documentaries are highly compressed—always loud. Part of it is the inability to pick up clues from the lips, facial expressions and so on when the talker is not facing the camera. Part of it is the mixture of music and sound effects emerging from all of the loudspeakers in the room, creating atmosphere and supporting on-screen action. The result is the same; more "noise," less intelligible speech. These extraneous, artistically and aesthetically justified sounds create problems.

Shirley and Kendrick (2004) investigated the effects of differing amounts of "extra" sound on listener impressions of dialogue clarity, overall sound quality and enjoyment. Clarity is a quality that corresponds to intelligibility, although it is not a direct quantitative measure of that parameter. Some of the listeners had measurably normal hearing, while others had varying degrees of impairment.

The test conditions involved only the three front channels, L, C and R, replaying those components of twenty 1- to 1.5-minute 5.1-channel film clips. The variable was the level of extraneous sounds delivered by the L and R loudspeakers, while the center loudspeaker delivered a constant level of dialogue. The first condition ran all three channels at reference level. The second condition attenuated the L&R channels by 3 dB, the third condition attenuated them by 6 dB, and the last condition ran the center channel alone.

The results shown in Figure 12.8 are interesting from several points of view; moving from left to right in the sequence of presentation styles results in a progressive increase in the dominance of the center channel. In terms of dialogue clarity, obviously *all* listeners thought this was a good idea, those with normal hearing and those with impaired hearing. This confirms that the structure of the soundtrack is a major factor in dialogue clarity and intelligibility (and remember, the surround channels are not involved in this test). The hearing impaired listeners never got to the high levels of "clarity" reported by the normal-hearing group, but they could appreciate the improvement achieved by attenuating the L and R channels. The lesson: if you want to experience clear speech, listen in mono using the center channel—turn all other channels off.

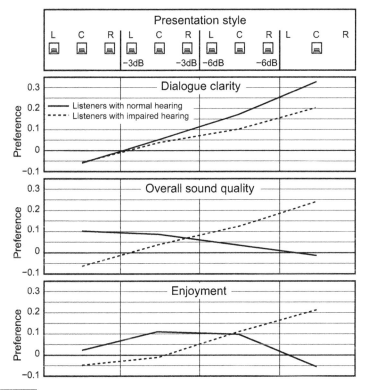

FIGURE 12.8 *The subjective preferences for various LCR front-soundstage presentations judged in the categories of "dialogue clarity," "overall sound quality" and "enjoyment." From left to right, the presentation styles progressively emphasize the center channel, ending with it in isolation. Listener responses were divided into two groups, listeners having normal hearing performance, and those having hearing impairments. Compiled from data in Shirley and Kendrick (2004).*

In terms of "overall sound quality," there is some disagreement between the groups. Those with normal hearing preferred all of the channels running at or close to reference levels; anything else is a degradation. Note that the difference between all three channels running at reference level, and the L and R loudspeakers attenuated by 3 dB, is trivial (and statistically insignificant). In total contrast, listeners with hearing disabilities, having difficulty understanding speech, clearly voted for progressively more monophonic sound. They obviously associated speech intelligibility with sound quality.

When we come to ratings of "enjoyment" the hearing impaired told us that if the dialogue is not clear, then the movie is not enjoyable. Those listeners with normal hearing had trouble with this category, because variations in the ratings were high, and differences between the averaged ratings shown here are not statistically reliable. To the extent that they may have meaning, it is interesting that these listeners entered *any* votes for turning the L&R channels down. But, they did.

Summarizing, listeners with normal hearing find themselves conflicted. In terms of "dialogue clarity" things improved as the L&R channels were progressively attenuated, and even turned off. However, in terms of "overall sound quality," 3 dB attenuation of the L&R channels was possibly acceptable, but more than that was rejected. Overall, these normal hearing listeners seemed to be saying that they could live with a system in which the L&R channels were running 3 dB below reference levels. More than that was better for dialogue, but worse from other perspectives. Listeners with impaired hearing were utterly predictable. Anything less than mono was degradation. This does not mean that they dislike the sound of a multichannel presentation; it means that they place a higher priority on the clarity of dialogue.

These data do not paint an attractive picture for multichannel audio for those among us with deteriorated hearing. The good news is that other tests done by these authors indicate an improvement in speech clarity when the person speaking is facing the camera; consciously or subconsciously, we read lips—all of us. Perhaps changes in cinematic technique can compensate somewhat for the reduced speech clarity caused by the invocation of distracting sounds in other channels. Otherwise it would seem that multichannel audio is a medium best suited for normal, usually younger ears. It is interesting to note that in their population of 41 subjects, with ages ranging beyond 75, those exhibiting hearing impairment began showing up in the 30–44 age group, with progressive deterioration at more advanced ages. The message here is that this is a factor to consider in all home theater installations. The challenge is to decide what to do about it.

This is not news to the motion picture industry. I. Allen (2006) mentions examples of dialog intelligibility problems traceable to both the director of the film and the operators of the theaters; if film sound tracks are too loud, they get turned down. Dialogue levels drop along with those of the special effects. There have also been conflicts caused by differences in the listening circumstances—dubbing stages and screening theaters.

In households having mixed populations of younger and older listeners, a convenient way to adjust the center-channel level balance would be an excellent feature. In the meantime, there are subtitles and closed captions, the reason many prefer watching at home.

12.5 RECORDING CONTROL ROOMS

Wherever the microphones respond to sounds, concert halls, jazz clubs or recording studios, those sounds are eventually delivered to a control room where they are signal processed and artistically crafted to satisfy musicians and their imagined target audience. Nowadays the control room can also be a performance space. Of course, sampled and synthesized sounds are common, so the only acoustical involvement is in the playback monitoring. In the end, what is heard through the monitor loudspeakers is powerfully influential on what ends up in the recordings.

Section 1.4 introduced the concept of the "circle of confusion," explaining the consequences of mismatched sound spectra in the control room and in the ultimate reproduction environment. It also has ramifications within the recording industry itself, as implied by Figure 5.17, where it shows that at a point the recorded master is handed over to a mastering engineer who attempts to configure it to be suitable for delivery through the chosen media, and to be acceptable to the maximum number of listeners. Bob Katz is one of these and he says: "Most times I can tell an [Yamaha] NS-10/near-field mix when it arrives for mastering" (Katz, 2002, p. 79). It contains telltale timbral cues introduced in compensation for the spectral errors of that loudspeaker. Katz goes into some detail explaining why for his purposes accurate, neutral, monitors are a fundamental requirement. However, the final mixes are auditioned through other loudspeakers to confirm that the program artistically "translates."

Monitor loudspeakers should be as neutral as possible, just as many consumer products clearly aspire to be. This is the way that the circle of confusion can be broken. However, professional loudspeakers are required to do things that consumer products may not have to: to play *very* loud, for *very* long periods of time, and not break. These products allow studios to earn money, and "dead air" is not an option.

TRANSLATION

Translation is the term used to describe the successful conveyance of the art to consumers, whoever and wherever they may be. It is used in all domains of sound reproduction from music to movies, and it is an apt description. However, as we know from linguistic translations, the original meaning is often changed, at least subtly and sometimes greatly. In sound reproduction the changes can take countless forms, and they are inevitable, so the goal is to try to anticipate what can go wrong and still maximize listener satisfaction. Frequently, even generous tolerances in translation are exceeded.

In an ideal world, the test would not be "translation" but "reproduction"—a perfect copy—but that is an impossible task, because sound reproducing systems take on so many forms in so many venues, including headphones. Therefore translation, a less demanding criterion, is applied, sometimes being reduced to conveying the most basic elements of the original art, or of the creator's "vision," whatever that may be.

Obviously the closer we can get to "reproduction," the better things can be. It may seem impossible, but as the following discussion shows, there are ways to simplify the process and still deliver an artistic product that can have wide appeal. Modern technology is making the job easier, in headphones and with loudspeakers in rooms.

There are three categories of studio monitor loudspeakers:

- Main monitors—large, usually in-wall-installed, powerful full-bandwidth systems capable of very high sound levels.

- Mid-field monitors—medium-sized loudspeakers that may be full bandwidth, or may use subwoofers, located a moderate distance in front of the console, positioned to minimize reflections from the working surface.

- Near-field monitors—small loudspeakers placed on the meter bridge of the recording console. The reflection from the working surface is part of the sound heard from these loudspeakers, and their locations may cause them to interfere with what is heard from main or mid-field monitors. Listeners are in the acoustical near field of the source (Section 10.5), meaning that small changes in head location cause changes in the sounds arriving at the ears.

In general, these loudspeakers are designed to have relatively high sensitivity, so less amplifier power is needed, meaning that the transducers run cooler and may last longer. On the more distant loudspeakers some directional control is desirable to deliver sound to the listening area, not the room boundaries. Horns are common.

Some mid- and near-field monitors are found in both the consumer and professional worlds. In general, though, professional loudspeakers lack cosmetic appeal, being mostly rectangular and black. There is one more point of differentiation: they may be active, with on-board or dedicated outboard power amplifiers and digital processing.

Some active units can be incorporated into a digital network making setup level calibration, multichannel configuration, subwoofer management and equalization uncommonly good and convenient.

Those professional loudspeakers with dedicated electronics have a huge advantage over passive loudspeakers. Consumers in general, especially high-end audiophiles, have not caught up with the advantages that technology has to offer. Good loudspeakers and amplifiers can deliver good sound, but merging them with dedicated digital crossovers, equalizers and amplifiers designed for those specific loudspeaker components, in that specific enclosure, can yield even better sound.

Traditional recording control rooms and studios are elaborately constructed and equipped spaces that musical groups rent at great expense. They exist for good reasons, but nowadays, numerous recordings are made in converted bedrooms and garages. The multichannel mixing and signal processing can now be done on site using powerful digital processing algorithms instead of massive consoles. The acoustics of small ordinary rooms has become a factor in the music industry so everything that has been learned from home listening rooms and home theaters is relevant to these small studio/control-room venues.

12.5.1 Old-School Monitoring

Although there is the potential for sound reproduction at its highest level in control rooms, one still sees examples of loudspeakers known not to be very good in the audiophile context. The rationale is that because so many consumers listen through "imperfect" loudspeakers, at least some of the monitoring needs to be done through "imperfect" speakers to ensure that the mix "translates" to the real world. The melody, rhythm, lyrics and basic harmonies may be communicated, but little else can be guaranteed. The reason is that there are so many ways to be "imperfect," and the audio industry has done a good job of exploring all possibilities. The expansion of audio into small wireless portable speakers and the popularity of headphones and earbuds of all descriptions add significant complication.

A benchmark in "imperfect" loudspeakers was the Auratone 5C. It consisted of a single full-range 5-inch (127 mm) driver of a kind popular in 1960s radios and television sets, mounted in a small closed box. Figure 12.9a shows on- and off-axis frequency responses of this tiny loudspeaker. Also shown is a similar set of data on a much larger and more expensive loudspeaker that shares many timbral and directivity characteristics, albeit with more bass and more power. Both of these loudspeakers were popular in their times, and some studios still advertise that they are in their equipment repertoire. In terms of sound quality, they were both highly colored; the UREI earned scores around 5.5/10 in double-blind listening tests conducted using professional recording engineers and producers around 1984 (loudspeaker C in figure 6, Toole, 1985).

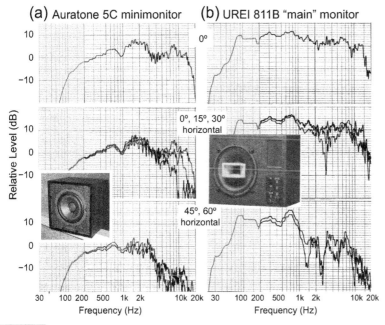

FIGURE 12.9 *Anechoic measurements on two loudspeakers that became popular in spite of, or because of, their measurable and audible imperfections. The measurements were at 2 m and are progressively less accurate below about 200 Hz because the chamber was not calibrated at that time. From Toole (2008), figure 2.6.*

The UREI 811B and other similarly bad off-axis performers of that period were reason to employ massive absorption on surfaces responsible for early reflections (see Section 18.3). Figure 7.5 shows the optional acoustical treatments, but when the sounds radiated toward the side walls, floor and ceiling resemble the 45° and 60° curves in Figure 12.9b, eliminating them is the best choice. The "truckload of fiberglass" dead-end control room designs were inevitable. Some went even further, damping the entire room, putting listeners in a dominant direct sound field over most of the frequency range. However, diminishing the influence of the room can be done much less expensively with small loudspeakers perched on the meter bridge of consoles or, even better, in mid-field locations just behind the console.

As near-field monitoring became more widespread, the search for new "sounds" from the small loudspeakers expanded. One of the loudspeakers that became popular was the Yamaha NS-10M. The original product was aimed at the consumer market and was designed to be listened to at a distance, in a normally reflective domestic room, and to be placed close to a wall, thereby reinforcing the bass. The designer (also of the NS-1000) visited me at the NRCC, to discuss our research and see our facilities. Unfortunately he designed for a flat radiated sound power and Figure 12.10a shows that he succeeded very well. Consequently, the on-axis frequency response suffered considerably because of the inherent directivity of the two-way design. The woofer becomes progressively more directional as frequency rises, then the system directivity falls above crossover to the small tweeter that, in its turn, becomes more directional. The opposite situation applies to the JBL 4301 in Figure 12.10b. This was designed for close listening and had a flat on-axis response. However, predictably, the sound power is uneven because of the DI. In (b) both DIs are superimposed, showing that this is a parameter dictated by the similar woofer and tweeter sizes chosen by these designers. Using these anechoic data each of these loudspeakers could be equalized to sound very much like

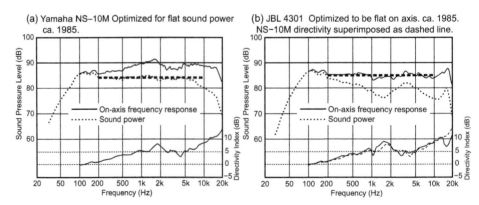

FIGURE 12.10 *Two small 7–8 inch two-way loudspeakers with similar directivity indexes, each one optimized to a different target. (a) the Yamaha NS-10M was designed to have flat sound power that was (erroneously) assumed to be desirable for distance listening in domestic rooms. (b) the JBL 4301 was designed to have flat on-axis frequency response that was (correctly) desirable for near-field listening.*

the other in near-field/direct-sound-dominant listening. The NS-10M was admired by some because of its "tight" bass, a consequence of the deficiency of bass fundamental frequencies. Woofers in their enclosures function as minimum-phase systems so the amplitude response defines the time response and equalization can change both—up to sound levels where reflex ports become non-linear.

The interesting part of the story is that both found use as near-field monitors, but only one was designed for that purpose, the JBL. When listeners complained about the NS-10 sounding too bright, it became fashionable to hang facial tissue in front of the tweeter. Looking at the on-axis response in (a) it is evident that the spectrum from about 1 kHz upwards is significantly higher than lower frequencies. Of course it sounded bright; it was the wrong loudspeaker for the job.

While recording engineers debated what brand of tissue was most effective, Yamaha took the product back to the lab and modified it, calling it NS-10M Pro. This new design is shown in Figure 12.11. In (a) it is seen that the woofer level was raised to match that of the tweeter at high frequencies, but the mid-frequencies were left as a prominent hump. But this is something we have seen before. In (b) measurements of three samples of Auratones are superimposed on the on-axis curve from (a). In near-field listening the new NS-10M Pro will have the basic spectral signature of the Auratone, but with more extended bass. It is also evident that the consistency of the low-cost driver in the Auratone was a problem.

This distorted spectrum shape was imitated by other manufacturers of small monitors. Even now it is a switchable option in some active monitor loudspeakers, amusingly called "old school" (www.barefootsound.com). However, the NS-10M is history, and it is significant that Yamaha has joined the "new school" of flat and neutral on-axis sound as shown in Figure 12.11c. The target is now marketed as "exceptionally flat response across the sound spectrum." Retaining a white woofer honors the legacy of the NS-10.

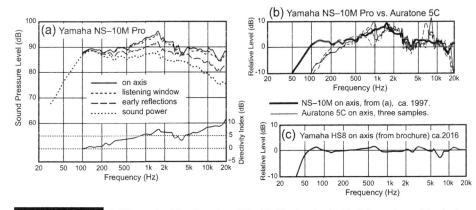

FIGURE 12.11 *(a) The revised "pro" version of the NS-10, showing that in this version nothing is close to being flat. (b) shows that the direct sound, important in near-field monitoring, now closely resembles that of the Auratone 5C, but with extended bass. (c) shows the published on-axis response of a current model.*

12.5.2 Modern Monitoring

Recording engineers still need to evaluate their mixes on "imperfect" loudspeakers to hear how they translate, so what can be done? My own perspective is colored by having evaluated, by measurements and listening tests, hundreds of consumer loudspeakers of all prices. This started in the 1960s (Figure 18.3) and by the time I formalized and published my findings in 1985–1986 (Figure 5.2), it was very clear that loudspeaker designers had a very simple performance objective: flat on-axis response. These were popular consumer products.

Not all loudspeakers were this good, but the variations were largely random—the designs failed in many different ways. In other words, "bad" sound cannot be standardized, as was attempted with the Auratone and continued in the NS-10M and others.

In 2000 Harman engineers benchmarked a collection of popular entry-level consumer mini-systems. They incorporated cassette and/or disc players, AM/FM tuners, amplifiers and loudspeakers in plastic cabinets, and ranged in price from $150 to $400. Figure 12.12 shows the results. All the systems had problems, but they were different, and when averaged the resulting unsmoothed curve was within ±3 dB tolerance from about 80 Hz to 20 kHz. This suggests that a flat on-axis target performance for control room monitors would please a high percentage of entry-level consumers.

The most common features of small, low-cost loudspeakers are (1) a lack of bass and (2) an inability to play loud. The obvious solution for modern monitoring, therefore, is to start with the most accurate, most neutral, widest bandwidth loudspeakers that can be afforded, and introduce a high-pass filter into the signal path and turn the volume down. That way loudspeakers of any price or size can be simulated—not any one in particular, but the "average" loudspeaker that might be heard by consumers. Another advantage is that the control room can be cleared of some amount of acoustical clutter. Figure 12.13 illustrates the principle. Obviously this function can be implemented with already existing controls with the added possibility of imitating "boomy" loudspeakers.

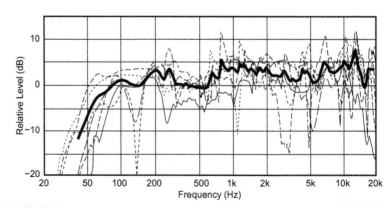

FIGURE 12.12 *On-axis frequency responses for six entry-level mini-systems, with the average shown as a heavy curve. From Toole (2008), figure 2.5.*

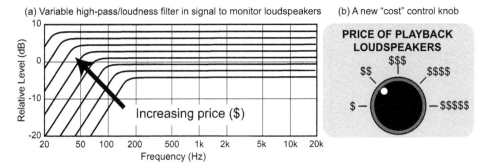

FIGURE 12.13 *A stylized view of how high-quality monitor loudspeakers might be modified to simulate the sound of average loudspeakers of various prices and sizes.*

Nowadays there are numerous high performing monitor loudspeakers in the marketplace. Figure 5.12 is one of them, but there are smaller, less expensive products that perform at a level that renders them effectively neutral reproducers. Basic neutrality in loudspeakers can be recognized in accurate and comprehensive anechoic data. When loudspeakers with similar-looking spinoramas are compared in double-blind listening tests, they sound very similar. It is common for the best loudspeakers to end up in a statistical tie in which the residual variable is the program material. Audio professionals and consumers both have the same problem: reliable anechoic measurements that can identify such loudspeakers are hard to find.

In Figure 12.14 loudspeakers (a) and (b) sound *very* different from each other, and not very good compared to a lot of consumer and professional loudspeakers in the marketplace. If one is looking for "personality" in a monitor loudspeaker, these are some choices. I have seen these in use in "notable" facilities and I can only wonder how the engineers manage to work with such timbrally distorted sounds. In one facility there were discussions of the importance of ultra-wide bandwidth, which was incongruous because their monitors didn't even make it to 20 kHz. The lack of measurements, or of a trust in measurements, is a serious handicap.

In contrast, loudspeakers (c) and (d) sound fundamentally similar. They could be described as "neutral," like the consumer loudspeakers in Figure 12.1. How do users find such loudspeakers? I show Harman products because the data are available to me, and for some of the monitor loudspeakers, the spinoramas are on the website. In a review of several manufacturers' websites the technical descriptions follow the uninformative traditions of the consumer loudspeaker market illustrated in Figure 12.3. Genelec is one notable exception, and although their anechoic data are not in spinorama format, there is enough of it to make an informed decision. One might argue that revealing such information should be a requirement of catering to a professional market.

But, one keeps hearing comments along the theme: "neutral is boring; I want monitors that excite (or pick a word) me." The visual parallel to this is "clear glass is boring; I want my views to be tinted." The problem with both is that the coloration, acoustical or visual, is fixed—it affects everything in the same way. Surely that too becomes

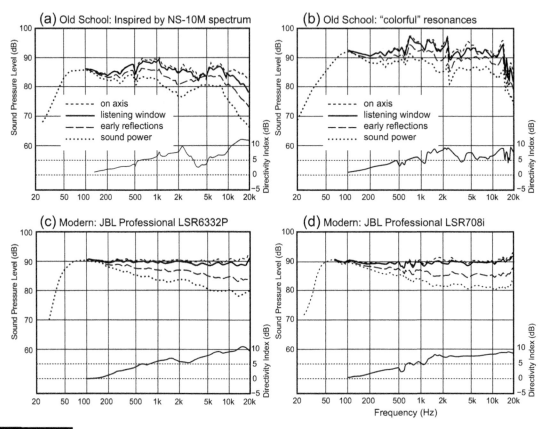

FIGURE 12.14 *Examples of old-school (a) and (b) and modern (c) and (d) monitor loudspeakers. (c) is a cone/dome design, and (d) a cone/horn design as can be discerned from the directivity indexes.*

boring. Whether one is a sound recording engineer judging the product through monitor loudspeakers, or a visual artist creating images behind glass, the artist is in control of the art. If it is boring, then add excitement with the numerous timbral and spatial algorithms available to mixers these days. That way the "excitement" is incorporated into the recording and has a chance of being communicated to the customers. If accentuating a part of the spectrum provides audible insights during the mix, do so with an equalizer, not a different loudspeaker.

Idiosyncratic monitor loudspeakers should not become part of the art, because then only listeners having the identical loudspeakers can appreciate the art. Audio professionals must fight the tendency to select loudspeakers that flatter their own recordings. That simply perpetuates whatever flaws were in earlier generation monitors. That may result in comfortable familiarity, but it impedes progress. Besides, I, as one of an increasing number of customers with relatively neutral loudspeakers, want to hear the art, unmodified.

The most perfect, most neutral, loudspeakers cannot sound good with all program material because all program material was not "born" equal. A study by Mäkivirta and Anet (2001) measured identical Genelec monitor loudspeakers in many recording control rooms. The variations in the room curves, especially at low frequencies, were substantial—more than enough to motivate recording engineers to adjust the timbral balance of recordings. These people at least started with potentially good sounding loudspeakers. However, it is interesting as an aside to note that the target steady-state room curve was flat. This is consistent with some recording/broadcast industry standards and, like the low-frequency portion of the X-curve, it runs counter to normal listening experience, and listener preferences when they are tested—see Section 13.2.2. It is not a virtue.

CHAPTER 13

A Rational Approach to Designing, Measuring and Calibrating Sound Reproducing Systems

Nothing about the detailed acoustics of the spaces we live and listen in is constant, yet humans cope remarkably well, adapting to the varying sound fields while maintaining a stable perceptual impression of the sound sources. Changing the venue does not alter the inherent timbral character of voices and musical instruments. They are merely those voices and instruments in different venues. In terms of sound quality, we consider almost any live performance to be a "reference" listening experience. This is normal—natural acoustics at work. Putting it in audio terms, the "room equalization" happens between the ears, cognitively and subconsciously.

We accept the varying bass boosts resulting from cumulative reflected sound as part of the character of the rooms, not so much as a description of the sound source. If a symphony orchestra sounds "thin," lacking in bass, it is blamed on the hall, not the instruments or musicians. If a male voice sounds "rich and sonorous," it is likely to be attributed to the reflective character of the space within which the sound is communicated—for example, singing in the shower. That works well in real-time situations, but if sounds with these qualities were captured, stored and reproduced in our listening spaces we might object, and be motivated to reach for a tone control or equalizer if one is available. This is precisely what recording engineers do to restore what they perceive as a preferable balance while listening through their monitors in their control rooms. So, the circle of confusion is always a factor in recorded sounds.

The recognition of the character of the sound source as distinct from that of the room is well described in Figure 5.16, and it applies to recordings of musical instruments and to the live events. The separation, the perceptual streaming, of the sound source and the room, is not perfect though. Adaptation to anything, including listening circumstances, has limits, and one of the requirements of this discussion and any measurement system is to identify the key variables and define the limits within which human listeners accommodate variations, and where intervention is required.

Generalized notions that rooms are fundamental problems to be eliminated are faulty, but the exaggeration is often used to promote sales of consulting services and acoustical materials. No doubt we need some amount of these to achieve the best possible sound, but used selectively, in moderation, and targeted to specific issues.

For the mass market, it appears that there is a generous happy middle ground, which is delivered by a lot of normally furnished domestic rooms. The key is to start with good loudspeakers—something not easy to identify in this culture (Section 12.1). If that is accomplished, refining the acoustical performance of normal listening spaces or designing a custom one from a clean sheet of paper is not difficult—except for the bass frequencies.

13.1 LOW FREQUENCIES—
THE UNIVERSAL PROBLEM

Chapters 8 and 9 of this book were devoted to audible variations in the quantity and quality of bass sounds in rooms caused by resonances and the associated standing waves, and the influence of adjacent boundaries. Below the transition frequency, the room, in one way or another, is the dominant factor in what we hear. It is not possible to completely generalize the effects of these factors because rooms themselves and the physical layout of loudspeakers and listeners within them are individualistic. The analysis and treatment must be done in the venue, after the audio system is operational.

At low frequencies, steady-state in-situ measurements are the measurement of choice. Here 1/6-octave resolution, or even higher, is useful as it provides insight into the origin of the resonances if one chooses to be analytical; see Figure 8.8 for an example. It is useful to know which modes are causing problems so that targeted remedies can be employed to attenuate them. Indiscriminately filling a room with bass traps can have positive effects, but being relatively broadband treatments, there may be excessive attenuation of beneficial sounds. There are also the obvious visual-aesthetic consequences. It is therefore good to have information than can guide their deployment within the space. In the end, revealing measurements are useful guides to insightful analysis of the situation, optimizing solutions to bass problems, assuming the necessary understanding of room mode and adjacent boundary effects. Figure 8.9 illustrates the clarity of high-resolution, low-frequency measurements.

Having adequate resolution also permits the identification of the center frequency and Q of room resonances that need to be equalized. The better the match of the equalizing filter to the resonance, the better the resonance is damped, reducing the ringing duration (Figure 8.12 and discussion in Section 8.3).

At low frequencies it has long been believed (the author included) that ringing in bass frequencies might be more audible than at high frequencies simply because of the longer durations. Section 8.3 discusses experimental evidence pertaining to this, and it is a complex situation. A general conclusion seems to be that the ringing is less of a problem to our perceptual processes than the spectral bump. The implication is that, even though waterfall diagrams provide visually compelling confirmation of the existence of resonances, they are not required to detect their presence or to confirm the success of remedial action. The necessary information is in a high-resolution steady-state response curve. Presumably because the ringing was not a major factor, it was found that significant improvement could be achieved with 1/3-octave amplitude equalization, at least for a single listener.

Invoking multi-subwoofer mode-control methods (Sections 8.2.6–8.2.8) to reduce seat-to-seat variations has a profound effect in reducing ringing. One study found that, based on the evidence of ringing-duration thresholds, these methods worked better than was necessary. However, that ignores the primary purpose of these techniques, which was to reduce seat-to-seat variations, expanding the listening area, delivering similarly good bass to several listeners.

A challenge for all equalization, whether it is manually implemented or executed by an algorithm, is to avoid boosting narrow destructive-interference dips. These are highly localized in the space and attempting to fill them simply adds a resonance to the system that is heard everywhere else. Boosting the sounds that are involved in an acoustic cancellation does not eliminate the cancellation.

Finally, as discussed in Chapter 9, adjacent boundaries generate acoustical interference that affects the sound delivered to individual seats, and, more broadly, that modifies the sound power radiated into a room. It is a component in all measurements made in rooms, superimposed on the effects of room resonances. Addressing the sound-power modification can be done with spatial averaging and equalization, as discussed in Section 9.2.1.

13.2 SOUND ABOVE THE TRANSITION FREQUENCY

At low frequencies the loudspeaker and the room operate as a closely coupled system. But at middle and high frequencies the wavelengths are shorter; "ray" acoustics can be applied to understanding what acoustical events follow the radiation of the sound. It is possible to separate individual direct and reflected sound components. New perceptual dimensions are added to the basic sound quality: spaciousness, envelopment and the like. In the following, we will review some relevant history, look at some relevant standards, and attempt to arrive at a sensible approach to measurements for the future.

13.2.1 Thirty Years—Some Things Change, Some Don't

For the entire history of sound recording and reproduction, a flat and smooth on-axis frequency response has been the dominant performance target for loudspeaker designers. The logic was that whatever the spectrum of the original captured sound was, the loudspeaker should reproduce it as faithfully as possible. My earliest blind listening tests, in the mid-1960s, indicated that listeners highly rated the sound from loudspeakers with flattish on-axis responses (see Section 18.1). Later, more rigorous research confirmed this and added smoothness and an absence of resonances to the requirements (Figure 5.2). Still more observation revealed that it was important to maintain those attributes in the far off-axis sounds that generate early reflections in rooms. Figure 13.1 provides an interesting comparison of data separated by 30 years.

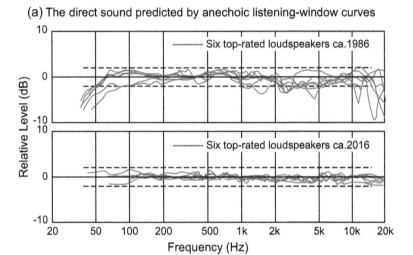

(a) The direct sound predicted by anechoic listening-window curves

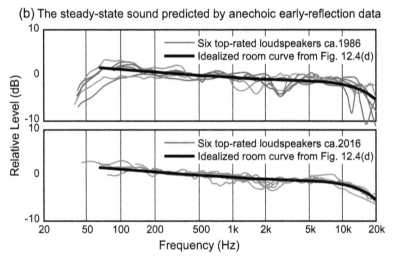

(b) The steady-state sound predicted by anechoic early-reflection data

FIGURE 13.1 *Thirty years of improvement. A comparison of measured performances of six loudspeakers that were rated highly in double-blind listening tests done in 1986 and now. Historical data are from Toole (1986): listening window data from figure 8 and mean front hemisphere data (comparable to early reflections in the spinorama) from figure 22. In (a) the dashed lines denote the 4 dB tolerance window allowed by ITU-R BS.1116–3 and EBU Tech 3276 for 1/3-octave smoothed on-axis curves. The curves shown are spatially averaged over the listening window, and have 1/20-octave resolution thereby showing more detailed and larger variations than spectrally smoothed data would. In (b) the predicted steady-state room curves are compared to the idealized room curve for present day highly rated loudspeakers shown in Figure 12.4d.*

The reduced variations indicate that loudspeakers have clearly improved, but the performance objectives of their designers have not changed. The dashed lines in Figure 13.1a are the 4 dB window limits suggested in ITU-R BS.1116–3 and EBU Tech 3276 for the 1/3-octave filtered on-axis response of an acceptable loudspeaker. Even the

30-year-old winning loudspeakers would suffice, and would look even better if the data were 1/3-octave smoothed.

An on-axis curve is subject to position-sensitive acoustical interference. Small irregularities mostly caused by diffraction may be seen that change or disappear when the microphone is moved, consequently a better indicator of loudspeaker performance is a spatial average over the listening window, preferably with high resolution. The ITU and EBU recommendations examine performance at off-axis angles up to 30°, which addresses direct sound only. In the spinorama these data and more are combined into a nine-curve average covering 0°, ±10°, ±20° and ±30° horizontal, and ±10° vertical, thereby including locations for an audience, or persons within a working area (Figure 5.6). In the 1986 National Research Council of Canada (NRCC) data, it was a simpler five-curve average: 0° and ±15° horizontal and vertical. With well-designed loudspeakers, the difference between on-axis and listening window curves is minimal; see Figure 12.1 and examples in Chapter 5. If the on-axis and listening window curves are not closely aligned, then there is likely to be sufficient directivity to cause different sounds to arrive at listeners seated next to each other or an engineer moving laterally along a recording console. This is clearly not desirable for stereo imaging or sound quality.

In Figure 13.1b the predicted room curves tilt downward because the frequency-dependent directivity of loudspeakers results in progressively more sound energy being delivered to the room at lower frequencies, room reflectivity normally increases at lower frequencies, and high frequencies are reduced by air attenuation. This is confirmed by the measurements in Figure 12.4b. If listeners are attracted to loudspeakers with flat direct sound performance, it means that they also approve of the accompanying downward tilted steady-state room curves. It is not possible for both the direct sound and the steady-state curves to be flat except in an anechoic space. This physical reality—bass rising after the arrival of the direct sound—exists in live listening experiences and in reproduced sound; see Section 10.8 and Toole (2015b).

13.2.2 The Wrong Room Curve Target?
As discussed earlier, both ITU-R BS.1116–3 and EBU Tech 3276 require loudspeakers to exhibit quite flat on-axis frequency responses, which is consistent with decades of loudspeaker design tradition and double-blind listener preferences. It would therefore be logical for the corresponding steady-state in-room "operational" frequency responses to be tilted. But that is not the case.

Figure 13.2 illustrates the situation, with (a) illustrating the previously shown desirable direct-sound performance of loudspeakers. In (b) the required "operational room response curve" (ORR) tolerances are shown. A steady-state, pink-noise, 1/3-octave analyzer measurement at the listening position is expected to fall within these bounds. If not, equalization is permitted to achieve compliance.

The desired "target" ORR (operational room response) curve is stated to be a horizontal line in EBU Tech 3276 and is assumed to be in the ITU document. I added such a target and the idealized room curve that has been demonstrated to result from measuring steady-state room curves for loudspeakers meeting the requirement of (a). (See

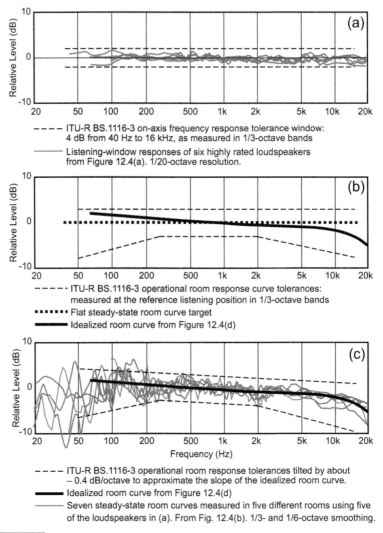

ITU-R BS.1116-3 on-axis frequency response tolerance window: 4 dB from 40 Hz to 16 kHz, as measured in 1/3-octave bands

Listening-window responses of six highly rated loudspeakers from Figure 12.4(a). 1/20-octave resolution.

ITU-R BS.1116-3 operational room response curve tolerances: measured at the reference listening position in 1/3-octave bands

Flat steady-state room curve target

Idealized room curve from Figure 12.4(d)

ITU-R BS.1116-3 operational room response tolerances tilted by about − 0.4 dB/octave to approximate the slope of the idealized room curve.

Idealized room curve from Figure 12.4(d)

Seven steady-state room curves measured in five different rooms using five of the loudspeakers in (a). From Fig. 12.4(b). 1/3- and 1/6-octave smoothing.

FIGURE 13.2 *Explanations are in the figure and text.*

Figure 13.1.) The curves are not the same, one is flat, the other tilted; one cannot have both at the same time except in an anechoic chamber. If the operational room response curve takes precedence, the performance of the loudspeakers must be altered—the bass must be attenuated and the treble boosted. If the loudspeakers had high subjective ratings in their original state, those ratings will now go down because the on-axis curve would be tilted upwards: less bass, more treble.

The threshold of audibility of a spectral tilt is about 0.1 to 0.2 dB/octave (Section 4.6.1) and this difference is of the order of 0.4 dB/octave. This is a problem, especially for broadcasters (the EBU audience) who are creating audio programs for playback through typical domestic systems.

Judgments provided by critical listeners in decades of double-blind tests indicate that the direct sound should be flat and the operational room response curve tolerances should be tilted by about −0.4 dB/octave, to be better aligned with the idealized room curve, as shown in Figure 13.2c. Then the measured room curves of loudspeakers with flat on-axis responses and well-behaved directivity will fit within the tolerances, without equalization above about 200 Hz, and allow listeners to hear the excellent sound these loudspeakers are capable of. This is as they are heard in homes, where in most instances the sound propagation path from the loudspeakers to the listeners is relatively clean.

However, in professional recording and broadcasting circumstances, the engineer is sitting behind a large reflecting console, possibly with computer screens jutting upwards, and near-field monitors on the meter bridge. The loudspeakers are often close to room boundaries or other large reflecting objects. All of this adds complication to the measured room curve at the operating position, the ORR. It will exhibit acoustical interference undulations, some of which may be audible, and others that are not. Falling outside the ORR tolerance range is not a certain indicator of an audible problem, especially when the target is an inappropriate one.

So, there is an obvious conflict when such ITU- and EBU-compliant systems are used in the creation of broadcast content, music recordings or movies. The circle of confusion is being aggravated, not alleviated.

A preferable approach may be to use the anechoic measurement of on-axis and/or listening window performance as the criterion of excellence for frequencies above about 300 Hz, the transition frequency. Loudspeakers failing to meet the anechoic requirement portrayed in Figure 13.2a can be equalized to deliver a flattish, smooth direct sound. The remaining question is relates to the off-axis/early reflected sounds.

Requiring additional off-axis anechoic data, possibly including sound power, could verify the requirement for relatively uniform directivity. Data in a spinorama format would provide a direct estimate of the steady-state room curve (the early-reflections curve) in an acoustically "clean" listening venue (see Section 5.6). A measured steady-state room curve in an acoustically "clean" room would also confirm the excellence of the loudspeakers and provide guidance for treating and adjusting low-frequency problems, as discussed in Chapters 8 and 9.

Obviously, equalizing inferior loudspeakers or good loudspeakers in problematic acoustical settings to match the ORR target curve cannot guarantee good sound. It is hoped that revisions can be made to ITU-R BS.1116 and EBU Tech 3276 when they come up for review.

13.2.3 "Room Correction" and "Room Equalization" Are Misnomers

Currently, "room correction" processes are widely dispersed throughout the audio industry. They are the basis of cinema sound calibration and have found their way into receivers and surround processors, and some exist as stand-alone devices. The claim is that measurements of the steady-state sound field using an omnidirectional microphone, and signal processing by an algorithm, can repair imperfections in unknown loudspeakers in unknown rooms. There is no doubt that such a process can yield improvements at low frequencies for a single listener, but above the transition frequency to claim that a smoothed steady-state room curve derived from an omnidirectional microphone is an adequate substitute for the timbral and spatial perceptions by two ears and a brain is absurd. However, it clearly can be good business.

Section 12.2.3 discusses this matter. All that can be stated with reasonable certainty at this moment is that, given a non-aberrant "typical" listening room, if a *spatially averaged* steady state room curve looks like the ones in Figure 13.2c, the customer has probably purchased a very good loudspeaker. Nothing needs to be done above about 300–400 Hz. In these examples, the anechoic data on the loudspeakers indicated that they were not responsible for the small acoustical interference irregularities seen in the curves. The "room correction" process is based on the belief that a room curve is the definitive statement of sound quality. Consequently, the processors perform equalization corrections, including non-minimum-phase acoustical interference irregularities, in order to hit the specified target curve. Doing so arbitrarily modifies the loudspeaker, conceivably degrading a very good one. If a high-resolution "corrected" room curve looks superbly smooth, there is a possibility that something inappropriate has been done. As has been said many times before, much of this could have been predicted if one had adequate specifications on the loudspeakers (Figure 12.3). As described in Chapters 8 and 9, prudently used equalization is one of the tools that can alleviate low-frequency problems.

If the loudspeaker is well designed—that is, a good spinorama—and the room curve is significantly different from the prediction—the early reflection curve—something in the propagation path must be suspected. In that case the solution is to find and fix the acoustical problem by physical means. Equalization is not the remedy.

Human nature is such that sometimes simply hearing a difference is enough to believe that there has been an improvement. That is what has allowed so many indifferent loudspeakers and room correction algorithms to exist.

13.2.4 Automotive Audio

Consumers spend considerable time listening to music in their cars, through audio systems that nowadays can be quite good. These systems can only be measured after installation in the acoustically complex automobile cabin. Figure 13.3 shows results from subjective evaluations by Olive and Welti (2009), Clark (2001), and Binelli and Farina (2008). For comparison are shown measurements in five multinational luxury cars, and a target curve used by Dirac, a popular design-aid algorithm. All boost the bass, in part to compete with the substantial road, aerodynamic and mechanical noise at low frequencies (in some vehicles the bass boost can vary with speed and/or background noise). The distinctly non-resonant character of the automobile cabin at low frequencies may also be a factor. There is substantial agreement through the middle frequencies, including the idealized small-room curve from Figure 12.4d. Apparently, designers wanted cars to sound like good home systems. Differences in how the in-cabin automotive data are gathered lead to some disagreements in curve shape at very high frequencies; the perceived sounds are likely to be more similar than these measurements indicate.

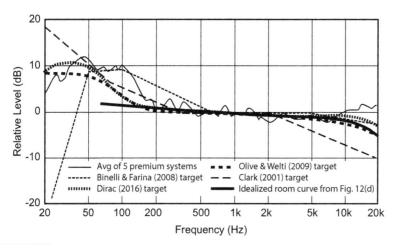

FIGURE 13.3   *Four in-cabin steady-state target curves used in the design of car-audio systems, compared to some real-world examples and to the idealized steady-state curve for loudspeaker reproduction in listening rooms.*

13.2.5 Headphones

Headphones entertain masses of people. Professionals occasionally mix through them when conditions demand it. Both rely on some connection to normal sound reproduction, that is, loudspeakers in rooms, because that is how stereo is intended to be heard. Stereo recordings are mixed using loudspeakers.

Headphone measurements have been and still are a debated issue. What form of mechanical-acoustical microphone-to-headphone "coupler" best mimics the eardrum-to-headphone coupling that we listen through? Then there is the second issue: what is the target curve for the measurement?

Sounds that reach eardrums in normal listening are modified differently for every incident angle by the elaborate geometry of the external ear, and the head and shoulders. The ear canal resonance is superimposed on all of these. Measurements of these effects are called head-related transfer functions (HRTFs) that can be measured at the entrance to the ear canal with it blocked (called blocked meatus) or at the eardrum, which is a risky operation. The latter includes the 2 to 3 kHz ear canal resonance. These frequency responses are anything but smooth and flat, but they are part of us, from birth, and we compensate for these variations when making judgments of sound quality, and use the variations when deciphering, or localizing sounds in, complex multidirectional sound fields.

When wearing headphones, these directional cues are lost, and we must rely on cues in the recordings. That is why binaural or dummy head recordings exist: to try to deliver the missing information by capturing it in an anatomically accurate dummy head microphone. There are two problems: (1) the dummy ears are not your ears, and (2) moving your head on playback moves the band that is playing—the brain is puzzled and defaults to in-head localization. Head-tracking schemes that synchronize modifications to the binaural signals greatly improve the externalization of sound images.

Stereo recordings are mixed using loudspeakers, and are just simply "different" in headphone listening. Popular music recordings often are perceived as left-ear, right-ear and middle of the head. It is intimate and entertaining, but not at all realistic.

When I started measuring and subjectively evaluating headphones in the 1970s, sound quality was hugely variable. Supplying measurements and subjective evaluations for reviews in *AudioScene Canada* magazine provided my database, and over a few years I assembled measurements from headphones that listeners liked and disliked, looking for a pattern. For practical reasons, the subjective evaluations were sighted, which is a weakness. However, the listeners in these tests were very familiar with double-blind loudspeaker evaluations, and with the poor correlation between subjective ratings and brand, price, and technology. They had developed a healthy skepticism.

My colleague at the NRCC, Dr. Edgar A.G. Shaw had been doing pioneering work on measuring the acoustical performance of the external ear (precursors to HRTFs), so I took advantage of this data. To me it seemed logical that human listeners would pay most attention to sounds arriving from the front hemisphere, so I assembled the free-field to blocked-meatus transformations for these angles, thereby establishing a "prediction" of what might be heard from a perfectly flat on-axis loudspeaker with constant, wide-dispersion, directivity in a reflective room. It was a huge simplification, but it was a start.

For the measurements I used a flat plate coupler fitted with a rubber pinna replica borrowed from a KEMAR anthropometric mannequin and a 1/4-inch pressure-response microphone at the entrance to the ear canal—a blocked-meatus measurement.

As product evaluation results accumulated, it was clear that the better sounding headphones had measurements that were in the general area of the predictive target. The variations were enormous, as can be seen in figure 8 of Toole (1984). In Figure 13.4a can be seen a selection of the "good" sounding headphones. Among these one product was highly praised by all listeners, and this is shown in (b) to fit very well into the shaded area embracing sounds arriving from the front hemisphere. It is interesting that starting in 1974 the magazine published review data in the format of (b) showing the coupler measurements, with and without air leaks, superimposed on the "ideal" loudspeaker/room target area. It was a pioneering effort.

There was reason to believe that this kind of measurement, used with this kind of target, could be useful in designing headphones that have a chance of sounding good. But such was not to be the case. A small number of brands took the hint, and some visited me at the NRCC, but most companies apparently had other "marketing" priorities. Headphone sound has continued to be characterized by large variations.

Some research was done on targets, concluding that something resembling the response to a diffuse sound field was appropriate. This evolved into an IEC recommendation, but it failed to have much effect on products. Olive and Welti (2012) provide background.

Many years later, my colleagues Sean Olive and Todd Welti focused their attention on the currently flourishing market for headphones, asking the same basic questions that I had 38 years earlier. The passage of time had provided new measurement apparatus

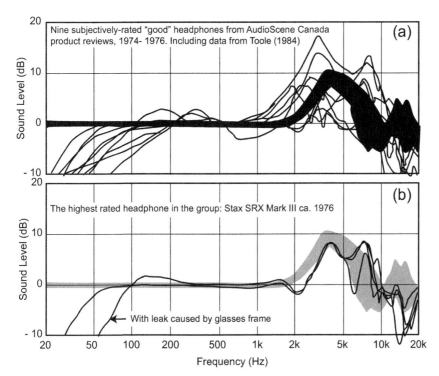

FIGURE 13.4   *Blocked-meatus measurements on headphones compared to a shaded area representing the free-field-to-blocked-meatus transformation for sounds arriving from the front hemisphere (horizontal only).*
From Shaw (1975) (used with permission of the Acoustical Society of America).

(a GRAS 43AG) delivering more trustworthy data at the eardrum, and powerful signal processing permitted new kinds of experiments. Several informative papers resulted (e.g., Olive et al., 2012, 2013, 2014, 2016). It was gratifying when they found that listeners gave the highest ratings to headphones that were equalized to mimic, at the eardrums, the sound of highly rated loudspeakers in a good listening room (Olive et al., 2012, 2013b, 2016)—an evolved form of the conclusion in Figure 13.4.

Conclusion: the best sounding headphones sound like good loudspeakers in a good listening room. Because that is where recordings usually originate, is anyone surprised? The good news is that it is possible to anticipate sound quality using the right measurements, interpreted in the right manner.

13.2.6 Cinemas

Chapter 11 discusses this in detail. A synopsis of the present situation is that the current cinema calibration target curve, the X-curve, is in significant disagreement with practices in the rest of the audio world. Within the closed dubbing-stage-record-to-cinema-playback loop, an arbitrary target could conceivably work. The present target artificially distorts the spectrum, requiring soundtrack mixers to "pre-equalize" based on subjective judgments. This is not consistently done. A calibration process that yields inconsistent results is an additional problem. There are many variables.

Movie soundtracks make their way into our homes and personal devices with increasing frequency, and some cinemas stimulate business by exhibiting programs generated in the music and television contexts. Compatible audio, not needing repurposing, would be logical. A first step at a psychoacoustical investigation of the cinema sound situation indicates a preference for a flattish direct sound radiated on the audience side of the screen. This is consistent with the rest of audio.

13.3 IS THERE A COMMON FACTOR— A GENERALIZABLE TARGET?

Those who have read the book to this point already know the answer to the title question in the heading for this section. From my earliest primitive subjective evaluations, through three decades of careful technical and subjective evaluations of hundreds of loudspeakers, to the present, it has been evident that listeners in double-blind comparisons routinely identify loudspeakers with flattish and smooth on-axis/listening window performance as their preferred products. Those with similarly good off-axis performance receive even higher ratings.

As discussed in Chapter 11, investigations into cinema and large venue sound starting around 1965 indicated that listeners preferred steady-state room curves with an overall downward tilt—a rising bass. Compared to the mid- and high frequencies, the amount of the rise depends on the reflectivity of the room, which tends to be higher in large commercial venues than in domestic rooms or recording control rooms. The rise is also a function of loudspeaker directivity, which compared to mid-high frequencies results in a more rapid rise in bass power output in large cinema loudspeakers than occurs in smaller domestic or monitor loudspeakers.

Figure 13.5, a repeat of Figure 11.10, summarizes the present situation, showing the best evidence of optimum steady-state target curves for movie sound venues and small listening rooms and home theaters.

> Both fit the same pattern. Both are the result of loudspeakers radiating flattish on-axis, direct, sound into rooms. That is the common factor. The true target is a flat direct sound. The resulting room curves adopt a slope based on the frequency-dependent directivity of the loudspeakers, the frequency-dependent reflectivity of the venue and the high frequency roll-off due to air attenuation associated with different listening distances. A room curve without reasonably comprehensive anechoic loudspeaker data is unreliable evidence of performance.

As explained, differences are traceable to two factors at low frequencies: the reflectivity of the room and the directivity of the loudspeaker. At high frequencies the differences are traceable to the air attenuation associated with the propagation distances to the listeners. Each is the subjectively preferred room curve target for the venue.

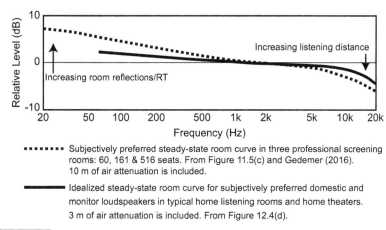

.... Subjectively preferred steady-state room curve in three professional screening
rooms: 60, 161 & 516 seats. From Figure 11.5(c) and Gedemer (2016).
10 m of air attenuation is included.

———— Idealized steady-state room curve for subjectively preferred domestic and
monitor loudspeakers in typical home listening rooms and home theaters.
3 m of air attenuation is included. From Figure 12.4(d).

FIGURE 13.5 *A comparison of subjectively preferred steady-state room curves derived from evaluations done in movie-sound venues and in small listening rooms and home theaters.*

As has been discussed elsewhere, a pristine direct sound is where auditory perception begins in normal listening. It defines direction, much about timbre, and is the reference for evaluating later arriving sounds that provide spatial cues. In the complex sound field of a concert hall the direct sound, even diluted by reflections, remains a strong identifier of timbre. If a recording has captured the essence of voices and musical instruments, it is not surprising that the preferred mode of sound reproduction preserves the direct sound.

The bass rise is an accompaniment to normal listening, as all practical sound sources trend to omnidirectional radiation at low frequencies (Figure 10.15). The amount of bass rise will vary according to the reflectivity of the reproduction space at low frequencies and the directivity of the woofer(s). There is no doubt that humans have substantial ability to separate the sound quality of the source, live or reproduced, from that of the listening space. We expect reflected sounds. When they are missing, as in an anechoic chamber, it is strange; one may feel disoriented or uneasy. Room sound provides context for what we hear. Common experience is that we adapt, and it is background information. But adaptation has limits, and it is important to appreciate that truly excessive reflections and reverberation are detrimental to sound reproduction, including the prime requirement of movies—speech intelligibility. The key is to identify the point at which acoustical intervention is necessary, and accept the fact that some of what we measure is simply not a problem to human listeners; we don't want to "eliminate" rooms, we want to optimize them.

I won't pretend that there is a definitive answer at this stage. However, it is reasonable to assume that loudspeakers delivering a high-quality, flat, neutral direct sound is a persuasive starting point. Above the transition frequency that, along with confirmation of similarly well behaved directivity may be sufficient to predict good sound for human listeners.

As measured by a microphone in a room some amount of bass rise is natural and expected. It remains to be determined if there is an aesthetic limit, and what it is. The apparent subjective preference for enhanced bass below about 100 Hz, as seen in Figure 12.7 is something deserving of more research.

1. Is it a fundamental characteristic of human preference? The fact that listener experience was a factor suggests that this has some validity.

2. Is it a loudness compensation for playback at lower levels than were used during the mix? The comparison in Figure 4.6e suggests that this may be a factor.

3. Is it related to the "circle of confusion," meaning that different programs might yield different levels of preferred bass and/or treble? As I listen to vast collections of music through uncompressed streaming sources, it is evident to me that this is part of the picture.

It is difficult to escape the reality that, for fussy listeners, easily accessible old-fashioned tone or tilt controls are useful.

CHAPTER 14

Measurement Methods

14.1 ALTERNATIVE VIEWS OF FREQUENCY RESPONSE

The preceding chapters have provided persuasive evidence that, if good sound is the objective, frequency response is the dominant factor. Non-linear distortion is audible if sufficiently loud, but in the world of conventional, "full-sized" loudspeakers, it has not been a frequent problem at normal playback sound levels. Professional use can be more demanding. Impulse response and time-domain ringing of resonances are already accounted for over most of the frequency range because loudspeaker transducers are minimum-phase devices—a flat, smooth frequency response indicates an absence of such misbehavior. Rise time is a function of the high-frequency limit—for example, the "speed" of a subwoofer is limited by the 80 Hz low-pass filter. Phase shifts, polarity concerns and group delays are below the thresholds of detectability in normal listening. These variables are discussed in Chapter 4.

In the end, the frequency response we need is in comprehensive anechoic information about the loudspeaker. Without that, one is severely handicapped. With it, one is able to anticipate sound quality as perceived in rooms (Section 5.7), and to predict in-room measurements with useful precision (Section 5.6). The latter is more a matter of interest than importance, because room curves alone are not definitive data. The exception is at low frequencies, where there are problems intimately related to the arrangements of loudspeakers and listeners within the boundaries of the room and the acoustical properties of the venue itself. Here it is necessary to do in-situ measurements of and corrections for the problems, as explained in Chapters 8 and 9.

As has been discussed, the assumption that a steady-state room curve is sufficient to be definitive of perceived sound quality is questionable. What we measure is simply not all of what we hear. It is obviously related, but equally obviously, binaural humans make perceptual differentiations among incident sounds from different directions at different times that a "monaural" microphone cannot. Plug one ear with a finger and listen to how the sound in a room changes. The situation is further compromised when a time-blind measurement system (i.e., steady-state) is used. However, decades of precedent exist in the audio industry, and one must do due diligence in collecting and examining the evidence before deciding whether there may be a superior method.

First we need to be clear what is meant when we say "frequency response." There are a few options, and they are not all equal.

- Anechoic spinorama and predictions of sound fields in rooms.

- Compromised in-situ measurements of direct sound.

- Steady-state room curves.

At frequencies below the transition frequency, there is no substitute for steady-state room curves. Above the transition frequency, an understanding of a playback situation requires at least two of them. In the following each of these is examined.

14.1.1 Prediction of the Direct Sound and Room Curves from Anechoic Data

The first sound to arrive at the listening location, the direct sound, defines the localization of sound images, initiates the precedence effect that permits localization in reflective spaces, and provides an initial reference for comparison to later arriving sounds that, combined, result in perceptions of both spatial effects and sound quality.

From the very beginning of my subjective/objective examinations, listeners were unambiguously attracted to loudspeakers with flat and smooth anechoic on-axis (direct sound) performance. Figure 13.1 shows results over the last 30 years. This has not changed for small rooms and evaluations of sound in cinemas came to a similar conclusion, as discussed in Section 11.5. Figure 5.15 showed that what listeners described as the perceived spectrum was most closely associated with the direct sound. Queen (1973) emphasized the importance of the direct sound as the prime identifier of timbre. It is reasonable to begin with a neutral direct sound—an accurate depiction of the sounds in recordings and sound tracks.

Predictions of direct sound are based on anechoic loudspeaker on-axis or listening window data (Figure 5.6). This describes the sound radiated from the loudspeaker, which would need to be corrected for screen loss (Figures 10.13 and 10.14), if one is present. Expectations of direct sound at listening positions must also acknowledge air attenuation in the propagation path to listeners (Figure 10.12).

Because we are dealing with anechoic measurements, high resolution is possible over the entire audible frequency range. Resonances can be identified and evaluated from spinorama data, thus providing fundamental insights into potential sound quality (Figures 4.4 and 4.13).

However, as explained in earlier sections (e.g., Sections 5.6 and 11.7), data from the spinorama can be used to estimate steady-state room curves at middle and high frequencies. It is instructive to compare the predicted and measured curves because it reveals how much acoustical interference "noise" there can be in room measurements. Most of these variations cannot be corrected by equalization, and in fact most are not problems for human listeners.

If sufficient anechoic data are available describing the loudspeaker, sound quality in a venue can be anticipated with some confidence at middle and high frequencies. However, only in-situ steady-state measurements can reveal what is happening at low frequencies.

14.1.2 In-Situ Measurement of the Direct Sound

In the absence of anechoically measured data on the loudspeakers, it is possible to perform quasi-anechoic measurements in the venue. These use time-windowed digital measurement techniques to capture the first arriving sound and ignore reflected sounds. The compromise is that the frequency resolution is the inverse of the reflection-free time window, for example, 10 ms = 100 Hz (1 ÷ 0.01). With the normal logarithmic frequency scale, this resolution provides more than enough detail at high frequencies, but not enough at low frequencies. To maximize the time window when measuring in a fixed installation like a cinema, it is suggested that the microphone or, even better, multiple microphones at different vertical angles, be located in the free space above the seats at a distance of 5–7 m from the screen. Some acoustical interference in the cross-over region(s) can be anticipated and multiple microphones permit some spatial averaging. Locating the microphone(s) on a line between the high-frequency horn and the reference seat at the 2/3-distance, the measurement would capture the direct sound on its flight to the prime listener. Figure 14.1 shows the arrangement used in the SMPTE

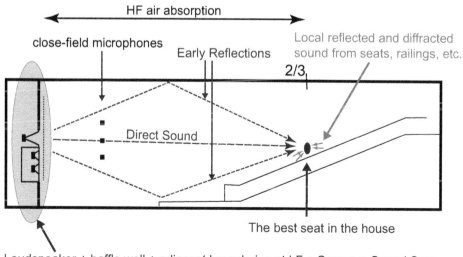

FIGURE 14.1 *The physical arrangement for some measurements in cinemas showing the standardized reference location at the 2/3 distance from the screen to the back of the venue, and a set of three close-field microphones to capture the sound as it propagates to the reference location.*

TC-25CSS, 2014 B-chain study, results of which can be seen in figures 8 and 9 of Toole (2015b). There it can be seen that there is useful agreement between the 50 ms windowed data measured at the 2/3 distance and that from the close-field microphones. The agreement is especially good above about 500 Hz because there the direct sound is the dominant sound field in cinemas.

Methods using a variable time window, expanding as frequency falls, are a trade-off in that the frequency resolution is better maintained at lower frequencies, but reflected sounds can be included in the expanded time window. The choice and setup of such a measurement is very much dependent on the size and reflectivity of the venue. It is well described as fractional-octave filtering. In practice, different parts of the spectrum may be more usefully represented by different smoothing algorithms.

All of these are useful techniques, especially when, as often happens, the loudspeakers are undocumented. If the loudspeakers are portable, it is possible to greatly improve the frequency resolution of these pseudo-anechoic measurements by transporting them to a situation where the reflection-free time window can be extended. Elevating the loudspeaker and microphone and separating them from reflecting surfaces or objects can be done in very large rooms, a warehouse or a hangar, or outdoors (measure at an indoor temperature, out of direct sunshine, so that transducers behave normally). These restrictions and the need for access to power and to avoid wind and rain are significant deterrents.

The difficulties just described can be substantially alleviated for many loudspeaker measurements by adopting the ground-plane technique described by Gander (1982). In this, both the loudspeaker and microphone are placed on a large flat surface such as a parking lot, a flat rooftop and so forth, thereby reducing the number of reflecting surfaces that need to be avoided. It is widely used, especially for the lower frequencies that are the most problematic for time-windowed measurements.

All of these techniques are used by manufacturers and consultants who lack anechoic chambers, or who lack one large enough to permit the 5–7 m measuring distances demanded by large loudspeakers and arrays. The challenge is that the microphone should be in the acoustical far field (2 m or more for domestic and monitor loudspeakers, and much more for large cinema and array loudspeakers). The greater the microphone distance, the greater the challenge to maintain a long reflection-free time window. If this is done it is possible to rotate the loudspeaker on two axes—on its bottom and on its side—obtaining a useful approximation to the anechoic spinorama (Figure 5.5).

Modules used in large sound reinforcement line arrays are measured on many axes at fine angular resolution that are used in computer models. Various physically shaped and electronically tapered arrays of the modules can be used to deliver satisfactory audience coverage for different venues. This is important in tour sound. In contrast, cinemas are somewhat standardized venues and the loudspeakers are more like scaled-up hi-fi systems. Arrays would likely be better, but cost is a significant factor.

Clearly, life is a lot easier if one can begin with anechoic data from the manufacturer. Because the directivity index is stable, even substantially predictable, significant design or sample flaws should be visible in the quasi-anechoic frequency response.

14.1.3 The Steady-State Room Curve

This is the result of an accumulation of energy from the direct sound and reflected sounds arriving at the microphone from all directions at all times. Traditionally it has been measured using 1/3-octave analysis of pink noise by a real-time analyzer. Now there are several alternative methods that yield accurate data, with selectable frequency resolution, curve smoothing and data manipulation. Obviously, data need to be accumulated over a sufficient time window to permit the sound field to stabilize, which varies with the room, but typically 500 ms or more would be used, especially at low frequencies.

For small rooms, Section 5.6 explains that it is possible to obtain a reasonable prediction of a steady-state room curve at middle and high frequencies from anechoic data on the loudspeaker. However, the reverse is not true: a steady-state room curve provides unreliable information about the loudspeaker, which is the prime factor in sound quality. Once the sound from a wide dispersion loudspeaker has been launched into a three-dimensional, somewhat reflective space, there are some aspects of loudspeaker performance that can no longer be determined, but that may matter to binaural human listeners. By itself, a steady-state room curve is incomplete data and it is a bold assertion (often repeated) that it is a definitive statement of sound quality as perceived by two ears and a brain (see Section 13.2.3).

Cinemas present a special situation, as explained in Sections 11.2 and 11.7. The directional loudspeakers, combined with an acoustically absorptive venue, result in a dominant direct sound field above a few hundred hertz, meaning that the steady state and direct sound fields are essentially identical. In all circumstances, though, prior knowledge of the anechoic loudspeaker and screen performances provide important insight.

As shown in Figure 12.4, the best-sounding loudspeakers, which all have flattish on-axis frequency responses, exhibit very similar room curves in different typical listening rooms, which is an interesting finding. The high-resolution curves exhibit only small fluctuations around a clear central tendency. Differentiating between ripples in the frequency response caused by resonances (audible) and those caused by acoustical interference (not normally an audible problem) is a major challenge. A microphone output displays both simultaneously, and identifying the origins of peaks and dips is difficult. Acoustical interference is displayed as comb filtering, which in limited circumstances can be a serious problem. However, in rooms, the direct and reflected sounds arrive from different directions, and binaural human listeners perceive this as a form of spaciousness; it provides information about the room.

Disguising the distracting visual effects of interference while preserving useful information is the problem. Two remedies for this situation are commonly employed:

- **Spatial averaging**: smoothing the curve by averaging measurements made at several locations. Seat-to-seat variations can be huge in small rooms, a fact that is hidden in spatial averages. In some common instances it is a labor-intensive statistical exercise in reducing the visibility of small spectral variations that were not problems to begin with, yielding data that could have been predicted (e.g., see Figure 11.11(d)). Nevertheless, such measurements can be used to evaluate certain kinds of problems that are common to all or most locations—for example, adjacent boundary effects (Section 9.2.1) and sizeable low-Q resonances and spectral imbalances in loudspeakers—the latter also being evident in anechoic loudspeaker data (Figure 4.13).

■ **Spectral averaging**: smoothing the curve by summing energy over a range of frequencies, usually expressed as a fraction of an octave. The greater the frequency range, the smoother the curve looks. For anechoic loudspeaker measurements, it is common to use 1/20-octave resolution. In-room measurements are often done using 1/12- or 1/6-octave resolution, although Olive (2004b) showed that sound quality predictions were more accurate with 1/20-octave resolution. Most people prefer to look at smoother curves, which partially explains why traditional RTA (real-time analyzer) room curves have been done with 1/3-octave resolution (see Section 4.6.5 for perspective). Full-octave-band resolution is used for background noise measurements and is useful to reveal spectral trends and balance in rooms while minimizing distracting fluctuations caused by acoustical interference.

Both methods reduce the visible effects of interference. As is seen in Figure 10.6 the very highest frequencies arrive at listening positions as predominantly direct sound. No significant interference is present and the performance of the loudspeaker can be seen in the room curve. However, through the middle frequencies room reflections generate interference in an amount dependent on loudspeaker directivity and room reflectivity. This is where spatial and/or spectral averaging can be useful. At low frequencies true steady-state measurements are needed (Chapters 8 and 9), and high resolution allows for the identification and treatment of room mode problems.

Figure 11.3 explains that in cinemas the direct sound is dominant above a few hundred hertz, and Figure 11.11 shows that high resolution room curves are highly consistent, usefully revealing, and predictable from anechoic data. Gedemer (2015a) showed that in a 516-seat cinema, above the low-frequency range there was very little difference between the spatial averages of four microphone locations and 109 locations.

Therefore, access to the intrinsic anechoic loudspeaker performance is greatly advantageous. The processes discussed in Sections 14.1.1 and 14.1.2 substantially avoid problems with acoustical interference, while still revealing important high-resolution information about the loudspeakers.

Some of the acoustical interference originates with the traditional practice of placing the microphone at listener ear level. It is a practice based on the notion that an omnidirectional microphone is a reliable substitute for two ears and a brain. Measurements will then include sounds reflected or diffracted from seats causing irregularities from phenomena that appear not to be bothersome to listeners. Human heads and torsos are variable, but rows of seats and the locations of measurement microphones relative to them tend to be unvarying, leading to repetitive patterns in the data. Section 11.7 discusses this.

If system equalization is to be based on these data, careful interpretation is necessary. When it is done by inadequately trained technicians or by an automatic algorithm, there is a real risk that equalization may be applied to non-minimum-phase irregularities that should be left alone. Good loudspeakers can thereby be degraded. This is the explanation why many people find that the best setting for a room equalizer is "off." The exceptionally linear and smooth room curves seen at the end of some room correction exercises are good advertising, but they are clues that some inappropriate manipulations may have been done. Automatic equalization algorithms are especially prone to these errors. In these situations, especially above the transition frequency, reducing the spectral resolution to 1/2-octave or even full-octave smoothing is advantageous. In this way, broad spectral trends can be seen, but distracting irregularities are not.

14.2 MEASURES OF LOUDNESS AND SYSTEM-LEVEL CALIBRATIONS

There has been a long quest for a perfect "loudness" meter that yields a single number perfectly representing the subjective loudness effect of any sound, as heard in any environment. The practical problem is that in the frequency and time domains, music, movies and mixed broadcast programs are not stable. The sounds continuously change in bandwidth, spectrum, level and transient/temporal structure. If we add the psychoacoustic complications of loudspeakers with different directivities located in different rooms, at different incident angles relative to listeners, at different distances, the situation becomes very much more complicated. In movies the informational content and emotional impact of sounds can be biasing factors in perceived loudness and annoyance. An omnidirectional microphone in a somewhat reflective room cannot collect data for processing the way two ears do. Even if one invokes a dummy-head recording device, we lack the brain that performs the elaborate and adaptive binaural processing. Yet, this is the situation in audio recording control rooms, home theaters, movies, dubbing stages and cinemas.

It is important to examine what matters and what does not matter. Even though the perfect solution may not exist, we do have measures that permit us to achieve our practical needs.

The first simple approach to a single number loudness rating derived from the equal-loudness contours shown in Figure 4.5a. The A-weighting curve shown in Figure 14.2, for example, resembles the shape of an inverted 40-phon loudness contour, and thereby was thought to be appropriate for measuring low-level sounds. B- and C-weighting curves were thought to be more representative of progressively higher sound levels. It didn't work as intended and, over the years, C-weighting came to be used as an approximation to flat, but with some discrimination against very low and very high frequencies. A-weighting has become widely accepted as a general purpose measure where low frequencies are not an issue, including assessment of hearing damage risk for occupational hearing conservation programs (Chapter 17). B-weighting has been all but forgotten, and is no longer a standard feature in sound-level meters.

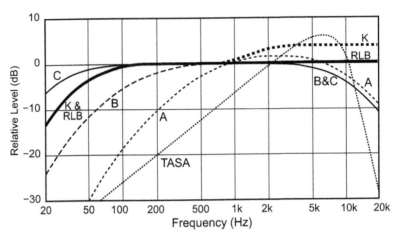

FIGURE 14.2 *Frequency-weighting curves that are available in sound-level meters (A, B and C) and the RLB curve proposed by Soulodre and Norcross (2003) for providing a single-number measure of the perceived loudness of typical audio signals. K-weighting is the RLB curve with a high-frequency boost, as used in the ITU-R BS.1770–3 (2012) loudness recommendation. TASA (2013) is the weighting used to evaluate the loudness and/or potential annoyance of film trailers.*

It really should not have been surprising when schemes based on equal-loudness contours should be found deficient. After all, they were derived from comparisons between pairs of pure tones in an anechoic environment, not a continuous broadband program with continuously varying temporal and spectral content in a semi-reflective normal room.

Substantial research over many years resulted in much more elaborate methods for computing loudness, as described in Fastl and Zwicker (2007) and Moore et al. (1997). They begin with narrow-band spectral analysis and loudness-summing procedures based on psychoacoustic models of some of the frequency, amplitude and masking properties of the hearing process. The parameters of the models are based on relatively limited experimental data, which are all averages of the responses of many listeners, all of whom exhibit large statistical variations. As stated by Moore et al. (1997) in section 3, describing limitations of their model, "the predictions of the model should not be expected to be accurate for individual listeners or even small groups of listeners." That paper gives an excellent overview of the variables considered in these models, showing just how complex the process is. It is still a work in progress, but it is clear that much is quite well understood, at least for relatively well-defined sounds. The real-world problem is that so many variables insist on varying.

14.2.1 Evaluating Relative Program Loudness Levels

Fortunately most of the reasons for attempting loudness evaluations are not based on a need for an *absolute* measure, but on a need to track or evaluate *relative* loudness. A very common need is to maintain relatively stable loudness experiences from recordings of many kinds from many sources. One's own music library or streaming audio are examples where occasional level adjustments are necessary unless loudness stabilization has been done at the source, or automatically at the point of playback. Broadcast programming is another, and there is a long history of devices created for monitoring these program levels—the long row of meters on the meter bridge of a recording or broadcast console, for example. Other devices in the signal path can automatically monitor and adjust outgoing program levels. One critical need is purely technical: to avoid overloading something in the signal path for which a peak signal level indicator is needed. The other is to prevent listeners from being presented with inappropriately loud or quiet passages.

In these cases the variables related to the loudspeaker(s) and room are relatively constant for individual listeners. An amount of adaptation will have taken place, so factors affecting loudness measurement are primarily those that exist in the electrical signal path.

Given the complexities and uncertainties of the options, it was interesting when research suggested that a much simpler solution might be sufficient (Soulodre, 2004; Soulodre and Norcross, 2003). It is easy to implement: a high-pass characteristic somewhere between B- and C-weighting but with no high-frequency roll off. Figure 14.2 shows the standard A-, B-, and C- weighting curves, along with the new proposal, the RLB (Revised Low-Frequency B) curve.

The RLB curve provides the basis for K-weighting used in ITU R-REC-BS.1770–3 (2012), which adds a 4 dB boost above about 1 kHz to accommodate the acoustical gain added by a spherical head. This standard is aimed the radio and TV broadcast industries and embraces reproduction in mono, double-mono and multichannel formats.

Because the RLB filter attenuates very low frequencies, and the original subjective/objective tests employed a small loudspeaker with a rapid rolloff below 50 Hz, the author has concerns about the ability of this measurement to adequately represent some of the very loud movie LFE (Low-Frequency Effects) in the total numerical approximation of loudness in home theater or cinema contexts. This became an issue with several recent movies being considered too loud by patrons, but the question is, was this reaction due to loudness or annoyance, clean loud sound or distorted loud sound? These perceptual dimensions are, of course, correlated, but some sounds can be loud without being annoying. In movies, some sounds must be loud to be convincing.

The TASA (2013) curve is particularly interesting. It was created to address audience complaints about excessively loud movie trailers in cinemas. But, rather than measure traditional loudness, which involves all frequencies, this group focused on what they call "annoying volume." Dolby figured prominently in this activity, and they fell back on a metric for background noise in communication systems, the ITU-R 468 curve that had its origins as the CCIR 468–4 curve—this is the TASA weighting curve. Dolby employed this to evaluate its Dolby-B noise reduction system, widely used in audio cassette systems. Here, though, instead of measuring the annoyance of noise close to the threshold of hearing, it is being used to evaluate something closer to the threshold of pain. Because the measure responds mainly to high-frequency sounds, it is not surprising that mixers repurposing soundtracks for trailers end up liberally boosting the bass because it does not show up in the measurement. Success is evaluated by the level of audience complaints, which guides the decision of what level is acceptable. I know of no scientifically conducted verification of the method.

It is important to remember that in evaluating loudness measurement schemes, the reference has been a purely subjective judgment. So, in the absence of any instrumentation, an unbiased subjective comparison of loudness is irrefutably correct.

14.2.2 Multichannel Sound System-Level Calibration

In the calibration of multichannel sound systems, one can adopt one of two approaches.

1. Deliver a defined sound level from each channel, as measured by a microphone located at the prime listening location.

2. Deliver a sound of equal perceived loudness from each channel to a person located at the prime listening location.

The first of these is a straightforward physical measurement, and implicit in it is the assumption that if the sound mixer and the recreational listener have similar loudspeaker/room setups, that the art will be well communicated. Any subjective, perceptual considerations will be incorporated in the mix, and these will be reproduced for customers. This is the current practice in movie sound and home theater calibrations. However, if the home listening situation is very different from the mixing situation, the sounds will be different. The question is: are the differences sufficient to degrade the entertainment value of the program?

The second approach requires measurements that are able to simulate human perceptions of loudness. These human responses may be different for sounds arriving from different directions, from loudspeakers that may not all have the same directional characteristics, and that may not be at the same distance from the listener. All of these factors need to be taken into account by the measurement scheme, which seems like a substantial challenge. Probably there will be errors, and the question is again, are these errors sufficient to degrade the entertainment value of the program?

If the loudspeakers being compared have very different frequency responses, a loudness balance achieved with one kind of signal may not apply to a signal with a different spectrum. Fortunately, as loudspeakers have improved and are now more similar, the problem has lessened, although it has not disappeared. The frequency-dependent directivity of the loudspeakers, and the frequency dependence, directional behavior and placement of absorbing, diffusing and reflecting surfaces relative to the loudspeakers and listeners, must also be factors. All of this requires a lot of detailed examination, and more technical information than is commonly available from manufacturers of loudspeakers or acoustical materials. The widely accepted recommendations for setting up listening evaluations are seriously lacking in all of these respects.

Elaborate experiments performed by several persons in several locations in Europe attempted to provide answers to some of these questions. I will summarize highlights from the work, but recommend that the original papers be read by those with a serious interest in the topic.

All of them share what I consider to be a combined virtue and liability—the constant radius, free-standing, ear level, five-channel loudspeaker arrangement defined by ITU-R BS.775–3 (2012) (similar to Figure 15.8a). The authors were aware of the limitations of this idealized setup, which can be found in some professional mixing environments, some academic facilities and homes of a few uncommonly dedicated audiophiles. The real world of consumer multichannel layouts is significantly different in random ways, even in custom home theater installations as shown in Figure 15.8. Because multichannel music is not widely available, the bulk of multichannel programming is movies, with some TV, including sports events, offering surround effects.

Typical movie dubbing stages and cinemas have the front and surround loudspeakers mounted on or in the room boundaries and the surround channels are each reproduced through multiple loudspeakers mounted on walls, thereby adding acoustical interference to the loudness and timbre complications. In the film-sound world, few of the loudspeakers are at constant radius and not all are at ear level—even for those in the center of the cinema. This means that attempts at equal-loudness calibrations for home audio systems, however successful they may be, will not and cannot reproduce what was heard on the dubbing stage with any precision. This does not preclude great entertainment value, but only because humans are marvelously adaptable—and, frankly, some of these factors are of little consequence. The ever-advancing world of mass entertainment now embraces many more channels. With additional "immersive" loudspeakers in elevated and overhead locations, and conflicting channel counts and loudspeaker configurations from Atmos, Auro-3D and DTS, the variables have multiplied.

By controlling the experimental circumstances as they did, the authors have made it possible for others to replicate the tests and perhaps to elaborate on them to include more of the real-world circumstances. All that said, the experiments were well done and yielded interesting results, most of which can be usefully applied to present day circumstances. The following experiments focused on loudspeakers in the horizontal plane.

Perspective on the series of experiments can be seen in Zacharov (1998) and Bech (1998). More details are reported in a series of five AES papers:

- Part I (Suokuisma et al., 1998) describes the test signals that are used.

- Part II (Zacharov et al., 1998) considers the influence of signals and loudspeaker placement.

- Part III (Bech and Zacharov, 1999) considers the effects of loudspeaker directivity and reproduction bandwidth.

- Part IV (Zacharov and Bech, 2000) investigates the correlation between measurements and subjective judgments.

- Part V (Bech and Cutmore, 2000) looks at reproduction level, the room, step size and symmetry.

Consistent throughout the experiments was a comparison of the following test signals (my descriptions):

- 700 Hz narrow-band noise;
- 250–500 Hz bandpass-filtered noise;
- 500 Hz–2 kHz bandpass-filtered noise (as used in many consumer products);
- Pink noise (as used in cinema calibrations);
- B-weighted noise;
- Four broadband noise spectra based on Zwicker and Moore loudness models.

Altogether it was an impressive undertaking. There are many details in the results, but some substantive conclusions were:

- In **symmetrically arranged systems** with identical forward-firing loudspeakers, the choice of test signal for subjective adjustments of loudness was not a factor. This was interesting given the huge variations in the options. Listeners adjusted loudness to match that of the forward-firing center channel. Repeated adjustments for the L and R and surround channels had standard deviations in the range 0.4 to 0.6 dB (Part II).

- In **asymmetrical setups**, loudspeaker distance was a strong factor. Only the narrow-band noise signal deviated from the pattern of the other test signals. Listeners were able to match loudness levels among the channels with standard deviations of under 0.4 dB (Part II).

- A comparison of two forward-firing loudspeakers with significantly different frequency-dependent directivities (Genelec 1030A and B&O BL6000) and a Quad ESL63 dipole yielded no significant influence on the level adjustments for the front channels, and no significant influence when used as surround channels when the prime axis of the Quad was aimed at the listeners. When the dipole was at 90° (with the null facing the listener) there was, predictably, a significant level difference between it and the forward-firing reference center channel. The technical and subjectively evaluated level differences were similar (3.2 dB measured, 3.4 dB subjectively judged, averaged across all nine test signals). It needs to be noted that the Quad ESL63 is a true dipole with a broadband null at 90°, not the bidirectional-out-of-phase pseudo-dipoles sold as on- or in-wall surround loudspeakers (see Figures 15.11 and 15.12) (Part III).

- Again, the signal type was not a factor for any of the normal bandpass filtered or unfiltered pink noise signals. Only one of the signals tailored to a computed loudness function yielded limited differences (Part III).

- Pulling together all of the preceding data and more, correlations between subjective and objective measures were performed to find the combination of test signals and measurements that best predict subjective responses (Part IV). The results indicate the following as being optimal:

- Test signal: constant specific loudness signal according to the Zwicker free-field model.

- Metric: Moore or Zwicker (diffuse or free field) loudness, or B- or C-weighted sound pressure levels.

- 500 Hz high-pass filtering any of the test signals improved the correlations slightly.

- It is important to know that even the less successful test signals and metrics were not in error by an amount that is likely to degrade entertainment value. This is especially so when the numerous variables in the production of programs are included. The spread of average values was typically in the range of 1 to 2 dB.

- Reproduction sound level was not a significant factor, at least over the range that was explored: 15 to 25 sones (about a 7 dB change in phons or dB).

- The listening room had no significant influence on the calibrations of the front L, C, R channels, but in some situations, with some signals the level of the surround channels may be displaced by 1 dB or so.

It is clear from these conclusions that the perception of comparative loudness levels is a very durable phenomenon, showing considerable consistency, at least among the trained listeners used here. While there were occasional differences in playback levels determined by measurements, they are, in my opinion, rather small—not likely to be noticed in, much less to degrade, recreational listening experiences. However, one physical factor was a significant variable in asymmetrical listening setups: loudspeaker distance. This will be discussed in the following section.

Since these tests were done more work has been done on loudness evaluations of program, discussed in Section 5.2.1. These are now embodied in ITU R-REC-BS.1770–3 (2012). Given the range of test signals employed in the experiments just described, it is not likely that using this metric would have changed the results.

The bottom line to these investigations is that there are several readily available metrics for measuring steady-state sound levels that offer good correlation with subjectively judged loudness. The errors, when they exist, are small enough that it is highly unlikely that listeners would be aware of them while listening to a program, and even less likely that enjoyment of music or movies would be compromised. Where one sits in the playback venue may be the dominant factor. This is good news.

14.2.3 The Effect of Propagation Distance— A Side-Channel Challenge

As we move farther from a sound source, the sound level diminishes. The rate at which it diminishes, and how it diminishes at different frequencies, determines what we hear in loudness and timbre. So, if one is to standardize playback sound levels, the location of the microphone is a determining factor.

In music, there are no reference levels because the industry is not at all standardized. In stereo playback, the only requirement is that the left and right loudspeakers are as similar as possible, that the balance control is centered so signals sent to those loudspeakers are the same, and that the prime listener is seated equidistant from the loudspeakers. It would require gross acoustical asymmetry in the room to degrade the stereo soundstage to the point where pleasure was compromised, although there can be small effects. Play a variety of monophonic source materials, stereo switched to mono will do nicely; sit in the sweet spot, and if there is a clearly defined sound image floating midway between the loudspeakers the job is essentially done. Feel free to measure at the listening position, but in this case the perceptual illusion is the definitive test.

In movies, standardized playback sound levels apply to the dubbing stages in which soundtracks are mixed and cinemas in which they are reproduced (SMPTE RP200–2012). The measurements are made using a broadband pink noise test signal and a C-weighted sound pressure level measurement, with the meter set to slow response. The reference level is 85 dBC, for the screen channels, and 82 dBC for each of the surround channels. All of these levels assume 20 dB headroom. This means that L, C, R screen channels must each be capable of 105 dBC, and each surround channel 102 dBC. These are not trivial sound pressure levels, so it is important to be able to predict in advance what is demanded of the loudspeakers and power amplifiers. High-end home theaters are often set up to meet cinema sound level requirements, even though many listeners find them excessive, both in home theaters and cinemas. Let us explore how one might do this.

First, let us look at sound propagation in rooms. The well-known inverse-square law tells us that the sound from a source that is small compared to the wavelength being propagated is reduced in amplitude by a factor of 6 dB for each doubling of distance from the source (−6 dB/dd where "dd" is double distance). This is obviously true in reflection-free (anechoic) environments. Otherwise it applies only to the direct sound: the first sound to arrive at a listening location in a room, because reflected sounds very quickly add to the sound level that is measured and heard. Section 10.4 discusses this in detail.

In terms of perceptions, the first sound arrival dictates where the sound will be localized to have come from and initiates the precedence effect. It, combined with later sounds, will determine how loud the sound will be perceived to be, and the steady-state sound levels that are measured with the slow-responding sound level meters specified in the SMPTE recommendation. If the direct sound is reduced by 6 dB/dd, the steady-state sound will be attenuated at a reduced rate, depending on the reflectivity of the listening venue and the directivity of the loudspeakers. Figure 10.8 shows that the steady-state sound declines at a rate of about −3 dB/dd. It also shows that the physical measurement is in agreement with the subjective evaluations of relative loudness conducted by Zacharov et al. (1998). This is important.

The information in Figure 10.8, while correct, is a little misleading because of the logarithmic distance scale. Figure 14.3 shows the data on a linear scale, where it is

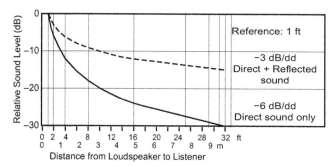

FIGURE 14.3 *The essential information from Figure 10.8 plotted on a linear distance scale.*

obvious that the sound levels change very rapidly close to the sound source, and less rapidly at a distance. This is especially significant for the surround loudspeakers, which can be localized—a distraction—by listeners who are seated close to the side walls in home theaters and cinemas. The direct sound is responsible for localization (the precedence effect), and it climbs in amplitude very rapidly at the shorter distances experienced in typical home theaters.

The fact that localization in complex sound fields is substantially determined by high-frequency transients means that one method of reducing the localization distraction is to reduce the level of high frequencies by aiming the speakers so that the more directional high frequencies fire over the heads of listeners, or to use loudspeakers that attenuate the high frequencies—the so-called dipole surround loudspeaker being one example.

A more elegant solution is to use loudspeakers having less than inverse-square loss per double distance, such as line sources, see Section 10.5.2 or other array designs such as CBTs shown in Figure 9.13f and g, which would logically be inverted to use the ceiling as the adjacent boundary. Other CBT configurations can be found in the JBLPro product line, and these have found use as surround loudspeakers. Suitably configured, these loudspeakers can deliver a relatively constant sound level over much of an audience.

This capability moves us to a new level of surround sound performance. The perception of "envelopment" is the impression of being in an acoustic space as determined by the movie or music recording engineer. It is an important ingredient in human satisfaction with both live and reproduced performances. It is one of the principal determinants of concert hall quality. It is maximum when the interaural cross-correlation coefficient is low—that is, the sounds arriving at both ears are similar in level and very different from each other. Sounds are maximally different when they arrive from the sides, but there are problems delivering sounds of similar level, as illustrated in Figure 14.4. The person seated midway across the room will always perceive good envelopment if the information is in the recording. However, at seats away from center the interaural amplitude differences increase, envelopment is degraded and as the side wall is approached, the side loudspeaker can be localized and the effect is completely lost.

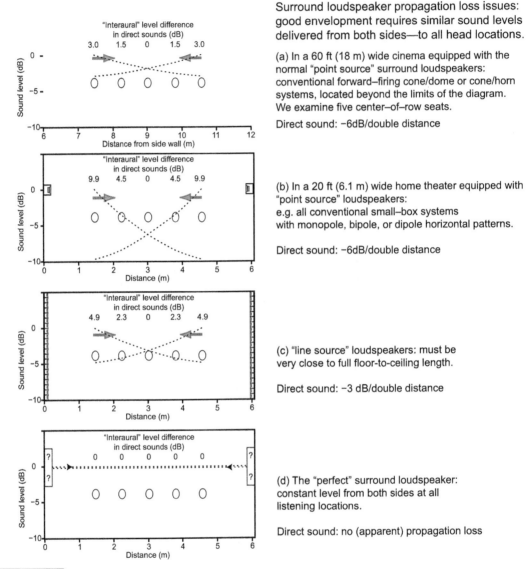

Surround loudspeaker propagation loss issues: good envelopment requires similar sound levels delivered from both sides—to all head locations.

(a) In a 60 ft (18 m) wide cinema equipped with the normal "point source" surround loudspeakers: conventional forward–firing cone/dome or cone/horn systems, located beyond the limits of the diagram. We examine five center–of–row seats.

Direct sound: −6dB/double distance

(b) In a 20 ft (6.1 m) wide home theater equipped with "point source" loudspeakers:
e.g. all conventional small–box systems with monopole, bipole, or dipole horizontal patterns.

Direct sound: −6dB/double distance

(c) "line source" loudspeakers: must be very close to full floor-to-ceiling length.

Direct sound: −3 dB/double distance

(d) The "perfect" surround loudspeaker: constant level from both sides at all listening locations.

Direct sound: no (apparent) propagation loss

FIGURE 14.4 *(a) the anticipated sound levels arriving at five listening locations halfway across the width of a cinema, from conventional surround loudspeakers to the left and right. (b) the comparable situation in a home theater. (c) the effect of changing the surround loudspeakers to floor-to-ceiling line sources. (d) what we really want.*

In the days of a monophonic surround channel, this was a great affliction, leading to the introduction of electronic decorrelation in the signals fed to the left and right surround loudspeakers in Shure HTS surround processors, and later in THX licensed processors "to diffuse the rear image and discourage localization at the closer surround loudspeaker" (Julstrom, 1987). With the discrete multichannel audio we now have, this decorrelation can be, and should be, incorporated into the master recording. No additional processing is required at playback.

This is the situation that led to the promotion of multidirectional surround loudspeakers as a means of sending more of the sound in other directions—the notion being to enhance the "diffuse" sound field. The problem is that there is no diffuse sound field in small, normally furnished rooms—there is too much absorption—and sounds reflected from the front and rear walls arrive from the wrong directions to be effective at generating impressions of space. If the design follows the well-known dipole surround loudspeaker configuration—more correctly called a bidirectional out-of-phase loudspeaker—then there is the additional consideration that the null is oriented in the direction that is most productive for creating envelopment. This has the effect of turning down the level of the useful (lateral) direct sounds and replacing them with less useful (front-back) indirect sounds. The loudspeaker may be less easily localized, but the desirable spatial effect, even for the sweet spot, has been compromised. There is another problem with this loudspeaker configuration: sound quality (Section 15.8.3).

14.3 MEASUREMENT MICROPHONES

Measurements in anechoic chambers are straightforward. The only sound of interest is that traveling a straight line from the loudspeaker to the microphone. If that axis of the microphone is calibrated to be flat, the data will be good. However, the situation is very different in reflective rooms. Sounds arrive at the microphone from multiple directions simultaneously, and in multichannel systems loudspeakers are located at various angles to the measurement point. The directivity of the microphone therefore is important.

Measurement microphones are pressure devices, and high-quality ones are delivered with calibration curves. Such microphones are omnidirectional at frequencies up to those that have wavelengths comparable to the diaphragm diameter. Above that they become progressively more directional.

However, all microphones exhibit nominally flat frequency response over a useful frequency range on specific axes. For microphones calibrated as "free field," that axis is perpendicular to the diaphragm—straight ahead. Such microphones should be aimed at the loudspeaker being measured. For "random-incidence" or "pressure" calibrated microphones, that axis is about 90° off-axis. Such microphones can be aimed at the ceiling in order to measure sound from the front L, C and R loudspeakers, and very likely the lower level surround loudspeakers. Loudspeakers in elevated positions will exhibit high-frequency rise in amounts depending on the size of the microphone and the elevation angle of the loudspeaker. For this reason, small 1/4-inch microphones are a popular choice, providing usefully accurate data over most of the frequency range from loudspeakers anywhere in the upper hemisphere. Nevertheless, without reorientation of the microphones, there will be small errors at the highest frequencies. These errors will be relevant to sound quality, but will not affect sound level calibrations (Section 14.2.2).

Good data can be gathered using 1/2-inch microphones if care is taken to ensure that the direct sound from the loudspeaker arrives along an axis of flat frequency response.

Always consult the manual for the microphone being used. Be very careful to keep the microphone diaphragm far away from reflecting and diffracting objects such as microphone stands and clamps. It is best if the microphone can be suspended using fine wires or attached at the end of a tube having diameter similar to the wired end of the microphone.

CHAPTER 15

Multichannel Audio

In the beginning, there was mono. Everything heard was stored in and reproduced from a single channel. In those early days, listeners enthused and critics applauded the efforts of Edison, Berliner and others as being the closest possible to reality. They were wrong, but a revolution in home entertainment had taken place.

It would be 50 years before stereo, motivated in part by movies, would emerge. With two channels came dramatic improvements in the impressions of direction and space. Once we got past the exaggerated "ping-pong," "hole-in-the-middle" problems of many early recordings, listeners enthused and critics applauded the efforts of artists and recording engineers as being the closest possible to reality. They were wrong again, but clearly another revolution in home entertainment had taken place.

Now, after another 50 years, multichannel audio is a reality, this time truly motivated by movies, not music. We started with 5.1 channels and now many more are possible. Is this the solution that we have been searching for?

15.1 A FEW DEFINITIONS

Monaural: listening through one ear. This term is widely misused, as in "monaural" power amplifier, a single-channel amplifier that, of course, can be listened to binaurally or, with a finger in one ear, monaurally.

Binaural: listening through two ears. Natural hearing is binaural. When the ears are exposed to the sounds in a room, we can enjoy any number of channels, binaurally. However, there is another audio interpretation of the word, and that narrowly applies to "binaural" recordings made with an anatomically correct dummy head, a mannequin, capturing the sounds arriving at each ear location, so that subsequently these two signals can be reproduced at each of the two ears. This is most commonly done through headphones, which offer excellent separation of the sounds at each ear, or through two loudspeakers using a technique called *acoustical crosstalk cancellation*. The idea is that the listener hears what the dummy head heard.

Monophonic: reproduction through a single channel.

Stereophonic: *Stereo* as a word has the basic meaning of "solid, three-dimensional." It seems that, in the early days of our industry, some influential people thought that two channels were enough to generate a three-dimensional illusion. Now, stereophonic, or just stereo, is firmly entrenched as describing two-channel sound recording and reproduction. In its original incarnation, the intent was that stereo recording would be reproduced through two loudspeakers symmetrically arrayed in front of a single listener. Nowadays, stereo recordings are enjoyed by multitudes through headphones. What is heard, though, is not stereo; it is mostly inside the head spanning the distance between the ears, with the featured artist placed just behind and maybe slightly above the nose. There may be a kind of "halo" of ambience in some recordings. This is sound reproduction without standards, but the melodies, rhythms, and lyrics get through.

Multichannel: This is an ambiguous descriptor, because it applies to two-channel stereo and to systems of any higher number of channels. At the present time, in the mass market, that number is five, plus a limited-bandwidth channel reserved for low-frequency special effects in movies. Together they are known by the descriptor 5.1. This quickly evolved to 7.1. In the upmixed systems, some sounds can be delivered by height channels derived from the 5.1 channel mix—they are artificial.

Immersive sound: More recently the system has expanded into what is called immersive sound, delivering many independent or relatively independent channels. The goal is to permit sounds to be panned horizontally and vertically around listeners. Naturally, more channels also permit much more realistic reproductions of ambience and envelopment of music in a jazz club, concert hall or cathedral. Some systems with higher channel counts—for example, 10.2 (Holman, 1996, 2001) and 22.2 (Hamasaki et al., 2004)—have been around for several years, but are not (yet) commercial. I have heard examples that were impressive. Currently there are commercial systems from Dolby Atmos, Auro-3D and DTS-X. Each has several levels of complexity, ranging up to 62 channels in elaborate cinema setups and up to 32 channels in home theaters. I have heard breathtakingly realistic concert hall and cathedral renderings in Auro-3D, even while moving around the home theater. We can only hope that the other systems will incorporate the music domain in addition to augmenting blockbuster movies. The situation is still somewhat fluid at time of writing, with compatibility issues in the delivery and processing of the digital streams and the number and placement of loudspeakers. These ambiguities exist in both professional and consumer systems.

Bass management: a signal processing option in surround processors with which it is possible to combine the low frequencies from any or all of the channels, add them to the (level corrected) low-frequency effects (LFE) channel and deliver the combination signal to a subwoofer output. Bass sounds will be stripped from all channels in which the loudspeakers have been identified as "small," and will be reproduced through those identified as "large," as well, of course, as the subwoofer(s). See Section 8.5 for details. Holman (1998) gives a good history.

Downmixer, downwards conversion, down-converter: an algorithm that combines the components of a multichannel signal, making it suitable for reproduction through a smaller number of channels. It is also widely used to store the multichannel signal in a smaller number of channels. This is not a "discrete" process, there is inevitable cross-channel leakage when the signal is subsequently upmixed. Dolby Surround/ Dolby Stereo is an example of a specific kind of downmixer, processing four channels for storage in two, as Lt + Rt Dolby Digital, a 5.1 channel signal, can be downmixed by the Dolby Digital decoder into mono, stereo, or Lt + Rt outputs. Lt + Rt can then be upmixed to 5.1, 6.1 or 7.1 channels by Dolby ProLogic, or any of several other competing algorithms.

Upmixer, upwards conversion, up-converter (aka surround processor): an algorithm or a device that processes a signal making it suitable for reproduction through a larger number of channels. Two-channel signals can be upmixed for reproduction through five or more channels, or five channels can be upmixed for reproduction through more. In one common application, multichannel recordings are downmixed (encoded) with a specific form of upmix decoding in mind, as in the case of Dolby Surround (which generates Lt + Rt composite signals) and Dolby ProLogic (which upmixes those signals into 5.1 channels). Other upmixers are designed to operate on these same encoded two-channel signals, arguing that they have a superior strategy to generate a multichannel result for listeners. Finally, upmixers may also be optimized to convert standard stereo music recordings into multichannel versions. Mostly, the front soundstage is minimally altered, and uncorrelated sounds are extracted and sent, usually with delays, to surround channels. These are known as "blind" upmixers because the stereo recordings were not made with this processing in mind and, consequently, there is no way of predicting the result. Some recordings inevitably will work better than others. In the uncontrolled reality of life, it is highly probable that any two-channel signal will be upmixed in some fashion. It is widely used in car audio.

15.2 THE BIRTH OF MULTICHANNEL AUDIO

Monophonic reproduction conveys most of the musically important dimensions: melody, harmony, timbre, tempo and reverberation, but no sense of soundstage width, depth or spatial envelopment—of being there. In the 1930s, the essential principles by which the missing directional and spatial elements could be communicated were understood, but there were technical and cost limitations to what was practical. It is humbling to read the wisdom embodied in the Blumlein-EMI patent (Blumlein, 1933) applied for in 1931, describing two-channel stereo techniques that would wait 25 years before being popularized. It is especially interesting to learn that the motivation for stereo came in part from a desire to improve cinema sound (Alexander, 1999, pp. 60, 80). Cinema played a central role in the development of present-day incarnations of multichannel audio and now immersive audio. It is curious to read purist audiophiles belittling movie sound when all of the soundstage, space and imaging effects they so treasure would be unlocked if they were to embrace multichannel audio. The sad part of that story is that the mainstream music recording industry has dragged its feet at every stage, beginning with stereo, resisting efforts to add channels.

The benefits of more channels were recognized very early in the history of audio, notably when the Bell Telephone Laboratories investigated the reproduction of a realistic auditory perspective (Steinberg and Snow, 1934). They concluded that there were two alternative reproduction methods that would work: binaural and multichannel.

Binaural (dummy head) recording and headphone reproduction is the only justification for the "we have two ears, therefore we need two channels" argument. Two-channel stereo as we have known it is the simplest form of multichannel reproduction—it is not binaural.

Multichannel loudspeaker reproduction is more obvious, because each channel and its associated loudspeaker creates an independently localizable sound source, and interactions between multiple loudspeakers create opportunities for phantom sources and impressions of movement. Inevitably, the question "how many channels are necessary?" must be answered. Bell Labs scientists assumed that a great many channels would be

BINAURAL RECORDING AND REPRODUCTION

This has been a great tease to the audio industry. It is a true encode/decode system. Timbral, directional and spatial cues are "encoded" into the two recorded channels, one for each ear, by the sounds approaching the dummy head from different directions, arriving at the ears at different times and with different amplitudes because of the acoustical interaction with the head and torso. All of this is captured at the entrances to the mannequin's ear canals or at eardrum locations. If the appropriate sounds can be delivered to the same reference points on playback, one should hear what would have been heard by a human listener seated where the recording mannequin was placed. The "decoding" is done by the ears and brain of the listener.

Does it work? In the beginning it really never had a chance. Microphones were large, noisy, cumbersome things—impossible to fit into a modeled ear canal. Headphones were rudimentary transducers designed for delivering the beeps of Morse code or basic voice communication, not for reconstructing a Beethoven concert. Things improved, of course, but a lingering bias against being tethered to and isolated under a pair of headphones posed resistance in the marketplace. The current "portable music" generation has no such problem, and headphone listening is widespread. But there was another negative bias, the fact that most of the time, for most listeners, the sound was perceived to

be inside or very close to the head. The lack of externalization was a problem.

For every problem there is a solution. Now, with the benefit of exceptionally accurate and sensitive small microphones and superb headphones, the sound quality can be impressive. Adding head tracking is the final step. In scientific work, a Binaural Room Scanning (BRS) system characterizes the sound of a venue using a dummy head that is rotated in small angular increments, measuring at every stop. On playback, headphones fitted with head tracking hardware follows head movements, changing the sounds at the ears to what would have been heard in the original situation. The result is that sounds are perceived to be external, when the head moves the room stays still, and the listener perceives the sound sources to be where they should be. Even simplified versions of this system substantially alleviate the front-back reversal and in-head localization problems. With, or even without, customization for individual ears, the result is a credible reconstruction of a three-dimensional acoustical event. These are systems likely to be used in virtual reality schemes. Nowadays it is possible to digitally synthesize binaural sounds that are localized in various positions around a listener.

necessary to capture and reproduce the directional and spatial complexities of a musical front soundstage—not even attempting to recreate a surrounding sense of envelopment. Their goal was to capture a performance in one hall using a row of microphones across the front of a stage, and then reproduce that "wavefront" in another hall. One loudspeaker would be used for every microphone channel in a similar position arrayed across the front of the performance stage. There was no need to capture ambient sounds, as the playback hall had its own reverberation. Being practical people, they investigated the possibilities of simplification, and concluded that, while two channels could yield acceptable results for a solitary listener, three channels (left, center and right) would be a workable minimum to establish the illusion of a stable front soundstage for a group of listeners (Steinberg and Snow, 1934).

By 1953 ideas were more developed and, in a paper entitled "Basic Principles of Stereophonic Sound," Snow (1953) describes a stereophonic system as one having two *or more* channels and loudspeakers. He says: "The number of channels will depend upon the size of the stage and listening rooms, and the precision in localization required." He goes on:

For a use such as rendition of music in the home, where economy is required and accurate placement of sources is not of great importance if the feeling of separation of sources is preserved, two-channel reproduction is of real importance.

So, two channels were understood to be a compromise, "good enough for the home" or words to that effect, and that is exactly what we ended up with. The choice had nothing to do with aesthetic ideals, but with technical reality that, at the time stereo was commercialized, nobody knew how to store more than two channels in the groove of an LP disc.

Vermeulen (1956) had a superb understanding of what stereo could and could not do, saying:

Although stereophonic reproduction can give a sufficiently accurate imitation of an orchestra [the soundstage], it is necessary to imitate also the wall reflections of the concert hall, in order that the reproduction may be musically satisfactory. This can be done by means of several loudspeakers, distributed over the listening room, to which the signal is fed with different time-lags. The diffused character of the artificial reverberation thus obtained [creating envelopment] seems to be even more important than the reverberation time.

He was absolutely right on all counts. The numerous spatial enhancers over the subsequent years, up to and including contemporary stereo-to-multichannel upmix algorithms, absolutely support his insight.

Around that same time, the film industry managed to succeed where the music side of the audio industry failed, and several major films were released with multichannel surround sound to accompany their panoramic images. These were discrete channels recorded on magnetic stripes added to the film.

Although they were very successful from the artistic point of view, the technology suffered because of the high costs of production and duplication. Films reverted to monophonic optical sound tracks, at least until the development of the "dual bilateral light valve." This allowed each side of the optical sound track to be independently modulated, and two channels were possible. However, optical sound tracks were notable for their noise and distortion forcing severe high-frequency rolloffs on playback, which made magnetic, and subsequent digital sound tracks vastly preferable.

As we will see, the two-channel mode did not last long, even with upmixing, and, ironically, it has been the film industry, not the audio industry or audiophiles, that has driven the introduction of discrete multichannel sound reproduction in homes—and is still doing so. One hopes that digital streaming can liberate multichannel audio for musical applications. When well done, the results are very impressive, and from my perspective, the ability to hear good music in a realistic spatial reconstruction is high entertainment, especially when it is accompanied by a large-screen video of the concert.

15.3 STEREO—AN IMPORTANT BEGINNING

In contrast with binaural audio, stereophony is not endowed with an underlying encode/decode system. It is merely a two-channel delivery mechanism. Yes, there have always been some generally understood rules about setting up the playback loudspeakers and about sitting in the symmetrical "stereo seat." But everybody knows that these simple rules are routinely violated. In professional recording control rooms there was an attempt to adhere to standard playback geometry (loudspeakers at ±30°, or so), but otherwise it was simply wide open for creativity.

Over the years, the struggle to capture, store and reproduce realistic senses of direction and space from two channels and loudspeakers has been a mighty one. There has been no single perfectly satisfactory solution, even after all these years. Professional audio engineers have experimented with many variations of microphone types and techniques, trying to capture the directional and spatial essence of live musical events. In the end, most have apparently given up trying and create totally artificial illusions. This is acceptable because it is art. Numerous analog and digital signal processors are available to expand the soundstage. Even simplified binaural crosstalk-cancellation processing has been used to place sounds outside the span of the loudspeaker pair, even though it can only work for a person in the stereo seat.

At the playback end, countless loudspeaker designs have attempted to present a more gratifying sense of space and envelopment. What does one say about a system that accommodates loudspeakers having directional characteristics ranging from omnidirectional, through bi-directional in-phase (so-called bipole), bi-directional out-of-phase (dipole), predominantly backward firing, and predominantly forward firing? The nature of the direct and reflected sounds arriving at the listeners' ears from these different designs runs the entire gamut of possibilities, and yet one finds intelligent advocates for them all.

From this perspective, stereo seems less like a system and more like a foundation for individual experimentation. Older audiophiles may remember the rudimentary "four-dimensional system" employing four loudspeakers sold by Dynaco as the QD-1

Quadapter. It delivered a sum of both channels to a center-front loudspeaker and a difference signal to a center-rear loudspeaker. David Hafler, the inventor, proposed a quadraphonic multichannel recording system to complete the package (Hafler, 1970).

Taking a different approach, the "Sonic Hologram" (Carver, 1982), was a simplified version of binaural crosstalk cancellation in the electronic signal path, while Polk built it into the SDA-1 loudspeakers. Lexicon's digitally implemented "panorama" mode allowed for individual setup adjustments to cater to different loudspeaker/listener geometries. The goal of all of these was to expand the soundstage beyond the stereo loudspeakers, potentially out to ±90°. None was doing anything that was intended by the recording artists, but all were attempting to reward listeners with a more enveloping listening experience. A host of digital "hall" and other artificial-reverberation effects came along in this period; they came to be known as "DSP" effects. Most were not very good, and some were laughably bad. All of these devices and processes existed because listeners found a dominant direct sound field from two loudspeakers to be insufficiently rewarding.

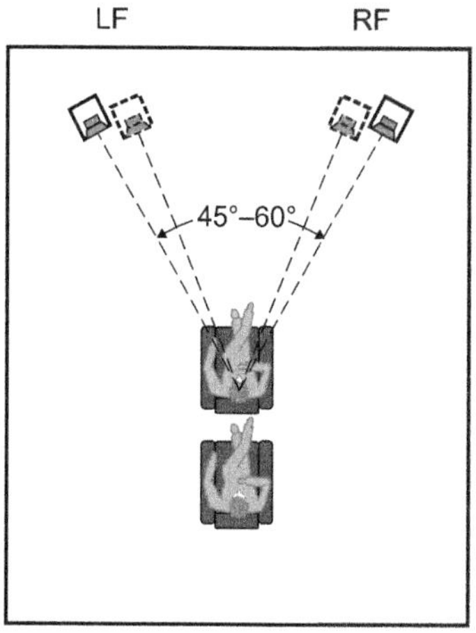

Eickmeier (1989) proposed a multidirectional loudspeaker that employed room boundaries as reflecting surfaces to create a scaled-down collection of reflections that mimicked the incident angles of those from an orchestra in a rectangular concert hall. Undoubtedly there will be some spatial enhancement, but perfection will be elusive because all of the direct and reflected sounds are identical, originating in the same two loudspeakers, and the delays in reflected sounds in a domestic room will be tiny compared to those in concert halls. Listener envelopment, arguably the most important perceptual factor in concert halls, is created by late sounds arriving from the sides (e.g., Bradley et al., 2000). Delays exceeding 80 ms are common and these don't happen in small room reflections. This is a justification for multichannel audio.

An ongoing attempt to extract the maximum from legacy stereo recordings is Ambiophonics (Glasgal, 2001, 2003 and www.ambiophonics.org). It has gone through several phases of evolution, incorporating binaural techniques and complex synthesis of spatial effects to enhance sound delivery.

Added to these fundamental issues is the inconvenience of the "stereo seat." Because of the stereo seat, two-channel stereo is an antisocial system—only one listener can hear it the way it was created. If one leans a little to the left or right, the featured artist flops into the left or right loudspeaker and the soundstage distorts. When we sit up straight, the featured artist floats as a phantom image between the loudspeakers, often perceived to be a little too far back and with a sense of spaciousness that is different from the images in the left and right loudspeakers (see Figure 7.1 and the associated discussion).

This puts the sound image more or less where it belongs in space for the person in the stereo seat, but then there is another problem: the sound quality is altered because

FIGURE 15.1 *Two-channel stereo. The ±30° arrangement is a widespread standard for music recording and reproduction, although many setups employ a smaller separation, especially those associated with video playback. To hear the phantom center image, and any other panned images between the loudspeakers, correctly located listeners must be on the symmetrical axis between the loudspeakers. Away from the symmetrical axis, in cars and through headphones, we don't hear real stereo; we hear a spatially distorted, but still entertaining, rendering.*

of the acoustical crosstalk, as described graphically in Figure 7.2. As shown in that figure, the audibility of this significant dip in the frequency response depends entirely on how much reflected sound there is to dilute the effect; room reflections are beneficial. In the reflection-controlled environment of a typical recording control room it is very audible. If the recording or mastering engineer attempts to compensate for this effect with equalization, another problem is created. When such recordings are played through an upmix algorithm, and the featured artist is sent to a center channel loudspeaker, the sound will be too bright.

15.3.1 Loudspeakers as Stereo Image Stabilizers

Moving away from the sweet spot, the phantom center image—the featured artist—moves with you, quickly jumping to the nearer loudspeaker. Alleviating this problem has been a long-standing challenge. Early on, listeners figured out that the balance control could be used to compensate somewhat for an off-center listening position. But this worked only for a single listener.

Also in the early days of stereo, Hugh Brittain, notable for other audio achievements while he was at GEC in England, promoted the idea of crossing the axes of left and right loudspeakers in front of the listener. His name became associated with that setup. Knowing that loudspeakers are somewhat directional, especially at high frequencies, the idea was that as one moved away from center, the more distant loudspeaker would get louder and the nearer one would be reduced in level, thereby achieving a crude amplitude panning to compensate for the delay error; the phantom image would be perceived to move less. Obviously it only works at mid- to high frequencies for typical cone/dome loudspeakers.

In the early 1980s I visited Stig Carlsson in Stockholm. He and his not-quite-omnidirectional loudspeakers had a following in Europe. These "ortho-directional (OD)" designs aimed at creating an active reflected sound field in rooms and, with the directional asymmetry, attempted to stabilize the stereo image over a larger listening area. I heard several existing and developmental models and all had varying degrees of success, depending greatly on the recording. All of them generated an active reflected sound field that solved the interaural crosstalk coloration. The randomized sounds arriving at the ears softened the images, which made movements of them less obvious. Steady-state room curves were quite smooth and flat. Readers may recall my own experience with an almost omnidirectional loudspeaker shown in Figures 7.19 and 7.20. If the reflected sounds are abundant and timbrally matched to the direct sounds, the subjective effects can be pleasant indeed.

A few years later, Mark Davis (1987) applied science to developing the dbx Soundfield loudspeaker. This was an arrangement of 14 drivers on the four faces of an enclosure that, in acoustical summation, radiated a 360° sound field in which the asymmetrical directivity had been carefully tailored. The purpose was to deliver direct sounds to off-center listeners that stabilized the location of the phantom center image. It was based on empirical data relating delay and amplitude panning, and differed from other attempts in that it was effective over a wide bandwidth. In addition to this direct sound feature, the loudspeaker radiated generously in all directions, delivering what many listeners found to be a pleasant spaciousness. My 1986 exposure to the product was limited to the Soundfield 10 model in which the transducer count was reduced to eight. In any array more transducers are better, but cost and complexity are limiting factors. Unfortunately the spatially averaged computer data has been lost, leaving only a single listener-axis frequency response that is not very smooth because of acoustical interference from the sparse transducer array. Polar plots confirmed Davis's published goals. Listener notes indicated that it generated an attractively spacious illusion, with soft, and reasonably stable phantom images. Even simpler implementations followed, and then the product disappeared.

Addressing the problem in a different way, the recently demonstrated Lexicon SoundSteer SL-1 uses a digitally controlled array of transducers to generate a uniformly shaped sound field over most of the frequency range. The sound pattern can be altered from omnidirectional to a narrow beam, which can then be digitally steered. Using a wireless interface, the listener can move to different locations, and the loudspeakers will deliver direct sounds of the necessary amplitude and timing to create a new sweet spot, or several of them stored for convenient recall.

There are numerous additional possibilities for adaptable array loudspeakers including creating multiple independent beams aimed at reflecting surfaces to simulate additional channels as was done in 2003 by Pioneer's PDSP-1 digital sound projector loudspeaker (254 individually controlled and amplified transducers), and a year later in Yamaha's 40-driver YSP-1. In this way a "stereo" pair of loudspeakers can provide a gateway to multichannel audio for those with the right configuration of listening room. The reducing cost of digital processing and small power amplifiers allow for impressive combinations of state-of-the-art sound quality and hitherto impossible acoustical and auditory manipulations.

15.4 QUADRAPHONICS—STEREO TIMES TWO

In the 1970s, we broke the two-channel doldrums with a misadventure into four-channel, called quadraphonics. The intentions were laudable—to deliver an enriched sense of direction and space. The key to achieving this was in the ability to store four channels of information in the existing two channels, on LPs at that time, and then to recover them.

There were two categories of systems in use at the time, matrixed and discrete. The matrixed systems downmixed four signals into the bandwidth normally used for two channels. In doing this, something had to be compromised and, as a result, all of the channels did not have equal channel separation. In other words, information that was supposed to be only in one channel would appear in smaller quantities in some or all of the other channels. The result of this "crosstalk" was confusion about where the sound was coming from and an inordinate sensitivity to listener position; leaning left, right, forward or back caused the entire the sound panorama to exhibit a bias in that direction.

Various forms of signal-adaptive "steering" were devised to assist the directional illusions during the playback process. The "alphabet soup" is memorable: SQ from

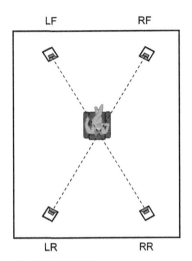

LF   RF

LR   RR

A quadraphonic listening arrangement showing the side-to-side/front-to-back restricted seating caused, mainly, by acoustical crosstalk, signal leakage among the channels.

CBS, QS from Sansui, E-V from ElectroVoice and others. Peter Scheiber, a musician with a technological bent, figures prominently as a pioneer in matrix design, with his patented encoder and decoder ideas being incorporated into many of the systems. The best of these systems were remarkably good in creating the illusion of completely separate, or discrete, channels when an image was panned around the room. However, this clear separation breaks down when there is a demand for several simultaneously occurring discrete images. In the limit, the steering ceases and we listen through the raw matrix, with its generous crosstalk leakage.

Ultimately, four discrete channels were needed. However, achieving this on the vinyl LPs required that the recorded bandwidth be extended to about 50 kHz—quite a challenge. Nevertheless, it was accomplished, as CD-4 from JVC and, although this quadraphonic format was short-lived, the technology necessary to achieve the expanded bandwidth had a lasting benefit on the quality of conventional two-channel LPs. Among these were half-speed cutting processes, better pressings, and playback cartridges with high compliance, low-moving mass and exotically shaped styli combined to yield wider bandwidth and reduced tracing and tracking distortions. All of these developments had a continuing positive influence on the vinyl disc industry.

Discrete multichannel tape recordings were available, but open reel tape was a nuisance to say the least, and high-quality packaged tape formats (e.g., cassettes) were not ready for true high-fidelity multichannel sound.

Years passed, with the industry unable to agree on a single standard, which was an intolerable situation from a business perspective. There were issues with mono, stereo and broadcast compatibility (Crompton, 1974). Eventually, the whole thing dissolved (Torick, 1998). The industry lost money and credibility, and customers were justifiably disconcerted after many of them invested in the soon-to-be-useless hardware.

Looking back on this unfortunate episode in the history of audio, one can see another reason for failure—the system was not psychoacoustically well founded. Lacking an underlying encode/decode rationale, the problems of two-channel stereo were simply compounded. There were even notions of "panning" images front to back using conventional amplitude-panning techniques, something that Ratliff (1974) and others have found problems with—the ears are in the wrong locations for that to work. The quadraphonic square array of left and right, front and rear, was still an antisocial system, with even stricter rules. The sweet spot now was constrained in the front-back and the left-right direction. Most importantly, there was no center channel: a basic requirement needed to eliminate the stereo seat.

Placing the additional channels symmetrically behind the listener is now known not to be optimum for generating a sense of spaciousness. Figure 15.3 shows a comparison from experiments looking at the effect of loudspeaker number and placement on the creation of a sense of spaciousness, which is correlated with a measurement of interaural cross-correlation coefficient (IACC). The solid line curve is the target.

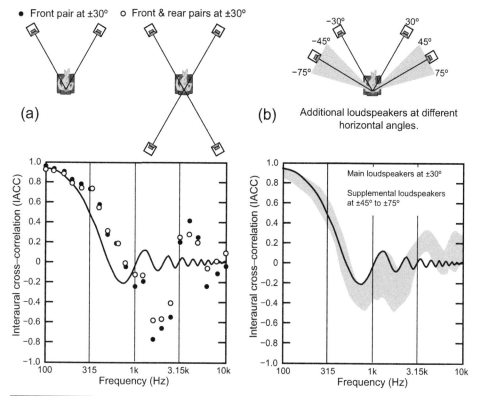

FIGURE 15.3 *(a) A comparison of IACC measured for a stereo pair of loudspeakers at ±30° and with a second pair of loudspeakers added at the same angles behind the listener. Signals delivered by the individual loudspeakers were narrow bands of uncorrelated noise, as might be recorded in a perfectly diffuse sound field. The target curve, the solid line, is the IACC of a perfectly diffuse sound field. (b) As (a), but with the second loudspeaker pair located at different angles within the range ±45° to ±75°. All of the measurements fall within the shaded area.*
Adapted from Tohyama and Suzuki (1989), copyright Acoustical Society of America.

Tohyama and Suzuki (1989) looked at a few arrangements of two and four loud-speakers, comparing measured IACCs to those found in a truly diffuse sound field—an all-but-unachievable goal in the real world. Results shown in Figure 15.3a indicate, not surprisingly, that two-channel stereo did not come close to replicating the diffuse-field IACC. The real news, though, is that doubling the number of channels by adding a pair of loudspeakers behind the listener at the same angular separation did not really change anything. The solid (two channels) and open (four channels) dots in the figure are very similarly distributed, and neither matches the target curve. This was the original layout for quadraphonic sound—obviously not an optimum concept.

When the extra pair of loudspeakers was deployed at several angles between 45° and 75°, as shown in Figure 15.3b, the IACC results all move closer to the target curve. Two things have changed: all four loudspeakers are in front of the listener, and they are at different horizontal angles. Which is responsible for the improvement? Later it will be seen that it is the difference in the angles that is the key. Whether the loudspeakers are in front of or behind the listeners is less important. One could debate the value of being able to place musicians behind listeners, but there is no debate that in movies one needs to be able to perceive flyovers. Therefore in future developments a rearward bias was preferred.

The important message in these data is that the placement of loudspeakers may be as important as the number, and placement more to the side is advantageous. Sounds arriving from the rear are extremely rare in the standard repertoire of music, but the need for a credible spatial impression is common. Ironically, a 1971 publication entitled *Subjective Assessment of Multichannel Reproduction* (Nakayama et al., 1971) showed that listeners preferred surround loudspeakers positioned to the sides, compared to ones placed behind, awarding subjective rating scores that were two to four times higher. It seems that nobody with any influence read it.

Fortunately, much of the clever technical active matrix innovation that went into quadraphonics was not wasted; it went to the movies where, finally, a center channel was a component.

15.5 MULTICHANNEL AUDIO—CINEMA TO THE RESCUE

The key technological ideas underlying quadraphonics were:

- Four audio channels stored in two channels

- The ability to reconstruct them with good separation by using adaptive matrices—electronically enhanced steering.

Dolby Laboratories Inc. was well connected to the real multichannel pioneers, the moviemakers, in the application of its noise reduction system to stereo optical sound tracks. Putting the pieces together, Dolby rearranged the channel configuration to one better suited to film use: left, center and right across the front, and a single surround channel that was used to drive several loudspeakers arranged beside and behind the audience. All of this was stored in two audio-bandwidth channels. With the appropriate adjustments to the encode matrix, and to the steering algorithm in the active decoding matrix, in 1976 they came up with the system that has become almost universal in films and cinemas: Dolby Surround or, as it is also known in the movie business, Dolby Stereo.

This system was subject to some basic rules that have set a standard for multichannel film sound: well-placed dialogue in the center of the screen, and music and sound effects across the front and in the surround channel. Reverberation and other ambience sounds are steered into the surround channel, as are various sound effects. At times the audience can be surrounded by crowd sounds as in a football game, or it can be transported to a giant reverberant cave or gymnasium, or it can be inside the confines of a car engaged in a dramatic chase, or it can be treated to an intimately whispered conversation between lovers where the impression is that of being embarrassingly close. Because the optical film sound track was relatively noisy, even with Dolby noise reduction, and relatively distorted, occasional "splatters" of vocal sibilants would leak into the surround channel and be radiated by the surround loudspeakers causing them to be localized. Consequently the surround channel was attenuated above about 7 kHz, eliminating the annoying misbehavior, but also degrading the sound quality.

To achieve this dynamic range of spatial experiences requires a flexible multichannel system, controlled-directivity loudspeakers, and some control over the acoustics of the playback environment. When it is done well, it is remarkably entertaining. There are still better and worse seats in the house, but there are numerous acceptably good seats toward the center.

It is important to note that the characteristics of the encoding matrix, the active decoding matrix, the spectral, directional and temporal properties of the loudspeakers and room all are integral parts of the functioning of these systems. Fortunately, the film industry recognized the need for standardization and so, for many years, it has tried to ensure that sound dubbing stages, where film sound tracks are assembled, resemble cinemas, where audiences are to enjoy the results.

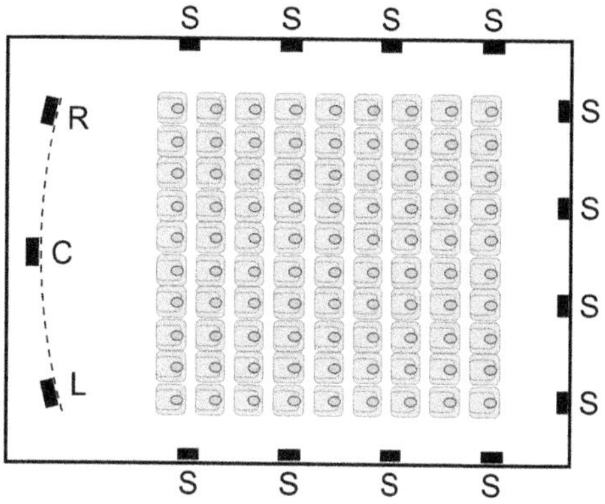

FIGURE 15.4 *Dolby Stereo as it was developed in 1976: three channels across the front and a single surround channel delivered identically to loudspeakers surrounding the audience. This is the foundation from which the home theater and multichannel music technologies we know today evolved.*

15.6 MULTICHANNEL AUDIO COMES HOME

With the popularity of watching movies at home, it was natural that Dolby Surround made its way there on videotape, laserdisc, broadcast television and all of the streaming delivery methods that have followed. Adapting it to the smaller environment required only minor adjustments to the playback apparatus. Reducing the number of surround loudspeakers to two ensured greater consumer acceptance, and recommending the placement of these loudspeakers to the sides of the listeners ensured that they would be most effective in creating the required illusions of space and envelopment (see Figure 15.5). The slight rearward bias in the surround loudspeakers is to allow for credible flyovers; it is not a requirement for envelopment.

The situation depicted in Figure 15.5a is straightforward: a sound, usually a sound effect, is hard-panned to one of the surround loudspeakers and is simply heard to come from that location. The situation in Figure 15.5b is more complex. Here the surrounds, following a strong direct sound from one or more front channels, deliver sounds intended to describe an acoustic space. Delaying the sounds to the surround speakers uses the precedence effect to ensure that, even in a small room, the surround sounds are perceptually subordinate to the front channels. The solid lines running from the side surround loudspeakers indicate sounds that are useful in reducing IACC, thereby augmenting the sense of space. They arrive at the listener from the side, which maximizes the

FIGURE 15.5

Reflected sounds for two usages of surround loudspeakers: (a) as an independent sound source and (b) as a supplement to the front loudspeakers reproducing delayed sounds intended to yield impressions of distance, spaciousness or envelopment. Delays and sound levels shown are relative to the direct sound, assuming perfect reflections. For a 21.5 ft (6.5 m) × 16 ft (4.9 m) room.

(a) Discrete sounds panned to a side surround channel. *The perception: sound coming from that loudspeaker.*

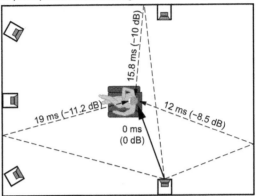

(b) Surround channels provide delayed "reflections" following the direct sound from any or all front channels. Delays are as shown in (a) plus delays introduced in the recording process between the front channels and each surround. *The perception: spaciousness, envelopment.*

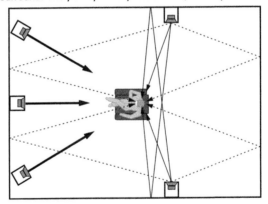

interaural decorrelation. The dotted reflections, on the other hand, are not very useful as they arrive from directions close to the median plane of the listener.

There is a clue here about the utility of bidirectional surround loudspeakers. Those configured as bipoles have wide dispersion and will deliver strong direct sounds to a large audience, including the money seat as shown in (b). Those configured as dipoles will attenuate the direct sounds, leaving the less useful front and rear reflections. This will be discussed later.

At the outset, a simple fixed-matrix version was available in entry-level consumer systems. The fixed-matrix systems exhibited so much crosstalk among the channels that listeners were surrounded by sound most of the time, even when it was inappropriate.

As I recall, it was Fosgate and Shure HTS who brought the first active-matrix decoders to the home-theater market. Julstrom (1987) describes the HTS device, which had an innovative feature, an "image-spreading technique . . . to diffuse the rear image and discourage localization at the closer surround loudspeaker." It also avoided the mono-phonic "in-head" localization effect that could be heard by listeners seated on the center line of the room. This was done with a complementary-comb-filter technique to intro-duce differences between the left- and right-side surround loudspeakers, decorrelating the otherwise monophonic surround channel. Decorrelation of the surrounds was included as a feature of Home THX a few years later. These were discrete-component products, and they came at premium prices. When low-cost silicon chips incorporating the active-ma-trix Dolby ProLogic decoder hit the market, home entertainment entered a new era.

Having enjoyed the spatial illusions in movies, it was inevitable that listeners would play conventional stereo recordings through a Dolby ProLogic processor. The results were spotty; some recordings worked quite well, and others didn't. The translation of a phantom center to a dedicated center loudspeaker was unreliable, often being overly prominent. The high-frequency rolloff in the surround channel was also noticeable as dullness in the surround sound field. The active matrix steering could sometimes be heard manipulating the music. Recordings made specifically for Dolby Surround were better, but they failed to establish a significant following in the music recording indus-try. There was work yet to be done.

15.6.1 THX Embellishments

In a natural succession to their THX program for certifying cinema sound systems, around 1990 Lucasfilm established a licensing scheme for certain features intended to enhance, or in certain ways regulate, the performance of home theater systems based on Dolby ProLogic decoders. Home THX, as it was called, added features to a basic ProLogic processor and to the loudspeakers used in home theater systems, and set some minimum performance standards for the electronics and loudspeakers. At a time when the market was being inundated with small, inexpensive add-on center and surround loudspeakers and amplifiers, THX made a clear statement that that would not do; all channels had to meet high standards. Tomlinson Holman deserves credit for assembling this amalgam of existing and novel features into what became an early benchmark for consumer home theater. Arguably the most positive lasting contribution to the industry has been the certification program for components that ensures that their specifications are adequate for satisfactory real-world home theaters, that their performances live up to those specifications, and that all of the functions in the ever more complex surround processors actually work as they should.

The first generation of THX certified components also embodied some features unique to the THX program. Much has changed since those early days, so not all of the original THX embellishments continue to be relevant and some have been phased out. The THX embellishments of relevance to this discussion are shown in the following list. It is important to spend a little time discussing them because their influence is still felt within the home-theater industry. Comments have been added that attempt to put them into a present-day context:

1. High- and low-pass filters in the surround processor to approximate a proper crossover between the subwoofer and satellite loudspeakers. A glaring omission

from previous systems was any consideration of how the outputs of subwoofers and satellite loudspeakers merged within the crossover region. It was a matter left entirely to chance, and inevitably there were many examples of really bad upper bass sound. However, preset electronic filters cannot do it all, because the loudspeakers and the room greatly influence the final result and these effects cannot be predicted in advance. The high-pass characteristics of loudspeakers vary substantially (although THX-approved loudspeakers were somewhat controlled). More importantly, the room is part of the system; as has been explained in Chapters 8 and 9, the listening room is powerfully influential in this (80 Hz) frequency region. The subwoofer and satellite loudspeakers, and the listener of course, are in different locations, so there can be no assurance that the low-pass and high-pass slopes will add as intended. Only in-situ acoustical measurements and equalization can do that. At the time this was impractical to consider, but the application of *any* high-pass filter to the satellite loudspeakers will prevent them from trying to duplicate the job of the subwoofer, and the entire system should be able to play louder, with less distortion—so the basic idea was constructive.

2. Electronic decorrelation between the left and right surround signals. Picking up the Shure HTS idea (Julstrom, 1987), THX recommended decorrelation of the signals supplied to the left and right surround loudspeakers. A pitch-shifting algorithm was suggested but in fact they approved other forms as well. Today's upmixers incorporate decorrelation or at least delays in the process of subdividing the surround signal(s) into multiple channels. With current discrete formats, it is up to the recording engineers to decide how much decorrelation is appropriate for the two main surround channels; no further processing is needed.

3. "Timbre matching" of the surround channels. All loudspeakers in a surround system should radiate similarly good sound. To evaluate this one must face each loudspeaker in turn. It is wrong to think that sounds from all directions should sound the same when the listener is facing the front center channel. The differences are built into the head-related transfer functions (HRTFs) in the binaural hearing process, and give us information about where sounds originate. Because home systems are now expected to reproduce momentary discrete left, right, front, back and elevated sounds in movie, all channels need to be as similar as possible in performance.

4. "Re-equalization" of the movie sound tracks: it is stated that this is a compensation for excessive treble that is sometimes found in film soundtracks as a result of the sound systems used in large cinemas and the manner in which they are calibrated. A single high-frequency correction curve was chosen. As discussed in detail in Chapter 11, cinema sound is not a constant, and the variations involve both bass and treble. It is easy to understand the original intent, but the present-day reality is that for listeners who are fussy about timbre and spectral balance, traditional tone controls are the answer.

5. Limited vertical dispersion of front (L, C, R) loudspeakers. It was a flawed concept from the outset, and it now seems that this directivity requirement has been removed from the THX specifications.

6. Dipolar radiation pattern for the surround loudspeakers. The argument for this is often expressed as a means of enhancing the diffuse sound field in the room. Such loudspeakers are often described as "diffuse" sound sources. The reality is that in domestic home theaters there is no consequential diffuse sound field—there is too much absorption—and loudspeakers cannot "diffuse" sound. In fact, we would not want it if it could exist; the sensations of space are in the multichannel recordings. The real effect of the dipole null is to attenuate the high frequencies for those listeners close to the sides of the room to reduce the tendency to localize the loudspeakers. This is not a bad idea, but unfortunately it leads to compromised sound quality, as will be discussed in a following section. These designs are not recommended for immersive audio, where the loudspeakers are intended to be localizable sound sources with high sound quality.

15.7 HOW MANY LOUDSPEAKERS AND WHERE?

Let us begin by defining the duties of a surround-sound system. It will be assumed that such a system is intended to cater to more than one listener, although it is likely to work best for the "money" seat.

- **Localization.** The perception of direction: where the sound is coming from. The minimum number of locations would be the number of discrete or steered channels/loudspeakers in the system. Beyond that, we would need to rely on phantom images floating between pairs of loudspeakers, those across the front being familiar because of stereo. With a center channel these are even more stable. Other opportunities exist, for example, between the front and sides. These are mainly useful to convey a brief sense of movement. Fixed-position panned images can have different locations, depending on where one is sitting, and the locations of the ears means that front-to-back panning is grossly imprecise. In reality movement is conveyed by a simple fade from one loudspeaker to another. Anyone seated away from the sweet spot would hear a distorted panorama of phantom sound images and degraded spaciousness. Conclusion: more discrete channels driving more loudspeakers are desirable; phantom images are not useful if perceived location is important and there are multiple listeners.

- **Distance.** By the appropriate addition of delayed, reflected, sounds in the recordings it is possible to create impressions of distance, moving the apparent locations well beyond the loudspeakers themselves.

- **Spaciousness and envelopment.** The sense of being in a different space, surrounded by ambiguously localized sound. This is a very important function.

The possibilities for improving localization are limited by the practical number of channels the industry feels customers will buy and install in their homes. For most people, it seems that 5.1 is difficult to accommodate in their living spaces. However, for enthusiasts with the space and money, there are immersive systems up to 32 channels available for home installations. These can deliver superheroes and aliens flying around the room, and as much ambience as one can tolerate.

However, the number of screen channels remains at three for most systems. In spite of that for most of the time all on-screen sounds are delivered by the center channel. I get up from my chair from time to time just to satisfy my curiosity because, for sounds associated with on-screen action there is the powerful "ventriloquism" effect. We hear the sound coming from where we see the lips move, the gun flash, or the door slam. The hours spent by moviegoers listening to a monophonic center channel as action moves around on the screen suggests that it works very well. Only big-budget blockbuster movies do much panning of sounds across the screen. Why? It costs more to do and complicates the editing process that can sometimes go on up to hours before a film is released. When the locations of sounds are few and stable, editing becomes mostly a visual and storytelling operation.

Sounds originating off-screen are frequently brief sound effects for which no real precision is demanded (nor delivered in the cinema situation). Other off-screen sounds fall into the broad "ambience" category where, if anything, ambiguity of location is desirable. A sense of distance is an important factor in "transporting" listeners out of the listening room, but this factor is difficult to separate from the essential perception of envelopment, the sense of being in a different, larger, space.

15.7.1 Optimizing the Delivery of "Envelopment"

As discussed earlier, interaural cross-correlation coefficient (IACC) is a strong correlate of a perception of apparent source width (ASW), image broadening, spaciousness and envelopment. The more different the sounds are at the two ears, at certain frequencies and delays, the greater the sense of these spatial descriptors. The locations of the ears then determine that sounds arriving from different directions generate different amounts of IACC and perceived ASW. Sounds from the sides are most effective; those from front and back, least effective. It is also known that diffusion in a sound field is a contributing factor, but that diffusion—or at least directional diversity in many reflections, is not a requirement for the perception of spaciousness.

Here it is necessary to emphasize that reflections occurring in small rooms cannot *alone* generate a sense of true envelopment. Envelopment requires delays (more than about 80 ms) that can only be supplied by recorded signals reproduced through multiple loudspeakers. Additional room reflections of those greatly delayed signals may enhance the impression, but the initial delay and the appropriate directions must be provided in the recorded sound and an arrangement of playback loudspeakers. How many channels and where do we put the loudspeakers? Two important experiments provide important insights, one from subjective evaluations, the other from measurements. They both arrive at the same conclusion.

Hiyama et al. (2002) conducted subjective evaluations of how closely the sound of a reference diffuse sound field, generated by a circular array of 24 loudspeakers, could be approached by arrays of smaller numbers of loudspeakers. Listeners were required to judge the degree of impairment in perceived envelopment (LEV) for each of the loudspeaker configurations, when compared to the 24-loudspeaker reference array. All loudspeakers radiated uncorrelated noise.

The first experiment they conducted examined the performance of different numbers of loudspeakers *equally spaced* in circular arrays. The results indicated that arrays of 12, 8 and 6 loudspeakers did well at imitating the perceived envelopment of the 24-loudspeaker array. Arrangements of 4 and 3 loudspeakers did poorly, however. It was a very clear delineation, suggesting that equally spaced arrays of 4 and 3 loudspeakers should be avoided.

Figure 15.6 shows some excerpts from the results of their numerous experiments in which two to five loudspeakers in different arrangements attempted to imitate the perceived envelopment of the 24-loudspeaker reference array (score = 0). The closer to 0 is the "difference grade," the better the subjective performance of the array. There is much more data in the paper; these selections related most closely to multichannel sound reproduction options. Results for three test signals are shown: 100 Hz to 1.8 kHz noise (the frequency range over which there appears to be the strongest correlation with envelopment), and dry recordings of cello and violin to which have been added

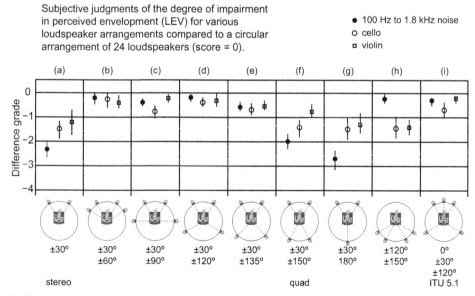

FIGURE 15.6 *Results of subjective comparisons in which listeners judged the degree of impairment in perceived envelopment (LEV) when an array of a smaller number of loudspeakers attempted to imitate the performance of a circular array of 24 loudspeakers.*
Adapted from the data in Hiyama et al. (2002).

convolved and simulated early and late reflections from the appropriate directions for a concert hall. The results are interesting.

- Two-channel stereo does not fare well (a), and neither does the "quad" arrangement (f), which performed very similarly, strongly confirming the results of Tohyama and Suzuki, discussed earlier.

- Symmetrical front-back arrays, it seems, contribute nothing to envelopment, but only add two more locations for special-effects sounds in movies and voices or instruments in music.

- A center-rear loudspeaker is worse (g).

- All combinations of a pair of loudspeakers at ±30° and another pair of loudspeakers at angles from ±60° to ±135° perform superbly (b), (c), (d), (e). Avoid ±150° (f), or whatever angle identifies the spread of the front loudspeakers.

- Four loudspeakers behind the listener (h) do not perform as well as four in front, at the same reflected angles (b).

- The five-channel arrangement described in ITU-R BS.775–2, shown in (i), performed about as well as any other configuration.

The last statement is obviously very good news, as it is the basis of conventional 5.1 systems.

Muraoka and Nakazato (2007) used a measurement of frequency-dependent inter-aural cross-correlation (FIACC) as a measure of the successful "sound-field recomposition." The idea was simple: FIACC was measured in four large spaces: a large lecture hall and three concert halls. An omnidirectional loudspeaker on the stage was the source. Recordings were made at the measurement locations using a circular array of 12 equally spaced microphones. These recordings were reproduced through different numbers of these channels using loudspeakers placed at 2 m from a measurement mannequin in an anechoic chamber. The FIACC of the sound field reproduced by each of the loudspeaker configurations was measured, and a "square error" metric was computed, describing the degree of difference between the FIACC at the original location, and that reproduced by the test arrangement of loudspeakers. The difference was computed over a "full" bandwidth (100 Hz to 20 kHz) and over a "fundamental" bandwidth (100 Hz to 1 kHz), the frequency range believed to be most related to the perception of envelopment.

Figure 15.7 shows selected data from the experiments. With the exception of (e), the arrangements shown in the top row are the same as those shown in Figure 15.8. That particular arrangement was not tested in this study, so the space (e) has been filled with results for what should be the best possible configuration: 12 channels at 30° intervals.

This is a very different kind of experiment from that of Hiyama et al., and yet the conclusions are almost identical to those listed earlier. There are some minor differences, as might be expected, because here the goal was to replicate the sound field of real rooms, not a mathematical ideal, or a synthesized approximation.

Comparisons of frequency-dependent interaural crosscorrelation (FIACC) measurements in four large halls (A,B,C,D) with FIACC measurements of reproductions of those spaces through different multichannel loudspeaker arrangements in an anechoic space. The results are expressed as a "square error" computed over the frequency range.

▬▬▬ "Full" bandwidth: 100 Hz to 20 kHz
▬▬▬ "Fundamental" bandwidth: 100 Hz to 1 kHz, the frequency range most related to the perception of spaciousness.

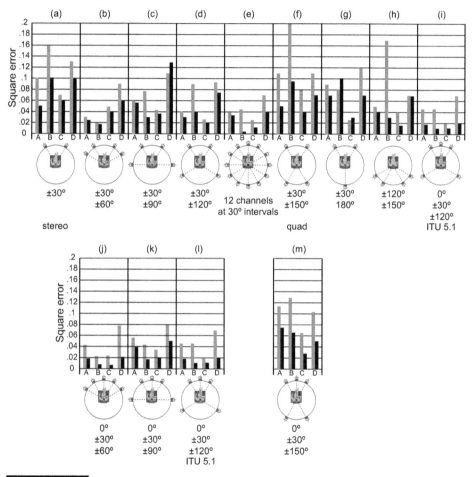

FIGURE 15.7 *Selected data from Muraoka and Nakazato (2007) showing how closely various configurations of reproduction channels and loudspeakers, (a) through (m), can replicate the FIACC measured in four large venues: a large lecture hall "A" and three concert halls "B," "C" and "D." The shorter the vertical bars, the better the reconstruction of the original sound field. It is probable that the black bars are more meaningful.*

Of special note in the top row results is the great similarity between (e), the 12-channel system, and (i), the widely used five-channel "home theater" arrangement. All of the combinations of a front pair of loudspeakers at ±30° and another pair of loudspeakers at angles from ±60° to ±120° performed reasonably well, as did the front or

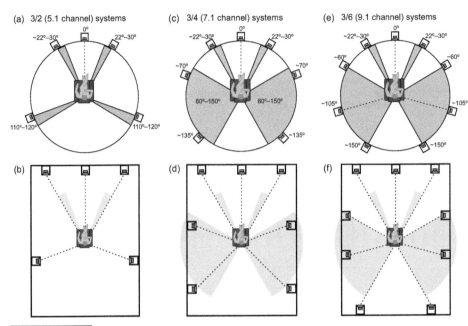

FIGURE 15.8 *Suggested layouts for common multichannel systems as viewed in the "academic" circular arrangements and in the real-world rectangular schemes.*

CHANNEL NUMBERING SCHEMES

To keep track of how many loudspeakers there are in a multichannel system, and to help in understanding where they are located, the industry originally adopted a simple designation. It consists of two numbers: one is the number of front channels, and the second is the number of surround/side/rear channels. So, 2/0 is stereo. 3/1 is the original Dolby Stereo/Surround system, with a single surround channel. 3/2 is conventional five-channel (5.1) surround, with L, C and R across the front, and two surround channels. 3/4 has four surround channels, which is called 7.1 in the consumer world. However, 7.1 in the cinema world is more likely to be interpreted as 5/2, which is Sony's SDDS system with five channels across the front and two surround channels. With the arrival of competing immersive sound schemes this numbering system has become more confused. To the horizontal more-or-less ear level channels we have become familiar with, there are added height channels. Some are below ear level, others are above ear level and still others are in the ceiling. An unambiguous descriptive formula for these new formats has yet to be worked out, but the present situation seems to be a numbering system that adds some number of ceiling loudspeakers to the basic 5.1 or 7.1 "surround" systems making them, for example, 5.1.2 or 7.1.4, where the last number indicates the number of ceiling loudspeakers. When the channel count gets to 16 or 32, as it can in elaborate systems, simplistic number systems lose their meaning because different immersive algorithms use different combinations of the installed loudspeakers.

rear combinations of four channels (b) and (h). Stereo (a) performed poorly, as did the front-back symmetrical "quad" arrangement (f).

The lower row of results show the effect of adding a center channel to (b), (c), (d) and (f), creating optional five-channel configurations, including a repetition of the ITU arrangement as (l). The already good performance of the four-channel versions is improved, with (j), (k) and (l) all exhibiting highly attractive results. The front-back symmetrical arrangement (f) is slightly improved by the addition of the center channel (m), but it is still not an attractive option. The lesson: avoid symmetrical front/back arrangements of left/right loudspeakers.

15.7.2 Summary

There is *very* good news. Large numbers of channels are not necessary to provide excellent facsimiles or reconstructions of enveloping sound fields. This is true whether the evaluating metric is subjective or objective. The optimal selection of four or five channels and loudspeakers can provide performance very similar to circular arrays of 12 or 24 loudspeakers.

Two elaborate studies, one subjective, one objective, concluded that the existing popular 5.1 channel arrangement of L, C, R loudspeakers spanning a 60° arc across the front, combined with two surround loudspeakers at ±120° performed superbly (arrangements "i" and "l" in Figures 15.6 and 15.7). However, it is also clear that there are other five-channel options, (j) and (k), that work comparably well. There is no requirement for loudspeakers behind the listener. Only if one wishes discretely panned localizable images at other locations, as in immersive audio, would one need more than five channels surrounding a listener.

NOTE: the previous sentence ended with "a listener." If there is an audience of multiple listeners located away from the symmetrical sweet spot, more channels and/or different kinds of loudspeakers will be necessary to generate similarly enveloping effects over the enlarged listening area. This is where the more elaborate immersive systems have advantages. In the demonstrations I have heard, it was impressive how stable the spatial illusions were as I moved around the listening room. More channels clearly are able to deliver a more natural sound field.

All three studies provide persuasive evidence that front-back left-right symmetrical arrangements should be avoided, in that they contribute little or nothing to the perception of envelopment over simple stereo, only adding directional options for panned sounds. Morimoto (1997) did some experiments in which he concluded that sounds originating behind listeners are important to LEV; in situations exhibiting similar IACC, listener envelopment was improved when a greater proportion of the sound arrived from the rear. Unfortunately, the loudspeaker arrangement used in the experiments was symmetrical left-right/front-back, a situation already disadvantaged in the generation of convincing LEV. Because all of the practical loudspeaker arrangements for home theater require rear loudspeakers for localized sounds, this is an issue that is settled automatically.

A caution: all of these experiments were conducted in anechoic listening circumstances. Reflections within the listening room will have some effect on the conclusions, but it is highly probable that the direct sounds from the loudspeakers will have the dominant effects and, the more channels, the greater that dominance is likely to be.

15.8 SURROUND SYSTEM LAYOUTS

Ideally all loudspeakers in a surround sound system should be comparably good and similar in design. That is a luxury few can afford and many cannot accommodate in their domestic situations. The common solution is to use smaller on- or in-wall loudspeakers that are less conspicuous and that provide adequate sound output down to the bass-managed crossover frequency to the subwoofer(s), about 80 Hz (Section 8.5). These are placed with the tweeters about 2 ft (0.6 m) above ear level; perhaps higher if there are seats close to the side walls to discourage localizing the loudspeakers.

There are several variations for laying out conventional multichannel systems, and even more with the new immersive systems. Figure 15.8 shows the basic layouts that I recommend. Small variations will not destroy the entertainment. It is the angles from the prime listener's perspective that matter, not the walls on which the loudspeakers are mounted; it depends on the shape of the room. For example, in (d) the rear loudspeakers could be located on either the side or rear walls of the room. The shaded areas are those within which variations are possible without seriously compromising the results. In rectangular rooms, delays in the setup process compensate for differing distances to the reference seat.

15.8.1 Loudspeaker Directivity Requirements

Sound from loudspeakers is of little value if it is not delivered to listeners. In stereo one aims the loudspeakers at a listener and the job is done. In home theaters it is more difficult, depending on the room layout. Figure 15.9 shows a difficult situation, a crowded room. Ideally one would not want any listeners close to walls, and to surround loudspeakers that can be distractingly localized when the sounds they radiate are intended merely to add ambience, audience sounds and so on. The first task is to ensure that all listeners are delivered a strong direct sound from all loudspeakers. This localizes the sound event, which is especially important in immersive systems, and provides high-quality spatial cues as required.

It can be seen that any competently designed loudspeaker easily meets the task of the front channels when the left and right loudspeakers are angled inward, which should be normal practice. The surround channels are more challenging. If the loudspeakers are on adjustable brackets, they can be aimed to maximize coverage, but such mounting schemes are unsightly and place the loudspeakers at a distance from the walls causing a boundary interaction. As shown in Figures 9.8 and 9.10 this is not ideal. In-wall

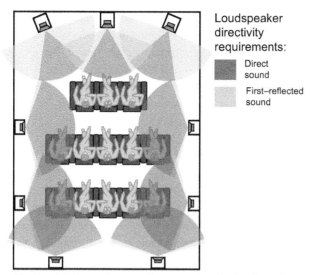

Loudspeaker directivity requirements:

■ Direct sound

░ First–reflected sound

Horizontal dispersion requirements for the loudspeakers
in this crowded room with a large audience:
LF, RF: ±30° for the direct sound
±87° for first–reflected sounds
Center: ±30° for the direct sound
±70° for first–reflected sounds
Surrounds: ±70° for direct sounds
±87° for first–reflected sounds
Loudspeakers are manufactured with symmetrical left/right
dispersion whether it is needed in a particular location or not.

FIGURE 15.9 *A summary of the horizontal-plane angular dispersions required of the loudspeakers to deliver direct sounds of comparable quality and level to all listeners and to deliver sounds to the wall surfaces from which the first reflections occur. The propagation loss due to the inverse-square law will inevitably cause differences in level at different distances. The criterion of excellence for direct sounds (the darker shaded angular range) is that they should all be as similar as possible to the on-axis performance of the loudspeaker. This is obviously a challenge for the surround loudspeakers because of the very large, almost 180° dispersion required of these units. Here the surround loudspeakers are all non-steerable in-wall or on-wall designs.*

designs or on-wall designs of the kind shown in Figure 9.13 are much better, as they avoid the acoustical interference problem with the adjacent boundary. The challenge is to find loudspeakers with the necessary horizontal dispersion to address the audience. In general, this is satisfied by bipole—bidirectional in-phase—designs. There are other variants with clever names but most amount to no more than loudspeakers having very wide dispersion: a good thing in this case. Dipoles should be avoided.

15.8.2 Mission-Oriented Acoustical Treatments

Most discussions of room acoustical treatments adopt the perspective that the room is a problem to be eliminated. That is simply not true, of course, even though rooms present difficulties, especially at low frequencies. However, if one thinks about it, there are certain areas where room surfaces can be put to good use.

From wide dispersion loudspeakers there will be a lot of sound that misses listeners and encounters room boundaries, so attention must be paid to those surfaces. If the loudspeakers are of high quality, with very uniform directivity over a wide frequency range, these reflections typically have minimal negative effect. However, if these off-axis sounds are of inferior quality, they are better absorbed. This was discussed in Chapter 7. However, in a surround system the sounds traveling side-to-side have a special value in that they are responsible for delivering the highly desired impressions of space and envelopment—of being in a space different from the one the eyes see. They do this by reducing the interaural cross-correlation coefficient (IACC). Figure 15.10 shows how room acoustical treatments can be used to assist this illusion especially for those seated toward the sides of the room. Diffusers used as shown here should be two-dimensional devices, scattering sound in the horizontal plane; that is, the notches and geometric features should run vertically.

15.8.3 Surround Loudspeaker Options

Because adding more loudspeakers to home theater systems is frequently difficult, there is a challenge to maximize the utility of each one. In the very early days of a monophonic surround channel, two conventional loudspeakers in a home theater clearly were not able to mimic the perceptions of arrangements of multiple surround loudspeakers in a cinema. It was thought that by spraying the sound around the room using a bidirectional surround loudspeaker that the situation would be alleviated. That and polarity inversion in left and right surrounds in normal rooms helped out. But in those early days the single surround channel was low-pass filtered at 7 kHz, to avoid distracting sibilant splashes that leaked into the surrounds because of leaky matrix decoders. All that changed when the surround channels became broadband, electronically decorrelated, and somewhat independent due to steering. It was transformed completely by digital discrete multichannel delivery. No more special loudspeaker "technology" is necessary. But some of it lingers on.

Figure 15.11 shows a surround loudspeaker with switchable directivity, the product shown in Figure 9.12. Here are shown spinoramas in each of its modes. It is seen to be very well behaved in the measurements shown in (a) and (b) but not in (c). The product was optimized for bipole, that is, uniformly wide dispersion, operation, as is required in most home theater applications. The degradation of the on-axis curve in dipole mode (c) is profound and seriously affects the sound quality.

To be fair to the promoters of dipoles, Figure 15.12 shows four products that were designed for this configuration only, and three of them were approved by THX. Note that two of them have insufficient bass extension to match the 80 Hz crossover frequency to a normal bass managed subwoofer. They are all different in measurements and sound, and by comparison to the subjectively highly rated loudspeakers shown in Chapter 5, not very good. It is time to retire the concept. The new immersive systems require conventional loudspeakers, which can include the bipole-configured bidirectional designs as

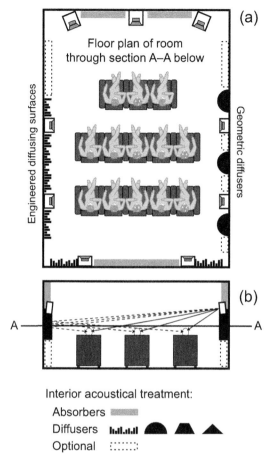

FIGURE 15.10 *(a) shows the horizontal plan for room acoustical treatment based on the notion of assisting spatial illusions. The materials described here apply to a horizontal band around the middle of the room, around and above seated ear height. The front side walls are "optional" territory. If stereo listening is to be part of the entertainment in the room, Chapter 7 discusses the choice of reflecting, absorbing or diffusing the first reflections from the side walls. For dedicated multichannel/home theater use, absorption is advised. The absorbers on the front and rear walls avoid reflections within the angular ranges that contribute little to envelopment. The diffusers along the side walls provide reflections of sounds from surround loudspeakers on the opposite walls, from directions that aid in the perception of envelopment by listeners seated away from the center of the room. Envelopment is most influenced by sounds in the 100 Hz to 1 kHz frequency range, the lower frequencies. Therefore these diffusers should be about 12 inches (0.3 m) deep if they are geometric shapes, and about 8 inches (0.2 m) for well-engineered surfaces. The corners are available for low-frequency absorbers if they are needed, or for subwoofers in multi-sub configurations. Figure 15.10b shows the room in elevation, with the diffusers placed at the appropriate elevation to redirect sound to the listening locations. A height of 36 to 48 inches (0.9 to 1.2 m) should be sufficient for this treatment. The space above can have patches of absorption. The space below is "optional" because the carpet or the seating will likely capture sounds reflected from this portion of the wall.*

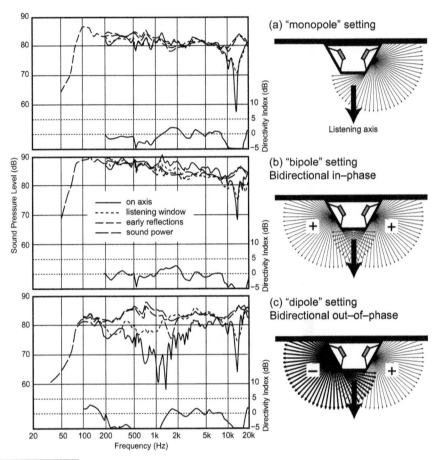

FIGURE 15.11 *The Infinity Beta ES250, a bidirectional on-wall loudspeaker measured in each of its switch-selectable directivity modes.*

shown in Figure 15.11b. They are simply very wide dispersion loudspeakers that usefully deliver sounds to widely distributed listeners in rooms as shown in Figure 15.9.

For those listeners in the single row in the null of the dipole surround loudspeakers (they don't work for multiple rows) the lack of high frequencies will make the loudspeakers less attention getting, and therefore less likely to be localized by listeners seated close to them. The origin of this problem is discussed in Section 14.2.3, and there really is no satisfactory solution that employs small-box loudspeakers. Line and CBT (constant bandwidth transducer) arrays offer advantages and can deliver state-of-the-art sound quality.

For those willing and able to install elaborate immersive sound systems it is likely that, with the right recordings, these issues will cease to exist. However, not all movies of interest are in immersive formats.

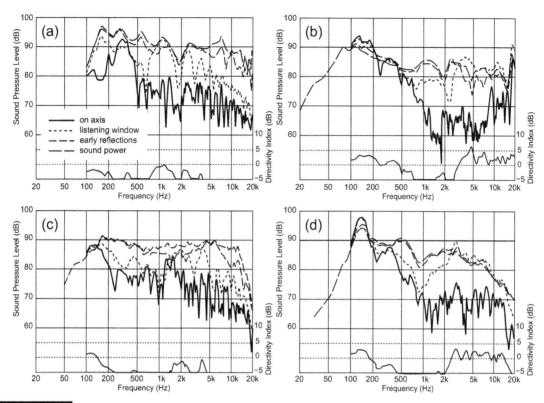

FIGURE 15.12 *Four on-wall bidirectional out-of-phase "dipole" surround loudspeakers.*

15.9 THE AMBISONICS ALTERNATIVE

There are two parts to the Ambisonics premise. The first is that, with the appropriate design of microphone, it is possible to capture (record in some number of channels) the three-dimensional sound field existing at a point. The second part is that, with the appropriate electronic processing, it should be possible to reconstruct a facsimile of that sound field at a specified point within a square or circular arrangement of four or more loudspeakers. Therefore, this system distinguishes itself in that it is based on a specific encode/decode rationale. Several names are associated with the technology. The basic idea for this form of surround sound was patented first by Duane Cooper (Cooper and Shiga, 1972). Patents were also granted to Peter Fellgett and Michael Gerzon (1983), who were working simultaneously and independently in England. Peter Craven contributed to the microphone design and, with the support of the NRCC, the UK group commercialized the Ambisonics record/reproduction system. See also www.ambisonic.net/ for historical perspectives and enthusiastic support.

It is an enticing idea, and the spatial algebra tells us that it should work. And it does, up to a point, at a point in space. Ambisonics remains a niche player in the surround-sound industry. Most people know little to nothing about it, although there are some encoded recordings (www.ambisonic.net/), and the Soundfield microphone continues to be used in some recordings. The scarcity of playback decoders is a clear problem. However, there are other considerations that may be significant.

The author has heard the system several times in different places, including a precise setup in the NRCC anechoic chamber in which he participated. There, theoretically, it should have worked perfectly because there were no room reflections to contaminate the delivery of sounds to the ears. In general, Ambisonics seems to be most advantageous with large spacious classical works with which the system creates an attractively enveloping illusion for a listener with the discipline to find and stay in the small sweet spot. It tolerates a certain amount of moving around, but leaning too far forward results in a front bias, leaning too far backwards creates a rear bias, leaning too far left and so on. Big, spatially ambiguous, reverberant recordings are more tolerant of listener movements of course. All of this should be no surprise in a system in which the mathematical solution applies only at a point in space, and then only if the setup is absolutely precise in its geometry, and the loudspeakers are closely matched in both amplitude and phase response. Room reflections absolutely corrupt the theory. So, what did it sound like in the anechoic chamber? Like an enormous headphone; the sound was inside the head. When the setup was moved to a nearby conventional listening room, the sound externalized and all previous comments apply.

In order to reconstruct the directional sound intensity vectors at the center of the loudspeaker array, some amount of sound may be required to be delivered by many, or all, of the loudspeakers simultaneously—that is the way the system works. A practical problem then arises, because we listen through two ears, each at different points in space, and both attached to a significant acoustical obstacle, the head. If a head is inserted at the summation point, then it is not possible for sounds arriving from the right loudspeakers to reach the left ear without timing errors and large head-shadowing effects, and vice versa. The system breaks down, and we hear something other than what was intended. What is heard can still be highly entertaining, but it is not a "reconstruction" of the original acoustical event. For that, one would need to generate individual sound fields at two points, one for each of the ears—binaural audio.

There are numerous ways to encode and store the Ambisonics signals, and even more ways to process the signals into forms suitable for reproduction from different numbers of loudspeakers in different arrangements. Higher-order Ambisonics is held by some as the real solution, but this means more channels, more paraphernalia and more cost. Ambisonics algorithms have found their way into various multichannel recording schemes. Whatever happens, having multiple digital discrete channels within which to store data can only be an advantage.

15.10 UPMIXER MANIPULATIONS: CREATIVITY AT WORK

Some of the criticisms of the first-generation Dolby ProLogic stemmed from it having been designed to deal with the imperfections of optical sound tracks on films. When delivered to homes on formats not having these problems, the limitations were obvious, especially when the upmixer was used for music; the surrounds were a dull, and there was too much emphasis on the center channel—the stereo soundstage seemed to shrink in width.

Recognizing an opportunity to improve on a good thing, inventors over the years have found great satisfaction manipulating the parameters of the matrixes, with delays and with steering algorithms, all in attempts to finesse the multichannel decoders either to be more impressive when playing movies, or to be more compatible with stereo music, or both. Most of them allowed for full bandwidth surround channels, and the more adventurous ones augmented the system with additional loudspeakers behind the listeners.

It needs to be emphasized that, when playing stereo music through these algorithms, we are hearing "ambience extraction," not reverberation synthesis. All of the reflected and reverberated sound that is reproduced in the surround channels was in the recording. It is just redirected to the side and/or rear loudspeakers rather than being reproduced exclusively through the front channels. Consequently, it generally sounds more natural, although it can also sound exaggerated with some recordings; stereo recordings were not designed for this form of reproduction. In order to get a sense of spaciousness in stereo reproduction through two loudspeakers, more "ambiance" (poorly correlated sound) is often recorded than would have been required if surround channels had been anticipated. The solution: use the remote control and turn the surrounds down.

Willcocks (1983) provides a good overview of surround decoder developments in the fruitful period of the 1980s. More recent developments have been made by Fosgate, Lexicon, DTS, Dolby, Harman and others. All exist in various digital-domain incarnations, in various products, available at the touch of an icon. Many people, the author included, find upmixed stereo to be a mostly rewarding experience, although some recordings still sound best when left in conventional stereo. It all depends on the microphone technique and electronic processing used in the mix.

Rumsey (1999) conducted controlled listening tests on some upmix alternatives, comparing all of them to the stereo original. There were substantial variations among the upmixers, strongly depending on program, which would be expected, and significantly on the listeners and their accumulated listening experience (all were either students in a sound recording program or active in the recording/broadcasting industry). In general the upmixers degraded the front soundstage. These expert listeners (presumably having grown up and worked with stereo) preferred the stereo original to any upmixer. Nevertheless, the best upmixers were given only slight demerits. Opinions about spatial impression were different, with some listeners giving upmixed versions substantial bonus points, while others thought the opposite. In the end, some of the "expert" listeners wanted to be left with their stereo systems, but others thought the new formats had some interesting and engaging things to offer. Clearly, there is a cultural component to tests of this kind and, as Rumsey points out, it would be interesting to conduct similar kinds of tests with a wider population of listeners, asking more basic, "preference" oriented questions.

Choisel and Wickelmaier (2007) compared mono, stereo, wide stereo, three matrix upmixers and five-channel discrete playback formats. The results are worth looking at, as they indicate that a good upmixer can be as, or occasionally even more, rewarding than a five-channel original. But, there are also some unrewarding upmixers. Stereo appears to provide much of what listeners want, but it loses in terms of perceptions like width, spaciousness and envelopment, very much as one would expect. Stereo also suffered in terms of sound quality (brightness) of the phantom center image, due to acoustical crosstalk—discussed in Section 7.1.1.

Personally, I enjoy upmixed stereo for many, but not all, programs. My system has identically performing loudspeakers in seven horizontal locations (7.1), and the resulting spatial impression is quite seamless. Switching back to stereo results in a diminished sense of envelopment and a shrinking of the soundstage.

15.11 MULTICHANNEL AUDIO GOES DIGITAL, DISCRETE AND COMPRESSED

The few samples of discrete multichannel recordings from the quadraphonics era were sufficient to generate a lasting interest, if not an outright lust, to develop a viable format that did not suffer from crosstalk among the channels. Today we are experiencing several versions of that dream. There are the expected pro and con arguments about which ones sound better, but under the bluster and ballyhoo, all of the systems, so far, have sufficient sonic integrity that our entertainment is not likely to be compromised. Purist audiophiles have pushed for systems of such bandwidth and dynamic range that even the most fastidious superhumans, dogs and aliens will be pleased. Professional audio needs extra "space" in recordings to cope with inevitable artist excesses, mistakes, multiple overdubs and so forth, but at the point of delivery to consumers we are adequately served by several of the popular media. One of the joys of digital encoding and decoding is that it is all possible for a price—bandwidth, or data rate—and that price is dropping.

Data rate, in audio terms at least, is abundant and inexpensive, but not limitless. There are situations where it is restricted, as when Dolby squeezed packages of digital data between the sprocket holes in movie film. The consequence was that there was only a certain amount of digital data space into which 5.1 channels could be stored. It was not enough to allow uncompressed audio, so Dolby incorporated an audio codec, a perceptual encode/decode scheme that devoted less data space to recording those sounds or components of sounds that were estimated not to be audible due to masking. The codec they developed, AC-3, proved to be highly durable as the basis for Dolby Digital.

All audio codecs are scalable, meaning that the amount of data compression can be adjusted according to bandwidth/data-rate availability, so when evaluating the performance of any codec it is important to know which version is being used.

There is a point of diminishing returns in the performance of all codecs; as the data rate is reduced each codec has a different data rate at which it begins to exhibit audible problems—artifacts. And, to make things more complex, different codecs exhibit different kinds of artifacts. It depends on the strategy used to identify components of the sound that are to be encoded more simply, or discarded. *All* codecs can be made to misbehave, but the best of them misbehave in a manner that is revealed as a momentary "difference" in sound, not a gross distortion or lapse in information. Most of the time, competent codecs are transparent. Those people involved with the subjective evaluation of codecs have found it necessary to train listeners to know what to listen for. It is not common for these problems to be discovered in casual listening.

Obviously, if the data rate is drastically reduced, no codec can perform flawlessly, and there is ample evidence of this in Internet and streaming audio. Check the data rate.

Audiophile paranoia suggests that all perceptually encoded systems are fatally flawed, alluding to the discarded musical information. Well, it is only lost if it could have been heard. Auditory masking is a natural perceptual phenomenon, operating in live concert situations just as it does in sound reproduction. It has assisted our musical enjoyment by suppressing audience noises during live performances and, over several decades, by rendering LPs more pleasurable. If we talk here about compressing data, it would be fair to say that LPs perform "data expansion," adding unmusical information in the form of crosstalk, noise and distortions of many kinds. More comes off of the LP than was in the original master tape. However, because of those very same masking phenomena that allow perceptual data reduction systems to work, the noises and distortions are perceptually attenuated. This perceptual noise and distortion masking is so successful that good LPs played on good systems can still sound impressive.

Serious subjective evaluations by experienced and trained listeners have been involved with the optimization of these encode/decode algorithms to ensure that critical data are not deleted. These are in tests where listeners can repeat musical phrases and sounds as often as necessary for them to be certain of their opinions. Having participated in comparative listening tests of some of these systems, the author can state categorically that the differences among the good systems at issue here are not "obvious." Even in some of the aggressive data reduction configurations, audible effects were quite infrequent, and limited to certain kinds of sounds only. It was helpful to know what to listen for. And then, the effects were not always describable as better or worse, sometimes they could only be identified as being "not quite the same."

Finally, there is lossless compression, in which nothing is discarded. Taking advantage of redundancy, quiet moments and so forth, the data rate can be reduced but the reconstructed data are complete; there is no loss. In formats and delivery systems with adequate bandwidth, such systems are attractive in that they simply eliminate all discussions about possible degradations.

As bandwidth has expanded in the delivery formats the need for dramatic compression has been reduced, so all in all the future looks bright.

15.12 THREE-DIMENSIONAL SOUND— IMMERSIVE AUDIO

At each stage of audio evolution, from mono to stereo to multichannel, there have been arguments that it is not necessary. The arguments rarely come from listeners, the customers, but from the established industry for whom any change is costly, complicated and risky. The unfortunate venture into quadraphonics left a bad taste in many mouths and some significantly damaged bank accounts. Adding more channels was not a bad idea, but the manner in which it was implemented was seriously compromised. It was not an optimum arrangement of channels, but the biggest problem was that it was really premature. LPs were an inadequate delivery format.

All along it has been the movie industry that has promoted more channels. They want audiences to be more persuasively drawn into the spaces portrayed in the visual images. Three-dimensional visuals have been tried more than once, and while impressive they still have not caught on in a commercial sense. The fact that only blockbuster films can truly take advantage of it has been a restraining factor.

Adding 3D sound is the latest attempt to provide a higher form of entertainment, and I would like to think that it might succeed. We have had five- and seven-channel film sound for many years, and at its best it is very rewarding, at least for those seated toward the center of the audience area. Adding more channels can expand the "sweet spot" for the perception of ambience and envelopment, while providing the means to steer specific sounds to locations around and above listeners. They can be dramatic sounds in action movies or they can be no more than birds and wind in overhead trees. The reconstruction of sounds in three-dimensional spaces like concert halls, opera houses and massive popular concerts are, to some people, also great attractions. Cinemas can play such music performances in addition to movies for people without access to the real thing. Personally, I find video concerts in my home theater to be highly entertaining. A few—very few—are audibly rewarding; often the audio is timbrally and spatially mediocre. Too many ignore or abuse the center channel. We need to do better, and perhaps this is the motivation. The same has been true in existing cinemas, especially when attempting to deliver classical repertoire.

The professional and consumer sides of the industry are still feeling their way. Differences in the digital audio streams and in the number and placement of loudspeakers are problems to be resolved, if indeed resolution is possible. Each of the currently competing systems—Dolby Atmos, Auro-3D and DTS-X—has several levels of complexity in its implementation, but even in home theaters the numbers can be quite high. Visits to the respective websites will yield up-to-date information.

I am in the process of installing such a system in my home theater, just to be able to experience what exists and to explore what might be possible. In the end, there is no substitute for many discrete channels driving many loudspeakers.

15.12.1 The Perception of Elevation

Decades of living in a two-dimensional sound reproduction world has given us a good understanding of the properties of binaural hearing and sound localization. Now, we are being given height channels and overhead channels to augment spatial presentations.

Because human ears are in the horizontal plane, we can localize with great precision in that plane. Differences in the timing (interaural time difference, ITD) and amplitude (interaural amplitude difference, IAD) of sounds at the two ears provide us with the basic binaural cues. This is well-explored territory, and I won't go into it. However, localization in the vertical plane, in elevation, is different. Here the traditional cues can still contribute information, but only when the head is in motion and the sound source is stationary. The amount of movement to demonstrate the effect can be substantial (e.g., ±30° swings in the Perrett and Noble (1997) experiments). For home and cinema entertainment, head movements are likely not to contribute much, and the sound sources of interest are often moving.

In the important median plane we rely on spectral cues derived from the way sounds interact with our external auditory apparatus: the pinnae. When I was at the National Research Council of Canada, a colleague, Dr. Edgar A. G. Shaw, was investigating the acoustics of the external ear, exposing the acoustical reasons why sounds arriving from different angles had different timbral signatures at the eardrums. These were the precursors to what we now call head-related transfer functions (HRTFs). The mechanisms include acoustical interference and resonances in the folds of the pinna and the cavities within them. Several of them are sensitive to the direction of the incoming sound, exhibiting different levels of response for different directions.

Looking at ears, the differences we can see among subjects result in very different acoustical performances (Shaw, 1974a, 1982): peaks and dips of ever-changing shapes. Some researchers were attracted to the acoustical interference dips in the curves. Such dips are highly variable among subjects, and they also occur routinely as a result of interfering direct and reflected sounds associated with either the sound source or the listener; they are not stable, changing frequency with even slight head movements. Evidence from sound reproduction research indicates that acoustical interference dips are largely ignored.

However, underlying the untidy measured curves was evidence of relatively stable resonances. Humans pay attention to, and are very sensitive to resonances, as they are significant timbral cues for all the sounds we are interested in: voices, musical instruments and so on (Section 4.6.2 and Toole and Olive, 1988).

Figure 15.13 summarizes the conclusion that Shaw came to, showing that averaged across several subjects, there was a progressive increase in the energy in a resonance around 7 to 8 kHz and a decrease at frequencies around 12 kHz. The change is dramatic.

In order to localize sounds in elevation:

- The sound must contain frequencies above about 6 kHz, and listeners must be able to hear it.

- There must be wide bandwidth and a dense spectrum help to reveal the height cue.

- Familiarity with the sound is important to recognize what the spectrum means.

Vertical localization "blur" (uncertainty) is about 17° for an unfamiliar voice, about 9° for a familiar voice and 4° for white noise (Blauert, 1996). If we have no idea of the spectral content of a specific sound, there is no assurance that we would automatically perceive it at its true elevation.

We rely greatly on "plausibility" when localizing in complex situations. Aircraft are automatically assumed to be above us. Nothing is localized below the floor. Voices are where the lips are on the screen—the ventriloquism effect. If listeners are predisposed

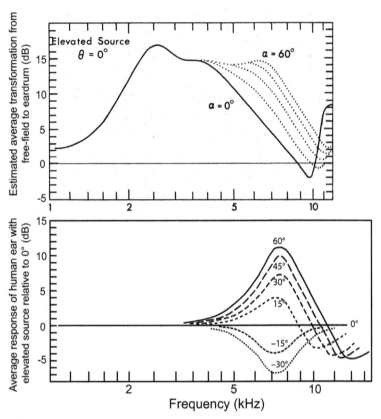

FIGURE 15.13 *(top) The mean curve from 10 subjects shows a progressive increase in energy around 7 kHz as a function of source elevation angle. Data from Shaw (1974b), copyright Acoustical Society of America. The lower curves show evidence of energy in one resonance being elevated and another decreasing as a function of source elevation.*
Data from Shaw, personal correspondence, 29 September 1972.

to think that sounds are coming from above, that is what they are likely to perceive; hence the emphasis on hovering helicopters and bird sounds in demonstrations of immersive sound systems. Even bouncing some high frequencies off the ceiling can assist the illusion, as has been shown with some current products. However, persuasive and durable localization in elevated and overhead locations requires real loudspeakers in those locations, and they must be able to deliver the necessary high frequencies to all listeners.

These directional cues are built into each of us and we are individualistic, so learning the elevation cues in-situ is part of the purpose of head movements. Whether we attend to resonant peaks, interference dips, or both, variations among listeners are likely to be permanent features.

CHAPTER 16

Loudspeakers and Power Amplifiers

I am not going to enter the debate about which amplifier or wire sounds best. In the context of loudspeakers and rooms, any real or imagined influences they have are of a much lower order than anything discussed in this book. However, there are a few factors that do have significant effects on sound quality, and in the interests of completeness we will discuss them here. These are hard electrical engineering issues, but they routinely get elevated to different planes of thought. The reason for talking a little about them here is because, as a result of them, frequency responses of loudspeakers get altered, and sometimes this can be heard.

16.1 CONSEQUENCES OF LOUDSPEAKER IMPEDANCE VARIATIONS

Impedance: 8 ohms. This is the kind of specification one sees for loudspeakers. It is an invented number. For a few, very, very few loudspeakers it is a good approximation, but for the vast majority it is a dreadful description of reality. Figure 16.1a shows a typical impedance that varies substantially with frequency and that crosses the rated impedance at a few places only. The variations are normally of no concern.

Most power amplifiers are designed to be constant-voltage sources so, unless an unfortunate interaction between amplifier and loudspeaker provokes limiting or protection, all is well. Sadly there have been some notable examples of high-end loudspeakers having impedances that dipped very low: to an ohm or less. This is a problem of incompetent loudspeaker design. However, sensing a market, amplifier designers responded with monster "arc-welder" devices that can drive these problem loudspeakers, but it is overkill for most circumstances. It was amusing at the time to read that these incompletely designed loudspeakers "revealed" differences between power amplifiers, as if it were a virtue. They were the *cause* of the differences.

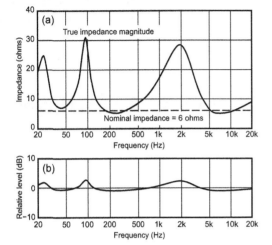

FIGURE 16.1 *(a) an impedance curve for a loudspeaker compared to the nominal impedance rating chosen by the manufacturer for it. (b) the change in frequency response of this loudspeaker caused by driving it with a tube amplifier having a large output impedance. Note that the shape of the frequency-response error is the same as the loudspeaker impedance curve.*

But there is a situation in which simple variations in impedance become an issue. Going straight to the problem, Figure 16.1b shows the kind of change in loudspeaker frequency response that can be caused by variable impedance—it is easily audible. The culprit? In this case a tube power amplifier with a large output impedance. The explanation is in Figure 16.2a and b. The output impedance of the power amplifier and the resistance of the loudspeaker wire are components in a voltage divider circuit. When combined with the frequency-dependent impedance of the loudspeaker it means that the "flat" frequency response voltage at location "A" inside the power amplifier acquires a shape following that of the impedance curve at location "B." Because this is the voltage driving the loudspeaker, the overall performance of the loudspeaker, that is, all of its frequency response curves, are modified by this amount. Different loudspeakers have different impedance curves; some are strikingly variable, others change little.

The amount of the change in frequency response depends on the total voltage drop across the combined amplifier output impedance and wire resistance, meaning that minimizing both of these is desirable. For solid-state power amplifiers output impedances tend to be very small: typically 0.01–0.04 ohms. Those for tube power amplifiers are much higher: typically 0.7–3.3 ohms, but occasionally even higher, which is inexcusable. These numbers come from a survey of *Stereophile* magazine amplifier reviews over several years—thank you, John Atkinson, for doing useful measurements.

To reviewers these are moderately discomfiting numbers, because the inevitable conclusion is that tube power amplifiers, as a population, cannot allow loudspeakers to perform as they were designed. Different reviewers handle it in different ways. Some ignore it; others have danced around the issue concluding that it is just one more uncertainty in sound reproduction. Rarely is it acknowledged to be what it is.

Loudspeakers can be designed to exhibit almost constant impedance, although it is rarely done. Such loudspeakers can perform with remarkable consistency in spite of significant losses in the upstream signal path. Rarely, though, is impedance ever discussed as a virtue or a problem. One well-known high-end loudspeaker specified that it should be used with wire having resistance less than 0.2 ohms. This conscientious advice is admirable, but it means that the restriction is violated as soon as any tube amplifier is connected, no matter what wire is used. It is sobering to think that 10 ft (3 m) of the much despised lamp cord has 0.148 ohms resistance—much less than typical tube amplifiers (See Table 16.1).

Minimizing wire resistance is easy: use large wire, having low gauge numbers (see Table 16.1)—or, better yet, use less wire. If there is a risk of radio-frequency signal pickup, it is important to know that unshielded wires act as antennas. A great deal

TABLE 16.1 Resistances per unit length of two-conductor stranded copper wire, stated for *both* wires in the circuit—so just measure how long the two-conductor wire is, and multiply by these numbers. For reference, common lamp cord is typically 18 gauge. If you do not see a gauge rating for a loudspeaker wire, be very suspicious. Some exotic cables use small wire for seriously mistaken reasons.

AWG wire gauge	Resistance per ft (ohms) BOTH conductors	Resistance per m (ohms) BOTH conductors
10	0.0020	0.0067
12	0.0032	0.0106
14	0.0052	0.0169
16	0.0082	0.0268
18	0.0148	0.0483

of mystique has evolved around loudspeaker wires, attempting to elevate this simple device to impossible heights of importance. Notions that they behave as transmission lines persist, but Greiner (1980) offers persuasive arguments that this is unrealistic. There are other beliefs, some of which are impossible (e.g., directional wires), or irrelevant (e.g., skin effect, which is significant only at frequencies much above the audio frequency band).

At prices that can exceed $20,000 for a pair of 8-foot (2.4 m) loudspeaker wires, one expects a lot. Enough said. Wire is a good product for the industry: totally reliable, inexpensive to manufacture, highly profitable and, if you like what you hear, an excellent investment, so long as you did not pay more than you needed to—"aye, there's the rub."

16.2 THE DAMPING FACTOR DECEPTION

One of the universal compliments attached to audio products, including wires, is that it results in "tighter bass." In the case of loudspeaker wire, it seems as though there might be some truth to it because of its role in the loudspeaker/amplifier interface and damping. Damping unwanted motion of a loudspeaker diaphragm is undoubtedly a good thing.

In 1975, I wrote an article for *AudioScene Canada* called "Damping, Damping Factor and Damn Nonsense." I still like the title because is a succinct statement of reality. The point of the article is summarized in Figure 16.2c. The internal impedance of the power amplifier is used to calculate something called the damping factor (DF) of the amplifier (DF = 8 divided by the output impedance); the number 8 was chosen because it is the nominal load (resistive) used to measure the power output capability of amplifiers. The logical inclination is to think that larger is better. Solid state amplifiers have damping factors ranging from about 200 to 800, using the impedances quoted earlier in this section. Tube amplifiers in my survey ran from 2.4 to 11.4 because of their high output impedances.

Figure 16.2c shows the complete circuit involved in the electrical damping of loudspeakers—it does not mysteriously stop at the loudspeaker terminals. Current must flow through components and devices inside the enclosure. After flowing through the wire,

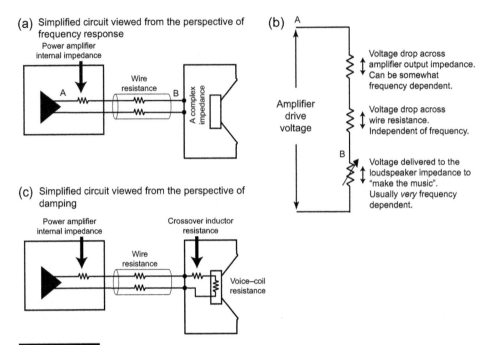

(a) Simplified circuit viewed from the perspective of frequency response

Power amplifier internal impedance

Wire resistance

A complex impedance

(b) A

Voltage drop across amplifier output impedance. Can be somewhat frequency dependent.

Amplifier drive voltage

Voltage drop across wire resistance. Independent of frequency.

Voltage delivered to the loudspeaker impedance to "make the music". Usually *very* frequency dependent.

(c) Simplified circuit viewed from the perspective of damping

Power amplifier internal impedance

Crossover inductor resistance

Wire resistance

Voice–coil resistance

FIGURE 16.2 *Schematic diagrams showing (a) and (b) the electrical circuit explaining how amplifier and wire impedances cause variations in loudspeaker frequency response, and (b) how they affect loudspeaker damping.*

it typically passes through an inductor, part of the low-pass filter ahead of the woofer in a passive system. Then, inside the woofer there is the voice coil. The inductor resistance is commonly around 0.5 ohm, and the voice coil resistance can have different values but is commonly around 6 ohms. So, let us examine all of the resistances in the circuit to arrive at the following progression of damping factor changes:

Amplifier internal impedance: 0.01 ohm	DF = 800
Add wire resistance: 10 ft of 10-gauge	
Both conductors: 0.02 ohm	DF = 266
Add crossover inductor resistance:	
0.5 ohm (typical)	DF = 15
Add voice-coil resistance: 6 ohms (typical)	DF = 1.2

Obviously the resistances inside the loudspeaker are the dominant factors. Even eliminating the inductor and driving the woofer directly changes things only slightly. The article (Toole, 1975) shows oscilloscope photographs of tone bursts of various frequencies and durations while the damping factor of the amplifier was varied from 0.5 to 200. At damping factors above about 20 (internal impedance less than 0.4 ohms), no change was visible in any of the transient signals, and changes in frequency response were very much less than 1 dB, and then only over a narrow frequency range. On music no change in sound quality could be discerned, including attentive listening for "tightness." Because 0.4 ohms is at least a factor of 10 higher than internal impedances found in typical solid state amplifiers, it means that, from the perspective of damping the transient behavior of loudspeakers, the wire resistance can be allowed to creep up substantially. However, as shown earlier, doing so can change the frequency response of the loudspeaker and that, we know, *is* audible.

In summary, with tube amplifiers, the internal impedance is already so high that damage is done to the frequency responses of loudspeakers having normal impedance variations. Added losses in wire simply make the situation worse. Listeners do not hear the loudspeaker that the manufacturer created.

With solid-state amplifiers internal impedances are negligibly low, so wire resistance must be controlled in order to minimize corrupting the frequency response of loudspeakers. How low? It depends on the *variations* in the impedance of the loudspeakers being used, and how low those impedances are—wire resistance represents a higher percentage of low impedances. For example: a loudspeaker ranging from 3 ohms to 20 ohms (not unusual for consumer loudspeakers and a moderately demanding situation) would experience about 0.6 dB variations in a system with 0.2-ohm wire resistance. Section 4.6.2 shows that this is slightly higher than the detection threshold for low-Q spectral variations in quiet anechoic listening. Twelve-gauge wire would allow for a run of 0.2 / 0.0032 = 63 ft (19 m). Obviously this is not very restrictive.

Loudspeakers having nearly constant impedance (a few exist) can tolerate large wire losses, sacrificing only efficiency up to the resistance at which damping is affected. If compelled to do better than this suggestion, more copper, shorter runs, or higher-impedance loudspeakers are the solutions.

16.3 LOUDSPEAKER SENSITIVITY RATINGS AND POWER AMPLIFIERS

Years ago, loudspeaker sensitivity was rated as the sound level at a distance of 1 m for an input of 1 watt. Power input is voltage^2 / resistance. Because loudspeakers do not have the same impedance at all frequencies, a sensitivity rating would apply only at a single, or at most a few, frequencies. Figure 16.3 shows the impedance curve for a loudspeaker, specified by the manufacturer as an 8-ohm unit. It is 8 ohms at four frequencies, but looking at the curve it is generally hovering at a level slightly above 4 ohms, dropping to a minimum of 3 ohms. A more realistic rating would have been 5 ohms. But that is an "odd" number in the industry, and such numbers, however true, tend to be avoided. The 3-ohm minimum is important because many receivers and some standalone power amplifiers are unhappy driving these low impedances as they lack the current capacity to deliver the required power into the load.

The figure shows the actual power delivered at a constant input voltage of 2.83 V, and it ranges from a high of 2.7 W at the impedance minimum, to a low of 0.4 W at the highest impedance point. Obviously, rating sensitivity according to power input does not work well. The domination of solid-state amplifiers really provided the solution.

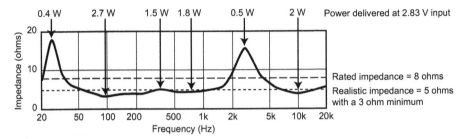

FIGURE 16.3 *A typical loudspeaker impedance curve, showing the rated impedance of the product and, at several frequencies, the input power for 2.83 V.*

These amplifiers are essentially constant-voltage sources, with power rated according to what they can deliver into an 8-ohm resistor. If the load impedance drops to 4 ohms, the power will double; at 2 ohms the power quadruples, and so on, until the amplifier can deliver no more current or exceeds some other limitation. At this point one can see an opportunity to "work the system" by dropping the impedance of the loudspeaker, thereby extracting more power from the amplifier, and elevating the sensitivity rating. Being the louder loudspeaker in a simple A vs. B comparison is not a bad thing at the point of sale. However, if there is truth in advertising, the customer will know to seek out a power amplifier that is content to drive a relatively low impedance.

Amplifiers that are optimized to meet a specification sheet deliver their rated power into 8 ohms, but may fail to deliver double power into 4 ohms. This is a major differentiating factor among power amplifiers. Those big, heavy monoblocks with massive heat sinks are the ones that are able to drive huge currents into very low impedances, and they tend to double their output into halved impedances.

Returning to the theme of sensitivity ratings, the present circumstances allow us simply to define an input voltage, not an input power. The selected standard voltage is 2.83 V, the voltage that delivers one watt into 8 ohms. All of the measurements shown in this book were made with loudspeakers driven with 2.83 V. Measurements were made at 2 m, a distance that safely represents the far field for small loudspeakers, although it is borderline for very large ones (see Section 10.5). The SPL is adjusted to show what it would have been at 1 m, which is the standardized distance (in this case 6 dB higher than the sound pressure level measured at 2 m). Not all manufacturers are accurate in their sensitivity ratings. However, as shown in Figure 16.3, at only four frequencies is the input power 1 watt—those where the curve crosses the 8-ohm line. That is why SPL @ 1 W @ 1 m is a specification relegated to history.

16.4 THE AUDIBILITY OF CLIPPING

How much amplifier power is necessary? Enough, is the correct answer. If the loudspeaker load is well behaved, well-designed inexpensive amplifiers can work just fine. Powered loudspeakers have a big advantage: the power amplifiers needed to drive individual transducers can be much less ostentatious devices because the details of the load they drive are known and well defined. It is the uncertainty of the load that forces us to buy amplifiers that are able to drive anything we connect to them. There is more to this tale than is revealed here. Benjamin (1994) and Howard (2007) add much more perspective.

If the amplifier does not have sufficient power to deliver a rewarding sound level it will likely be driven into clipping—the tops of waveforms are chopped off. If the amplifier is well behaved, this will result in clean clipping. At the 1987 AES Convention in London, there was a public test of the audibility of amplifier clipping. Although the results were never published, the result was that about 6 dB of clean clipping was rarely audible. I participated in the test and was surprised by the result. Voishvillo (2006, 2007) removed 50% of the waveform amplitude before it became an audible problem. This amount of clipping generates about 20% THD (total harmonic distortion). Interestingly, in comparison he found that this was less annoying than 3% of zero-crossing distortion. If the clipping is not clean or symmetrical, many forms of audible misbehavior can occur, something more likely to occur in low-cost amplifiers, especially those with inadequate power supplies.

Obviously, all such misbehavior should be avoided, but when attempting to deliver very high sound levels to listeners at a distance from the loudspeakers, the power requirements may rise more rapidly than is commonly thought. Examine the right-hand scale in Figure 4.3 for the relationship to sound level, and Figure 14.3 for the relationship to distance from the loudspeaker—remembering that each 3 dB increment is a doubling of power. In the real world it is likely that a lot of people are listening to clipped audio during loud passages.

CHAPTER 17

Hearing Loss and Hearing Conservation

This entire book is irrelevant if one cannot hear. It is only somewhat relevant if one is hearing impaired. Good hearing is important to our pleasure, but if one is professionally involved with the audio industry, it is essential to being able to do our jobs. Dr. Brian C.J. Moore is well known for his informative and insightful research and writings on the physiology and psychology of hearing. The abstract of a recent paper (Moore, 2016) states:

Exposure to high-level music produces several physiological changes in the auditory system that lead to a variety of perceptual effects. Damage to the outer hair cells within the cochlea leads to a loss of sensitivity to weak sounds, loudness recruitment (a more rapid than normal growth of loudness with increasing sound level), and reduced frequency selectivity. Damage to the inner hair cells and/or synapses leads to degeneration of neurons in the auditory nerve and to a reduced flow of information to the brain. This leads to poorer auditory discrimination and may contribute to reduced sensitivity to the temporal fine structure of sounds and to poor pitch perception.

To any person interested in music, sound recording or sound reproduction, this is an alarming summary.

For audio professionals, hearing loss is an occupational hazard, especially for those who also play in bands or have noisy hobbies. Anyone at work or at play who is exposed to loud sounds for prolonged periods, or to a single *very* loud sound, experiences some amount of hearing loss.

It is well known that temporary hearing loss sets in during loud recording sessions, rock concerts and so on. A ringing in the ears may result. With adequate quiet rest periods, we appear to recover, but if the process persists some of the temporary loss becomes permanent. Whether the hearing loss is temporary or permanent, it affects something as basic as one's ability to detect equalization changes (Kruk and Kin, 2015).

Section 3.2 shows that hearing loss well within the range of "normal" by standard audiometric criteria is sufficient to degrade one's ability to form good and consistent opinions about sound quality. Audiometry focuses on speech intelligibility, not on hearing small nuances in music and movies. So, what is "normal" to an audiologist can be significantly impaired to an audiophile or audio professional.

At the present time, the fact is that once the inner ears have been damaged, the damage is permanent. In audio parlance, the microphone is broken. Some of the audio information fails to reach the brain, and some that is delivered is corrupted. It is more than just not hearing small sounds; some higher-level aspects of perception occurring within the brain are modified, even when threshold shifts are not present. It is a truly insidious affliction.

Clearly it is important to do everything possible to prevent hearing loss, which is difficult as it has many causes:

- A single exposure to an extremely loud sound, such as an explosion, firing guns and many military weapons without protection, or automobile airbag deployment.

- Multiple exposures to very loud sounds, especially if sustained or repeated over a period of time that does not offer quiet periods for recovery.

- An accumulation over many years of exposure to sounds that would be considered "normally" loud, including sounds related to both work and recreation.

- Over a lifetime, one may experience any or all of the foregoing, and this accumulated hearing loss is called presbycusis—the statistical normal hearing loss as a function of age. When an audiologist tells you that your hearing is normal "for your age," you are basically being told that your hearing is not what it used to be.

- Ototoxic drugs, which include both prescription and over-the-counter medications. An Internet search will reveal many, including some that may be in your medicine chest.

- Diving, disease, head trauma, surgical procedures and so on.

Some of these may be avoidable, but not all. I was an active participant in listening tests through most of my career, and I was among the best: an observant and consistent listener. However, things change. In our tests we track not only the performance ratings of the products under test, but also the ratings themselves. The variations in judgments when the same product is evaluated multiple times, and the spread of the ratings, best to worst, are tracked. These are important statistics describing listener performance, as shown in Figure 3.9. Around age 60 I noted that my performance was deteriorating, and indeed had become aware that opinions did not come as quickly and easily as they used to. It was hard work, and I was not delivering consistent opinions. I retired from the listening panel. Figure 3.6a shows my hearing performance, which could have been worse, but the hearing threshold is a crude metric, ignoring much. Now I enjoy music and movies. I still have opinions, but they are my own. It was a bitter pill. Critical listening is a young person's game, but then only if the young person has functioning ears.

Now everything I hear has a background of tinnitus. In my case it is a high-pitched hissing sound, almost a whistle. It varies in level depending on factors I have not been able to identify, but it is always there. I also am more sensitive to loud sounds than when I was younger. It is called hyperacusis (check: www.hyperacusis.net). I know people for whom even reasonably loud sounds inflict pain. I still enjoy dynamic range in movies and music, but not with the same enthusiasm as I used to. I occasionally hear distortion that is not in the sound arriving at my ears.

All of this happened in spite of the fact that I have been very careful about wearing ear protection, starting in the 1970s. It turned out that the hard shell ear defender with liquid-filled cushions had been invented by two of my National Research Council of Canada (NRCC) colleagues. They provided persuasive arguments about hearing protection. I wore these devices faithfully while mowing grass, operating power tools in my wood shop, and the like. I wore earplugs when driving my noisy (but enormously entertaining) Lotus Elan+2 and later sports cars.

The appearance of musicians' earplugs, a development of Dr. Mead Killion at Etymōtic Research, revolutionized my hearing protection program (www.etymotic.com). These are custom-fitted earmold devices that simply turn the volume down, leaving the timbre relatively untouched. I always wear the 15 dB version when flying. I am in touch with my surroundings, but the sound level is reduced. Headphones work on top of them, and I can sleep comfortably. They fit into my shirt pocket. At the end of a long flight, when I take them out I realize just how much noise I had been avoiding. They are compulsory at loud concerts, where one notices that the lyrics are actually intelligible, but the bass still shakes the torso. Now, several companies make musicians' earplugs. All of them reduce sound levels, but some of them distort the timbre; the attenuation varies substantially with frequency. Shop around.

17.1 OCCUPATIONAL NOISE EXPOSURE LIMITS

I will begin by getting something off my chest.

> Occupational hearing conservation programs are almost totally irrelevant to audio professionals and serious audiophiles.

It is a bold assertion, but it is totally supportable. Figure 4.3 contains the OSHA occupational noise exposure limits. They suggest that a sound level of 90 dBA is acceptable for an eight-hour work day. But we know that anything above about 75 dB can have progressively greater effects on hearing. What is going on here?

It is essential to know that in existing national and international standards, the *only* criterion being considered is the preservation of the ability to understand speech. The

OSHA, NIOSH and similar occupational noise exposure criteria were created for manufacturing and other industrial workers, and the goal was not to *prevent* hearing loss, it was to *preserve* enough that at the end of a working life, conversational speech at 1 m distance was possible. Permanent damage of an important kind is inevitable. Hi-fi hearing, critical listening ability is not preserved.

Distilled to its essence, it is considered acceptable for hearing loss to accumulate up to 25 dB in both ears at the 1 kHz, 2 kHz and 3 kHz audiometric frequencies. In practical terms this translates to a loss of about 10% understanding of entire sentences, about 50% misunderstanding of monosyllabic "PB" words (words that are ambiguous because of similar sounding consonants) during conversation at normal voice levels, in the quiet, with persons one meter apart (Kryter, 1973). And this is considered to be an acceptable situation—"normal" hearing. Further losses from 25 to 40 dB are described as "slight." Really? For whom?

17.2 NON-OCCUPATIONAL NOISE EXPOSURE

Noise in the workplace is an understandable part of many occupations. However, many people underestimate the high sound levels that accompany many leisure activities. Hearing damage from recreational noises is sometimes called "sociocusis." Figure 4.3 shows a few examples of dangerously high sound levels, but much more extensive data and analysis are in Clark (1991). There are many everyday sounds that reach dangerous levels, some around the home, or home workshop, and numerous combustion-engine-powered machines. Of them all, though, Clark concludes that the most serious threat to hearing comes from recreational hunting or target shooting, where both ears experience loss, but the ear opposite to the one pressed against the gunstock (most often the left ear) can show thresholds that are higher by 15 to 30 dB.

17.3 BINAURAL HEARING IS ALSO AFFECTED

We think of hearing loss as an inability to hear small sounds, especially at high frequencies, but this is just part of the story. The sounds we are left with are modified; our perceptions are changed. One aspect is that we lose frequency resolution in our perceptual spectral-analysis capabilities, see Section 4.6.2. It modifies our perception of music itself. This can result in a reduced ability to separate sounds from one another, especially in a reflective or noisy environment. Speech intelligibility suffers (Leek and Molis, 2012).

Coupled with this is a recently discovered phenomenon, *hidden hearing loss*, that can exist with or without elevated audiometric thresholds. It involves reduction in binaural discrimination, that results in difficulty separating the components of an acoustically complex situation with multiple sound sources and/or reflections. Many of us recognize that carrying on conversations in restaurants gets more difficult with age and hearing loss. Movie mixers deal constantly with dialog competing with background music, sound effects and ambient sounds in the multichannel—that is, multidirectional—sound field. Are the artistic decisions they make likely to be affected by this kind of hearing loss? Will audience members with more normal hearing react differently to their artistic creations?

The disquieting thing is that this affliction can exist with little or no measured elevation of the hearing threshold. Kujawa and Liberman (2009) concluded:

It is sobering to consider that normal threshold sensitivity can mask ongoing and dramatic neural degeneration in noise-exposed ears, yet threshold sensitivity represents the gold standard for quantifying noise damage in humans. Federal exposure guidelines (OSHA, 1974; NIOSH, 1998) aim to protect against permanent threshold shifts, an approach that assumes that reversible threshold shifts are associated with benign levels of exposure. Moreover, lack of delayed threshold shifts after noise has been taken as evidence that delayed effects of noise do not occur. The present results contradict these fundamental assumptions by showing that reversibility of noise-induced threshold shifts masks progressive underlying neuropathology that likely has profound long-term consequences on auditory processing. The clear conclusion is that noise exposure is more dangerous than has been assumed.

Liberman et al. (2016) elaborates on the situation with evidence that college-age students can exhibit hidden hearing loss. Bernstein and Trahiotis (2016) conclude that "listeners whose high-frequency monaural hearing status would be classified audiometrically as being normal or 'slight loss' may exhibit substantial and perceptually meaningful losses of binaural processing." These are troubling findings.

Some additional recent results add to this picture. The tests involved evaluations of Spatial Release from Masking (SRM), which is a measure of the angular separation between a "target" voice and an interfering voice necessary for the listener to be able to spatially separate them. This relates to the ability to separate images in a soundstage, something of increased importance in immersive audio. Srinivasan et al. (2016) concluded that young-normal-hearing listeners could discriminate $\approx 2°$, older-normal-hearing listeners could discriminate $\approx 6°$, but older-hearing-impaired listeners found little advantage even at 30° separation. The latter is disturbing. The older-normal-hearing listeners exhibited more high frequency loss than the young listeners—presbycusis at the very least. In the end, age *per se* is a factor, but hearing loss, regular and/or hidden, is the main factor in being able to discriminate among multiple images in a panorama.

All of this should be as concerning to casual audiophiles as it is to professional audio people. These factors determine what we are able to hear in the complex sound illusions of movies and music. Could it be that audio professionals seek simplicity in the acoustical environment in which they work because of temporarily or permanently deteriorated hearing? Quite possibly. Absorbing strong early reflections diminishes the room effect, as does placing loudspeakers close to the listener on the meter bridge—that is, near-field monitoring. Both of these put the listener in a dominant direct-sound field.

Almost certainly this is a factor in personal preferences in loudspeakers and room treatments.

17.4 SOME OBSESSION CAN BE A GOOD THING

Most of us know people who rely on hearing aids. I know several, and none of them, not one, is pleased with the result. Hearing is not restored, it is rendered somewhat more functional than without the devices, but the true essences of the audible world are gone forever. I have some hearing loss and tinnitus, and I am obsessive about preserving what I have left. I still enjoy some loud music and movies; frankly, that is part of what I am saving it for. For me this is recreational listening, and my opinions are irrelevant.

However, for those professionally involved with the audio industry, opinions are important, if not crucial, to performing their tasks. I have met some professionals who are aware of the risks and take precautions. I also have met some who shrug it off, believing they can use their vast experience to compensate for degraded abilities. I have met one who seriously thought that at his advanced age his hearing was a good as it had ever been. He had not noticed any change (!). Indeed, many hearing-impaired mixers still can perform in a workmanlike manner; their experience has taught them the ritualistic settings for repetitive tasks. But in reality, they are handicapped, and some things will be overlooked and others not exploited to the fullest. This should not have happened.

The bottom line is that each of us is responsible for our own hearing. Occupational noise exposure limits are not sufficient for anyone who wants to enjoy the full panorama of music and movie sound. They are certainly not enough for audio professionals who earn their livings by listening, making judgments and offering opinions based on what they hear. It may not seem "macho" to wear musicians' earplugs while mixing, at least while working with routine high-level material. But surely that is better than wearing hearing aids later on.

CHAPTER 18

Fifty Years of Progress in Loudspeaker Design

Readers who have digested the book to this point will be well aware that anechoic measurements can reveal most of what we need to know about the potential sound quality of loudspeakers as they are heard in normal rooms. The exception is bass performance, where there is no substitute for in-situ measurements and corrective actions. Enough anechoic measurements can allow us to predict steady-state room curves above a few hundred hertz, and, more importantly, to anticipate subjective ratings in double-blind listening tests. This did not happen by accident.

I did not intend for this to be my life's work. Hi-fi was my hobby, but my research interest was sound localization; in particular, the manner in which sounds at the two ears are processed by the brain to yield perceptions of direction (Sayers and Toole, 1964; Toole and Sayers, 1965a, 1965b). All of the university experiments had been done with headphones, which allowed independent control of the signals to each ear.

A thrilling prospect of the job as a research scientist at the National Research Council of Canada (NRCC) was that there was an excellent anechoic chamber, within which the experiments could be extended to include natural listening, starting in a reflection-free environment, and then moving on to more complex circumstances. For this, loudspeakers were needed. When anechoic measurements were made on some highly rated audiophile loudspeakers of the time, the results were disturbing. The frequency responses were far from flat, and these were simple on-axis anechoic measurements made for the purpose of performing anechoic listening tests. Moreover, supposedly comparably good loudspeakers were very different from each other. Up to this point, the author had only seen "specifications" for frequency response and, if it were not for the unimpeachable pedigree of the measurement circumstances, it would have been possible to think that there had been a tragic error in making the measurements. Suddenly, claims that the loudspeaker was the "weakest link" in the audio chain rang true. But could these products really sound as bad as some of the curves looked?

18.1 MY INTRODUCTION TO THE REAL WORLD

A logical "Friday afternoon" experiment was to do a simple comparison listening test in one of the laboratory rooms. Having learned the basics of experimental psychology for the doctoral thesis work, it was obvious that this test had to be somewhat controlled. So, cotton sheeting was hung up to render the experiment "blind." The loudspeakers were compared in monophonic A/B/C/D comparisons, and were adjusted to be equally loud. There was no statistical imperative for listening in groups of four loudspeakers; it just seemed convenient. Interestingly, three- or four-way multiple comparisons have remained the norm in our subjective evaluations ever since. A supportive technician built a simple relay switch box. After that, I and a few interested colleagues took turns listening, forming opinions and making notes. A "Gestalt" impression, a summarized overall rating, was required: a number on a scale of 10.

The results surprised all of us. The audible differences were absolutely enormous, but there was general agreement about which ones seemed to sound good. It remained a topic of discussion for days. Where was the much-touted individuality of opinion?

The need for loudspeakers for my anechoic sound localization experiments remained, and the winner of this simple test, a KEF Concord, showed promise. It was dismantled; I found the cause of the misbehavior (flexure along the vertical axis of the B139 woofer). A phase plug eliminated the cancellation and a new crossover resulted in a much-improved loudspeaker—my first loudspeaker "design." The sound localization experiments proceeded.

Months passed before another listening test was staged in February 1966. By then I had learned that bed sheeting is not acoustically transparent and that music passages needed to be short and repeated. This meant that we needed a "disc jockey" to perform the tedious task; it was the LP era. Our agreeable technician did it. Word had spread and audio enthusiasts from within the organization lined up to participate, and in some cases bringing their personal loudspeakers to be evaluated. This test went on for days, and yielded enough subjective data to warrant rudimentary statistical analysis.

Again, there was good agreement about the products that were preferred and those that were not. The winning loudspeaker was the redesigned unit that was being used in my anechoic chamber tests. It also had the best looking set of measured data, assuming one puts any value in smooth and flat frequency responses on and off axis, Figure 18.1f. I still have the original hand written response sheets from these tests done 50 years ago.

The original Quad ESL Figure 18.1c may have set a standard for on-axis behavior, but the large radiating panels degraded off-axis performance, causing in-room performance to be compromised. The theoretically attractive "dual-concentric" Tannoy (a) radiated excruciatingly bright and colored high frequencies along with very boomy bass. The multi-driver large box Wharfedale (d) exhibited poor bass extension, numerous resonances and uncontrolled acoustical interference. It was very colored and lacked high frequencies. The original KEF (e) is what I purchased for my post-graduation hi-fi system, based on listening in tiny rooms at the 1964 audio show in the Hotel Russell in London. To my then uneducated ears, it simply sounded good. Raymond Cooke of KEF

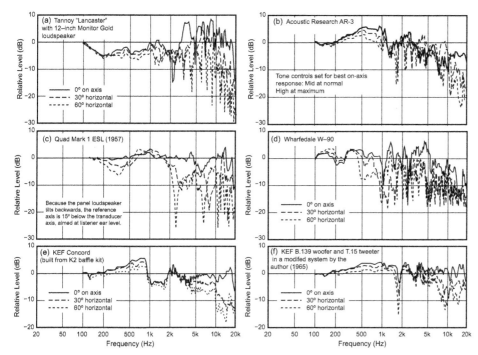

FIGURE 18.1 *On- and off-axis anechoic measurements on five loudspeakers that were highly respected in the mid-1960s. The sixth (f) was (e) as redesigned by the author. Low frequency data are truncated because the anechoic chamber had not yet been calibrated.*

gave me a discounted price because it would be for "research." I lied then, but it turned out to be true. I picked them up at the factory in Maidstone, Kent.

The Acoustic Research AR-3, which was already in the lab when I arrived at the NRCC, was highly regarded at the time. It embodied Edgar Villchur's innovative acoustic suspension woofer, offering deeper bass in a small enclosure than had previously been possible. It also had dome midrange and tweeter units with uncommonly good dispersion. AR had upgraded the AR-3 to the AR-3a, and when I enquired about an upgrade, I was invited to visit. The loudspeakers and I arrived in Cambridge, Massachusetts, and while I was being toured around the engineering area and factory, they were converted to the 3a version and tested.

Roy Allison was a gracious host and provided insight into what was behind the design and their measurement philosophies. The high-frequency rolloff was explained as being a necessary compromise between a tweeter that had the desirable characteristics of smoothness, bandwidth and dispersion, and one that sacrificed some or all in order to gain output. AR decided to let the high frequencies roll off. But then, it was explained that doing so brought them closer to their design goal: "To produce the same spectral balance at the ears of listeners in both concert halls and living rooms" (Allison and Berkovitz, 1972). The paper shows curves for a few halls, and the high frequencies roll off by different amounts. Noting the variability of this and also of the bass responses of the halls, the authors said in closing, "we think that home listeners should be encouraged to make more liberal use of amplifier tone controls." Forty-five years later, it is still good advice.

That said, the idea of rolling off the highs in the loudspeaker because mike placement can exaggerate the high frequencies in classical music seems odd. It is the role of the recording and mastering engineers to make whatever music is being recorded sound correctly balanced through the monitor loudspeakers. These almost certainly did not roll off the high frequencies. This is where equalization, if any, needs to be applied, not permanently in the customers' loudspeakers. However, because the recording industry lacked any standards for monitoring (and still does), that may have been a motivating factor. In the case of the AR-3a, the intended optimization only considered symphonic music, and then only for recordings that they argue would otherwise be unpleasantly bright. What about the ones that aren't bright, and what about the rest of the musical repertoire?

In any event the rolloff persisted in the 3a version, in which the crossover frequencies were lowered, moving the 1 kHz coloration down to about 400 Hz. Figure 18.2 shows a compilation of their measurements and mine. Given the possible sources of error, including production variations, the agreement is good.

According to listeners in my early blind listening tests, the significantly non-flat frequency response was a liability, with comments referring to mid-frequency coloration and high-frequency dullness. The program selections came from commercial LPs: Mozart's *Jupiter* (symphony), Chopin's waltzes (piano), Handel's *Messiah* (choral), a military band march, and Billy Strange playing acoustic guitar.

Some fans point to the AR "live vs. reproduced" demonstrations as absolute proof of reproduction accuracy. However, as explained in Section 1.8, such demonstrations employ recordings made especially for those events, reproduced through specific loudspeakers in concert venues. Such demonstrations have every reason to succeed and apparently all of them, from different companies, did. There is really no excuse for failure so long as the loudspeakers are not corrupted by gross distortion or resonances. What matters in the consumer world is different: How do commercial recordings sound in homes?

The room curve data in Figure 18.2c is especially interesting. From what is now known, at the listening position in normal rooms, the mid- to high frequencies consist substantially of early reflections and direct sound (Section 5.6). In this case, the directivity of the loudspeaker is very uniform, at least above 1 kHz, so the on-axis, off-axis and sound power curves will have similar shapes. Fortunately Allison and Berkovitz published enough data to permit retrospective analysis. Figure 18.2c shows the average of many in-room measurements, with the anechoic 60° off-axis curve and the measured sound power superimposed. As anticipated, there is substantial agreement. The single 60° curve represents one possible early reflection. As shown in Section 5.6 the sum of early reflections in a normal room provide a good estimate of the steady-state room curve. The spinorama measurement method averages several such curves and

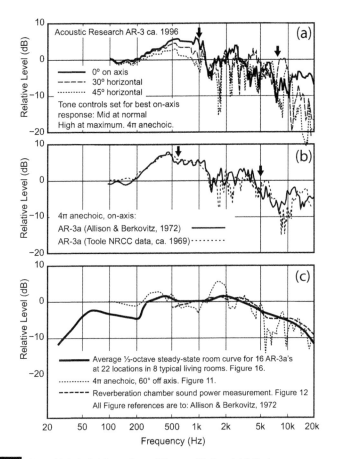

FIGURE 18.2   *Some historical data on the well-known AR-3 and AR-3a from my own anechoic measurements and manufacturer's data published in Allison and Berkovitz (1972). The arrows indicate the crossover frequencies for the AR-3 in (a) and the AR-3a in (b).*

such a spatially averaged curve would have been much smoother—not unlike the sound power curve. Allison (1974, 1975) correctly explained that the dip in the upper bass, around 100–200 Hz, was the result of adjacent boundary interactions; see Figure 9.4 and accompanying discussion. The acoustical interference roughness, the "grass," in the curves was the result of reflected and diffracted sound from the protruding grille-frame. Measurements of the raw transducers on plain baffles were much smoother. Acoustic Research deserves credit for being pioneers in attempting to relate in-room measurements to measurements pertaining only to the loudspeaker and having the candor to publish honest and accurate data.

My next encounter with this loudspeaker came a few years later in England, when AR was under new ownership and Tim Holl was planning a new engineering lab. One of his first projects was to create an AR-3a Improved. It was to have a high-frequency driver capable of delivering a flatter overall response; by then the high-frequency rolloff was regarded as a problem. Apparently neither it nor the lab was achieved.

18.2 TWO DECADES OF DOMESTIC LOUDSPEAKERS

It needs to be emphasized that I am not documenting the history of loudspeakers. In those years I was, and still am, a research scientist seeking answers to significant questions relating what we measure to what we hear. I had the opportunity to evaluate numerous loudspeakers, brought to the NRCC laboratory by audio enthusiasts, retailers and loudspeaker manufacturers. But by far the majority were submitted for evaluation by Canadian audio magazines: *AudioScene Canada, Sound Canada* and *Sound and Vision Canada*. They published reviews displaying anechoic measurements and subjective comments resulting from double-blind listening tests. I had no role in selecting the loudspeakers. Those were exceptionally informative reviews, but sadly the market could not support the activity.

There were interesting loudspeakers that simply did not come my way, and I had no budget to purchase them. Some readers of my first book thought I was being prejudiced for not having commented on certain brands or models. Even with the expanded collection of loudspeakers that follows, several noteworthy products are missing. From my perspective, though, the population of loudspeakers I was able to evaluate embraced enough of the "junk to jewels" sound quality range for patterns in the measurements that represented excellence to be recognized. I got what I needed, and the results are published in the public domain for all to see.

Figures 18.3 and 18.4 show some of the many loudspeakers that appeared in the years between my first tests and those published in Toole (1986). Measurements made after 1983 used a computer-controlled measurement system, and the anechoic chamber was calibrated so low frequency performance could be seen. Samples of the 1986 data are shown in Figure 5.2.

Several of these products performed well in double-blind listening tests at the time, but a few probably obvious ones had problems. Some were popularly priced and sold in large numbers. Others are products that showed evidence of serious engineering insight and effort. They came from the US, UK, Canada and Japan to indicate that nobody had a monopoly on making decent sounding loudspeakers—nor bad sounding ones either. I have not used valuable space to show poorly designed loudspeakers. They incorporate almost every conceivable flaw and a few display respected brand names. Price was an unreliable guide (see Figure 3.19). In the end, only data on specific models can be trusted. A good or bad result in one model conveys no dependable information about a brand so generalization is unreliable. Often, influential individuals insist on "voicing" the product, or marketing has seen what sold last year, and what is imagined to be the "sound of the season" gets delivered. In the early days, though, no loudspeakers were truly neutral; they all had distinctive sounds, and it became part of audio lore that one had to search for a loudspeaker that matched one's personal taste.

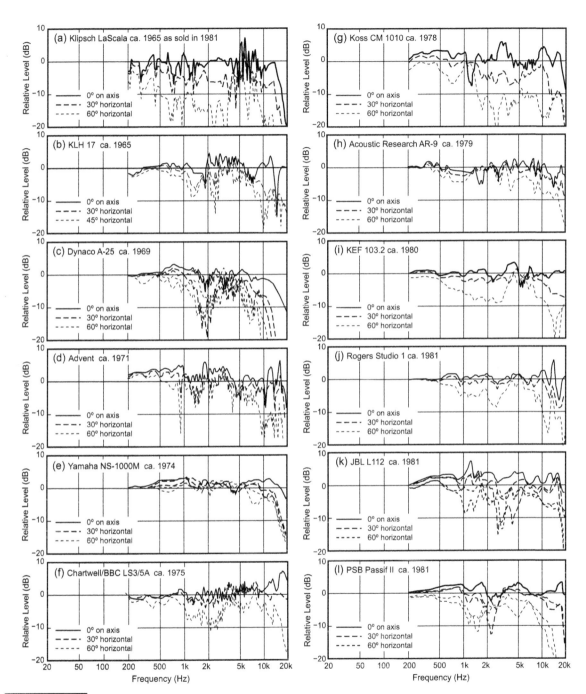

FIGURE 18.3 *Twelve consumer loudspeakers from the period 1965 to 1981. Those with asymmetrical driver arrangements were measured on both left and right sides, explaining why some have more curves. Curves truncated at 200 Hz were done using the first-generation analog measurement system that, for a variety of reasons, exhibited inconsistent errors at low frequencies. Sound levels have been normalized, relative sensitivity is not revealed.*

What is rarely understood is that the sound of a loudspeaker is primarily determined by its direct sound frequency response (on-axis and listening window) and its directivity (off-axis performance or DI) as a function of frequency. The former can be altered by equalization. The latter is built into the loudspeaker; equalization cannot change it. So, if one purchases loudspeakers having well behaved directivity (constant or smoothly and gradually changing over the mid- and high frequencies) then an equalizer or tone control can be freely used to create whatever sound is desired.

Equalizers can be changed at will to follow moods or variations in recordings. Most importantly they can be turned off. A compromised loudspeaker or one compromised by inappropriate equalization is a permanent timbral distortion for all reproduced sounds. This is understandable in consumer audio, but inexcusable in professional audio.

The problem has two components:

1. A fundamental mistrust of measurements.

2. A general absence of measurements worthy of one's trust.

Looking over the collections of curves, the one feature that stands out is that almost every designer set out to make a loudspeaker with a basically flat on-axis frequency response. Some succeeded better than others, but the underlying trend is there. They also differ in off-axis performance, and sometimes it was this, the spectrally uneven reflected sound, that differentiated how they sounded in a room.

In the very first evaluations shown in Figure 18.1, it was found that a loudspeaker exhibiting smooth and flat on- and off-axis performance consistently generated high subjective ratings. Looking at the data in Figures 18.1, 18.3 and 18.4, one can see that there was much progress toward this goal over the years. Figures 12.1 and 13.1 show that the progress in loudspeaker design and technology has continued, and the same pattern of subjective preference prevails. But, regrettably, everything in the marketplace is not better.

In Figure 18.3a a classic large horn loudspeaker exhibits evidence of acoustical interference and high directivity, both typical of the genre, but the underlying spectral balance was well maintained on and off axis. Very high sensitivity was a benefit at a time when amplifier power was limited. Loudspeaker (b), the Henry Kloss designed KLH 17 and (c), the Danish designed Dynaco A-25 were popular and affordable small loudspeakers of the period. Kloss also contributed the respected Advent (d), which, in a larger version, was often used in pairs—the "stacked Advents."

The Yamaha NS-1000M (e) was a serious effort, with beryllium dome midrange and tweeter units, mounted to minimize diffraction, to maximize and maintain constant directivity, and with obvious care in crossover design. They exhibited the lowest non-linear distortion of any consumer loudspeaker I had tested up to that point. The designer (also of the very different sounding NS10M) visited me at the NRCC, modestly explaining that it was simply "good engineering." It sounded good, but some listeners thought they could hear a "metallic" quality. In blind tests that problem went away. The curves show no evidence of resonances, metallic or otherwise, which was the point of using beryllium. The susceptibility of humans to biasing influences was and remains a major problem. A spectral balance that slightly favored the mid- and high-frequencies was audible, albeit easily correctable with some low frequency boost. It seems that the performance objective was a flat sound power curve for both the NS1000M and the NS10M, but that target was a better match for the almost constant directivity NS1000M than it was for the conventional small two-way NS10M, as illustrated in Figure 12.9. It was the wrong target, but an example of excellent engineering.

The BBC contributed several loudspeaker designs over the years, and the Chartwell licensed version of the LS3/5A was the "outside broadcast monitor" intended for use in small dead spaces. It was a small two-way, using KEF drivers, much like the midrange/tweeter combination in floor-standing loudspeakers. It sounded good at moderate sound levels, and acquired an enthusiastic following.

Figure 18.3h, the AR-9, was another serious effort, showing evidence of attention to both spectral balance and uniformity of directivity. Developed by Tim Holl, it embodied some control of room boundary interactions as advocated by Allison (see also Figure 9.12). This was another loudspeaker that anticipated the future: flat spectrum, constant directivity. It got high listener ratings.

The other loudspeakers in Figure 18.3, g, i, j, k and l, all show that the designers aimed at flattish on-axis frequency responses. Some came closer than others, but all lost control in achieving uniformity in off-axis radiation. In normally reflective rooms, these spectrally unbalanced reflected sounds affect the experience. Double-blind subjective ratings for the PSB Passif II (l) are in Figure 3.8 along with the AR-58s (Figure 18.4a) and the Quad ESL 63 (measurements in Figure 7.12).

Figure 18.4 continues the story, beginning with the AR 58s, which is dramatically different from the neutral spectral balance of the AR-9. Was it deliberately so? It was reminiscent of the AR-3a shown in Figure 19.2, so I superimposed the on-axis curve from that figure. Could it be that somebody wanted to revisit the sound of the legendary 3a in a new product? If so, it seems that the once desired AR-3a "improved" was achieved, because the tweeter now had more output.

The DM12 had good on-axis response, but exhibited the classic 6-inch (150 mm) two-way off axis misbehavior. The Infinity (c) had good directional control, but a mid-frequency hump determined the tonal balance of the product. The Energy (d) was uncommonly good in all respects: flat, smooth, almost constant directivity—it consistently scored well in listening tests. Others, as can easily be seen, were variations on a common theme.

The Ohm Walsh 2 shown in (f) is an interesting design, being horizontally omnidirectional up to a crossover to a conventional tweeter. It was a simplification of the original single broadband driver invented by Lincoln Walsh, which worked but was impractical. It had a relatively spacious presentation, and because the reflected sounds had the same spectrum as the direct sound, the room was less obvious than might be

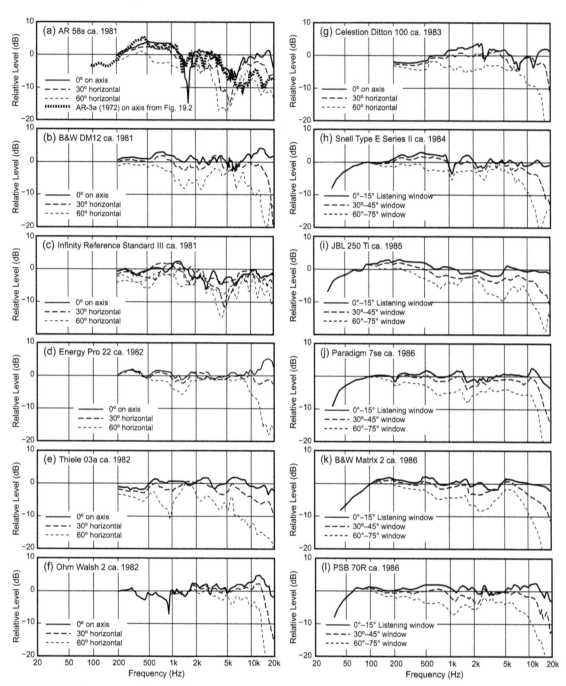

FIGURE 18.4   *Twelve loudspeakers that follow those in Figure 18.3, documenting progress in consumer loudspeaker design in the years 1981 to 1986.*

thought. This was one of the experiences leading me to recognize that humans have an ability to "listen through" rooms, to separate the sound of the source from the sound of the room. This ability seems to be degraded when the direct and reflected sounds have different timbral signatures.

Notably absent from the displayed products are those from Bose. All of the tested products, especially the distinctive direct-reflecting 901, radiated sound in several directions, from multiple drivers, some or most of which are aimed away from the listener. In the 901 eight drivers addressed the wall in front of the listener, and one was aimed at the listener. To reveal useful performance data on such loudspeakers requires the digital measurement system that can perform spatial averaging and sound power calculation; it did not exist at the time I was doing these tests. Measurements on single axes show acoustical interference that is disturbing to the eye but not necessarily evident to listeners in normally reflective rooms. So, rather than show misleading data, I show none. The same applies to the image-stabilizing dbx Soundfield loudspeaker discussed in Section 15.3.1. These designs added interest to the panorama of products in the marketplace, appealing to some, not to others. The bipolar but essentially omnidirectional Mirage M1 that ended up in one of my listening rooms (Section 7.4.6) is another in this class of products. Measurements are in Figure 7.20.

All multidirectional loudspeakers energize the reflected sound field in rooms, and all of these did. For some kinds of program and for some listeners, that is a positive attribute. The topic of early reflections is discussed in considerable detail in Chapter 7, and it is clear that there is not a single solution so having alternatives is good.

Reviewing the collection shows convincing reasons why loudspeakers sound different from one another—they *are* different from one another—but it is unlikely that the differences were always deliberate. Nevertheless, it can be seen that as the years passed they became increasingly similar. Today, loudspeakers can be even more similar than these (see Figures 12.1 and 13.1) and products with very similar measurements are sometimes difficult to distinguish in double-blind tests. Often it comes down to a decision between sounds that are slightly different, but comparably good. In those cases the variations in program material become the deciding factor, favoring one or the other at different times, and the end result is a statistical tie. There is a point of diminishing returns. The most perfect loudspeaker will not always sound perfect because recordings are not consistent. And, as mentioned several times throughout the book, humans are remarkably adaptable. We "break in" to accept many innocuous spectral aberrations, so long as gross resonances or non-linear distortion are not involved.

Given that bass is responsible for about 30% of one's overall assessment of sound quality, and that is primarily determined by the specific room and arrangement within the room (Chapters 8 and 9), it is clear that chance plays a large role in listening experiences. In controlled double-blind tests in which room effects have been rendered constant, it is possible to arrive at reproducible subjective evaluations that have a correlation with measurements such as these, and even better, with the spinorama style of data presentation. But this is a luxury that customers and reviewers do not have.

18.3 SOME EARLY PROFESSIONAL MONITOR LOUDSPEAKERS

Some audio professionals believe that monitor loudspeakers are designed to a higher performance standard than consumer products. I have heard it said that one can hear more things in a professional monitor than one can in a (lowly) consumer loudspeaker. If the criteria for excellence include the ability to play very loud for long periods without stress or failure, they are possibly right. If the criteria include bandwidth, timbral accuracy and an ability to revel the subtleties of a recording there is reason to reconsider.

If one focuses on conventional consumer loudspeakers from respected brands that aspire to a semblance of "fidelity," as was done in Figures 18.3 and 18.4, it is difficult not to be impressed by the overall uniformity of the performances and of the excellence of a few of the more recent ones. These products were mainly in the popular-price range because they were selected for readers of consumer audio magazines. Many consumers were listening to good sound. Today even more listeners are, including some bass and loudness limited small Bluetooth loudspeakers that sound better than they ought to.

Loudspeakers (e), (f) and (j) in Figure 18.3 are "crossover" products, having appeal to both consumer and professional users. My circumstances did not bring many large professional monitor loudspeakers my way, but over the years a few interesting examples were tested. Figure 18.5 shows 10 of them. It would be difficult to declare victory for the professional audio camp based on these data, although the 1987 product (j) did very well. The most serious concern is that these were, and—incredibly—some still are, the "windows" through which recording engineers and artists "view" their creations. Some recording engineers developed serious love affairs with certain of these products—another example of the power of human adaptation.

Harry Olson (1954, 1957) was a pioneer in sound reproduction, concluding in those early years that loudspeaker frequency response and directivity were the variables of greatest importance. However, measurement capabilities were limited, and Figure 18.4a shows an obviously artistically smoothed on-axis frequency response of the RCA LC1A, a 15-inch (380 mm) dual-cone loudspeaker in a low-diffraction enclosure. Directivity curves of both transducers are shown in the 1954 paper. Conical shapes on the woofer diaphragm broke up the regular interference effects that would have occurred from the central high-frequency driver. That driver was fitted with deflectors to broaden the dispersion above 10 kHz. The smoothing and the missing off-axis curves undoubtedly hide some sins, but it is evident that the designers had some of the right goals in mind.

Looking through the other loudspeakers over the 19 years from 1968 to 1987, there is a great deal of variation. The JBL 4310/L100 had a collection of strong resonances that were audible, but seemingly ignored or admired by many. The 4320, with its acoustic lens, was surprisingly well behaved, needing only an equalization tilt to make it competitive. To these should be added the Auratone (1960s onward) and the UREI 811B that appeared around 1979, shown in Figure 12.9. The UREI was based on the Altec 604(d), with a modified horn and crossover. These loudspeakers had strong

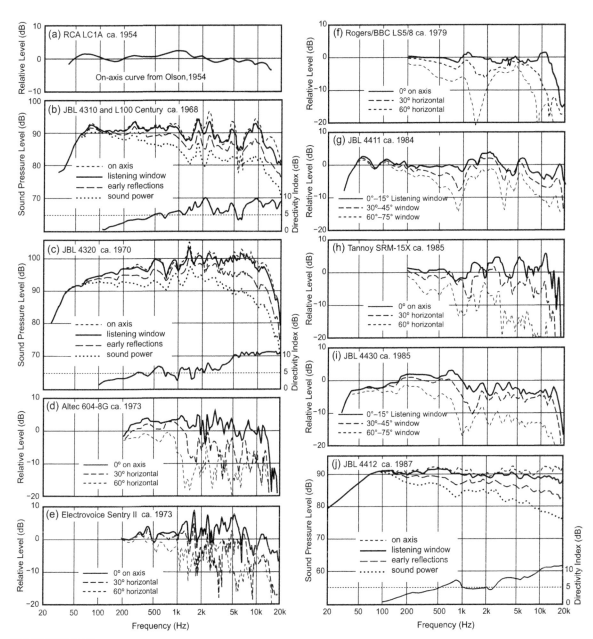

FIGURE 18.5 *A few of the medium and large monitor loudspeakers I tested. This collection shows the complete evolution of the NRCC measurements from the old analog system, to the digitally controlled system in which the anechoic chamber was calibrated at low frequencies: (g) and (i). I was able to test three of the JBLs after my arrival at Harman, so for these full spinorama data is shown: (b), (c) and (j).*

"personalities," and when inserted into double-blind comparison listening tests a couple of them were awarded the lowest subjective ratings ever noted.

Several can be seen to degrade very quickly as one moves away from the frontal axis. Such loudspeakers can only sound good in non-reflective rooms. Attempts at equalization using steady-state room curves cannot reliably succeed because of the frequency-dependent directivity exhibited by some of them. This, I suspect, had something to do with the tendency for many control rooms to be acoustically treated using what I call the "truckload of fiberglass" approach. Eliminate everything except the direct sound and equalize that. Directivity problems cannot be corrected by equalization. However, given the complexity of the problems in the direct sound from some of these loudspeakers it is evident that complete satisfaction would have been elusive.

The JBL 4412 (1987) breaks the pattern, exhibiting the spectral and directional control of what I have called modern monitors in Section 12.5.2. Figures 5.12 and 18.7(d) show the JBL Pro M2, an example of a modern mid-field or main monitor that is inherently well behaved and is then, based on anechoic data, equalized in a dedicated digital processor/power amplifier package.

These and others now in the marketplace provide relatively neutral timbral quality that allows mixers to estimate what customers might hear. Section 12.5.2 has more discussion on this topic. The conclusion: if it is necessary to boost or cut specific frequency ranges to gain insight into a mix, do it with an equalizer, not a different loudspeaker. The physical and acoustical clutter in the control room can be reduced.

18.3.1 A "Toole" Monitor Loudspeaker

For me, those were years of gaining experience and learning. Part of this was a brief career as a moonlighting studio designer—applying the knowledge I thought I had in the real world. I designed three complete studio complexes, one of which had a studio large enough to hold a 75-piece chamber orchestra, and contributed to others. The data shown in previous figures indicate that the professional monitors of that period, the mid-1970s, left something to be desired. The ones I had access to routinely did poorly in double-blind listening evaluations, although they almost always could play louder than the consumer products.

It bothered me when one of my clients wanted to install Altec 604–8Gs (Figure 18.5d), because that was what his lead engineer and some of his customers liked. I took the challenge of designing something better. In my opinion the same principles that guide the design of the best sounding consumer loudspeakers could be applied to large monitors. I borrowed a large selection of driver samples, tested them and selected what I thought were the best performers. It ended up being a four-way tri-amplified system with a 15-inch JBL 2215 woofer with a 2290 passive radiator (no port misbehavior), a JBL 2120 8-inch midbass, a JBL 2397 "Smith" horn driven by a 2440 compression driver, and an ElectroVoice ST-350A high-frequency horn. All were selected on the basis of smooth frequency responses and uniformly wide dispersion. The Smith horn was significantly less colored than most of the conventional horns of the time, and had uniformly wide dispersion. The result is shown in Figure 18.6; it was May 1978.

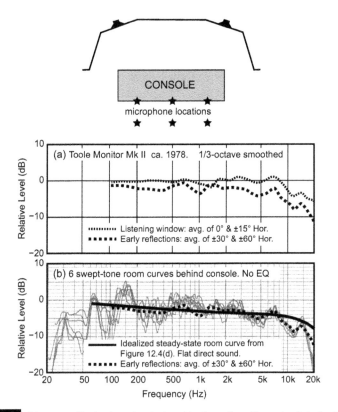

FIGURE 18.6 *A large monitor loudspeaker designed by the author. Discussion is in the text.*

Fortunately, in addition to swept-tone anechoic measurements I did some 1/3-octave curves. These, I have been able to combine by calculation into approximations of the spinorama listening window and early-reflections curves. As discussed in Section 5.6 steady-state room curves are often well estimated by the early-reflections curve.

Figure 18.6a shows a relatively smooth and flat direct sound loudspeaker with relatively constant directivity (no EQ). (b) shows pure-tone sweeps (no spectral smoothing) at six locations in one of the control room installations. The variations above about 1 kHz are relatively small and follow the shape of the curves in (a). The room was quite dead, RT around 0.2 to 0.25 s, and the early reflection surfaces were highly absorbing, so the evidence of accumulating reflected energy is mostly from the floor, the console surface, and other control room apparatus.

Figure 18.6b shows that the early-reflections estimate is a good match for the unsmoothed room curves. Also shown for interest is the idealized steady-state room curve from Figure 12.4d showing what can be expected from a state-of-the art, flat on-axis loudspeaker in a normally reflective room. If I had known then what I know now, I would have equalized the loudspeaker based on anechoic data, which would have delivered the idealized room curve almost exactly—except of course for the bass room-mode region.

After simple equalization to smooth the large undulations in the room curves, it sounded good enough for the senior recording engineer and the customers to give an enthusiastic go-ahead. In spite of the significant cost and complexity, six such systems were built into different projects. The results look much the same.

Lacking any entrepreneurial instincts, I never saw it as a product, only a proof of concept. It was just one more confirmation of trends that have continued since. Starting with a truly good loudspeaker, getting good sound is not difficult, except in the low-bass room-mode region.

Soon after this I published a deliberately provocative article (Toole, 1979) in which I challenged the professionals to move to the standard of sound quality being enjoyed by an increasing percentage of consumers; it was called "Hi Fidelity in the Control Room—Why Not?" I sensed that it went over like the proverbial lead balloon. I was rocking the boat.

This loudspeaker was just a basically neutral loudspeaker that differed from its hi-fi brethren in that it could play dangerously loud. Proof that I was right is that several present-day monitors make it possible to have the highest of fidelity in control rooms, if that is the choice.

A significant event in my career occurred when I conducted a massive monitor loudspeaker evaluation in collaboration with the Canadian Broadcasting Corporation (CBC) (Toole, 1985, 1986). The goal was to identify suitable large, medium and small monitor loudspeakers for use throughout the nationwide network. From the samples submitted by manufacturers and distributors 16 were selected as finalists. These were measured, and then subjectively evaluated by 27 listeners, 15 of whom were professional recording engineers and producers from the CBC. The others were colleagues and interested audiophiles well practiced in the double-blind listening procedure, which was an ongoing operation in my research. Subjective evaluations went on for two weeks. In the end the customer was satisfied that the results were conclusive. There was no major disagreement among the listeners about which were the most preferred products, which included loudspeakers from Figures 18.3, 18.4 and 18.5. The most memorable comment I heard was that some of the audio professionals thought that, during the tests, they had never heard such good sound in their lives. It turned out that most of the monitor loudspeakers with which the engineers were experienced did poorly in the tests, which, when the results were revealed, initiated a minor revolt. After some repeated tests using the engineers' own tapes, they were persuaded. It was a learning exercise for us all. The highest rated loudspeakers included both domestic and professional products.

As discussed in Section 3.2 and Toole (1985) it was this evaluation that yielded definitive evidence that hearing loss affects one's ability to form consistent opinions about sound quality, and introduces bias. Sadly it is an occupational hazard in the audio business (Chapter 17).

18.4 LOOKING AROUND AND LOOKING AHEAD

Loudspeaker design is now a mature technology. The trial-and-error, golden-ear phase is, or should be, long gone. Specialist manufacturers produce superb transducers, off the shelf or to order, and measurement and computer design aids provide enough trustworthy guidance to reliably yield neutral sounding loudspeaker systems. It requires skill and the desire to produce such loudspeakers. For some that is a problem. There is still a residue of the past, in which the sound of the loudspeaker is part of the magic, part of the art, even an emotional attachment to something real or imagined.

If we were perfectly successful in delivering uniform sound, neutral sound, from all loudspeakers professional and consumer, the art would be more likely to be communicated as it was created. This is good for the art, but from a loudspeaker marketing perspective this is a backwards step, because the "sound" of the loudspeaker has been a differentiating factor. People expect loudspeakers to sound different.

In spite of the folklore, electronics that are not misbehaving or driven beyond their limits are essentially transparent. Nobody builds electronics for an audio signal path that does not have a ruler flat frequency response over more than the audible bandwidth. They are becoming "invisible" parts of our systems, and if the future should include merging electronics with transducers in active loudspeakers that portion of audio tradition will fade. The result will be better sound, but something will have been lost.

To tour a state-of-the-art loudspeaker design lab is to see transducer and system measurement apparatus that did not exist until recent years. Computer-based simulation capabilities allow us to almost hear the product before a prototype exists. However, such a facility is expensive to build and expensive to populate with the skilled engineers necessary to operate the devices. Not everybody has it. It also requires a belief in science. Not everybody has that, either.

For some, loudspeaker design is portrayed as an art and the listening experiences that result are described in poetic terms. If the resulting sound is genuinely neutral, allowing listeners an opportunity to hear the real art—the recording—then all is well. If not, the customer is well advised to find another poet—or ponder reconsidering the value of science.

Figure 18.7 continues the review of interesting loudspeakers by showing two recent, highly regarded, loudspeakers: (a) a dipole panel radiator and (b) a conventional cone/dome design. (c) shows measurements on a fully active cone/dome prototype—multiple power amplifiers, multiple transducers and dedicated digital control electronics. It was indefinitely "postponed" because of cost, complexity and a belief that, at the anticipated price level, audiophiles might not be ready to cast off the entrenched tradition of separate components connected by wires. Professional audio people are more open minded, however, and (d) shows a recent manifestation of an active monitor. In this case the dedicated electronics are outboard, but several monitor loudspeakers have electronics incorporated into the package. These and other similar loudspeakers are transparent "windows" into the audio arts. Consumer audio needs to catch up. The new Lexicon SoundSteer SL-1 active array delivers quality sound and adds directivity control, steered beams, movable stereo sweet spot and more to the capabilities. The future should be very interesting.

For now, and for the foreseeable future, there will be excellent passive loudspeakers from which to choose. Figure 12.1 shows examples of some passive loudspeakers that approach the performance of active ones. Traditionalists will not be deprived.

The message is unmistakable. Looking back over the collection of measurements in this chapter, one sees evidence of considerable improvement. But one also sees that

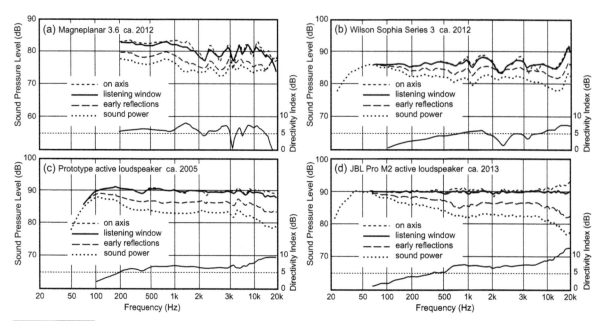

FIGURE 18.7 *Two highly regarded loudspeakers. (a) is a large panel loudspeaker. The low frequencies have been truncated because the anechoic chamber was not calibrated for large dipole radiators and they have unique ways of coupling to room modes in the listening room. (b) shows a conventional cone/dome design. (c) and (d) are loudspeakers having dedicated electronics.*

some years ago progress seems to have plateaued. Good-sounding loudspeakers were being made, and still are, but they all have residual flaws of one kind or other. They may sound comparably good, but just not exactly alike—variations on the same theme.

Some of them require monster monoblock power amplifiers to drive the current-hungry low impedances, sometimes the consequence of the complicated passive crossover networks required to smooth the frequency responses. This is much easier and better if done as equalization in electronics upstream of the power amplifiers, with the power amplifiers optimally designed to match the load conditions and power needs of individual transducers. And, active crossovers are vastly superior to passive versions. The monster monoblocks get replaced by a number of much smaller, simpler, amplifier modules. Modern equalizers can be as complex as needed, and because transducers are minimum-phase devices over their operating frequency ranges a smooth frequency response is a good indicator of freedom from transient misbehavior.

Although some people are still catching up, one needs only to look back over the decades to see that the clues to good sound have been with us for a long time. As I point out in Section 5.1 the "ancients," as I call them, had the right instincts as far back as 80 years ago (Brittain, 1936–37), even without the benefit of accurate measurements. Harry Olson (1954, 1957) provided more focused guidance accompanied by measurements. Gilbert Briggs (1958) of Wharfedale, assisted by Raymond Cooke (later of KEF), showed credible looking on and off-axis response curves and polar diagrams, and expounded with good sense and insight on enclosures, resonances, phase shift and other factors. Edgar Villchur (1964) of Acoustic Research, was an enthusiastic supporter of measurements, including the idea of integrating on and off axis radiated sound in a measure of the total sound power. He was right when he said: "Once a test has been validated as an index of performance, it can reveal information that might take many hours or even days of uncontrolled listening to discover."

However, validation is difficult, especially acquiring reliable subjective data. Looking back it is evident that all of the technical metrics existed, but not all were equally useful indicators of sound quality as heard in typical semi-reflective listening rooms. When the subjective and objective data existed and were analyzed, as discussed in Chapter 5, everyone could take some credit in having "anticipated" at least portions of the solution. But until the proofs were in, the divergent claims were simply opinions, and some of them did nothing to advance the science of audio or the interests of consumers. Sections 5.7.1 and 5.7.3 illustrate the dangers of placing too much trust in a single metric, the wrong one, and failing to test it with ongoing rigorous subjective evaluations.

In the intervening years many others have added innovations to loudspeaker design and aids to making engineering measurements. The list of contributors is long and the results have been significant. In every case the ability to measure has been the foundation. Woofers and their enclosures are now designed in computers, with predictable performance. Transducers can be modeled from the magnetic motor systems through the suspension systems and diaphragm flexural modes to predictions of the radiated sound field. Complete systems can be measured in several ways, from anechoic chamber and outdoor free fields, through time-windowed in-room measurements, to the new mind-bending near-field scanning system (Klippel and Bellman, 2016). The merging of science, engineering and art has been exciting to witness and to be a small part of.

> My contribution has mainly been to look, as dispassionately as possible, at the possibilities, to contrive and conduct experiments seeking answers to questions, and to offer results and explanations of what was observed. Sometimes this included evidence of who and what might have been right. Nobody has been completely wrong, as it turns out, and even then errors were most often associated with the incomplete information available at the time. Over the years this has often been the absence of unbiased subjective data. This is the scientific method.

Much has been learned about the human listener in terms of capabilities (considerable) and trustworthiness (only in blind listening tests). Chapter 5 shows what is needed to set the design objectives for neutral sounding loudspeakers, using measurement techniques that are available today. The problem for consumers, both domestic and professional, is the lack of useful measurements in specification sheets. Figure 12.3 compares where we are to where we need to be. The simplistic specifications offered by most manufacturers simply insult the intelligence. Choosing a loudspeaker is therefore consigned to the "listening test," which in the real world is almost always sighted, subject to preconditioning by sales people, colleagues, friends, reviews and so forth. Equal level comparisons of products are rare, and listeners may or may not have time to adapt to the peculiarities of the listening environment. And we must not forget that any listening test includes all of the upstream activities in the recording process. The recordings and the circle of confusion are part of the test. The results of such tests are subject to variation, so in the end the choice involves considerable opportunity for disappointment.

> This, I believe, is the true weak link in the audio industry. We know how to design, and to describe in measurements, neutral, accurate, high-fidelity loudspeakers for consumers and professionals. The problem is that the information is so rarely conveyed.

Marketing departments insist that graphical technical data are incomprehensible, so they deliver numerical data that are almost useless, but familiar. My assertion is that data in the form of spinorama or some other comparably revealing graphical format are easily comprehended with the simple guidance: flat and smooth are good, and the more similar all the curves look, the better the sound is likely to be. An engineering degree is not a requirement. However, I do understand why manufacturers are not rushing to reveal useful anechoic data in the ANSI/CTA-2034-A (spinorama), or any other, format. Some of them would be embarrassed, and all of them would have the commitment to maintain, in production, the performance of the "golden prototype." Manufacturing variations in driver sensitivity and frequency response can be significant, but directivity is unlikely to change. This is yet another advantage of active loudspeakers: the ability to make them functionally identical at the end of the production line.

However, even if we had the ability to identify good loudspeakers, not even the best of them can recreate the subjective impressions of listening to live performances in stereo, the default musical format. There is no magic tweak, no spikes, wires, or exotic electronics that can compensate for a directionally and spatially deprived format. As has been discussed in Chapter 15, multichannel and immersive formats provide welcome added dimension and space. They can be very persuasive, but the musical repertoire in these formats is limited. In any event I intend to enjoy what there is and will be in my soon to be upgraded immersive, listening room—and movies of course.

Finally, if one has the good fortune to acquire excellent loudspeakers, there is still the need to deal with low-frequency problems in small rooms. These have been addressed in Chapters 8 and 9. Equalization is likely to be a component in that corrective action, but outside the bass region equalization needs to be employed with great caution. The widely used "room correction" algorithms assume that the definitive information about sound quality is in the steady-state room curves generated using an omni-directional microphone. However, two ears and a brain are much more analytical, and in different ways. As has been discussed elsewhere in this book, there is a significant risk of these systems degrading good loudspeakers. Section 12.2.3 points out that some of these systems are installed as if they are program equalizers; adjust the curve until it sounds good. It is a subjective decision that includes the circle of confusion and is therefore biased by the music being listened to at the time.

Equalizers in the form of traditional bass, treble and tilt tone controls are useful to compensate for common spectral peculiarities in recordings caused by the circle of confusion, or to accommodate personal taste. They should be quickly accessible and easy to turn up, down, on or off, as required while listening. But, begin with resonance-free, low coloration—neutral—loudspeakers as the baseline. Audiophiles who reject tone controls in their electronics are simply ignoring the existence of the *very real* circle of confusion. They mistakenly assume that recordings are flawless.

In Conclusion: Science in the Service of Art

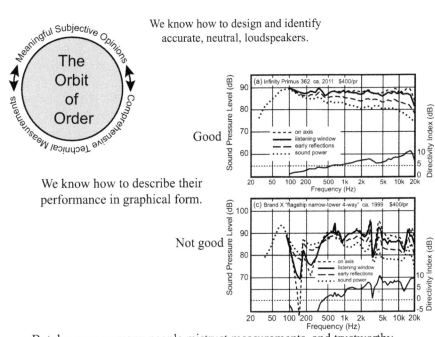

We know how to design and identify accurate, neutral, loudspeakers.

We know how to describe their performance in graphical form.

But, because so many people mistrust measurements, and trustworthy measurements are so scarce, the Circle of Confusion still exists.

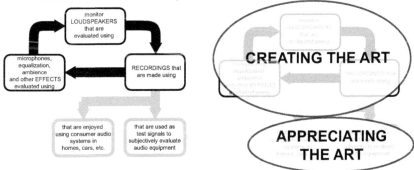

"The future is already here. It's just not evenly distributed yet."
William Gibson

FIGURE 18.8 *The "not good" loudspeaker was from a very well-known brand. This product was the brainchild of the US distributor, having nothing to do with the competent parent company. A consultant was hired to design it and one suspects that he phoned the design in. Current Internet forum chat indicates that one person is recommending bi-wiring as an upgrade.*

18.5 THE END

I could go on, but I won't. It is time to wrap it up. Due to space limitations not all topics have been covered and others have been simplified. However, there is a supporting website where additional information can be found on some of the topics. In Chapter 1 I described the "circle of confusion" as a fundamental problem in the audio industry. It still is. However, over the years we have learned how to design and describe loudspeakers that listeners approve of when listening to a wide range of commercial recordings. They do so in a variety of rooms, without elaborate acoustical treatments, or anything not achievable with suitably arranged normal domestic furnishings. The better the loudspeakers, the more easily the room fades into an innocent context. The higher the channel count, the less the room contributes to the experience.

A good listening experience starts with good loudspeakers. The highly rated loudspeakers have been neutral reproducers—clear "windows" through which to view the art. It is all very sensible, which makes it all the more painful to hear some of the products in the marketplace, portrayed as something they are not. It is unlikely that this will go away. But with education and patience more people may yet hear truly good sound. The discouraging Circle of Confusion has been supplemented with the optimistic Orbit of Order. Good things can happen.

Figure 18.8 summarizes my view of where we are. If you are among those fortunate enough to have a good sound reproducing system, I suggest you join me in a toast, as we listen to music as it should be heard: "here's to good sound, everywhere and always."

References

Alexander, R.C. (1999). *The Inventor of Stereo—the Life and Works of Alan Dower Blumlein*, Focal Press, Oxford, UK.

Allen, I. (2006). "The X-Curve: Its Origins and History", *SMPTE Mot. Imag. J.*, vol. 115, July/August, pp. 264–273.

Allison, R.F. (1974). "The Influence of Room Boundaries on Loudspeaker Power Output", *J. Audio Eng. Soc.*, vol. 22, pp. 314–320.

Allison, R.F. (1975). "The Sound Field in Home Listening Rooms, II", *J. Audio Eng. Soc.*, vol. 24, pp. 14–19.

Allison, R.F. (1982). "Rating Loudspeakers for Accuracy", *Stereo Rev.*, vol. 47, no. 6, pp. 60–61.

Allison, R.F. and Berkovitz, R. (1972). "The Sound Field in Home Listening Rooms", *J. Audio Eng. Soc.*, vol. 20, pp. 459–469.

Angus, J.A.S. (1997). "Controlling Early Reflections Using Diffusion", 102nd Convention, Audio Eng. Soc., Preprint 4405.

Angus, J.A.S. (1999). "The Effects of Specular Versus Diffuse Reflections on the Frequency Response at the Listener", 106th Convention, Audio Eng. Soc., Preprint 4938.

Angus, J.A.S. (2010). "Closed and Vented Loudspeaker Alignments That Compensate for Nearby Room Boundaries", 128th Convention, Audio Eng. Soc., Preprint 7990.

ANSI/ASA S12.2–2008. "Criteria for Evaluating Room Noise", Standards Secretariat of the Acoustical Society of America, Melville, NY.

ANSI/CTA-2034-A (2015). "Standard Method of Measurement for In-Home Loudspeakers", Consumer Technology Association, Technology and Standards Dept., www.CTA.tech.

Antsalo, P., Karjalainen, M., Mäkivirta, A. and Välimäki (2003). "Comparison of Modal Equalizer Design Methods", 114th Convention, Audio Eng. Soc., Preprint 5844.

Arau-Puchades, H. (1988). "An Improved Reverberation Formula", *Acustica*, vol. 65, pp. 163–180.

Ashihara, K. (2007). "Hearing Thresholds for Pure Tones Above 16 kHz", *J. Acoust. Soc. Am.*, vol. 122, pp. EL 52–57 (JASA Express Letters).

Atkinson, J. (1997). "Loudspeakers: What Measurements Can Tell Us—and What They Can't Tell Us!", 103rd Convention, Audio Eng. Soc., Preprint 4608.

Atkinson, J. (2011). "Where Did the Negative Frequencies Go?", 131st Convention, Audio Eng. Soc., Heyser Memorial Lecture.

Augspurger, G.L. (1990). "Loudspeakers in Control Rooms and Living Rooms", 8th International Conference, Audio Eng. Soc.

Avis, M., Fazenda, B. and Davies, W. (2007). "Thresholds of Detection for Changes to the Q Factor of Low-Frequency Modes in Listening Environments", *J. Audio Eng. Soc.*, vol. 55, pp. 611–622.

Ballagh, K.O. (1983). "Optimum Loudspeaker Placement Near Reflecting Planes", *J. Audio Eng. Soc.*, vol. 31, pp. 931–935. See also letters to the editor, vol. 32, p. 677 (1984).

Baskind, A. and Polack, J.-D. (2000). "Sound Power Radiated by Sources in Diffuse Field". 108th Convention, Audio Eng. Soc. Preprint 5146.

Bech, S. (1989). "The Influence of Room Acoustics on Reproduced Sound, Part 1—Selection and Training of Subjects for Listening Tests", 87th Convention, Audio Eng. Soc., Preprint 2850.

Bech, S. (1992). "Selection and Training of Subjects for Listening Tests on Sound-Reproducing Equipment", *J. Audio Eng. Soc.*, vol. 40, pp. 590–610.

Bech, S. (1998). "Calibration of Relative Level Differences of a Domestic Multichannel Sound Reproduction System", *J. Audio Eng. Soc.*, vol. 46, pp. 304–313.

Bech, S. and Cutmore, N. (2000). "Multichannel Level Alignment, Part V: The Effects of Reproduction Level, Reproduction Room, Step Size, and Symmetry", Audio Eng. Soc. 108th Convention, Preprint 5085.

Bech, S. and Zacharov, N. (1999). "Multichannel Level Alignment, Part III: The Effects of Loudspeaker Directivity and Reproduction Bandwidth", 106th Convention, Audio Eng. Soc., Preprint 4909.

Bech, S. and Zacharov, N. (2006). *Perceptual Audio Evaluation*, Wiley, West Sussex.

Benade, A. H. (1984). "Wind Instruments in the Concert Hall" Text of an Oral Presentation at Parc de la Villette, Paris; part of a series of lectures entitled "Acoustique, Musique, Espaces", 15 May 1984 (personal communication).

Benade, A. H. (1985). "From Instrument to Ear in a Room: Direct or via Recording", *J. Audio Eng. Soc.*, vol. 33, pp. 218–233.

Benjamin, E. (1994). "Audio Power Amplifiers for Loudspeaker Loads", *J. Audio Eng. Soc.*, vol. 42, pp. 670–683.

Benjamin, E. (2004). "Preferred Listening Levels and Acceptance Windows for Dialog Reproduction in the Domestic Environment", 117th Convention, Audio Eng. Soc., Paper 6233.

Beranek, L. L. (1986). *Acoustics*, Acoustical Society of America, New York.

Bernstein, L. and Trahiotis, C. (2016). "Behavioral Manifestations of Audiometrically-Defined 'Slight' or 'Hidden' Hearing Loss Revealed by Measure of Binaural Detection", *J. Acoust. Soc. Am.*, vol. 140, pp. 3540–3548.

Bilsen, F. A. and Kievits, I. (1989). "The Minimum Integration Time of the Auditory System", 86th Convention, Audio Eng. Soc., Preprint 2746.

Binelli, M. and Farina, A. (2008). "Digital Equalization of Automotive Sound Systems Employing Spectral Smoothed FIR Filters", Audio Eng. Soc. Convention Paper 7575.

Bissinger, G. A. (2008). "Structural Acoustics of Good and Bad Violins", *J. Acoust. Soc. Am.*, vol. 124, pp. 1764–1773.

Blauert, J. (1996). *Spatial Hearing: The Psychophysics of Human Sound Localization*, MIT Press, Cambridge, MA.

Blauert, J. and Divenyi, P. L. (1988). "Spectral Selectivity in Binaural Contralateral Inhibition", *Acustica*, vol. 66, pp. 267–274.

Blauert, J. and Laws, P. (1978). "Group Delay Distortions in Electroacoustical Systems", *J. Acoust. Soc. Am.*, vol. 63, pp. 1478–1483.

Blesser, B. and Salter, L.-R. (2007). *Spaces Speak, Are You Listening*, MIT Press, Cambridge, MA.

Blumlein, A. (1933). British Patent No. 394 325 "Improvements in and Relating to Sound Transmission, Sound-Recording and Sound-Reproducing Systems" granted to Alan Blumlein and EMI, 1933. Reprinted in *J. Audio Eng. Soc.*, vol. 6, pp. 91–98, 130 (1958).

Bolt, R. H. (1946). "Note on Normal Frequency Statistics for Rectangular Rooms", *J. Acoust. Soc. Am.*, vol. 18, pp. 130–133.

Bonello, O.J. (1981). "A New Criterion for the Distribution of Normal Room Modes", *J. Audio Eng. Soc.*, vol. 29, pp. 597–606.

Boner, C.P. and Boner, C.R. (1965). "Minimizing Feedback in Sound Systems and Room-Ring Modes With Passive Networks", *J. Acoust. Soc. Am.*, vol. 37, pp. 131–135.

Børja, S.E. (1977). "How to Fool the Ear and Make Bad Recordings", *J. Audio Eng. Soc.*, vol. 25, pp. 482–490.

Braasch, J., Blauert, J. and Djelani, T. (2003). "The Precedence Effect for Noise Bursts of Different Bandwidths. I. Psychoacoustical Data", *J. Acoust. Soc. Japan*, vol. 24, pp. 233–241.

Bradley, J.S., Reich, R.D. and Norcross, S.G. (2000). "On the Combined Effects of Early- and Late-Arriving Sound on Spatial Impression in Concert Halls", *J. Acoust. Soc. Am.*, vol. 108, pp. 651–661.

Bradley, J.S., Sato, H. and Picard, M. (2003). "On the Importance of Early Reflections for Speech in Rooms", *J. Acoust. Soc. Am.*, vol. 113, pp. 3233–3244, http://dx.doi.org/10.1121/1.1570439

Bradley, J.S., Soulodre, G.A. and Norcross, S. (1997). "Factors Influencing the Perception of Bass", *J. Acoust. Soc. Am*, vol. 101, p. 3135, http://dx.doi.org/10.1121/1.419017

Brandewie, E. and Zahorik, P. (2010). "Prior Listening in Rooms Improves Speech Intelligibility", *J. Acoust. Soc. Am.*, vol. 128, pp. 291–299.

Bregman, A.S. (1999). *Auditory Scene Analysis, the Perceptual Organization of Sound*, MIT Press, Cambridge, MA.

Bridges, S. (1980). "Effect of Direct Sound on Perceived Frequency Response of a Sound System", 66th Convention, Audio Eng. Soc., Preprint 1644.

Briggs, G.A. (1958). *Loudspeakers*, Fifth edition, Wharfedale Wireless Works, Ltd., Idle, UK.

Brittain, F.H. (1936–1937). "The Appraisement of Loudspeakers", *GEC J.*, Pt. 1, vol. 7, November, pp. 266–276; pt. 2, vol. 8, May, pp. 121–130 (1937).

Brittain, F.H. (1953). *Art and Science in Sound Reproduction*, The General Electric Co. Ltd., London.

Broner, N. (2004). "Rating and Assessment of Noise", Australian Institute of Refrigeration, Air Conditioning and Heating Conference, www.AIRAH.org.au

Büchlein, R. (1962). "The Audibility of Frequency Response Irregularities", reprinted in English in *J. Audio Eng. Soc.*, vol. 29, pp. 126–131 (1981).

Burgtorf, W. (1961). "Untersuchungen zur Wahrnehmbarkeit verzögerter Schallsignale", *Acustica*, vol. 11, pp. 97–111.

Cabot, R.C. (1984). "Perception of Nonlinear Distortion", 2nd International Conference, Audio Eng. Soc., Preprint C1004.

Campbell, D.M. (2014). "Evaluating Musical Instruments", *Physics Today*, vol. 67, no. 4, pp. 35–40.

Carver, R. (1982). "Sonic Holography", *Audio*, vol. 66, March.

Celestinos, A., Devantier, A., Bezzola, A., Banka, R. and Brunet, P. (2015). "Estimating the Total Sound Power of Loudspeakers", 139th Convention, Audio Eng. Soc., Paper 9463.

Celestinos, A. and Neilsen, S. (2008). "Controlled Acoustic Bass System (CABS), a Method to Achieve Uniform Sound Field Distribution as Low Frequencies in Rectangular Rooms", *J. Audio Eng. Soc.*, vol. 56, pp. 915–931.

Choisel, S. (2005). "Effect of Loudspeaker Directivity on the Perceived Direction of Panned Sources", Reports from the Sound Quality Research Unit, No. 32, Dept. of Acoustics, Aalborg Univ., Denmark, and PhD Thesis "Spatial Aspects of Sound Quality", Univ. of Aalborg.

Choisel, S. and Wickelmaier, F. (2007). "Evaluation of Multichannel Reproduced Sound: Scaling Auditory Attributes Underlying Listener Preferences", *J. Acoust. Soc. Am.*, vol. 121, pp. 388–400.

Clark, D. L. (1981). "High Resolution Subjective Testing Using a Double-Blind Comparator", 69th Convention, Audio Eng. Soc., Preprint 1771.

Clark, D. L. (1983). "Measuring Audible Effects of Time Delays in Listening Tests", 74th Convention, Audio Eng. Soc., Preprint 2012.

Clark, D. L. (2001). "Progress in Perceptual Transfer Function Measurement—Tonal Balance", Audio Eng. Soc. Convention Paper 5407.

Clark, W. (1991). "Noise Exposure From Leisure Activities: A Review", *J. Acoust. Soc. Am.*, vol. 90, pp. 175–181.

Cohen, E. and Fielder, L. (1992). Determining Noise Criteria for Recording Environments", *J. Audio Eng. Soc.*, vol. 40, pp. 384–402.

Cooper, D. H. and Shiga, T. (1972). "Discrete-Matrix Multichannel Stereo", *J. Audio Eng. Soc.*, vol. 20, pp. 346–360.

Corey, J. and Woszczyk, W. (2002). "Localization of Lateral Phantom Images in a 5-Channel System With and Without Simulated Early Reflections." 113th Convention, Audio Eng. Soc., Preprint 5673.

Cox, T. and D'Antonio, P. (2004). *Acoustic Absorbers and Diffusers*, Spon Press, London.

Cox, T. and D'Antonio, P. (2009). *Acoustic Absorbers and Diffusers, Second Edition*, Taylor and Francis, Abingdon.

Cox, T., D'Antonio, P. and Avis, M. R. (2004). "Room Sizing and Optimization at Low Frequencies", *J. Audio Eng. Soc.*, vol. 52, pp. 640–651.

Craven, P. G. and Gerzon, M. A. (1992). "Practical Adaptive Room and Loudspeaker Equalizer for Hi-Fi Use", 92nd Convention, Audio Eng. Soc., Preprint 3346. Also: UK DSP Conference, Paper DSP-12.

Cremer, L. and Müller, H. A. (translation by T. J. Schultz) (1982). *Principles and Applications of Room Acoustics*, Vol. 1 and 2, Applied Science Publishers, London.

Crompton, T. W. J. (1974). "The Subjective Performance of Various Quadraphonic Matrix Systems", BBC Research Department Report No. BBC RD 1974/29, www.bbc.co.uk/rd/pubs.

D'Antonio, P. and Konnert, J. (1984). "The RFZ/RPG Approach to Control Room Monitoring", 76th Convention, Audio Eng. Soc., Preprint 2157.

Dalenbäck, B.-I. (2000). "Reverberation Time, Diffuse Reflection, Sabine and Computerized Prediction", www.rpginc.com/research/reverb01.htm

Dash, I., Mossman, M. and Cabrera, D. (2012). "The Relative Importance of Speech and Non-Speech Components for Preferred Listening Levels", 132nd Convention, Audio Eng. Soc., Paper 8614.

Davis, D. and Davis, C. (1980). "The LEDE Concept for Control of Acoustics and Psychoacoustic Parameters in Recording Control Rooms", *J. Audio Eng. Soc.*, vol. 28, pp. 585–595.

Davis, M. F. (1987). "Loudspeaker Systems With Optimized Wide-Listening-Area Imaging", *J. Audio Eng. Soc.*, vol. 36, pp. 888–896.

Deer, J. A., Bloom, P. J. and Preis, D. (1985). "New Results for Perception of Phase Distortion", 77th Convention, Audio Eng. Soc., Preprint 2197.

DellaSala, G. (2016). "Bass Optimization for Home Theater With Multi-Sub + mDSP", www.audioholics.com/home-theater-calibration/bass-optimization-for-home-theater, March 3, 2016.

Devantier, A. (2002). "Characterizing the Amplitude Response of Loudspeaker Systems", 113th Convention, Audio Eng. Soc., Preprint 5638.

Djelani, T. and Blauert, J. (2000). "Some New Aspects of the Build-Up and Breakdown of the Precedence Effect", Proc. 12th Int. Symp. On Hearing, Shaker Publishing.

Djelani, T. and Blauert, J. (2001). "Investigations Into the Build-Up and Breakdown of the Precedence Effect", *Acta Acustica-Acustica*, vol. 87, pp. 253–261.

Eargle, J. (1973). "Equalizing the Monitoring Environment", *J. Audio Eng. Soc.*, vol. 21, pp. 103–107.

Eargle, J., Bonner, J. and Ross, D. (1985). "The Academy's New State-of-the-Art Loudspeaker System", *SMPTE Journal*, pp. 667–675.

EBU Tech. 3276. (1998). "Listening Conditions for the Assessment of Sound Programme Material: Monophonic and Two-Channel Stereophonic," Second edition.

Eickmeier, G. (1989). "An Image Model Theory for Stereophonic Sound", 87th Convention, Audio Eng. Soc., Paper 2869.

Evans, W., Dyreby, J., Bech, S., Zielinski, S. and Rumsey, R. (2009). "Effects of Loudspeaker Directivity on Perceived Sound Quality—A Review of Existing Studies", Audio Eng. Soc., 126th Convention, Paper 7745.

Fastl, H. and Zwicker, E. (2007). *Psychoacoustics, Facts and Models*, Springer, Berlin.

Fazenda, B.M., Avis, M.R. and Davies, W.J. (2005). "Perception of Modal Distribution Metrics in Critical Listening Spaces—Dependence on Rooms Aspect Ratios", *J. Audio Eng. Soc.*, vol. 53, pp. 1128–1141.

Fazenda, B., Stephenson, M. and Goldberg, A. (2015). "Perceptual Thresholds for the Effects of Room Modes as a Function of Modal Decay", *J. Audio Eng. Soc.*, vol. 137, pp. 1088–1098.

Fazenda, B., Wankling, M., Hargreaves, J., Elmer, L. and Hirst, J. (2012). "Subjective Preference of Modal Control Methods in Listening Rooms", *J. Audio Eng. Soc.*, vol. 60, pp. 338–349.

Fielder, L.D. (2012). "Frequency Response Versus Time-of—Arrival for Typical Cinemas", presented at the 2012 SMPTE Annual Technical Conference.

Fincham, L.R. (1985). "The Subjective Importance of Uniform Group Delay at Low Frequencies", *J. Audio Eng. Soc.*, vol. 33, pp. 436–439.

Fitzroy, D. (1959). "Reverberation Formula Which Seems to Be More Accurate With Nonuniform Distribution of Absorption", *J. Acoust. Soc. Am.*, vol. 31, pp. 893–897.

Flanagan, S., Moore, B. C. J. and Stone, M. (2005). "Discrimination of Group Delay in Click-like Signals Presented via Headphones and Loudspeakers", *J. Audio Eng. Soc.*, vol. 53, pp. 593–611.

Fletcher, H. and Munson, W.A. (1933). "Loudness, Its Definition, Measurement and Calculation", *J. Acoust. Soc. Am.*, vol. 5, pp. 82–108.

Fritz, C., Curtain, J., Poitevineau, J., Borsarello, H., Wollman, I. and Tao, F-C. (2014). "Soloist Evaluations of Six Old Italian and Six New Violins", *Proc. Nat. Acad. Sci.*, vol. 111, no. 20, pp. 7224–7229, www.pnas.org/cgi/doi/10.1073/pnas. 1323367111.

Fryer, P.A. (1975). "Intermodulation Distortion Listening Tests", *J. Audio Eng. Soc.* (abstracts), vol. 23, p. 402.

Fryer, P.A. (1977). "Loudspeaker Distortions, Can We Hear Them", *Hi-Fi News Rec. Rev.*, vol. 22, pp. 51–56.

Gabrielsson, A., Hagerman, B. and Bech-Kristensen, T. (1991). "Perceived Sound Quality of Reproductions With Different Sound Levels", Karolinska Institute, Stockholm. Rep. TA 123.

Gabrielsson, A. and Sjögren, H. (1979). "Perceived Sound Quality of Sound Reproducing Systems", *J. Acoust. Soc. Am.*, vol. 65, pp. 1919–1933.

Gander, M.R. (1982). "Ground-Plane Acoustic Measurement of Loudspeaker Systems", *J. Audio Eng. Soc.*, vol. 30 pp. 723–731. Correction in: *JAES*, vol. 34, p. 49.

Gardiner, B. (2010). "Yamaha's NS-10: The Most Important Speaker You've Never Heard Of", http://.gizmodo.com/5637077.

Gardner, M. (1968). "Historical Background of the Haas and/or Precedence Effect", *J. Acoust. Soc. Am.*, vol. 43, pp. 1243–1248.

Gardner, M. (1969). "Image Fusion, Broadening, and Displacement in Sound Localization", *J. Acoust. Soc. Am.*, vol. 46, pp. 339–349.

Gardner, M. (1973). "Some Single- and Multiple-Source Localization Effects", *J. Audio Eng. Soc.*, vol. 21, pp. 430–437.

Geddes, E.R. (1982). "An Analysis of the Low Frequency Sound Field in Non-Rectangular Enclosures Using the Finite Element Method", PhD Thesis, Pennsylvania State University.

Geddes, E.R. (2005). "Audio Acoustics in Small Rooms", a PowerPoint presentation available at www.gedlee.com

Geddes, E.R. and Lee, L.W. (2003). "Auditory Perception of Nonlinear Distortion—Theory", 115th Convention, Audio Eng. Soc., Preprint 5890.

Gedemer, L. (2013). "Evaluation of the SMPTE X-Curve Based on a Survey of Re-Recording Mixers", J. Audio Eng. Soc. Convention Paper 8996 (2013).

Gedemer, L. (2015a). "Predicting the In-Room Response of Cinemas From Anechoic Loud-speaker Data", Audio Engineering Society 57th Conference: The Future of Audio Entertainment Technology.

Gedemer, L. (2015b). "Subjective Listening Tests for Preferred Room Response in Cinemas—Part 1: System and Test Descriptions", 139th Convention, Audio Eng. Soc.

Gedemer, L. (2016). "Subjective Listening Tests for Preferred Room Response in Cinemas—Part 2: Preference Test Results", 140th Convention, Audio Eng. Soc.

Gedemer, L. and Welti, T. (2013). "Validation of the Binaural Room Scanning Method for Cinema Audio Research", 135th Convention, Audio Eng. Soc., Preprint 8974.

Gee, A. and Shorter, D.E.L. (1955). "An Automatic Integrator for Determining the Mean Spherical Response of Loudspeakers and Microphones", British Broadcasting Corp., Engineering Monograph 8.

Gerzon, M. (1983). "Ambisonics in Multichannel Broadcasting and Video", 74th Convention, Audio Eng. Soc., Preprint 2034.

Gilford, C. L. S. (1959). "The Acoustic Design of Talks Studios and Listening Rooms", *Proc. Inst. Electrical Eng.*, vol. 106, pp. 245–258. Reprinted in *J. Audio Eng. Soc.*, vol. 27, pp. 17–31 (1979).

Glasgal, R. (2001). "Ambiophonics", 111th Convention, Audio Eng. Soc., Preprint 5426.

Glasgal, R. (2003). "Surround Ambiophonic Recording and Reproduction", 24th International Conference, Audio Eng. Soc.

Goertz, A., Wolff, M, and Naumann, L. (2001). "Optimization of Sound Reproduction in Listening Rooms for Surround Sound Loudspeaker Setups", White paper for Klein+Hummel. English translation at: http://forums.klipsch.com/forums/storage/8/1485263/tmt2002_eng.pdf

Gover, B.N., Ryan, J.G. and Stinson, M.R. (2004). "Measurements of Directional Properties of Reverberant Sound Fields in Rooms Using a Spherical Microphone Array", *J. Acoust. Soc. Am.*, vol. 116, pp. 2138–2148.

Green, R. and Holman, T. (2010). "First Results from a Large-Scale Measurement Program for Home Theaters", 129th Convention, Audio Eng. Soc., Preprint 8310.

Greenfield, R. and Hawksford, M. (1990). "The Audibility of Loudspeaker Phase Distortion", 88th Convention, Audio Eng. Soc., Preprint 2927.

Greiner, R.A. (1980). "Amplifier-Loudspeaker Interfacing", *J. Audio Eng. Soc.*, vol. 28, pp. 310–315.

Greiner, R.A. and Melton, D. (1994). "Observations on the Audibility of Acoustic Polarity", *J. Audio Eng. Soc.*, vol. 42, pp. 245–253. And, letters to the editor: *JAES*, vol. 43, pp. 147–149.

Griesinger, D. (2009). "The Importance of the Direct to Reverberant Ratio in the Perception of Distance, Localization, Clarity and Envelopment", 126th Convention, Audio Eng. Soc., Preprint 7724.

Griffin, J.R. (2003). "Design Guidelines for Practical Near Field Line Arrays", www.audiodiy central.com/resource/pdf/nflawp.pdf

Haas, H. (1949, 1972). "The Influence of a Single Echo on the Audibility of Speech", Doctoral dissertation, University of Göttingen. Reprinted in: *J. Audio Eng. Soc.*, vol. 20, pp. 146–159, 1972. A reprint of a 1949 translation of Haas' PhD dissertation.

Hafler, D. (1970). "A New Quadraphonic System", *Audio*, July.

Hamasaki, E., Hiyama, K., Nishiguchi, T. and Ono, K. (2004). "Advanced Multichannel Audio Systems with Superior Impression of Presence and Reality", 116th Convention, Audio Eng. Soc., Preprint 6053.

Hansen, V. and Madsen, E.R. (1974a). "On Aural Phase Detection", *J. Audio Eng. Soc.*, vol. 22, pp. 10–14.

Hansen, V. and Madsen, E.R. (1974b). "On Aural Phase Detection: Part II", *J. Audio Eng. Soc.*, vol. 22, pp. 783–788.

Harvith, J. and Harvith, S. (1987). *Edison, Musicians and the Phonograph*, Greenwood Press, New York.

Hiyama, K., Komiyama, S. and Hamasaki, K. (2002). "The Minimum Number of Loudspeakers and Its Arrangement for Reproducing the Spatial Impression of Diffuse Sound Field", 113th Convention, Audio Eng. Soc., Preprint 5674.

Holman, T. (1996). "The Number of Audio Channels", 100th Convention, Audio Eng. Soc. Preprint 4292.

Holman, T. (1998). "The Basics of Bass Management", *Surround Professional*, 1, October.

Holman, T. (2001). "The Number of Loudspeaker Channels", 19th International Conference, Audio Eng. Soc. Paper No. 1906.

Howard, K. (2005). "Time Dilation, Part 1", *Stereophile*, vol. 28, January," Part 2" *Stereophile*, vol. 28, April, www.stereophile.com/features/105kh/index.html and www.stereophile.com/ reference/405time/index.html

Howard, K. (2006). "Wayward Down Deep", *Stereophile*, vol. 29, July, www.stereophile.com/ reference/706deep/index.html

Howard, K. (2007). "Heavy Load: How Loudspeakers Torture Amplifiers", *Stereophile*, vol. 30, July, www.stereophile.com/reference/707heavy/index.html

Hughes, R., Cox, T., Shirley, B. and Power, P. (2016). "The room-in-room effect and its influence on perceived room size in spatial audio reproduction". 141st Convention Audio Eng. Soc., Preprint 9621.

IEC 60268–13 (1998). *Technical Report—Type 3, Sound System Equipment—Part 13: Listening Tests on Loudspeakers*, International Electrotechnical Commission, Geneva, Switzerland.

ISO 226 (2003). *Acoustics—Normal Equal-Loudness Contours*, International Organization for Standardization, Geneva, Switzerland.

ISO 2969 (1987). *International Standard. Cinematography—B-Chain Electroacoustic Response of Motion-Picture Control Rooms and Indoor Theaters—Specifications and Measurements*. International Organization for Standardization, Geneva, Switzerland.

ITU-R BS.1116–3 (2015). *Methods for the Subjective Assessment of Small Impairments in Audio Systems*, International Telecommunication Union, Geneva.

ITU-R BS.1770–3 (2012). *Algorithms to Measure Audio Programme Loudness and True-Peak Audio Level*, International Telecommunication Union, Geneva.

ITU-R BS.775–3 (2012). *Multichannel Stereophonic Sound System With and Without Accompanying Picture*, International Telecommunication Union, Geneva.

Johnsen, C. (1991). "Proofs of an Absolute Polarity", 91st Convention, Audio Eng. Soc., Preprint 3169.

Julstrom, S. (1987). "A High-Performance Surround Sound Process for Home Video", *J. Audio Eng. Soc.*, vol. 35, pp. 536–549.

Karjalainen, M., Antsalo, P., Mäkivirta, A. and Välimäki, V. (2004). "Perception of Temporal Decay of Low-Frequency Room Modes", 116th Convention, Audio Eng. Soc., Preprint 6083.

Katz, B. (2002). *Mastering Audio*, Focal Press, Oxford, UK.

Katz, B. (2016). "PSI Audio AVAA C20—Equipment Review", *Stereophile*, vol. 39, no. 6, pp. 123–133.

Keele, D.B. and Button, D.J. (2005). "Ground-Plane Constant Beamwidth Transducer (CBT) Loudspeaker Circular-Arc Line Arrays", 119th Convention, Audio Eng. Soc., Preprint 6594.

King, R., Leonard, B. and Sikora, G. (2012). "The Practical Effects of Lateral Energy in Critical Listening Environments", *J. Audio Eng. Soc.*, vol. 60, pp. 997–1003.

Kishinaga, S., Shimizu, Y., Ando, S. and Yamaguchi, K. (1979). "On the Room Acoustic Design of Listening Rooms", 64th Convention, Audio Eng. Soc., Preprint 1524.

Klippel, W. (1990a). "Multidimensional Relationship Between Subjective Listening Impression and Objective Loudspeaker Parameters", *Acustica*, vol. 70, pp. 45–54.

Klippel, W. (1990b). "Assessing the Subjectively Perceived Loudspeaker Quality on the Basis of Objective Parameters", 88th Convention, Audio Eng. Soc., Preprint 2929.

Klippel, W. and Bellmann, C. (2016). "Holographic Nearfield Measurement of Loudspeaker Directivity", 141st Convention, Audio Eng. Soc., Preprint 9598.

Kommamura, M. and Mori, S. (1983). "New Measurement of Frequency Response Flatness Limen", *J. Acoust. Soc. Jpn.*, vol. 39, p. 463.

Krauss, G.J. (1990). "On the Audibility of Group Distortion at Low Frequencies", 88th Convention, Audio Eng. Soc., Preprint 2894.

Kruk, B. and Kin, M. (2015). "Perception of Timbre Changes vs. Temporary Threshold Shift", 138th Convention, Audio Eng. Soc., Preprint 9228.

Kryter, K.D. (1973). "Impairment to Hearing From Exposure to Noise", *J. Acoust. Soc. Am.*, vol 53, p. 1211.

Kuhl, W. and Plantz, R. (1978). "The Significance of the Diffuse Sound Radiated From Loudspeakers for the Subjective Hearing Event", *Acustica*, vol. 40, pp. 182–190.

Kujawa, S.G. and Liberman, M.C. (2009). "Adding Insult to Injury: Cochlear Nerve Degeneration After "Temporary" Noise-Induced Hearing Loss", *J. Neuroscience*, vol. 29, pp. 14077–14085. A summary can be found at http://newswise.com/articles/noise-induced-hidden-hearing-loss-mechanism-discovered

LaCarrubba, M. (1999). "The Wide-Dispersion Listening Space—a New Approach to Control-Room Design", *Mix*, November.

Lee, L.W. and Geddes, E.R. (2003). "Auditory Perception of Nonlinear Distortion", 115th Convention, Audio Eng. Soc., Preprint 5891.

Lee, L.W. and Geddes, E.R. (2006). "Audibility of Linear Distortion With Variations in Sound Pressure Level and Group Delay", 115th Convention, Audio Eng. Soc., Preprint 6888.

Leek, M. and Molis, M. (2012). "Hearing Loss and Frequency Analysis of Complex Sounds", *Acoustics Today*, vol. 8, no. 2, pp. 29–33.

Leembruggen, G. (2015a). "Equalizing the Effects of Perforated Cinema Screens", 57th Conference, Audio Eng. Soc., "The Future of Audio Entertainment Technology: Cinema, Television and the Internet".

Leembruggen, G. (2015b). "Low Frequency Issues: Cinema/Home/Internet", Soc., 57th Conference, Audio Eng., "The Future of Audio Entertainment Technology: Cinema, Television and the Internet".

Liberman, M., Epstein, M., Cleveland, S., Wang, H. and Maison, S. (2016). "Toward a Differential Diagnosis of Hidden Hearing Loss in Humans", *PLoS ONE*, vol. 11, no. 9, p. e0162726. http://dx.doi.org/10.1371/journal.pone.0162726

Linkwitz, S. (1998). "Investigation of Sound Quality Differences Between Monopolar and Dipolar Woofers in Small Rooms", 105th Convention, Audio Eng. Soc., Preprint 4786.

Linkwitz, S. (2007). "Room Reflections Misunderstood", 123rd Convention, Audio Eng. Soc., Paper 7162.

Linkwitz, S. (2009). "The Challenge to Find the Optimum Radiation Pattern and Placement of Stereo Loudspeakers in a Room for the Creation of Phantom Sources and Simultaneous Masking of Real Sources", 127th Convention, Audio Eng. Soc., Paper 7959.

Lipshitz, S. (1990). "The Great Debate—Some Reflections Ten Years Later", 8th International Conference, Audio Eng. Soc.

Lipshitz, S., Pocock, M. and Vanderkooy, J. (1982). "On the Audibility of Midrange Phase Distortion in Audio Systems", *J. Audio Eng. Soc.*, vol. 30, pp. 580–595. Comments by Shanefield, D. and authors' reply (1983), vol. 31, pp. 447–448. Comments by Moir, J. (1983), vol. 31, p. 939. More comments by van Maanen, H.R.E., Shanefield, D. and Moir, J. (1985), vol. 33, pp. 806–808.

Lipshitz, S. and Vanderkooy, J. (1981). "The Great Debate: Subjective Evaluation", *J. Audio Eng. Soc.*, vol. 29, pp. 482–491.

Lipshitz, S. and Vanderkooy, J. (1986). "The Acoustic Radiation of Line Sources of Finite Length", 81st Convention, Audio Eng. Soc., Preprint 2417.

Litovsky, R.Y., Colburn, H.S., Yost, W.A. and Guzman, S.J. (1999). "The Precedence Effect", *J. Acoust. Soc. Am.*, vol. 106, pp. 1633–1654.

Ljungberg, L. (1969). "Standardized Sound Reproduction in Cinemas and Control Rooms", *J. SMPTE*, vol. 78, pp. 1046–1053.

Lochner, J.P.A. and Burger, J.F. (1958). "The Subjective Masking of Short Time Delayed Echoes by Their Primary Sounds and Their Contribution to the Intelligibility of Speech", *Acustica*, vol. 8, pp. 1–10.

Long, B., Schwenke, R., Soper, P. and Leembruggen, G. (2012). "Further Investigations Into the Interactions Between Cinema Loudspeakers and Screens", Soc. of Motion Picture and Television Engineers, 2012 Fall Technical Conference.

Louden, M.M. (1971). "Dimension-Ratios of Rectangular Rooms With Good Distribution of Eigentones", *Acoustica*, vol. 24, pp. 101–104.

Mäkivirta, A.V. and Anet, C. (2001). "A Survey Study of In-Situ Stereo And Multi-Channel Monitoring Conditions", 111th Convention, Audio Eng. Soc., Preprint 5496.

Martens, W.L., Braasch, J. and Woszczyk, W. (2004). "Identification and Discrimination of Listener Percepts Associated With Multiple Low-Frequency Signals in Multichannel Sound Reproduction", 117th Convention, Audio Eng. Soc., Preprint 6229.

McMullin, E., Celestinos, A. and Devantier, A. (2015). "Environments for Evaluation: The Development of Two New Rooms for Subjective Evaluation", 139th Convention, Audio Eng. Soc., Paper 9460.

Meyer, E. (1954). "Definition and Diffusion in Rooms", *J. Acoust. Soc. Am.*, vol. 26, pp. 630–636.

Meyer, J. (2009). *Acoustics and the Performance of Music*, Fifth edition, Springer, New York.

Møller, H. (1974). "Relevant Loudspeaker Tests in Studios, in Hi-Fi Dealer's Demo Rooms, in the Home etc. Using 1/3-Octave, Pink Weighted, Random Noise", presented at 47th AES Convention, www.bksv.com/doc 17-197.pdf

Møller, H., Minnaar, P., Olesen, S., Christensen, F. and Plogsties, J. (2007). "On the Audibility of All-Pass Phase in Electroacoustical Transfer Functions", *J. Audio Eng. Soc.*, vol. 55, pp. 115–134.

Moore, B. C. J. (2003). *An Introduction to the Psychology of Hearing*, Fifth edition, Academic Press, London.

Moore, B. C. J. (2016). "Effects of Sound-Induced Hearing Loss and Hearing Aids on the Perception of Music", *J. Audio Eng. Soc.*, vol. 64, pp. 112–123.

Moore, B. C. J., Glasberg, B. R. and Baer, T. (1997). "A Model for the Prediction of Thresholds, Loudness, and Partial Loudness", *J. Audio Eng. Soc.*, vol. 45, pp. 224–240.

Moore, B. C. J. and Tan, C-T. (2004). "Development and Validation of a Method for Predicting the Perceived Naturalness of Sounds Subjected to Spectral Distortion", *J. Audio Eng. Soc.*, vol. 52, pp. 900–914.

Moore, B. C. J., Tan, C-T., Zacharov, N. and Mattila, V-V. (2004). "Measuring and Predicting the Perceived Quality of Music and Speech Subjected to Combined Linear and Nonlinear Distortion", *J. Audio Eng. Soc.*, vol. 52, pp. 1228–1244.

Morimoto, M. (1997). "The role of rear loudspeakers in spatial impression". 103rd Convention, Audio Eng. Soc., Preprint 4554.

Morton, D. (2000). *Off the Record*, Rutgers University Press, New Brunswick, NJ.

Moulton, D. (1995). "The Significance of Early High-Frequency Reflections from Loudspeakers in Listening Rooms", 99th Convention, Audio Eng. Soc., Preprint 4094.

Moulton, D. (2003). "The Loudspeaker as a Musical Instrument: And Examination of the Issues Surrounding Loudspeaker Performance of Music in Typical Rooms", Acoust. Soc. Amer., Nashville meeting, text available on www.moultonlabs.com.

Moulton, D. (2011). "Panoramic Power Response: A Fresh Approach to Loudspeaker Dispersion and Control Room Design", www.moultonlabs.com.

Moulton, D., Ferralli, M., Hebrock, S. and Pezzo, M. (1986). "The Localization of Phantom Images in an Omnidirectional Stereophonic Loudspeaker System", 81st Convention, Audio Eng. Soc., Preprint 2371.

Muraoka, T. and Nakazato, T. (2007). "Examination of Multichannel Sound Field Recomposition Utilizing Frequency Dependent Interaural Cross Correlation (FIACC)", *J. Audio Eng. Soc.*, vol. 55, pp. 236–256.

Nakayama, T., Miura, T., Kosaka, O., Okamoto, M. and Shiga, T. (1971). "Subjective Assessment of Multichannel Reproduction", *J. Audio Eng. Soc.*, vol. 19, pp. 744–751.

NARAS (2004). "Recommendations for Surround Sound Production", The Recording Acadamy's Producers and Engineers Wing of The National Academy of Recording Arts & Sciences, Inc.

Newell, P. (2003). *Recording Studio Design*, Focal Press, Oxford, UK.

Newell, P., García, J. G. and Holland, K. (2013). "An Investigation Into the Acoustic Effect of Cinema Screens on Loudspeaker Performance", *Proc. Inst. Acoust.*, vol. 35, pt. 2.

Newell, P., Holland, K, Torres-Guijarro, S., Castro, S. and Valdigem, E. (2010). "Cinema Sound: A New Look at Old Concepts", *Proc. Inst. Acoust.*, vol. 32, pt. 5.

Newell, P., Newell, J., Neskov, B. and Holland, K. (2016). "Cinema Sound and the Loudness Issue: Its Origins and Implications", *Proc. Inst. Acoustics*.

Newell, P., Santos-Dominguez, D., Torres-Guijarro, S. and Castro, S. (2015). "The Relationship Between Subjective and Objective Response Differences at Different Heights Above Cinema Seating", *Proc. Inst. Acoust.*, vol. 37, pt. 4.

NIOSH Occupational Noise Exposure (1998). www.cdc.gov/niosh/docs/98-126/pdfs/98-126.pdf

Nousaine, T. (1990). "The Great Debate: Is Anyone Winning", 8th International Conference, Audio Eng. Soc.

Olive, S.E. (1990). "The Preservation of Timbre; Microphones, Loudspeakers, Sound Sources and Acoustical Spaces", 8th International Conference, Audio Eng. Soc., Paper D1.

Olive, S.E. (1994). "A Method for Training Listeners and Selecting Program Material for Listening Tests", 97th Convention, Audio Eng. Soc., Preprint 3893.

Olive, S.E. (2001). "A New Listener Training Software Application", 110th Convention, Audio Eng. Soc., Preprint No. 5384.

Olive, S.E. (2003). "Difference in Performance and Preference of Trained Versus Untrained Listeners in Loudspeaker Tests: A Case Study", *J. Audio Eng. Soc.*, vol. 51, pp. 806–825.

Olive, S.E. (2004a). "A Multiple Regression Model for Predicting Loudspeaker Preference Using Objective Measurements: part 1—Listening Test Results", 116th Convention, Audio Eng. Soc., Preprint 6113.

Olive, S.E. (2004b). "A Multiple Regression Model for Predicting Loudspeaker Preference Using Objective Measurements: Part 2—Development of the Model", 117th Convention, Audio Eng. Soc., Preprint 6190.

Olive, S.E. (2009). "A New Reference Listening Room for Consumer, Professional and Automotive Audio Research", 126th Convention, Audio Eng. Soc., Paper 7677.

Olive, S.E., Castro, B. and Toole, F.E. (1998). "A New Laboratory for Evaluating Multichannel Audio Components and Systems", 105th Convention, Audio Eng. Soc., Preprint 4842.

Olive, S.E., Hess, S.M. and Devantier, A. (2008). "Comparison of Loudspeaker-Room Equalization Preference for Multichannel, Stereo, and Mono Reproductions: Are Listeners More Discriminating in Mono?", 124th Convention, Audio Eng. Soc., Preprint 7492.

Olive, S.E., Schuck, P.L., Ryan, J.G., Sally, S.L. and Bonneville, M.E. (1997). "The Detection Thresholds of Resonances at Low Frequencies", *J. Audio Eng. Soc.*, vol. 45, pp. 116–128.

Olive, S.E., Schuck, P.L., Sally, S.L. and Bonneville, M. (1995). "The Variability of Loudspeaker Sound Quality Among Four Domestic Sized Rooms", 99th Convention, Audio Eng. Soc., Preprint 4092.

Olive, S.E. and Toole, F.E. (1989a). "The Detection of Reflections in Typical Rooms", *J. Audio Eng. Soc.*, vol. 37, pp. 539–553.

Olive, S.E. and Toole, F.E. (1989b). "The Evaluation of Microphones—Part 1: Measurements. 87th Convention, Audio Eng. Soc., Preprint 2837.

Olive, S.E. and Welti, T. (2009). "Validation of a Binaural Car Scanning System for Subjective Evaluation of Automotive Audio Systems", 36th International Conference: Automotive Audio, Audio Eng. Soc.

Olive, S.E. and Welti, T. (2012). "The Relationship Between Perception and Measurement of Headphone Sound Quality", 133rd Convention, Audio Eng. Soc., Preprint 8744.

Olive, S.E., Welti, T. and Khonsaripour, O. (2016). "The Preferred Low-Frequency Response on In-Ear Headphones", Audio Eng. Soc. Int. Conf. on Headphone Technology, Paper 6–1.

Olive, S.E., Welti, T. and Martens, W.L. (2007). "Listener Loudspeaker Preference Ratings Obtained in Situ Match Those Obtained via a Binaural Room Scanning Measurement and Playback System", 122nd Convention, Audio Eng. Soc., Preprint 7134.

Olive, S.E., Welti, T. and McMullin, E. (2013a). "Listener Preference for In-Room Loudspeaker and Headphone Target Responses", 135th Convention, Audio Eng. Soc., Paper 8994.

Olive, S.E., Welti, T. and McMullin, E. (2013b). "Listener Preference for Different Headphone Target Response Curves", 134th Convention, Audio Eng. Soc., Preprint 8867.

Olive, S.E., Welti, T. and McMullin, E. (2014). "The Influence of Listeners' Experience, Age and Culture on Headphone Sound Quality Preferences", 137th Convention, Audio Eng. Soc., Paper 9177.

Olson, H.F. (1957). *Acoustical Engineering*, Van Nostrand, New York, republished in 1991 by Professional Audio Journals, Inc.

Olson, H.F., Preston, J. and May, E. (1954). "Recent Developments in Direct-Radiator High-Fidelity Loudspeakers", *J. Audio Eng. Soc.*, vol. 2, pp. 219–227.

OSHA Occupational Noise Exposure (1997). www.osha.gov/dts/osta/otm/new_noise/index.html

Pedersen, J.A. (2003). "Adjusting a Loudspeaker to Its Acoustic Environment—the ABC System", 115th Convention, Audio Eng. Soc., Preprint 5880.

Perrett, S. and Noble, W. (1997). "The Effect of Head Rotations on Vertical Plane Sound Localization", *J. Acoust. Soc. Am.*, vol. 102, pp. 2325–2332.

Perrot, D.R., Marlborough, K., Merrill, P. and Strybel, T.S. (1989). "Minimum Audible Angle Thresholds Obtained Under Conditions in Which the Precedence Effect Is Assumed to Operate", *J. Acoust. Soc. Am.*, vol. 85, pp. 282–288.

Pike, C., Brookes, T. and Mason, R. (2013). "Auditory Adaptation to Loudspeaker and Listening Room Acoustics", 135th Convention, Audio Eng. Soc., Preprint 8971.

Plenge, G. (1987). "On the Behavior of Listeners to Stereophonic Sound Reproduciton and the Consequences for the Theory of Sound Perception in a Stereophonic Sound Field", 83rd Convention, Audio Eng. Soc., Preprint 2532.

Pollack, I. (1952). "The Loudness of Bands of Noise", *J. Acoust. Soc. Amer.*, vol. 24, pp. 533–538.

Pulkki, V. (2001). "Coloration of Amplitude-Panned Virtual Sources", 110th Convention, Audio Eng. Soc., Preprint 5402.

Queen, D. (1973). "Relative Importance of the Direct and Reverberant Fields to Spectrum Perception", *J. Audio Eng. Soc.*, vol. 21, pp. 119–121.

Rakerd, B. and Hartmann, W.M. (1985). "Localization of Sound in Rooms, II: The Effects of a Single Reflecting Surface", *J. Acoust. Soc. Am.*, vol. 78, pp. 524–533.

Rakerd, B., Hartmann, W.M. and Hsu, J. (2000). "Echo Suppression in the Horizontal and Median sagittal planes", *J. Acoust. Soc. Am.*, vol. 107, pp. 1061–1064.

Ratliff, P.A. (1974). Properties of Hearing Related to Quadraphonic Reproduction", BBC Research Department Report No. BBC RD 1974/38. www.bbc.co.uk/rd/pubs.

Read, O. and Welsh, W.L. (1959). *From Tin Foil to Stereo*, Howard Sams, Indianapolis, IN.

Rettinger, M. (1968). "Small Music Rooms", *Audio*, October, pp. 25, 87.

Robinson, D.W. and Dadson, R.S. (1956). "A Re-Determination of the Equal-Loudness Relations for Pure Tones", *Br. J. App. Phys.*, vol. 7, pp. 166–181.

Robinson, P., Walther, A., Faller, C. and Braasch, J. (2013). "Echo Thresholds for Reflections From Acoustically Diffusive Architectural Surfaces", *J. Acoust. Soc. Am.*, vol. 134, pp. 2755–2764.

Roederer, J.G. (1995). *The Physics and Psychophysics of Music*, Third edition, Springer-Verlag, New York.

Rumsey, F. (1999). "Controlled Subjective Assessments of Two-to-Five-Channel Surround Sound Processing Algorithms", *J. Audio Eng. Soc.*, vol. 47, pp. 563–562.

Rumsey, F., Zielinski, S., Kassier, R. and Bech, S. (2005). "Relationships Between Experienced Listener Ratings of Multichannel Audio Quality and Naïve Listener Preferences", *J. Acoust. Soc. Am.*, vol. 117, pp. 3832–3840.

Saberi, K. and Perrot, D.R. (1990). "Lateralization Thresholds Obtained Under Conditions in Which the Precedence Effect Is Assumed to Operate", *J. Acoust. Soc. Am.*, vol. 87, pp. 1732–1737.

Sanfilipo, M. (2005). "Speaker Break In: Fact or Fiction", www.audioholics.com/loudspeaker-design/speaker-break-in-fact-or-fiction

Santillán, A. (2001). "Spatially Extended Sound Equalization in Rectangular Rooms", *J. Acoust. Soc. Am.*, vol. 110, pp. 1989–1997.

Sato, H. and Bradley, J. (2008). "Evaluation of Acoustical Conditions for Speech Communication in Working Elementary School Classrooms", *J. Acoust. Soc. Am.*, vol. 123, pp. 2064–2077.

Sayers, B. McA. and Toole, F.E. (1964). "Acoustical Image Lateralization Judgments With Binaural Transients", *J. Acoust. Soc. Am.*, vol. 36, pp. 1199–1205.

Schroeder, M.R. (1954). "Statistical Parameters of the Frequency Response Curves of Large Rooms", *Acustica*, vol. 4, pp. 594–600. Translated from the German original in *J. Audio Eng. Soc.*, vol. 35, pp. 299–306 (1987).

Schroeder, M.R. (1996). " 'Schroeder Frequency' Revisited", *J. Acoust. Soc. Am.*, vol. 99, pp. 3240–3241.

Schulein, R.B. (1975). "In Situ Measurement and Equalization of Sound Reproduction Systems", *J. Audio Eng. Soc.*, vol. 23, pp. 178–186.

Schultz, T.J. (1983). "Improved Relationship Between Sound Power Level and Sound Pressure Level in Domestic and Office Spaces", Report No. 5290, Am. Soc. of Heating, Refrigerating and Air-Conditioning Engineers (ASHRAE), prepared by Bolt, Beranek and Newman, Inc.

Self, D.R.G. (1988). "Science v. Subjectivism in Audio Engineering", *Electronics and Wireless World*, vol. 94, pp. 692–696, www.dself.dsl.pipex.com/ampins/pseudo/subjectv.htm

Seligson, L. (1969). "How Consumers Union Tests Loudspeakers", *Stereo Rev.*, February, pp. 62–66.

Sepmeyer, L.W. (1965). "Computed Frequency and Angular Distribution of the Normal Modes of Vibration in Rectangular Rooms", *J. Acoust. Soc. Am.*, vol. 37, pp. 413–423.

Seraphim, H.-P. (1961). "Über die Wahrnehmbarkeit mehrerer Rückwürfe von Sprachschall", *Acustica*, vol. 11, pp. 80–91.

Shaw, E.A.G. (1974a). "Physical Models of the External Ear", Proc. 8th Int. Cong. Acoust., London, vol. 1, p. 206.

Shaw, E.A.G. (1974b). "Transformation of Sound Pressure Level From the Free Field to the Eardrum", *J. Acoust. Soc. Am.*, vol. 56, p. 1848.

Shaw, E.A.G. (1975). "The External Ear: New Knowledge", presented at the VII Danavox Symposium, Denmark.

Shaw, E.A.G. (1982). "External Ear Response and Sound Localization", Chapter 2 in *Localization of Sound*, Gatehouse (ed.), Amphora Press.

Shinn-Cunningham, B.G. (2003). "Acoustics and Perception of Sound in Everyday Environments", Proc. 3rd Int. Workshop on Spatial Media, Aisu-Wakamatsu, Japan, http://cns.bu.edu/~shinn/pages/RecentPapers.html

Shirley, B.G. and Kendrick, P. (2004). "ITC Clean Audio Project", 116th Convention, Audio Eng. Soc., Preprint no. 6027.

Shirley, B.G., Kendrick, P. and Churchill, C. (2007). "The Effect of Stereo Crosstalk on Intelligibility: Comparison of a Phantom Stereo Image and a Central Loudspeaker Source", *J. Audio Eng. Soc.*, vol. 55, pp. 852–863.

Silzle, A., Geyersberger, S. Brohasga, G., Weninger, D. and Leistner, M. (2009). "Vision and Technique Behind the New Studios and Listening Rooms of the Fraunhofer IIS Audio Laboratory", 126th Convention, Audio Eng. Soc., Preprint 7672.

Smith, D.L. (1997). "Discrete-Element Line Arrays—Their Modeling and Optimization", *J. Audio Eng. Soc.*, vol. 45, pp. 949–964.

SMPTE ST 202 (2010). "For Motion Pictures—Dubbing Stages (Mixing Rooms), Screening Rooms and Indoor Theaters—B-Chain Electroacoustic Response".

SMPTE TC-25CSS (2014). "B-Chain Frequency and Temporal Response Analysis of Theaters and Dubbing Stages", Soc. Motion Picture and Television Eng., www.smpte.org/standards/reports

Snow, W.B. (1953). "Basic Principles of Stereophonic Sound", *JSMPTE*, vol. 61, pp. 567–587. Reprinted in *IRE Transactions-Audio*, vol. AU-3, pp. 42–53 (1955).

Snow, W.B. (1961). "Loudspeaker Testing in Rooms", *J. Audio Eng. Soc.*, vol. 9, pp. 54–60.

Somerville, T. and Brownless, S.F. (1949). "Listeners' Sound-Level Preferences—Part I", *BBC. Q.*, vol. III, no. 4, pp. 11–16.

Soulodre, G.A. (2004). "Evaluation of Objective Loudness Meters", 116th Convention, Audio Eng. Soc., Preprint 6161.

Soulodre, G.A. and Norcross, S.G. (2003). "Objective Measures of Loudness", 115th Convention, Audio Eng. Soc., Preprint 5896.

Soulodre, G.A., Popplewell, N. and Bradley, J.S. (1989). "Combined Effects of Early Reflections and Background Noise on Speech Intelligibility", *J. Sound Vib.*, vol. 135, pp. 123–133.

Srinivasan, N.K., Kasey, M.J. and Gallum, F.J. (2016). "Release From Masking for Small Spatial Separations: Effects of Age and Hearing Loss", *J. Acoust. Soc. Am.* (Express Letters), http://dx.doi.org/10.1121/1.4954386

Srinivasan, N.K. and Zahorik, P. (2012). "Prior Listening Exposure to a Reverberant Room Improves Open-Set Intelligibility of High-Variability Sentences", *J. Acoust. Soc. Am.*, vol. 133, pp. EL33–EL39 (JASA Express Letters).

Steinberg, J.C. and Snow, W.B. (1934). "Auditory Perspective—Physical Factors", *Elect. Eng.*, vol. 33, pp. 12–17.

Stelmachowicz, P.G., Beauchaine, K.A., Kalberer, A., Kelly, W.J. and Jesteadt, W. (1989). "High-Frequency Audiometry: Test Reliability and Procedural Considerations", *J. Acoust. Soc. Am.*, vol. 85, pp. 879–887.

Stevens, S.S. (1957). "Calculating Loudness", *Noise Control*, vol. 3, 11–22.

Stevens, S.S. (1961). "Procedure for Calculating Loudness: Mark VI", *J. Acoust. Soc. Am.*, vol. 33, pp. 1577–1585.

Suokuisma, P., Zacharov, N. and Bech, S. (1998). "Multichannel Level Alignment, Part 1: Signals and Methods", 105th Convention, Audio Eng. Soc., Preprint 4815.

Talbot-Smith, M. (editor) (1999). *Audio Engineer's Reference Book*, Focal Press, Oxford, UK.

TASA (2013). *The TASA Standard*, Trailer Audio Standards Association, www.tasatrailers.org

Temme, S. and Dennis, P. (2016). "In Vehicle Sudio System Distortion Audibility versus Level and Its Impact on Perceived Sound Quality", 141st Convention, Audio Eng. Soc., Preprint 9651.

Tervq, S., Laukkanen, P., Pätynen, J. and Lokki, T. (2014). "Preferences of Critical Listening Environments Among Sound Engineers", *J. Audio Eng. Soc.*, vol. 62, pp. 300–314.

Thompson, E. (2002). *The Soundscape of Modernity*, MIT Press, Cambridge, MA.

Tocci, G.C. (2000). "Room Noise Criteria—the State of the Art in the Year 2000", www.cavtocci.com/portfolio/publications/tocci.pdf.

Tohyama, M. and Suzuki, A. (1989). "Interaural Cross-Correlation Coefficients in Stereo-Reproduced Sound Fields", *J. Acoust. Soc. Am.*, vol. 85, pp. 780–786.

Toole, F. E. (1970). "In-Head Localization of Acoustic Images", *J. Acoust. Soc. Am.*, vol. 48, pp. 943–949.

Toole, F. E. (1972). "Understanding the Phono Cartridge", *Electron Magazine* (Canada), three parts, August, September and October.

Toole, F. E. (1973). "Loudness—Applications and Implications to Audio", *db The Sound Engineering Magazine*, part 1, May, pp. 27–29, part 2, June pp. 25–28.

Toole, F. E. (1975). "Damping, Damping Factor, and Damn Nonsense", *AudioScene Canada*, February, pp. 16–17.

Toole, F. E. (1976). "Measuring Headphones", *AudioScene Canada*, vol. 13, August, pp. 63–65.

Toole, F. E. (1979). "Hi Fidelity in the Control Room—Why Not", *db The Sound Engineering Magazine*, February, pp. 30–33.

Toole, F. E. (1982). "Listening Tests—Turning Opinion Into Fact", *J. Audio Eng. Soc.*, vol. 30, pp. 431–445.

Toole, F. E. (1984). "The Acoustics and Psychoacoustics of Headphones", 2nd International Conference, Audio Eng. Soc., Paper C1006.

Toole, F. E. (1985). "Subjective Measurements of Loudspeaker Sound Quality and Listener Preferences", *J. Audio Eng. Soc.*, vol. 33, pp. 2–31.

Toole, F. E. (1986). "Loudspeaker Measurements and Their Relationship to Listener Preferences", *J. Audio Eng. Soc.*, vol. 34, pt.1, pp. 227–235, pt. 2, pp. 323–348.

Toole, F. E. (1990). "Loudspeakers and Rooms for Stereophonic Sound Reproduction", AES 8th International Conference "The Sound of Audio", Paper 8–011.

Toole, F. E. (1991). "Binaural Record/Reproduction Systems and Their Use in Psychoacoustic Investigations", 91st Convention, Audio Eng. Soc., Preprint 3179.

Toole, F. E. (2003). *Art and Science in the Control Room*, Institute of Acoustics, Reproduced Sound 19, Oxford, UK.

Toole, F. E. (2006). "Loudspeakers and Rooms for Sound Reproduction—A Scientific Review", *J. Audio Eng. Soc.*, vol. 54, pp. 451–476.

Toole, F. E. (2008), "Sound Reproduction: The Acoustics and Psychoacoustics of Loudspeakers and Rooms", First & Second Editions, Focal Press, Oxford, UK.

Toole, F. E. (2015a). "The Measurement and Calibration of Sound Reproducing Systems", *J. Audio Eng. Soc.*, vol. 63, pp. 512–541. This is an open-access paper available to non-members at www.aes.org

Toole, F. E. (2015b). "Art and Science/Opinions and Facts", CIRMMT Distinguished Lectures in the Science and Technology of Music, McGill University, Monteal, www.youtube.com/watch?v=zrpUDuUtxPM

Toole, F. E. and Olive, S. E. (1988). "The Modification of Timbre by Resonances: Perception and Measurement", *J. Audio Eng. Soc.*, vol. 36, pp. 122–142.

Toole, F. E. and Olive, S. E. (1994). "Hearing Is Believing vs. Believing Is Hearing: Blind vs. Sighted Listening Tests and Other Interesting Things", 97th Convention, Audio Eng. Soc., Preprint 3894.

Toole, F. E. and Olive, S. E. (2001). "Subjective Evaluation", Chapter 13 in *Loudspeaker and Headphone Handbook*, Borwick, J. (ed.), Focal Press, Oxford, UK.

Toole, F. E. and Sayers, B. McA. (1965a). "Laterization Judgments and the Nature of Binaural Acoustic Images", *J. Acoust. Soc. Am.*, vol. 37, pp. 319–324.

Toole, F. E. and Sayers, B. McA. (1965b). "Inferences of Neural Activity Associated With Binaural Acoustic Images", *J. Acoust. Soc. Amer.*, vol. 37, pp. 769–779.

Torick, E. (1998). "Highlights in the History of Multichannel Sound", *J. Audio Eng. Soc.*, vol. 46, pp. 27–31.

Tsay, C.-J. (2013). "Sight Over Sound in the Judgment of Music Performance", *Proc. Nat. Acad. Sci.*, vol. 110, pp. 14580–14585, www.pnas.org/cgi/doi/10.1073/pnas.1221454110.

Van Keulen, W. (1991). "Possible Mechanism for Explaining Monaural Phase Effects of Complex Tones", 90th Convention, Audio Eng. Soc., Preprint 3065.

Vermeulen, R. (1956). "Stereo Reverberation", *IRE Trans. Audio*, July–August, pp. 98–105.

Vickers, E. (2009). "Fixing the Phantom Center: Diffusing Acoustical Crosstalk", 127th Convention, Audio Eng. Soc., Preprint 7916.

Villchur, E. (1964a). "High Fidelity Measurements, Science or Chaos", *Electronics World*, August.

Villchur, E. (1964b). "Techniques of Making Live-Versus-Recorded Comparisons", *Audio Magazine*, October.

Voelker, E.-J. (1985). "Control Rooms for Music Monitoring", *J. Audio Eng. Soc.*, vol. 33, pp. 452–462.

Voishvillo, A. (2006). "Assessment of Nonlinearity in Transducers and Sound Systems—From THD to Perceptual Models", 121st Convention, Audio Eng. Soc., Preprint 6910.

Voishvillo, A. (2007). "Measurements and Perception of Nonlinear Distortion—Comparing Numbers and Sound Quality", 123rd Convention, Audio Eng. Soc., Preprint 7174.

Walker, R. (1993). "Optimum Dimensional Ratios for Studios, Control Rooms and Listening Rooms", BBC Research Department Report No. BBC RD 1993/8, www.bbc.co.uk/rd/pubs.

Warnock, A.C.C. (1985). *Introduction to Building Acoustics*, Canadian Building Digest—CBD-236, published by National Research Council of Canada, Institute for Research in Construction. http://irc.cnrc.gc.ca/cbd236e.html

Waterhouse, R. V. (1958). "Output of a Sound Source in a Reflecting Chamber and Other Reflecting Environments", *J. Acoust. Soc. Am.*, vol. 30, pp. 4–13.

Watkins, A.J. (1991). "Central, Auditory Mechanism of Perceptual Compensation for Spectral-Envelope Distortion", *J. Acoust. Soc. Am.*, vol. 90, pp. 2942–2955.

Watkins, A.J. (1999). "The Influence of Early Reflections on the Identification and Lateralization of Vowels", *J. Acoust. Soc. Am.*, vol. 106, pp. 2933–2944.

Watkins, A.J. (2005). "Perceptual Compensation for Effects of Echo and of Reverberation on Speech Identification", *Acta Acustica* united with *Acustica*, vol. 91, pp. 892–901.

Watkins, A.J. and Makin, S.J. (1996). "Some Effects of Filtered Contexts on the Perception of Vowels and Fricatives", *J. Acoust. Soc. Am.*, vol. 99, pp. 588–594.

Welti, T.S. (2002a). "How Many Subwoofers Are Enough", 112th Convention, Audio Eng. Soc. Preprint 5602.

Welti, T.S. (2002b). "Subwoofers: Optimum Number and Locations", www.harman.com

Welti, T.S. (2004). "Subjective Comparison of Single Channel Versus Two Channel Subwoofer Reproduction", 117th Convention, Audio Eng. Soc., Preprint 6322.

Welti, T.S. (2009). "Investigation of Bonello Criteria for Use in Small Room Acoustics", 127th Convention, Audio Eng. Soc., Preprint 7849.

Welti, T.S. (2012). "Optimal Configurations for Subwoofers in Rooms Considering Seat-to-Seat Variation and Low-Frequency Efficiency", 133rd Convention, Audio Eng. Soc., Preprint 8748. Go to www.harman.com/innovation to download color versions of the plots in this paper.

Welti, T. S. and Devantier, A. (2003), "In-Room Low Frequency Optimization". 115th Convention, Audio Eng. Soc., Preprint 5942.

Welti, T. S. and Devantier, A. (2006). "Low-Frequency Optimization Using Multiple Subwoofers", *J. Audio Eng. Soc.*, vol. 54, pp. 347–364.

Welti, T. S. and Olive, S. (2013). "Validation of the Binaural Room Scanning Method Using Subjective Ratings of Spatial Attributes", 48th International Conference, Audio Eng. Soc., Paper 1–2.

Willcocks, M. E. G. (1983). "Surround Sound in the Eighties—Advances in Decoder Technologies", 74th Convention, Audio Eng. Soc., Preprint 2017.

Zacharov, N. (1998). "An Overview of Multichannel Level Alignment", 15th International Conference, Audio Eng. Soc., Paper no. 15–016.

Zacharov, N. and Bech, S. (2000). "Multichannel Level Alignment, Part IV: The Correlation Between Physical Measures and Subjective Level Calibration", 109th Convention, Audio Eng. Soc., Preprint 5241.

Zacharov, N., Bech, S. and Suokuisma, P. (1998). "Multichannel Level Alignment, Part II: The Influence of Signals and Loudspeaker Placement", 105th Convention, Audio Eng. Soc., Preprint 4816.

Zahorik, P. and Brandewie, E. (2016). "Speech Intelligibility in Rooms: Effect of Prior Listening Exposure Interacts With Room Acoustics", *J. Acoust. Soc. Am.*, vol. 140, pp. 74–86.

Index

Entries with an "f" indicate that there is a figure

Printed and bound by CPI Group (UK) Ltd, Croydon, CR0 4YY

22/10/2024

01777609-0001